Lecture Notes in Civil Engineering

Volume 44

Lecture Notes in Civil Engineering (LNCE) publishes the latest developments in Civil Engineering - quickly, informally and in top quality. Though original research reported in proceedings and post-proceedings represents the core of LNCE, edited volumes of exceptionally high quality and interest may also be considered for publication. Volumes published in LNCE embrace all aspects and subfields of, as well as new challenges in, Civil Engineering. Topics in the series include:

- Construction and Structural Mechanics
- Building Materials
- Concrete, Steel and Timber Structures
- Geotechnical Engineering
- Earthquake Engineering
- Coastal Engineering
- Hydraulics, Hydrology and Water Resources Engineering
- Environmental Engineering and Sustainability
- Structural Health and Monitoring
- Surveying and Geographical Information Systems
- Heating, Ventilation and Air Conditioning (HVAC)
- Transportation and Traffic
- Risk Analysis
- Safety and Security

To submit a proposal or request further information, please contact the appropriate Springer Editor:

- Mr. Pierpaolo Riva at pierpaolo.riva@springer.com (Europe and Americas);
- Ms. Swati Meherishi at swati.meherishi@springer.com (India);
- Ms. Li Shen at li.shen@springer.com (China);
- Dr. Loyola D'Silva at loyola.dsilva@springer.com (S-E Asia and Australia/NZ).

Indexed by Scopus

More information about this series at http://www.springer.com/series/15087

Ranjith Dissanayake · Priyan Mendis
Editors

ICSBE 2018

Proceedings of the 9th International Conference on Sustainable Built Environment

Springer

Editors
Ranjith Dissanayake
Department of Civil Engineering
University of Peradeniya
Kandy, Sri Lanka

Priyan Mendis
Department of Infrastructure Engineering
University of Melbourne
Parkville, VIC, Australia

ISSN 2366-2557 ISSN 2366-2565 (electronic)
Lecture Notes in Civil Engineering
ISBN 978-981-13-9748-6 ISBN 978-981-13-9749-3 (eBook)
https://doi.org/10.1007/978-981-13-9749-3

This Springer imprint is published by the registered company Springer Nature Singapore Pte Ltd.
The registered company address is: 152 Beach Road, #21-01/04 Gateway East, Singapore 189721, Singapore

Preface

It is with great pleasure that we present the proceedings of the 9th International Conference on Sustainable Built Environment (ICSBE) 2018. This is the ninth consecutively organized conference following a series of International Conference since 2010, keeping its tradition of adhering to engineering excellence.

Taking a step forward from the last seven events, the coverage of specialty areas in this conference has been diversified. This book contains the manuscripts of research work from many different sub-specialties. We expect that all these manuscripts will be presented at parallel sessions from 14 to 15 December 2018.

We would like to express our appreciation to all keynote lecturers for their invaluable contribution to the development of a sustainable world. We are also very grateful to the authors for contributing research papers of high quality. The manuscripts in this proceeding book have been reviewed by a panel of academic and professional experts who have vast expertise in their respective fields. The enormous work carried out by these reviewers is gratefully appreciated as well. We are also pleased to acknowledge the advice and assistance provided by the members of the international advisory committee and members of the editorial committee along with many others who volunteered to assist to make this very significant event a success. Furthermore, we acknowledge the financial sponsorship provided by many organizations that has been extremely supportive towards the success of this international conference.

It is the earnest wish of the editors that this proceeding book would be used by the research community and practicing engineers who are directly or indirectly involved in studies related to sustainability.

December 2018

Ranjith Dissanayake
Priyan Mendis
Editorial Committee

ICSBE 2018

Contents

Climate Change and Sustainable Transportation

Important Aspects of Evacuation Planning for the Coastal Communities in Sri Lanka

G. P. Jayasiri[1(✉)], O. P. C. Randil[1], G. M. C. A. Perera[1], C. S. A. Siriwardana[1], P. B. R. Dissanayake[2], and C. S. Bandara[2]

[1] University of Moratuwa, Moratuwa, Sri Lanka
gimhanjayasiri@gmail.com
[2] University of Peradeniya, Peradeniya, Sri Lanka

Abstract. Sri Lanka is a tropical island in the Indian Ocean which has suffered from various disasters in the recent past. Flood events which occurred in the last three consecutive years clearly shows that the country is affected from climate change. During a disaster event, evacuation of vulnerable and affected communities is necessary to reduce the number of deaths. In the Sri Lankan perspective, most of the recent disasters severely damaged the coastal belt of the country. This study was carried out to evaluate the different aspects of evacuation planning with respect to the coastal districts in Sri Lanka. Some of the aspects considered are early warning dissemination, evacuation routes, shelters, drills and training and the effects of having a family vehicle, disabled people, neighbours and domestic animals. During this study, an online questionnaire survey was distributed after carrying out an in-depth literature survey to gather data. The lack of trust in authorities who disseminate early warnings and limited knowledge on evacuation routes and shelters in the residential areas are some of the key issues which were identified during this study. Furthermore, it was identified that social media plays a pivotal role in disseminating the early warnings in the urban areas. Increasing the number of awareness campaigns related to evacuation and using social media wisely are some of the recommendations made, which will build the coastal community resilience via proactive measures.

Keywords: Evacuation planning · Resilience · Early warning

1 Introduction

Extreme weather events and natural hazards have been identified as the top two global risks in terms of likelihood (WEF 2018). When looking at Sri Lanka, it is visible that this statement is highly applicable considering the frequent disasters after the 2004 Indian Ocean Tsunami. During 2016, 2017 and 2018, the country witnessed extreme rains which triggered flooding in several urban areas. The number of affected people from the 2018 floods were around 150,000 which shows a significant decrease compared to the events in 2016 and 2017 (DMC 2018a, b). The 2017 flood event which

R. Dissanayake and P. Mendis (Eds.): ICSBE 2018, LNCE 44, pp. 3–10, 2020.
https://doi.org/10.1007/978-981-13-9749-3_1

damaged both the infrastructure and communities of the densely populated districts in Sri Lanka raised the number of affected families up to 630,000 (MoNPEA 2017a, b).

These three consecutive events show that Sri Lanka should undertake quick resilience measures in order to reduce the number of affected people. Evacuation planning is an aspect which should be given due recognition when building community resilience. Most of vulnerable communities face several problems during the evacuation process after an early warning is disseminated. There is always a set of individuals who will not respond to the EW messages and evacuation orders following the guidelines, but will act on their own (Sakamoto et al. 2016). Hence the respective authorities, in the Sri Lankan case, the tri-forces sometimes must persuade them. The lack of awareness and trust in authorities, unawareness on evacuation plans, shelters and shortest routes, minor experience in drills and training, increasing number of deaths and deficiencies in capacity has led to the need to carry out this research to find out the challenges and barriers in evacuation planning to provide recommendations to reduce the disaster risk of the coastal communities. The study is focused mainly on the evacuation scenarios of floods and Tsunamis since the coastal belt of the island is more vulnerable to these hazards.

2 Methodology

To evaluate the different aspects in evacuation planning for the coastal communities in Sri Lanka, initially, an in-depth literature review was carried out to identify the existing status by referring to several plans and programmes prepared by various national and Non-Governmental Organizations (NGOs). In addition, online databases like JSTOR, Google scholar ad ScienceDirect were used to search for peer-reviewed articles across the world. Combinations of keywords: evacuation planning, coastal hazards, emergency management, disaster resilience and climate change were employed in the searches. The preliminary materials were screened further to obtain a shortlist of materials relevant to the research. These were used for snowball sampling to find further literature. Finally, after a discussion with the Sri Lankan academics involved in the Disaster Management sector, three main themes, namely; early warning dissemination, evacuation routes, shelters, drills and training and the effect of having a family vehicle, disabled people, neighbours and domestic animals were selected to identify gaps.

Then, an online questionnaire survey was distributed to gather the community perception on evacuation planning in coastal communities. The Survey was distributed via social media and emails. This was done to gather the opinions of the people who have frequent internet access in the urban areas. Thematic content analysis method was used to analyse the data under the above aspects. Finally, after the analysis, several recommendations were provided to address the gaps and challenges identified in the above themes.

3 Results and Discussion

3.1 General Information

From the online survey, 97 filled questionnaires were collected which led to the assessment of the existing problems and challenges in evacuation planning. Most responses were from the younger generation (age group: 15–30 years) as they are the demographic that uses social media mostly.

Out of the 14 coastal districts, information was collected from 9 districts. Most of the data is concentrated in the Western province and the Southern province as the conclusions will be more focused on those two provinces. Most of the people who are responded were from the engineering and higher education sectors. 34% were female respondents. More than 70% of the families consist of 4 to 5 members and at least one person has a mobile phone. Furthermore, when looking at the education level of the families who responded; more than 90% of them have at least one member who has diploma level or university level education. Facebook is the most commonly used social network by the respondents. According to Fig. 1, when looking at the public transport facilities and road conditions in the respective areas, the respondents were satisfied with them.

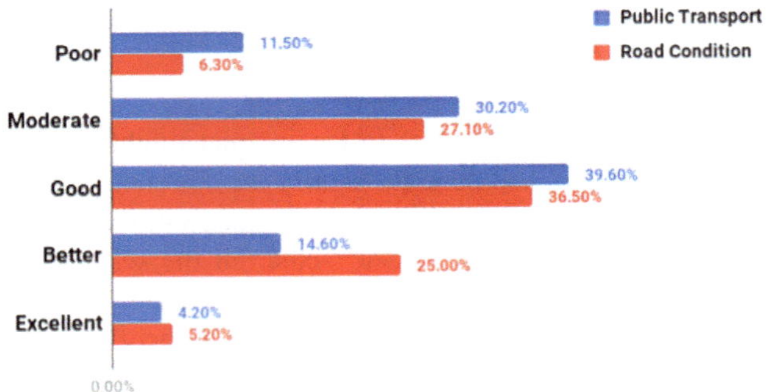

Fig. 1. Road condition and quality of public transport according to respondents

Looking at the general opinions about hazards, more than 40% of the respondents think that storms, floods and lightning strikes will affect their area while 15% were confident that their area is hazard free. This fact confirms that most of the respondents are aware that they are vulnerable to floods. When rating the trust placed on the authorities towards the notices issued by the authorities to evacuate for disasters more than 70% rate it below 3, on a scale of 1 to 5 (1 being worse and 5 being excellent). This is a critical issue which needs to be addressed, which shows that communities do not take evacuation notices seriously.

3.2 Early Warning Dissemination

The means of receiving Early Warnings (EW) also plays a pivotal role during evacuation. At present, Sri Lanka has a Multi-Hazard Early Warning Dissemination System and an Early Warning Coordination Network from the national level down to the grass root level (DMC 2015). In addition, the National Emergency Operations Plan (NEOP) has identified the role of the media and the military for an effective EW dissemination (DMC 2015). Analysis of the online survey shows that most respondents get early warnings via Facebook, TV and Mobile phone SMSs as shown in Fig. 2.

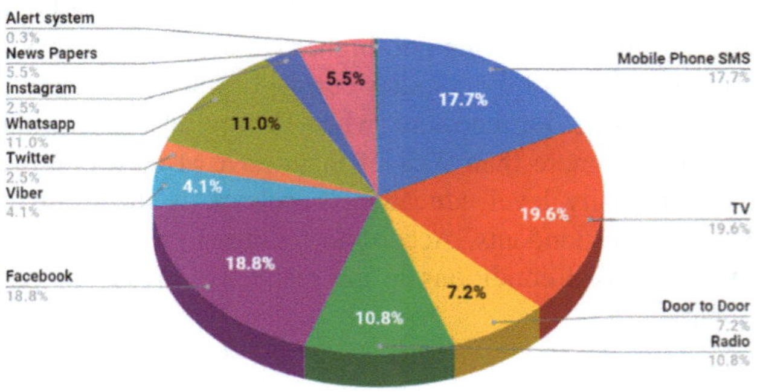

Fig. 2. Way of getting EW messages

During a disaster, sometimes, access to the internet might not be possible. Furthermore, a statistical analysis was done to identify the correlations between the above responses. According to the Shapiro – Wilk normality analysis done to find a correlation between the respondent's district and social media, all the selected variables have a significance less than 0.05, which depicts that the data significantly variates from a normal distribution. From the analysis, it was visible that there aren't any correlations between the respondent's districts and preferred way of getting EW messages. But it was visible that there are weak positive correlations between Facebook, WhatsApp, Twitter and other social media. This shows that a selected individual respondent has a slight similarity on their opinion of social media. Furthermore, WhatsApp and media channels have a weak negative correlation which shows a tendency in the respondents to prefer either one of them to get EWs. In addition, from the District and Social Network cross tabulation, it was observed that more than 80% of the respondents either use Facebook or WhatsApp as a social media network. Hence, this shows that when using social media to disseminate EW messages Facebook and WhatsApp can be more effective. The Facebook page which was launched recently and is managed by the Disaster Management Centre shows that they have undertaken this initiative (DMC 2018b).

Maintaining the trust on authorities who give evacuation instruction is a challenge (CEDMA 2017). When analysing the comments made by the respondents there were some serious responses regarding the EW mechanisms. Some believe that announcements by local authorities are reliable as those are issued right prior to the disaster, but the predictions issued by Department of Meteorology (DoM) are inaccurate. During a recent study, it was noted that DoM needs advanced technology to do local area modelling which is needed to produce accurate results. (Jayasiri et al. 2018). Untimed EW and unnecessary exaggeration were also noted as critical issues. Some people felt that it is very important to respond to disaster warnings from the respective authorities even though the level of trust is low, because otherwise, the loss will be severe. Furthermore, one respondent said that Sri Lanka has improved surveillance and other facilities regarding Disaster Management by introducing new communication methods. It was also noted that the authorities may be paying special attention to areas near the capital as well.

During the recent monsoon rains, several early warnings were disseminated via social media which helped the authorities inform people about the upcoming disaster risk. Respondents have also mentioned that conveying early warnings via Mobile Phones and door to door methods are very effective while the use of newspapers to distribute early warnings was not as effective. In addition, around 70% of the respondents are capable of passing the early warning messages they get to more than 10 people.

3.3 Evacuation Routes, Shelters, Drills and Training

According to NEOP, Administrative heads of provincial, district, divisional and GN levels, DMC coordinators and fire service departments of local authorities are some of the responsible stakeholders for evacuation (DMC 2015). In addition, Non-Governmental Organizations like Red Cross and Sarvodaya also come to aid during the evacuation. Apart from the national level programmes, awareness and knowledge on evacuation planning should be incorporated to the school level as well. More participation of school children in local evacuation planning is necessary to increase their preparedness (Vásquez et al. 2018). At present, the country has developed guidelines for the school's disaster safety which identifies the importance of evacuation planning as well (DoE 2008). In the post-disaster recovery plans, the importance of using schools as evacuation shelters was also identified (MoNPEA 2017b).

During the analysis, it was noted that 74% of the respondents have not undergone any evacuation drills. Most of them were from Colombo, Galle, Gampaha, Kalutara and Matara districts. 26% of the respondents who said that they have experience of drills noted that it was mostly Tsunami evacuation drills, fire drills at their workplaces and first aid drills.

Furthermore, almost every sector responded to having at least 10% of respondents with evacuation drill experience as in Fig. 3.

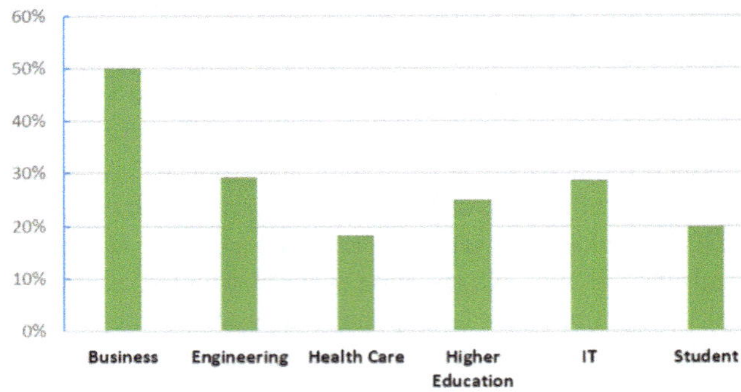

Fig. 3. Respondents experience in evacuation drills

Knowing the safest and fastest ways to the evacuation shelters helps the residents to get out of their houses and reach a more convenient place in time. In addition, identification of the safest place in the relevant area which can provide shelter is important. During the survey, 37% of the people said that they know the evacuation routes while others were either not sure or thought there aren't any. Out of the 37% who said they know about the evacuation routes, 74% firmly said that there are no evacuation signs along the routes. Due to the significant damage caused by the 2004 Indian Ocean Tsunami, most of the evacuation signs in the country are for Tsunamis. But for other disasters, there are hardly any.

In addition, only 18% of the respondents know the evacuation shelters in the area. This is a significantly a low value which the authorities should take immediate action to remedy by conducting awareness programmes. Religious places, schools and hospitals can be taken as evacuation shelters since most of them are on a higher ground. According to the survey, respondents said that it will take around 5–10 min to reach the nearest shelter in their areas. All of the 18% who said they know the evacuation shelter; were confident that they know the shortest routes as well.

Furthermore, 60% of the respondents said that they don't have any personal experience of a disaster. Most of the remaining 40% have experiences related to the Tsunami, strong winds and floods. More than 70% of them rated their evacuation experience by giving a value greater than 3 within a scale of 1–5 where 1 being poor and 5 being excellent.

3.4 Effect of Having a Family Vehicle, Disabled People, Neighbours and Domestic Animals

Sometimes, having a family vehicle helps a speedy evacuation for residents except during a flood. 81% of the respondents said that they have a family vehicle and most of them were cars which have the capacity to carry around 5 five passengers. In comparison to the total number of family members of the respondents, only 60% of them have the capability of carrying their entire family at once during an evacuation by their vehicle.

During an evacuation, protective care of disabled people is necessary, supported by advance preparation (Owens et al. 2013). Only 7% of the respondents had disabled people and most of them need assistance in walking.

During an evacuation, looking after relatives and neighbours sometimes becomes a priority rather than evacuating your own self. 80% of the respondents said that they will put an extra effort to take their neighbours with them if they are not capable. 6% of the respondents believed that they do not have enough space in their vehicles to carry them while 12% believe that they are capable to evacuate by themselves.

People who have pets and other domestic animals (people with farms) tend to care for them during an emergency. During this survey, the importance of this issue was analysed and out of the respondents 47% had domestic animals and 78% mentioned that they will put an extra effort to carry their pets during an evacuation while the others believe that the animals can survive by themselves.

4 Conclusions and Recommendations

The analysis of the information led to several conclusions regarding the evacuation planning for coastal communities in Sri Lanka. Even though Sri Lanka has a properly established EW mechanism, the lack of trust in authorities who disseminate early warnings is an issue. Social media plays an important role when informing the general public about evacuation notices. DMC and other relevant authorities should consider this fact and should use social media wisely for EW processes. If the authorities can develop a mechanism where only vulnerable people get an evacuation notice to their mobile phones as an SMS, that would be a perfect scenario. Limited knowledge and awareness on evacuation routes and shelters in residential areas are some of the key issues which were identified during this study. Local authorities should undertake necessary actions to establish sign boards which will guide people to evacuation shelters quickly. In addition, they can conduct smaller-scale awareness campaigns and scenario-based evacuation drills. Most people who live in towns have a family vehicle which helps them evacuate quickly. For those who need transportation assistance, local authorities, with the help of NGOs, can arrange those facilities during an evacuation process. Even though we should always care about our neighbours and other vulnerable people we have to make sure that we are safe first.

Proper evacuation notices, well-established guidelines and clear signboards to evacuation shelters will always help affected communities evacuate safely which will reduce the number of deaths during a disaster.

Acknowledgement. This research was conducted in collaboration with the Faculty of Engineering of University of Moratuwa and University of Peradeniya with the financial aid from CABARET (Capacity Building in Asia for Resilience Education), a project of the European Union's Erasmus+ programme – Key action 2 – Capacity building in the field of higher education.

Disclaimer. The European Commission support for the production of this publication does not constitute an endorsement of the contents, which reflects the views only of the authors, and the Commission cannot be held responsible for any use, which may be made of the information contained therein.

References

Center for Excellence in Disaster Management and Humanitarian Assistance (CEDMA) (2017) Sri Lanka disaster management handbook

Department of Education (DoE) (2008) National guidelines for school disaster safety. http://www.preventionweb.net/files/9252_srinatguilines.pdf

Disaster Management Centre (DMC) (2015) National emergency operations plan, vol I

Disaster Management Centre (DMC) (2018a) Summary situation report - southwest monsoon

Disaster Management Centre (DMC) (2018b) DMC Sri Lanka. https://www.facebook.com/dmc.gov.lk/

Jayasiri GP, Siriwardena CSA, Hettiarachchi SSL, Dissanayake PBR, Bandara CS (2018) Important aspects in building community resilience of Coastal Districts in Sri Lanka. In: International conference on disaster management

Ministry of National Policies and Economic Affairs (MoNPEA) (2017a) Floods and landslides. Integrated risk assesment

Ministry of National Policies and Economic Affairs (MoNPEA) (2017b) Post-disaster recovery plan

Owens JK, Warner Stidham A, Owens EL (2013) Disaster evacuation for persons with special needs: a content analysis of information on YouTube. Appl Nurs Res 26(4):273–275

Sakamoto M, Kuri M, Iguchi M, Maki N (2016) Disaster governance in disaster management planning—analysis of the evacuation planning process for Kuchinoerabujima Volcano Eruption. J Nat Disaster Sci 37(2):105–117

Vásquez A, Marinkovic K, Bernales M, León J, González J, Castro S (2018) Children's views on evacuation drills and school preparedness: mapping experiences and unfolding perspectives. Int J Disaster Risk Reduction 28(June 2017):165–175

World Economic Forum (WEF) (2018) The global risks report. ISBN 10.978-1-944835-15-6

Application of GPS/GIS Based Travel Mode Detection Method for Energy Efficient Transportation Sector

G. W. H. A. Chandrasiri(✉), K. A. G. G. K. Kumarasinghe, and H. K. Nandalal

Faculty of Engineering, University of Peradeniya, Peradeniya, Sri Lanka
hansinie404@gmail.com

Abstract. The planning of efficient transportation systems is a central topic of concern as it saves energy and contributes to economic development both at national and household level. The development and optimization of highway and transportation engineering models such as transportation networks, parking facilities and as well as reduction of traffic congestion etc., require information on travel modes. To overcome the burdens for the respondent and the other drawbacks of the conventional methods related to travel mode detection, combination of Global Positioning System (GPS) and Geographical Information System (GIS) were tested in this research. After validating the accuracy of the smartphone based 'My Tracks' GPS mobile application, GPS travel route with a manually entered travel mode diary was used to collect data from randomly selected days and students in two faculties in the University of Peradeniya. Out of 125 datasets, 99 accurate datasets were selected for the analysis using 'ArcGIS' software coupling with 'STATA' statistical software. Rule-Based algorithms and heuristic based methods are two main approaches that were used to detect four travel modes (Walk, Bus, Motor Vehicles and Motor Bicycles). Differentiation of Bus mode from the motor vehicles (except motor bicycles) was the most challenging part of the research and a country specific novel method was introduced in this research. Travel mode detection results were compared with relevant correctly completed travel mode diaries to validate the methodology. The overall travel mode detection accuracy of the new method is 96% for the accurate data sets compared to conventional travel mode diary method. The success rate of collecting an accurate data set is 79%. The GPS/GIS based travel mode detection could be applied in energy efficient development in the transportation sector. In future research, the bus mode detection method invented in this research could be used to detect driver behaviour.

Keywords: Travel mode · GPS · GIS · My tracks · University of Peradeniya

1 Introduction

An efficient transportation system is crucial for the development of a country. An energy efficient transportation system contributes to the development of the economy both at national and household level and the sustainability of a green environment and

R. Dissanayake and P. Mendis (Eds.): ICSBE 2018, LNCE 44, pp. 11–21, 2020.
https://doi.org/10.1007/978-981-13-9749-3_2

ecosystem. The detection of travel modes plays a key role while designing energy efficient transportation system. Household travel surveys have been used as a tool since 1970's to collect data to build transportation models such as transportation networks and parking facilities (Zong et al. 2016). The Travel diary method is one of the conventional methods to carry out travel surveys. It has many drawbacks as it is a more time consuming and respondent burdening practice. Consequently, the use of Global Positioning System (GPS) devices for collecting travel mode data became popular in the recent years. It reduces the respondent burden while researcher uses the respondent as a passive data collector. Though the GPS records accurate location, time, elevation information, it cannot record details like travel modes and trip purposes. Therefore, in this study Geographical Information System (GIS) was used with GPS data to detect travel modes. Gong and Chen (2011) have studied GPS/GIS method for travel mode detection in New York City using GPS devices. At present, smartphones have GPS facility and therefore accuracy and applicability of GPS facility in the smartphones has become an alternative to GPS devices-based travel mode detection method. However, information on use of smartphones-based GPS in travel mode detection is scarce. Therefore, a study has been designed to investigate the applicability of GPS/GIS method using mobile GPS applications for travel mode detection in the University of Peradeniya. The objectives of the study are; (i) to detect the travel mode using smart phone-based GPS applications (ii) to compare the results of modal shares (iii) to evaluate the success rate of smart phone-based travel mode detection compared to that of travel diary-based travel mode detection method (iv) identify and compare the travel modes of students in the Faculties of Engineering and Science using novel smartphone-based travel mode detection method.

2 Background Information on GPS Based Travel Surveys

GPS based travel surveys have many advantages over traditional paper-based travel surveys or travel diary method. Because, in traditional paper-based surveys, respondents have to fill many pages manually. It is reported that respondents often forget to enter the trip segment and it is a time-consuming procedure. In GPS based travel surveys, respondents are acting as passive data collectors and therefore respondent burden is comparatively lower. The GPS method shows the traveller's exact route and because of the accuracy of the data, sample size can be reduced. Moreover, travel speeds can be calculated for the particular route at particular time period. It is a good indicator to assess the level of service for a given transportation network. Due to the easiness of the method, length of the survey can be extended from traditional single day survey to multiple day survey. Accuracy can be validated for the daily routine of the same respondent through this multiple survey. Though the advantages of the GPS based travel survey are higher compared to traditional methods, it has few drawbacks like any other method. Since GPS traces the respondent's exact route, it minimizes the privacy concern. Signal loss and degradations in shopping malls, tunnels, and urban canyons are the other drawback in this method as well as the fact that GPS cannot directly identify travel mode and trip purposes. In this research, the main focus is on combining GPS and GIS methods to give a solution for travel mode detection.

3 Literature Review

Shalaby and Chung (2005) developed a trip reconstruction software tool that automatically identifies the travelled road links and modes used by the respondent using GPS traces. Conventional GIS map matching algorithm was adopted in this method to identify road links while rule based algorithms were used to identify four travel modes (walk, bicycle, bus and passenger car). GPS travel data and the multimodal transportation network model of downtown Toronto has been used for the evaluation of the developed software tool. It is reported that said method correctly matched 79% of all links travelled and 92% of all trip modes. However, they only allowed for one-purpose trips and maximum of two possible modes per trip.

Zong et al. (2016) developed a hybrid procedure for travel mode detection and it was presented using large-scale origin-destination survey conducted in Beijing in 2010. They determined the subway mode separately from other modes and used two methods to detect the subway mode. It is reported that 53.8% success rate for multinomial logit model while 97.8% success was for GIS algorithm. Results revealed that the GPS/GIS method provides higher accuracy for subway modes. Nested logit model was employed to determine the modes of Walking, Bicycle, Car and Bus. Combined success rate of the hybrid procedure was 86%.

Gong and Chen (2011) developed a GIS algorithm to automatically process data that was traced from GPS based travel surveys. They have detected five travel modes (Walk, Car, Bus, Subway and commuter rail) and those detected results from GIS algorithm were compared against conventional travel diaries. They have conducted two small GPS surveys and success rates of that combined survey was 82.6%. In their research, they allowed an unlimited number of mode transfers within a trip. Furthermore, they built connectivity into their multimodal transportation network in GIS to find the right link for matching the GPS traces, or interpolate the underground subway mode. In their multi modal transportation network, the success rate for the subway mode detection equals to 65.6%.

Clifford et al. (2008) used set of developed heuristic rules to determine both travel modes and trip purposes. Heuristics based on speed and route of travel, as well as some demographic information. In this study, four different travel modes (walk, bicycle, private vehicle and public transport) have been considered. They compared results with respective travel diaries.

4 Methodology

4.1 Study Area

The study area is University of Peradeniya, a state university in Sri Lanka. It has nine faculties and about 11000 students.

4.2 GPS Data Collection

Instead of handheld GPS loggers, smartphones with GPS facility were used for this research. Several GPS tracking applications such as *My tracks, Geo tracker, Track Me, GPS Trip recorder* were tested and *'My tracks' and 'Geo tracker'* were selected as GPS tracking applications based on the performance of said two applications. It was noticed that internet facility is needed for tracking paths in *Geo tracker* occasionally and GPS signal losses were higher than the *My tracks* application. Finally, 'My tracks' was selected as the best tracking application. Accuracy of the mobile GPS application was checked against a GPS receiver called etrex10. Obviously, the traced data points for a given route was low in the mobile phone compared to the GPS receiver. But we checked the accurate data points that were traced from the mobile phone. After three consecutive days of pilot travel survey with two different mobile phones, we found that the average accuracy of the traced data points of two mobiles phones were 88.6% and 93% respectively. Since the accuracy of the 'My Tracks' application is greater than 75%, it was selected for the study. It was noticed that the signal losses were occurred in both etrex10 GPS receiver and the two GPS enabled mobiles phones in shopping malls and the tunnels.

4.3 GIS Data Collection

In Sri Lanka, local agencies do not have transportation networks as GIS layers. As a result, to get an initial idea whether the traced GPS data points are correct or not, first travel routes were exported to the Google earth. Also, bus halts were identified using Google earth as an indicator to separate bus mode from the other modes.

4.4 Socio Demographic Data Collection

Peoples' mode choices are changed with their socio demographic information like age, gender, vehicle ownership, location of residence etc. Therefore, before selecting participants, questionnaire-based survey was conducted for total student population to collect socio demographic data.

4.5 Sample Selection

The target population could be stratified by three main attributes, namely, (i) Respondent type, (ii) Location of stay/residence wise and (iii) faculty wise. By respondent category, population can be classified as students, academic staff and non-academic staff. By location of stay/residence category, population can be classified as within the university and out of the university. By faculty, population can be classified into nine different faculties. The target group of this investigation was limited to the students of the Faculties of Engineering and Science. The reasons for selecting those faculties to this research was that the Faculty of Engineering is isolated from other faculties as well as it is the faculty of the research team while the Faculty of Science is inside the University premises and there is a higher number of students in that faculty. Academic and non-academic staff were not selected for this sample because of their busy life style. Under location of stay category both students in hostels and residential students were selected. The final sample of 100 students was selected using stratified random sampling technique from both Science and Engineering faculties. Of these 100 students, 79 students were selected for data collection as others have shown problems in data recording using mobile application.

4.6 Key Steps in Methodology

4.6.1 GPX Data

In 'My Tracks', the GPS routes are saved as GPX files. In that file there is no velocity data indicated directly. Velocity is the main parameter to separate travel modes. So that first GPX file was exported to the ArcGIS and route was projected from WGS 1984 to Kandawala, Sri Lanka grid. Example for a projected route is presented as Fig. 1. Then X, Y coordinates were created in the attribute table. After that, attribute table was exported to the MS Excel.

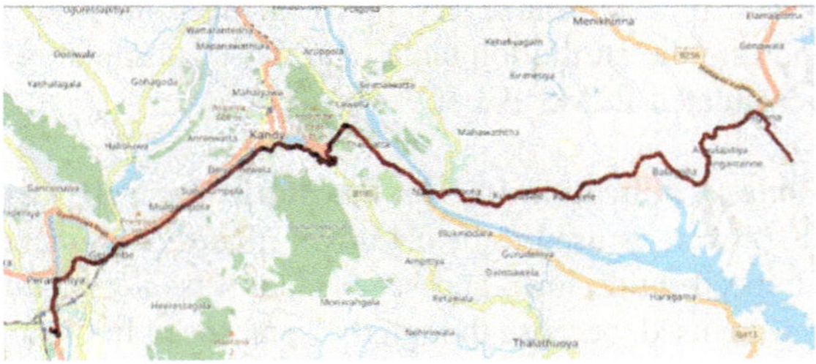

Fig. 1. Example for a projected route

4.6.2 Erroneous Data

After exporting the attribute table to the MS Excel, erroneous data were filtered by elevation. Erroneous elevation data were represented as −9999. The accuracy level of the data set was calculated using the Eq. 1 given below and data sets with accuracy more than 80% were selected to detect travel modes. Table 1 represents the accuracy results for ten datasets.

$$Accuracy = \frac{(a - b) * 100}{a} \tag{1}$$

Where,

a = Traced data points
b = Erroneous data points

Table 1. Accuracy results for ten datasets

Dataset	Traced data point	Erroneous data points	Accuracy %
1	1409	97	93.1
2	489	145	70.3
3	1615	123	92.4
4	752	299	60.2
5	3403	18	99.5
6	1168	19	98.4
7	485	14	97.1
8	1267	11	99.1
9	343	201	41.4
10	954	4	99.6

4.6.3 Velocity Calculation

X and Y coordinates for each and every point in the route were calculated in the ArcGIS. After exporting the attribution table with those coordinates, consecutive point velocities can be calculated in MS Excel.

4.6.4 Travel Modes Calibration in STATA Statistical Software

STATA statistical software is used to detect travel modes. Commands were generated according to local speed guide lines and set of data sets. Set of travel survey routes from 'My Tracks' were used, because though the local speed limits are indicated, actual level of service may be changed with the corresponding study area.

Data sets with calculated point velocities generated in Excel sheets were exported to STATA statistical software. Primarily, data points were categorized into three modes considering the velocities. Velocity ranges and corresponding travel modes are as below.

$$\text{Travel mode} = 0,\ \text{Standstill},\ \text{if Velocity} \leq 0.2\,\text{km/h}$$
$$\text{Travel mode} = 1,\ \text{Walk},\ \text{if } 0.2\,\text{km/h} < \text{Velocity} \leq 8\,\text{km/h}$$
$$\text{Travel mode} = 2,\ \text{Vehicle},\ \text{if } 8\,\text{km/h} < \text{Velocity} \leq 70\,\text{km/h}$$

Then the data set was further categorized into 300 s intervals and average velocities were taken in each 300 s intervals. Using those average velocities Vehicle mode was categorized as below.

$$\text{Travel mode} = 3,\ \text{Bus},\ \text{if } 8\,\text{km/h} < \text{Average velocity} \leq 30\,\text{km/h}$$
$$\text{Travel mode} = 4,\ \text{Motor Vehicle},\ \text{if Average velocity} > 30\,\text{km/h}$$

Most frequent travel mode for the points in the 300 s interval was used as the travel mode in that 300 s interval. Then detected travel modes for each 300 s intervals were plotted against time. Mode changes can then be identified from that plot. But the drawback in the plotting method is that sometimes due to traffic congestions motor vehicles can be categorised from time to time in the bus category as presented in Fig. 2. Therefore, to differentiate motor vehicles from bus category, a bus halt layer was created in Google earth around the University of Peradeniya area as presented in Fig. 3 and points representing standstill, walk and vehicle in each dataset were separated in ArcGIS. Then, that bus halt layer was exported to Arc GIS and 75 m buffer zones were created around the bus halt. After that, standstill and walk points inside the buffer zones were clipped as shown in Fig. 4. If more than 90% of standstill and walk points near the bus halts coincide with the buffer zone, it was filtered as a Bus Mode. Filled questionnaires were used to differentiate motor bicycles from the motor vehicles. Finally, one day mode changes in each respondent were tabulated.

Fig. 2. Travel mode vs. time

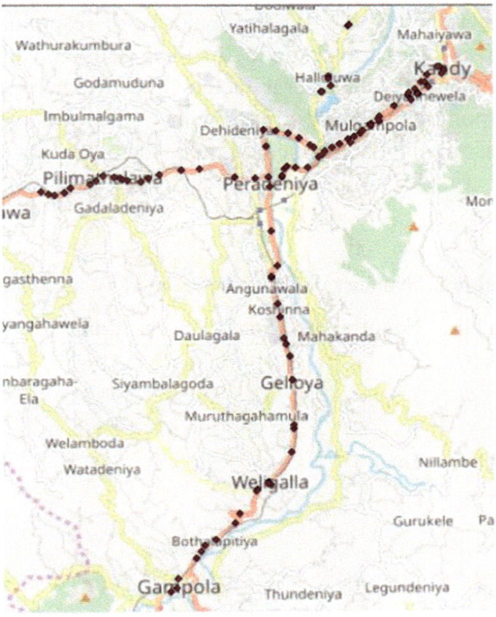

Fig. 3. Bus halt layer around Kandy area

Fig. 4. Clipped standstill and walk points

5 Results and Discussions

5.1 Travel Mode Detection

Analysis of data revealed that travel modes such as walk, bus, motor vehicle and motor bicycles could be detected using the smart phone-based GPS application. It was found that about 31% of the respondents use only one travel mode - for example walk mode – in their daily travel. About 61% of respondents use 2 travel modes and 6% use 3 travel modes in their daily travel (Table 2). The majority of respondents use Walk travel mode plus any one of other travel modes in their daily travelling. But their mode selections can be changed with several factors like distance, time, salary, level of service of the transportation networks etc. In this study we mainly focused on the applicability of smartphone-based method and the accuracy of the travel mode detection.

Table 2. Number of travel modes used by respondents

No. of modes	Freq.	Percent	Cum.
1	19	31.67	31.67
2	37	61.67	93.33
3	4	6.67	100
Total	60	100	

Table 3. Number of respondents and their selection of modes

Travel mode	No of respondent
Walk	79
Bus	39
Motor vehicle	13
Motor bicycles	8

The result in Table 3 indicates that all the respondents had walked in their daily routines. Among students in the Faculties of Engineering and Science, it was found that the bus mode usage in the Faculty of Science is higher than that of Engineering (Fig. 5). Common reason for that may be the distance from the location of stay to the Faculty of Engineering is short compared to the Faulty of Science.

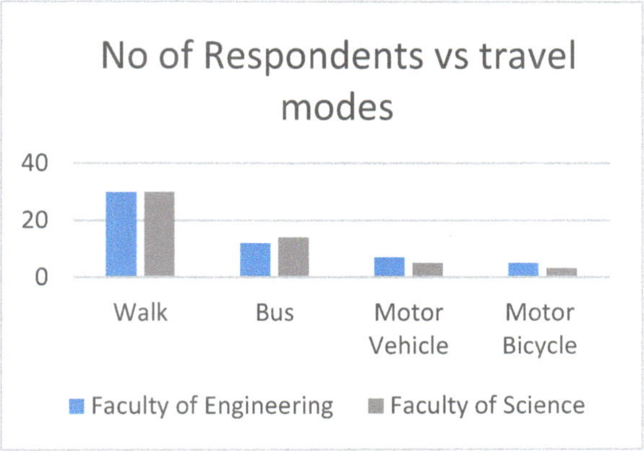

Fig. 5. Comparison of travel modes used by responded of two faculties

5.2 Modal Shares

Modal share percentage per day was identified for each respondent.

Figure 6 shows the results of the comparison of modal shares. It indicates that walk and bus combination is the highest share followed by walk and walk with motor vehicle combination.

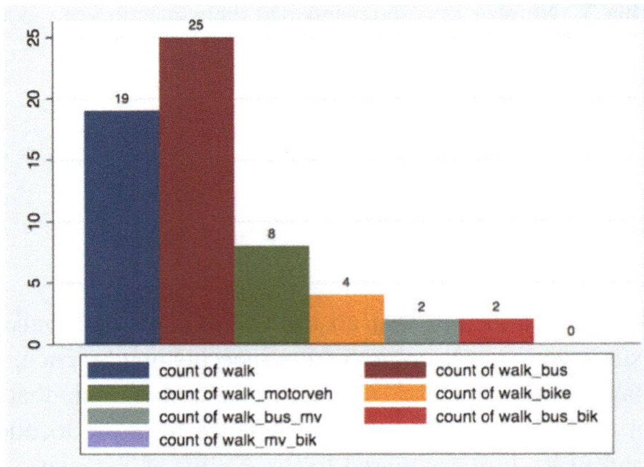

Fig. 6. Modal share of 60 students in both faculties

5.3 Success Rate of Travel Mode Detection Using Smart Phone-Based GPS Application

The travel modes detected using smartphone-based GPS/GIS method was compared with the travel diaries of relevant responded. It was found that out of 79 dataset results 76 were matched with the relevant travel diary. Therefore, the success rate of detecting travel modes in this research is 96%.

Among routes with more than 80% accurate data sets, overall travel mode detection accuracy is 96% and the success rate of collecting accurate dataset is 79%.

6 Conclusion

The new method of smartphone-based GIS/GPS application could be used to detect travel modes accurately. About 61% of the study population use two travel modes in their daily travel. The new method has a success rate of 96% compared to the conventional travel diary method. In this study location - University of Peradeniya - the high share associated with the walk mode is because more than half of the respondents are from university residences. Due to lack of owned vehicles, students had used buses to complete their trips. Overall, this new method is accurate, faster and less respondent burden and hence it can be used to detect travel modes in an institute or a city. Those results can be used to develop transportation facilities (Access roads, Parking, Shuttle services, etc.) in that area.

Travel planning using smart phone-based GPS/GIS application helps develop energy efficient transport facilities and would therefore contribute to sustainable management of green energy, environment and ecosystem services.

Acknowledgements. The authors wish to thank respondents who were willing to do the daily travel survey, evaluation panel for their guidance and comments to complete this research successfully.

References

Clifford E, Zhang J, Stopher P (2008) Determining trip information using GPS data. Institute of Transport and Logistics Studies, Sydney

Zong F, Yuan Y, Liu J, Bai Y, He Y (2016) Identifying travel mode with GPS data. Transp Plan Technol 40(2):242–255

Gong H, Chen C (2011) A GPS/GIS method for travel mode detection in New York City. Comput Environ Urban Syst 36(2):131–139

Shalaby A, Hwan Chung E (2005) A trip reconstruction tool for GPS-based personal travel surveys. Transp Plan Technol 29(1):381–401

Selection of Optimum Junction Operation Strategy for Gatambe Intersection Using VISSIM Simulation

K. L. D. Maduranga[1(✉)], R. G. N. Yasamali[1], I. M. S. Sathyaprasad[1], and H. U. Weerakoon[2]

[1] Department of Civil Engineering, University of Peradeniya, Peradeniya, Sri Lanka
lathindu07@gmail.com
[2] Central Engineering Services (Pvt) Limited, Digana Villege, Rajawella, Sri Lanka

Abstract. At present, traffic congestion is a major issue in the Kandy city area. It is observed that most of the congestion occurs at junctions and intersections. One of the most congested locations in the Kandy road network is the Gatambe junction. Under this research, the aim is to simulate the Peradeniya – Gatambe corridor using VISSIM simulation software and identify the optimum junction improvement strategy for the Gatambe intersection. Gatambe intersection is a four-way intersection that connects the Colombo – Kandy road (A1), William Gopallawa Mw (AB42) and Gannoruwa road (B365) directly. At present, the Gatambe intersection operates as a roundabout and at peak hours it is controlled manually by traffic police. Because of the heavy upstream traffic congestion near the Gatambe intersection, queues propagate upstream of the junction. Although many solutions including those needing heavy capital investment have been proposed to reduce congestion at Getambe junction, it is important that all possible solutions are tested on a simulation model before any physical implementation. In this study VISSIM is used as the simulation model. VISSIM is a microscopic simulation model that can predict network performance at the operational level, optimize control systems, and predict their behaviour before implementation. For simulation the necessary geometric and traffic data were collected as the first step. Then, the existing situation was simulated using VISSIM and the results validated with the observed survey queue length results. After that, the model was used to simulate three improvement strategies; signalization without geometric improvement, signalization with geometric improvement and grade separation. According to the delay results, the benefit of each alternative was estimated and a cost benefit analysis was conducted to identify optimum junction improvement strategy.

Keywords: Microscopic simulation · PTV VISSIM · Optimum junction operation

R. Dissanayake and P. Mendis (Eds.): ICSBE 2018, LNCE 44, pp. 22–35, 2020.
https://doi.org/10.1007/978-981-13-9749-3_3

1 Introduction

Traffic congestion is a major problem in many cities all around the world. Motorists and passengers waste their time on the streets due to heavy traffic congestion and it increases air pollution. In Sri Lanka this is a common issue in many cities. Kandy city is severely impacted with the lack of road facilities. There are many identified difficulties in expanding the existing road network in Kandy due to geographical aspects and historical value. Kandy is located in a mountainous terrain 116 km away from Colombo in the central province of Sri Lanka. It is recognized as the second busiest city of the country. The Kandy city area road network consists of a few approaches and the western approach is one of its main approaches. The road link, which directly connects Kandy and Colombo, belongs to the western approach. In this research, the aim is to simulate the road network to get an idea about the behaviour of traffic in the western corridor. Data on traffic behaviour is obtained through a simulation model. When considering the simulation, it is important to understand the concept, which is a set of variables that contain sufficient information to describe the evaluation of the system over a time period. Traffic simulation models are useful for both microscopic and macroscopic perspectives. This simulation is done by VISSIM and it is a microscopic simulation model. It is concerned with the individual elements of transportation systems, such as individual traveler behaviour and individual vehicle dynamics.

1.1 Microscopic Simulation

Transportation models can be generally classified into microscopic, mesoscopic, macroscopic models. Microscopic models study individual elements of transportation systems, such as individual vehicle dynamics and individual traveler behaviour. Mesoscopic models analyse transportation elements in small groups, within which elements are considered homogeneous. Macroscopic models deal with aggregated characteristics of transportation elements, such as aggregated traffic flow dynamics and zonal-level travel demand analysis. Microscopic simulation is related to small scale operations and it is designed to run actual traffic operations as well. There are functions based on the number of behavioural models such as car following model, lane changing model and gap acceptance model. Microscopic simulation is best suited to evaluate operational and management strategies to address congestion.

1.2 VISSIM

VISSIM is a part of the PVT software and it provides microscopic simulation methods. It helps to assess and solve a wide range of transportation problems. VISSIM primarily contains scientifically approved models for car –following, lane-changing and pedestrian moments. VISSIM consists of road users, public transport, pedestrians and their interactions with each other. This network is used by cars, HGVs, buses, trains and trams as well as cyclists and pedestrians. A wide selection of evaluations are available for online and offline analysis. VISSIM is capable of animated visualization in 2D or 3D and it fills the space between technical expertise and non – technical audience.

VISSIM can be used to analyse traffic of various alternative road designs which include both urban and highway configurations.

1.3 Importance of Finding a Strategic Solution for Gatambe Intersection

Gatambe intersection is located at 7.270577 latitude and 80.6047636 longitude and is a four-way intersection. This is connected with Colombo – Kandy road (A1), William Gopallawa Mw (AB42) and Gannoruwa road (B365) directly. The Gatambe intersection is within 111–112 km. Though the Gatambe intersection is a roundabout nowadays it is operated manually at peak hours using traffic police. Because of the high congestion near Gatambe intersection the queue propagates both upstream and downstream areas. So it has become a major issue in present.

Strategic solutions means a long-term solution. The solution should have the ability to adjust with different variations of the parameters and situations. The importance of finding a strategic solution for the Gatambe intersection is mainly to reduce the traffic congestion. The Peradeniya Botanical Gardens are located near the Gatambe intersection and it has a high influence on the traffic congestion. Hence, the propagation of the queue is rapidly increasing around it. There are two bridges near the Gatambe intersection and it also influences congestion. If the road network has been blocked near the Gatambe intersection, it will make sense to the congestion near Gannoruwa Bridge, Railway crossing at William Gopallawa Mw, Dangolla junction and towards the old Kandy road. Due to the heavy congestion, the travelers have to suffer from delay, high fuel consumption and CO2 emission. There are many proposed solutions for the reduction of congestion near the Gatambe intersection. It is essential to test the most suitable solution through a simulation software before construction. That will help to reduce expenses and save time by selecting most suitable solution (Fig. 1).

Fig. 1. Gatambe junction

2 Literature Review

The literature review was divided into two main topics. They are researches based on VISSIM simulation models and based on selection of optimum junction improvement strategy.

2.1 VISSIM Based Simulation Models

Dong (2013) analysed the road network of CBD in Beijing, China by using a VISSIM based simulation and it was used to analyse the traffic influences in future years. Here, four different traffic organization plans are compared. Since the driving behaviour in Beijing are significantly different from the USA or Europe, behavioural studies are conducted to calibrate parameters before building up the model. This paper uses travel time, travel speed, queue length and delays as evaluation indicators to conduct the comparisons. They used O-D matrices and some default values in VISSIM as input.

Natalia and Leszek (2017) intended to develop a simulation model to analyse impact of heavy transport generated by the port to the city traffic. They analysed the traffic conditions of individual and collective traffic. They considered peak hours when collecting the data. They mainly focused on traffic patterns, changing of time varying; pedestrian behaviour and impact of signalized systems. They used VISSIM simulation software to simulate the model and analyse the results. The results were presented based on time of travel and delay.

Heuristic algorithm based application tool is used to calibrate the microscopic simulation parameters (Mohammad Rayeeduk Kalam 2017). This described that simulation parameters in VISSIM are based on three heuristic optimization algorithms. This research explained that VISSIM calibration offers four objective functions such as speed, flow, delay and multi-objective criteria and VISSIM has the ability to calibrate all urban or rural driving behaviours. The results show that compared to other algorithms, simultaneous perturbation stochastic approximation generally obtained an acceptable set of parameters within a short period.

Papageorgiou (2006) conducted a research on a microscopic simulation model for traffic management. It mainly presents the microscopic simulation model development of a major highly congested traffic network in Cyprus. This paper explains about traffic flow theory and how a traffic flow simulation model is developed for an effective transport planning.

Kun and Lei (2007) presented a microscopic traffic emission simulation platform by using VISSIM and the study shows that vehicles' emissions are strongly dependant on vehicles operation modes(acceleration). Further, it shows that optimization of signal timing plans can improve the traffic operation as well as reduce the traffic emission.

Papageorgiou (2006) conducted a computer based approach that goes through various stages which include problem identification, model objectives, model development, model calibration, model validation, scenario preparation, simulation experiments and simulated results evaluation. The proposed approach is applied in the case of developing a microscopic traffic simulation model.

Brugmann et al. (2014) intended to develop a simulation model for real –world microscopic traffic simulation. This model consists many elements such as road network data, several subject specific model families for vehicle types and turning elements. VISSIM is a simulation tool for the design of traffic actuated control systems, mainly used to analyse traffic of both urban and highway configurations.

2.2 Selection of the Optimum Junction Improvement Strategy

Yun and Ji (2013) studied the difference between the stop sign intersection and yield sign intersection from the aspects of delay analysis. The traffic efficiency of the yield sign intersection and the stop sign intersection has been explained by using VISSIM.

Huanyun and Fei (2013) presented a development of traffic safety evaluation method based on simulated conflicts at signalized intersections. In this study, simulated conflicts were used to pre-evaluate the safety performance of signalized intersections. Simulated conflicts were generated by the VISSIM simulation. It was found that the safety performance of this intersection was improved after reducing the speed limit.

Zhixia and Madhav (2014) intended a modelling reservation based autonomous intersection control in VISSIM. This model was introduced and implemented in VISSIM. Comparison was done with the results obtained. The comparison was done within optimized signalized control, significantly reduce delays, intersection capacity and with lower volume to –capacity ratios under various traffic demand conditions.

Lee Vien and Mohammed Bally (2014) presented microscopic simulations on the design and operational performance of diverging diamond interchange. Results from traffic impact studies indicated that most of the existing intersections in urban areas have already exceeded the capacity. This paper elaborates the microscopic simulation model developed using the VISSIM software to investigate the design and operational performance of a diverging diamond interchange for different traffic scenarios (Fig. 2).

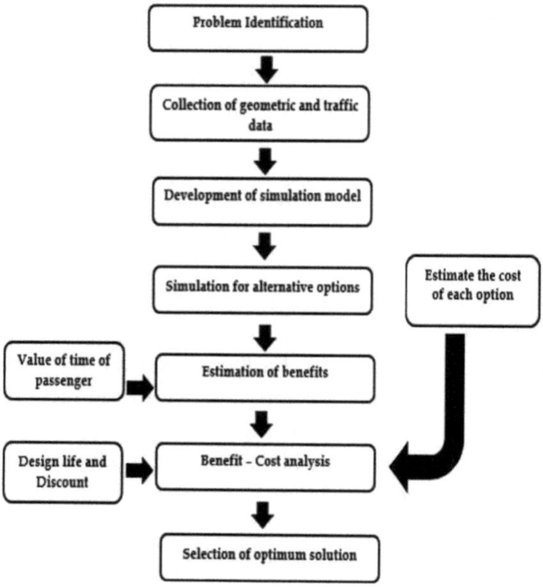

Fig. 2. Flow chart of the methodology

3 Methodology

3.1 Data Collection

A survey was conducted from 6.00 A.M to 8 P.M during a weekday and the below mentioned activities were conducted to collect the necessary data to identify the problem.

- Classified Turning Movement Count at 4 junctions. (CTMC)
- Queue length measurements of each road.
- Rail gate operation survey (Fig. 3).

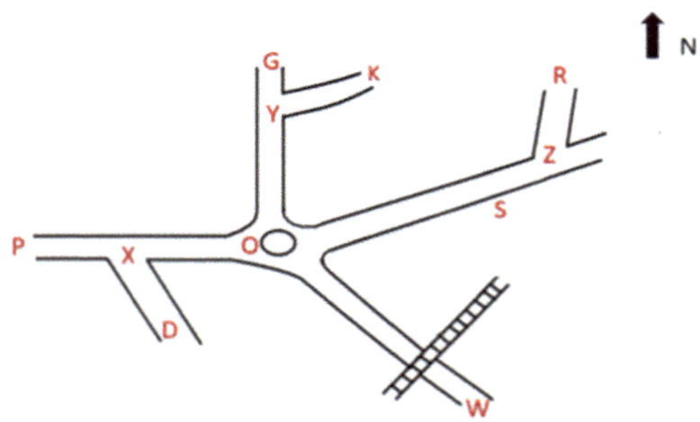

P – Peradeniya Road
G – Gannoruwa Road
D – Dangolla Road
W – William Gopallawa Mawatha
S – Sirimawo Bandaranayake Road
K – Katugasthota Road
R – Deveni Rajasinghe Mawatha
O –Getambe Junction
X – Dangolla Junction

Fig. 3. Layout of the conducted survey

Geometric data was collected to draw the roundabout. Following figure shows the dimensions of the existing roundabout. In addition to this, all necessary dimensions of the road network were measured for input in the simulation software (Fig. 4).

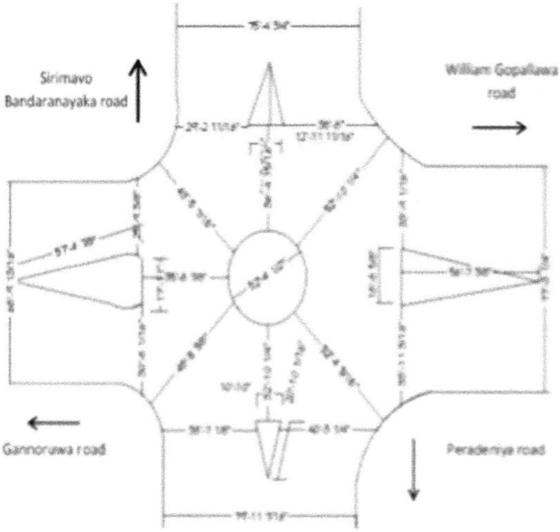

Fig. 4. Dimensions of roundabout

Through classified turning movement count peak hour and off peak hour was identified (Fig. 5).

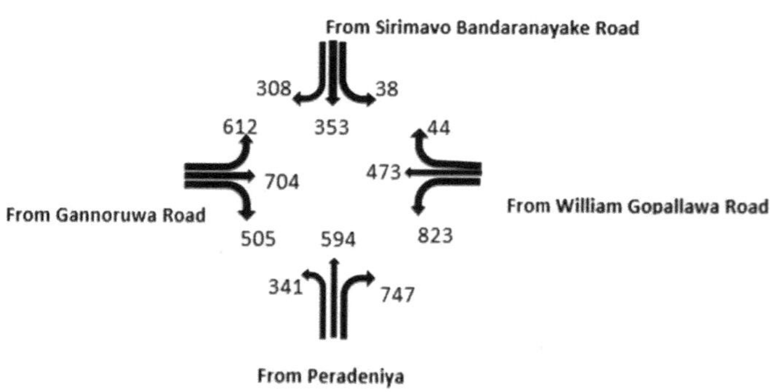

Fig. 6. Turning movements at the junction

Peak hour – 07:15–08:15

Off peak hour – 14:00–15:00

Following figure shows the turning movements of the roundabout during the peak hour (Fig. 6).

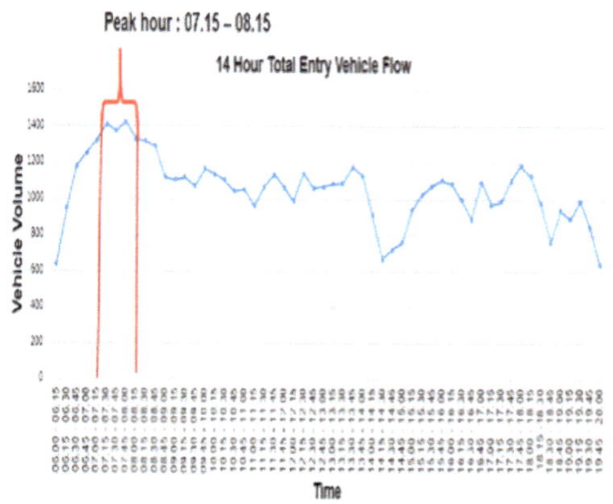

Fig. 5. 14 h total entry vehicle flow

4 Alternative Solutions

To find the optimum junction improvement strategy, different options were tested using VISSIM simulation software. The alternative solutions were;

1. Option A: Roundabout
2. Option B: Signalization without geometric improvement
3. Option C: Signalization with geometric improvement - 3 phase and 4 phase
4. Option D: Grade separation

From each option, the average delay per vehicle during peak and off peak hour was taken to find the optimum results with comparing the benefit cost ratio.

4.1 Option A

The first option was the existing situation, Roundabout option. To run the VISSIM program we needed to draw the roundabout with the road network in the VISSIM interface including correct geometric data. Google maps was used as the VISSIM interface and then road network was drawn in the VISSIM interface, traffic data was fed to the model considering the vehicle composition. Following figure shows the simulated model in VISSIM interface (Fig. 7).

Fig. 7. Existing condition

This is the reference model for the simulation. The model was validated using the simulated queue length results and with actual queue lengths which were taken from the conducted survey.

4.2 Option B

Under this option, four phase signalization was used without the geometric improvement. First the four phase signal time was calculated. Then validated road network model was used and the signal time was input to the model (Figs. 8 and 9).

Cycle time – 217 s

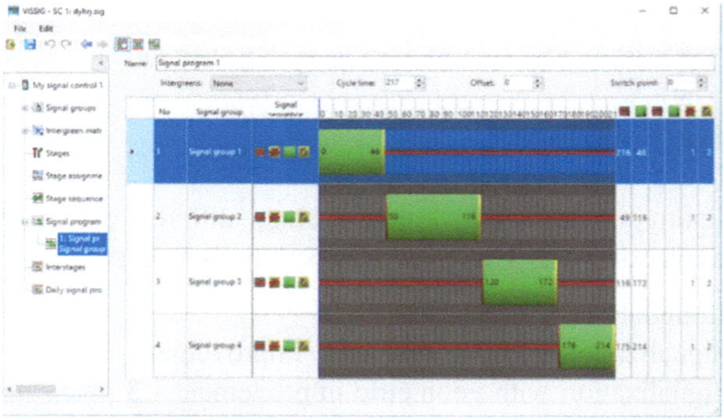

Fig. 8. Four phase signal time

Fig. 9. 4 – phase signalization without geometric improvement

4.3 Option C

Under the geometric improvement, an additional lane is provided for the right turn movement of the Peradeniya road and Gannoruwa entry road is widened to a three lane road. The length of the additional lane of the Peradeniya road was given by considering maximum queue length which can be accommodated and with the maximum queue length results which were taken from the simulation of the existing condition. The additional lane of the Gannoruwa road is provided along the bridge (Figs. 10, 11 and 12).

3- Phase signalization

Cycle time – 120 s

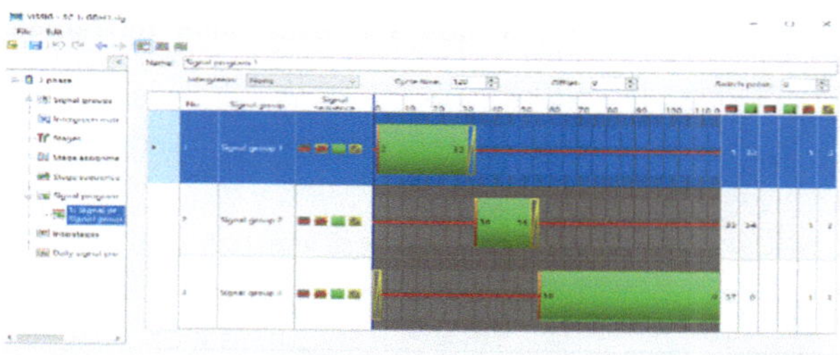

Fig. 10. Three phase signal time

Fig. 11. 3 – phase signalization with geometric improvement

4 – Phase signalization

Cycle time – 217 s

Fig. 12. 4 – phase signalization with geometric improvement

4.4 Option D

A conceptual proposal to grade separate movements PW and WP is considered as option D. When considering the 14-h flow and the peak hour flow, it is observed that there is no predominant movement. However, percentage turning movements of the William Gopallawa Mawatha to Peradeniya road (WP) and Peradeniya road to the William Gopallawa Mawatha (PW) sums up to 28% of total. This conceptual proposal incorporates grade separation of PW and WP over other movements and also the railway (Fig. 13).

Fig. 13. Grade Separation

Design details of grade separator

- Carriageway width (3.65 m * 2) = 7.3 m
- Median Width = 0.8 m
- Shoulder Width (0.3 m * 2) = 0.6 m
- Total Width of the Flyover = 8.7 m
- Gradient = 0.06
- Clearance Height = 7.5 m

5 Economic Analysis

5.1 Outputs from VISSIM Simulation

The results observed after running VISSIM simulation for the four options A, B, C and D are given in the Table 1.

By only looking the delay results of these alternatives, it is difficult to get the optimum results due to high cost of some options. Because of that, carrying a Benefit Cost analysis was a better way to find the optimum solution. Following table shows the Benefit Cost analysis which was done for each option.

Table 1. Average individual delay per vehicle during peak and off peak hour from simulation

Average individual vehicle delay/(s)		
Option	Peak hour	Off peak hour
A	268	157
B	212	138
C – 3 phase	88	64
C – 4 phase	176	114
D	22	13

5.2 Estimation of Benefits

According to the standards practice suggested for the assessment of the transportation section public investments, following are the identified benefits,

1. Travel time saving
2. Reduction of VOC
3. Reduction of emission
4. Improved safety

For the Benefit, only the travel time saving is considered to compare the alternatives. Assuming that night travel time saving is not a benefit to the alternative options, the benefit was calculated considering 3 peak hours and 9 off-peak hours per day. To calculate the benefit considered CCPI 2018 values based on CCPI 1999 values (Table 2).

Table 2. Estimated benefit for each alternative

Option	Benefit/(Rs. Million)
Option A	–
Option B	63.48
Option C – 3 phase	251.56
Option C – 4 phase	123.38
Option D	370.90

5.3 Estimation of Cost

Following cost components are considered in the economic analysis

1. Land acquisition cost
2. Construction cost
3. Maintenance cost

Construction costs were taken from the BOQ provided by the SCDP and land acquisition cost is calculated considering the land price and land area to be acquired. Summarized cost for four option are as follow (Table 3).

Table 3. Estimated cost of each alternative

Option	Land acquisition cost/(Rs. million)	Construction cost/ (Rs. million)	Signalization cost/ (Rs. million)	Total cost/ (Rs. million)
A	–	–	–	
B	–	50	30	80
C – 3 phase	70	114	30	214
C – 4 phase	70	114	30	214
D	60	513	–	573

Finally, the present value for the benefit and the cost were calculated to take the benefit to cost ratio for each option assuming the design life of 20 years and discount rate 10%. Maintenance cost for flyover is Rs 5 Million and for rest of the options maintenance cost is Rs 4 Million.

Table 4. Benefit – cost analysis for the alternative options

	Option					
	A	B	C		D	E
			3 Phase	4 Phase		
Benefit/Rs. million	–	63.48	251.56	123.38	370.90	240.97
PV of benefit/Rs. million	–	540.44	1912.98	938.24	2513.97	1832.45
Cost/Rs. million	–	80	214	214	573	269
PV of cost/Rs. million	–	103.15	221.66	221.66	520.22	274.96
Benefit/cost	–	5.24	8.63	4.23	4.83	6.66

6 Conclusion

The purpose of this study is to review various options that can be implemented to reduce the traffic congestion at the Gatambe intersection and find the optimum solution. Table 1 shows the average individual delay per vehicle during peak and off peak hour which were taken from the simulation. It shows that all the alternative options have low individual delay per vehicle than the existing situation. Through this simulation, the effect of each method can be implemented before construction. Therefore, this simulation method is very useful for all the practical situations. However, by considering only results of individual delay per vehicle, the best solution cannot be found. A Benefit –Cost analysis was done to get the best solution which gives the least delay for a low cost. To get the benefit, travel time saving was considered.

From Table 4, the maximum benefit to cost ratio is given by the option C, 3 – phase signalization with geometric improvement. This option consists of a 3 – phase signalization system with geometric improvement which has an additional right turn lane for the Peradeniya road and the Gannoruwa entry road is widened to a three lane road.

Although the travel time saving of the flyover is maximum, due to the high cost it got a lower value in the benefit-cost ratio. Also, when considering the 4 – phase signalization without geometric improvement, it gives a considerable delay. Therefore, the benefit is less. By analysing these benefit-cost ratios, the geometric improvement with a 3 – phase signalization is the optimum solution.

Acknowledgements. The authors wish to express their sincere thanks to Eng. (Mrs) T.N. Siyambalapitiya, Director (Planning), RDA, Engineering staff of planning branch of RDA, Traffic survey team of RDA and all the members in the survey team for the great support in conducting the survey successfully. A special thank goes to Eng. Manjula Kularathne of CMC, Mr. Chathura and Mr Nadeeka Jayasooriya of University of Moratuwa for their advice on VISSIM simulation software.

References

Brugmann J, Schreckenberg M, Luther W (2014) A variable simulation model for real-world microscopic traffic simulations. Simul Model Pract Theory 48:58–92

Dong L, Xiaokuan Y, Chao G (2013) VISSIM - based simulation analysis on road network of CBD in Beijing, China. Procedia Soc Behav Sci 96(2013):461–472

Huanyun Z, Fei H (2013) Development of traffic safety evaluation method based on simulated conflicts at signalized intersections. Procedia Soc Behav Sci 96:881–885

Kun C, Lei Y (2007) Microscopic traffic-emission simulation and case study for evaluation of traffic control strategies. J Transp Syst Eng Inf Technol 7(1):93–100

Lee Vien L, Mohamed Bally M (2014) Microscpic simulation on the design and operational performance of diverging diamond interchange. Transp Res Procedia 6:198–212

Mohammad Rayeedul Kalam S et al (2017) Heuristic algorithm based application tool to calibrate microscopic simulation parameters. traffic and transportation engineering (English edition)

Natalia S, Leszek S (2017) The use of a simulation model to analyze the impact of heavy transport generated by the port to the city traffic. J KONBiN 42:25–42

Papageorgiou GN (2006) Towards a microscopic simulation model for traffic management: a computer-based approach. IFAC Proc Volumes 39:403–411

Weerakoon H, Weeraman T (2017) Use of microscopic traffic simulation model as a test bench to assess improvement options for Gatambe Intersection. S.N., Kandy

Yun M, Ji J (2013) Delay analysis of stop sign intersection and yield sign intersection based on VISSIM. Procedia-Soc Behav Sci 96:2024–2031

Zhandong Z, Shaohui C, Yanquan Y, Aixiu H (2016) VISSIM simulation based expressway exit control modes research. Procedia Eng 137:738–746

Zhixia L, Madhav V (2014) Modelling reservation - based autonomous intersection control in VISSIM. Veh - Highway Autom Committee Peer 2381:81–90

Gap Assessment of Warning and Dissemination Process of Early Warning System in Coastal Areas of Sri Lanka

H. M. S. S. Hippola[1(✉)], E. M. S. D. Jayasooriya[1], G. P. Jayasiri[2], Chameera Randil[2], Chamal Perera[2], K. K. K. Sylva[1], A. K. Kulathunga[1], C. S. Bandara[1], C. S. A. Siriwardena[2], and P. B. R. Dissanayake[1]

[1] University of Peradeniya, Peradeniya, Sri Lanka
samhipo@gmail.com
[2] University of Moratuwa, Moratuwa, Sri Lanka

Abstract. Multi Hazard Early Warning Systems (MHEWS) issue warnings to help communities safely evacuate from hazardous areas. Studies showed that the response of coastal communities to warnings issued by the existing MHEW system in Asian countries including Sri Lanka is insufficient. In this study, the current context and gaps of MHEW system related to the "warning and dissemination" process in the coastal regions of Sri Lanka were identified using a proper research methodology. Survey was carried out through oral questionnaires in selected coastal areas of Sri Lanka. By interpreting the output of the interviews, along with a deep literature review, the current context and gaps of the MHEW systems related to warning and dissemination in coastal areas of Sri Lanka were listed out. From the identified problems, key areas which need special attention of the Disaster Management Centre, were highlighted. Among them, introducing a better Nation-wide Emergency Communication system using more advanced technologies to directly disseminate warnings to vulnerable communities instantly, automating possible components of the warning dissemination process, increasing the number of early warning towers, increasing the human resource in the disaster management centre and establishing a specialized media department at the DMC to coordinate with the media were prominent. In addition, the necessity of improving response capability was highlighted.

Keywords: Disasters · MHEW · Warning and dissemination

1 Introduction

Sri Lanka, being an island situated in the Indian Ocean, is prone to coastal hazards. Among those hazards, floods, cyclones, thunder storms and coastal erosions occur at more frequent intervals, a tsunami had occurred once in the recent past while there are risks of earthquakes as well. Sri Lanka, already ravaged by the floods in 2003, was then hit by the monstrous Indian Ocean tsunami in 2004 which engulfed approximately 35,000 Sri Lankan lives. These conversions of coastal hazards in to massive disasters

R. Dissanayake and P. Mendis (Eds.): ICSBE 2018, LNCE 44, pp. 36–44, 2020.
https://doi.org/10.1007/978-981-13-9749-3_4

led the government of Sri Lanka to strengthen legislature and institutional arrangements for disaster risk management. An efficient Multi Hazard Early Warning (MHEW) system was required urgently to reduce the risk of destruction of lives and property. Subsequently, Sri Lanka enacted the Disaster Management Act No. 13 in 2005. Under this Act, Disaster Management Centre (DMC) of The Ministry of Disaster Management, Sri Lanka was established with the responsibility of preparing and implementing of a National Disaster Management Plan and a National Emergency Response Plan.

Despite these legislative and institutional reformations, Sri Lanka has experienced several coastal hazards with considerable damage since 2005 (Disaster Management Centre 2016). Flood events were reported every year affecting 19 districts on average. Each flood event has caused considerable loss of lives with the flood events in 2008 and 2011 being those with highest number of death tolls, 44 and 69 respectively. In addition, high winds have caused a considerable loss of lives in the years 2011 and 2013. Furthermore, the most recent southwest monsoon weather condition has caused 22 deaths. It is to be noted that these numbers are quite high for hazards such as floods and high winds and the consequences are much more severe if another tsunami happens. These facts reflect either the inefficiency of MHEW or the reluctance of the vulnerable communities to respond to early warnings. Therefore it is timely to identify gaps of the existing MHEW system in Sri Lanka and make the required improvements. Furthermore it is equally vital to address and remedy the problem of communities being reluctant to obey early warnings.

The United Nations International Strategy of Disaster Reduction (UNISDR 2006) points out four key elements of a people centred early warning systems. They are Risk Knowledge, Monitoring and Warning Service, Dissemination and Communication and Response Capability. Risk knowledge involves assessing the dynamic nature of hazards and vulnerabilities due to climate change and change in demography. Monitoring and Warning services should consist of a sound scientific basis to continuously predict and forecast hazards, timely and accurately while dissemination and communication processes should ensure receipt of clear, meaningful messages containing critical and useful information for vulnerable communities, with sufficient lead time to evacuate safely. The final key element focuses on the response capability of vulnerable communities. These communities should understand their risk situation and act accordingly obeying and respecting the early warnings issued.

The whole MHEW process can be discretized in to the aforementioned key elements to study the current context and gaps. Hettiarachchi et al. (2017) have investigated the efficiency and effectiveness of the existing disaster management frameworks in Sri Lanka in a more general sense. This paper presents a discussion that focuses on the current context and gaps of the Dissemination and Communication processes of existing MHEW in Sri Lanka under the subtopics, Communication Channels, Warning Messages and Coordination with Media and Military. Furthermore the reasons for the poor response from the vulnerable communities for early warnings issued are discussed towards the end. Dutta et al. (2014) have done a similar study to assess gaps and strengthen early warning System in Cambodia.

2 Existing Warning and Dissemination System in Sri Lanka

An efficient early warning system should focus to generate and disseminate timely and meaningful warning information to enable individuals, communities and organizations threatened by hazards to take necessary prepared measures and act appropriately with sufficient time to reduce the possibility of harm or losses. The process of delivering early warnings, originated from the technical agencies, to end users, involves several pre-planned steps. A schematic diagram of the warning and dissemination system of Sri Lanka is given in Fig. 1.

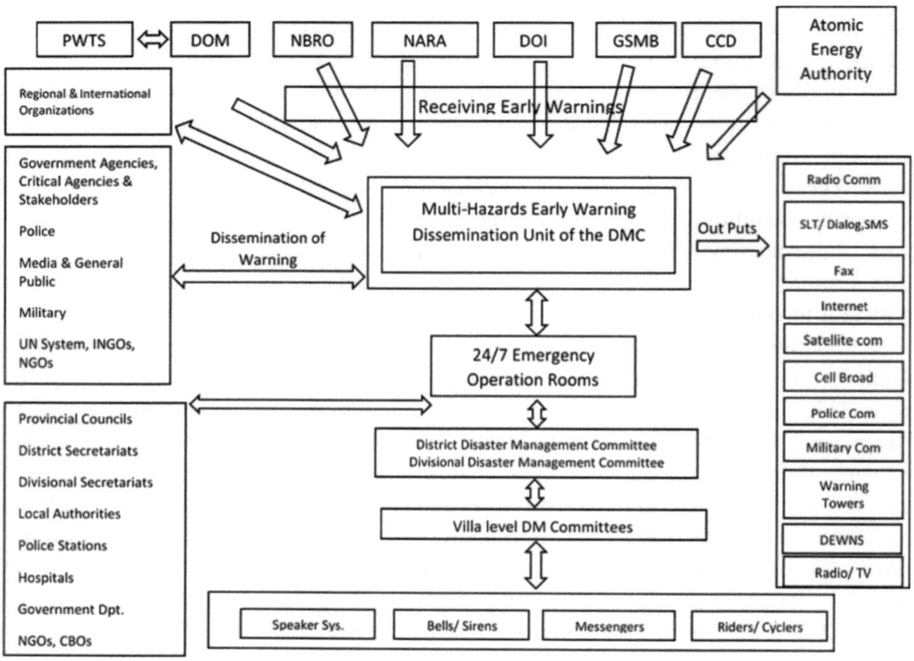

Fig. 1. Multi hazard warning dissemination system (Extracted from the DMC official website)

At a time of an impending disaster, both local and international technical agencies analyse the data they receive, to predict the occurrence of the concerned disaster. It is their responsibility to forecast the scale and issue warning alerts for different types of hazards. In Sri Lanka, local technical agencies involved are the Department of Meteorology (DOM), National Building Research Organization (NBRO), The National Aquatic Resources Research and Development Agency (NARA), Department of Irrigation (DOI), Geological Survey and Mines Bureau (GSMB), Coast Conservation and Coastal Resource Management Department and Marine Environment Protection

Authority (MEPA). The DOM generally monitors the occurrence of tropical cyclones and tsunami while the DOI monitors floods. The international technical agencies that assist the DMC are also listed in Fig. 1. It is the responsibility of these local and international technical agencies to inform the Director General of DMC and Emergency Operating Centre (EOC) if a disaster is likely to happen. It should be highlighted that Probability of Detection (POD) and Probability of False Alarm (PFA) are decisive factors which governs the credibility of these early warnings issued.

Once the DMC is informed regarding an impending disaster, the MHEW dissemination unit of the DMC initiates the process of warning dissemination, after an analysis of the situation. They communicate with EOC that operates for 24 h every day which will later coordinates the whole warning dissemination process with the armed forces (army, navy, air force, coast guard and police), media and relevant government and non-government agencies. The outputs of the MHEW dissemination unit of the DMC are radio/television news, short message system (SMS), fax, internet, satellite, police and military communication, early warning towers and other media. It is important to note that efficiency, technology of these communication channels and outputs directly affect the speed of warning dissemination.

It is the responsibility of the EOC to disseminate the warnings to administrative divisions such as provincial councils, district secretariats, divisional secretariats and local authorities. In parallel to these communications, the EOC disseminates the early warnings to disaster management committees at district, divisional and village levels. This is followed by the receipt of early warnings by the vulnerable community through speaker systems, bells/sirens, messengers and riders/cyclers. Police stations and the Grama Niladhari (village level administrative officers of the government) usually assist the warning dissemination process at the village level.

3 Methodology

The gap assessment of the warning dissemination process and assessing the response capability for coastal hazards was done by conducting separate interviews with disaster management officials, academics who are engaged in disaster related research and vulnerable communities in coastal areas of Sri Lanka. The technique used to select respondents in vulnerable communities was the snow ball sampling method. The oral interviews carried out on professionals and vulnerable people were based on carefully prepared questionnaires. The meetings were conducted with the aim of understanding warning and dissemination process and key elements governing response capability. The flow chart in Fig. 2 demonstrate the methodology.

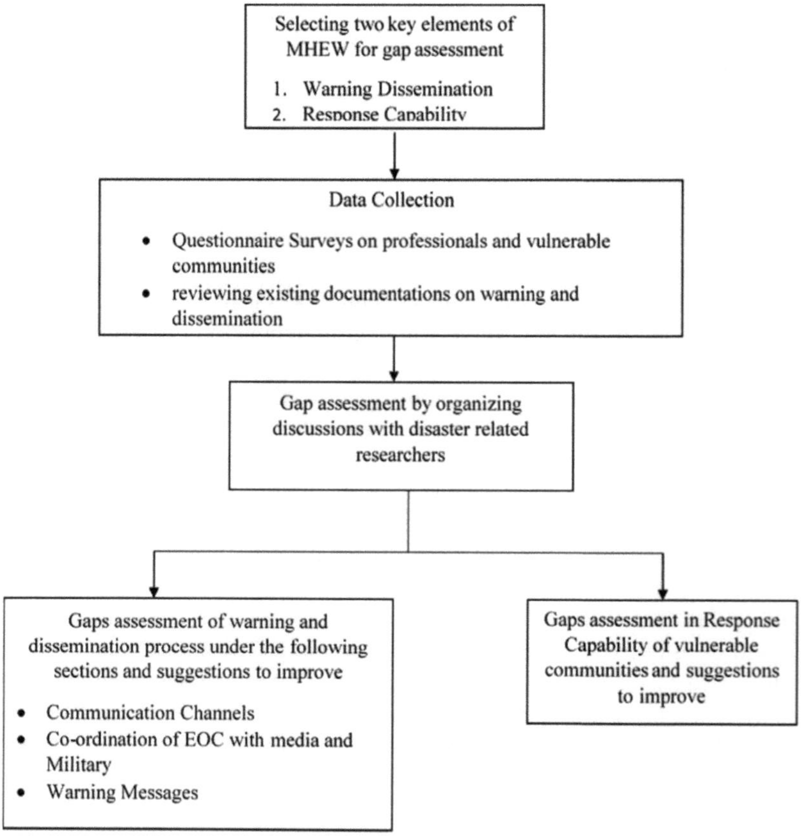

Fig. 2. Methodology

4 Results and Discussions

This section presents the current context, gaps and suggestions for improvement of the warning dissemination process for coastal communities in Sri Lanka under the sub-topics communication channels, coordination, warning messages and response capability of vulnerable communities.

4.1 Communication Channels

Communication of the early warnings generated by local and international technical agencies is currently carried out in four layers namely, National, District, Divisional and Grama Niladhari (village) levels. At the National level, Early Warning (EW) messages are disseminated to emergency response committees that are then passed to their relevant organizations. The District Disaster Management Centre Units (DDMCU) act as focal points of communicating EW messages to district, divisional and Grama Niladhari levels. It is the responsibility of the DDCU to disseminate EWs to

District Secretariat, Divisional Secretariat, stake holder agencies and relevant political authorities. In the meantime, the warnings are disseminated to the vulnerable communities through local authorities. At Grama Niladhari level, EW messages are disseminated to the vulnerable communities through last mile communication tools. The communication methods used in each of these levels are presented in Table 1. It is to be noted that a common set of communication methods are used by technical agencies, national level and district levels while last mile communication methods are used at the village level.

Table 1. Communication methods used in warning dissemination process

Technical agencies/national and district levels	Village level
Police and military communication	Police vehicles
Satellite and radio communications	Public address system
Intra government network	Sirens
Fax	Temple and church bells
Internet	Riders/messengers
Early warning towers	Early warning towers
Media	Media
Telephones, SMS	Telephones, SMS

The responses from the professionals suggest that overall communication network can be rated as moderate. However, some of the professionals have specifically rated the communication of EOC, with administrative levels and media as poor. An efficient warning dissemination process demands a seamless flow of warnings. Drawbacks in a few links might act as bottle necks to drastically reduce the efficiency of warning dissemination process. The following paragraphs discuss the possible gaps in the existing communication channels.

Most of the professionals rated the standard and conditions of the equipment as moderate and agreed that the equipment are maintained at frequent intervals. Among these equipment, early warning towers were identified as the most preferred method of warning receiving. The general public in coastal communities confirmed that the early warning towers are maintained and tested in frequent time intervals. However, they have pointed out that the number of existing early warning towers are insufficient to cover all the regions; thus there is a need of increasing number of early warning towers.

Majority of the professionals were not satisfied about the speed of communication and technologies used. It is to be noted that the speed of communication can be improved drastically by establishing a better Nation-wide Emergency Communication system with new technologies which can directly convey warnings from National level to end users. In addition, the conventional equipment can be updated and maintained at better standards to disseminate the warnings effectively. Furthermore, almost all professionals suggested that updating databases is as equally important as much as the standard of equipment.

Majority of the professionals pointed out that the lack of staff, is a major concern in disaster management institutions. It is true that in the current system, a considerable human involvement is required when disseminating a warning from national level to end users. Therefore the current system requires either the enrolment of new staff or introducing new technologies to automate possible sections of the warning dissemination process. In addition, enrolment of new staff can be encouraged by increasing salaries and other benefits.

4.2 Coordination of Emergency Operation Centre with Media and Military

At a time of an impending disaster, the DMC provides accurate warnings to the media through appointed media spokespersons. The media will disseminate the warning messages to last mile vulnerable communities. Therefore, it is the responsibility of DMC to ensure an accurate press release providing the required warning messages to media within the shortest possible time. Majority of the professionals had a moderate feeling regarding the coordination between the EOC and the media. However, some of the professionals pointed out that the media tends to act on their own will and thus reduce the efficiency of warning dissemination.

To address this issue, some professionals have suggested to establish a specialized media department at the DMC to build up a better understanding with the media. This specialized coordination will allow the DMC and the media to be efficiently prepared for a warning dissemination situation depending on the type of hazard and vulnerable community.

The military forces consists of the Army, Navy, Air Force and the Police. These forces have the inherent capability to disseminate early warnings to vulnerable communities and direct target population to safe locations. Therefore the DMC should disseminate the warning messages to the forces with maximum possible accuracy. Almost all of the professionals have rated the coordination between EOC and Military forces as good.

4.3 Warning Messages

Warning messages received by the vulnerable communities should convey the specific nature of the threat and its impacts. The UNISDR (2009) presents that warning messages should be tailored to the specific needs of those at risk, should be geographically specific, should in cooperate the understanding of the values, concerns and interests of those who will need to take action, should be well recognized and understood, should be consistent over time, and should include follow up actions.

Majority of the professionals rate the condition of warning messages as moderate. They point out that the current warning dissemination system has previously prepared warning messages with procedures to update them. However, the existing warning messages should be improved drastically to meet the standards suggested in the UNISDR (2006). The reduced quality of warning messages was reflected in the recent flood events where twenty people died. Therefore it is mandatory to create new meaningful warning messages accounting for the type of disaster, vulnerable community, communication method etc.

4.4 Response Capability

From the above analysis, it is evident that the warning and dissemination processes implemented in Sri Lanka are satisfactory despite the improvements suggested. However, the death tolls in recent hazardous events suggests that there is a serious issue when disseminating early warnings to vulnerable communities. If the warning dissemination process stands satisfactory, the problem lies with the response capabilities of the vulnerable communities. The interviews conducted with the coastal communities pointed out several root causes for this poor response capability. The UNISDR (2009) presents a check list to improve the response capability of vulnerable communities. This check list includes respecting warnings, disaster preparedness and established response plans, assessed and strengthened community response capacity and enhanced public awareness and education. This check list can be used to propose suggestions in order to improve the response capability of Sri Lankan coastal communities.

The business oriented life style of people in coastal communities has led to their lack of knowledge on disasters. This lack of knowledge was reflected in 2004 tsunami incident, where people went towards the sea when water level reduced prior to the huge waves. The surveys revealed that most of the people are unaware about the decision makers, how warnings are disseminated, simple information on hazards, vulnerabilities etc. This unawareness may lead to situations where people ignore warning messages and become victims of the disaster. Another aspect of this unawareness is, most of the people preferred to respond only to bells/sirens, messengers and early warning towers rather than warning messages from SMS, internet and media. To address this issue of not knowing, public awareness programs should be conducted highlighting the MHEW process, a general knowledge about disasters, reliable early warning sources and methods of responding to early warnings. In addition, the unawareness can be further reduced by introducing new academic curriculums with disaster management courses from the school level to the university level.

False alarms is another key factor which reduces the response of vulnerable communities for early warnings. One false alarm by the authorities will drastically reduce the credibility of the sources. For example, most of the people in Benthota area experienced false alarm before the storm event of November 2017 and faced the consequences of not obeying the next warning where the storm actually happened. Due to these facts, false alarms should be minimized as much as possible. However, this breach of trust can be restored by educating the general public on the real scenario of forecasting and early warnings.

The Sri Lankan culture is built around the concept of "home" where people live throughout their lives. The interviews with people in coastal communities exposed that some people show reluctance to evacuate leaving houses, personal belongings, vehicles and agricultural lands for the sake of one warning. Furthermore, this reluctance can increase due to the absence of a specific place to evacuate. To avoid these misconceptions, the awareness of people needs to be raised through education and awareness programs.

5 Conclusions

The paper proposes several improvements to the warning and dissemination process under the sections communication channels, warning messages and co-ordination with media and military forces. The main improvements proposed to the communication channels are, introduce a better Nation-wide Emergency Communication system using more advanced technologies to directly disseminate warnings to vulnerable communities instantly, automating possible components of warning dissemination process, improving the maintenance of communication tools, increasing the number of early warning towers and increasing the human resources in the disaster management centre. In addition, the possibility of establishing a specialized media department at the DMC to coordinate with the media was highlighted. Furthermore, it was proposed to make new meaningful warning messages, accounting for the type of disaster, vulnerable community and communication method. In the next stage of the methodology, it was identified that the moderate warning and dissemination system existing in Sri Lanka becomes useless when the vulnerable communities lack response capability. Finally, education and awareness programs were identified as a remedy to improve the response capability of coastal communities in Sri Lanka.

Acknowledgements. This study was conducted with the financial aid from CABARET (Capacity Building in Asia for Resilience Education), a project of the European Union's Erasmus + Programme - Key action 2 - Capacity building in the field of higher education.

References

Disaster Management Centre (2016) Impacts of disasters in Sri Lanka. https://app.adpc.net/sites/default/files/public/publications/attachments/Impacts%20of%20Disasters%20in%20Sri%20Lanka%202016.pdf

Dutta R, Basnayake S, Ahmed AK (2014) Assessing gaps and strengthening early warning system to manage disasters in Cambodia. J Integr Disaster Risk Manag 172 and 173

Hettiarachchi SSL, Siriwardena CS, Jayasiri GP (2017) Investigation of efficiency and effectiveness of the existing disaster management frameworks in Sri Lanka. In: 7th international conference on building resilience; using scientific knowledge to inform policy and practice in disaster risk reduction

Ministry of disaster Management - Sri Lanka (2013) Progress Report 2013. http://www.dmc.gov.lk/index.php?option=com_content&view=article&id=75&Itemid=161&lang=en

UN-ISDR Developing Early Warning Systems: A Check list (2006) The third international conference on early warning. https://www.unisdr.org/2006/ppew/info-resources/ewc3/checklist/English.pdf

UN-ISDR Terminology on Disaster Risk Reduction (2009). http://www.unisdr.org/files/7817_UNISDRTerminologyEnglish.pdf

Gaps in the Evacuation of People with Special Needs During Coastal Disasters in Sri Lanka

E. M. S. D. Jayasooriya[1(✉)], H. M. S. S. Hippola[1], C. S. Bandara[1], A. K. Kulathunga[1], K. K. K. Sylva[1], P. B. R. Dissanayake[1], G. P. Jayasiri[2], Chameera Randil[2], Chamal Perera[2], and C. S. A. Siriwardana[2]

[1] University of Peradeniya, Peradeniya, Sri Lanka
darshana.jayasooriya@eng.pdn.ac.lk
[2] University of Moratuwa, Moratuwa, Sri Lanka

Abstract. Coastal hazards taken place all over the world have ended up with major devastations such as loss of life, property damage and environmental degradation more often. Sri Lanka, being an island in the Indian ocean is much vulnerable to coastal hazards; the unprecedented human disaster that caused by the Asian Tsunami was one of the worst examples. Present studies show that there are lot of programs carried out by the government and non-government organizations to drag down the adversity towards the coastal communities in the country who are much vulnerable to coastal disasters, focusing all the segments of basically social, economic and environmental aspects. Evacuation of coastal communities safely in disastrous situations is one of the main requirements within the disaster recovery framework. Having effective multi-hazard early warning (MHEW) systems and procedures in place in the coastal regions of the country helps to reduce the disaster risk. Although there are policies and practices for the evacuation of coastal communities including MHEW systems, the present study shows that the attention and preparedness towards the evacuation of people with special needs in the coastal communities is more or less omitted by the responsible parties. Moreover, when disasters strike, people with special needs meet various discriminations in terms of evacuation to a shelter or relief points and more often are overlooked from the emergency preparedness, response and recovery stages. Therefore, the present research focuses on determining the gaps in the evacuation of people with special needs in the context of MHEW systems during coastal hazards in Sri Lanka. Open interviews and telephone interviews were conducted in few major coastal cities and understood the likelihood of the obstacles, issues and the perception towards the evacuation of people with special needs. A questionnaire was developed according to the open interviews and fine-tuned in three sandpit events employing various local and foreign professionals in academics. The questionnaires were distributed online and the responses were analyzed. Eventually, a substantial number of gaps were identified in supporting the prevailing systems in Sri Lanka and was arrived at conclusions to impact the government through the findings in order to make the evacuation of people with special needs during coastal disasters more efficient.

R. Dissanayake and P. Mendis (Eds.): ICSBE 2018, LNCE 44, pp. 45–53, 2020.
https://doi.org/10.1007/978-981-13-9749-3_5

Keywords: Coastal hazards · Evacuation · People with special needs · Multi hazard early warning systems

1 Introduction

The ongoing studies from the National Council for Persons with Disabilities in Sri Lanka professes a deficiency of research regarding the guidelines and understanding about the preparedness in the early stage before a disaster, emergency response stage during a disaster and recovery stage for people with special needs; pointedly people who are living with disabilities. When it comes to world scenarios, disasters such as Asian Tsunami 2004 jeopardise the past approaches such as "wait-for-help" in an evacuation of people with mobility deficiencies. Asian Tsunami that brought severe damage to Sri Lanka even showed that the above method resulted in lot of loss of lives due to its irregularities. In the Sendai Framework for Disaster Risk Reduction, people with special needs have been addresses more or less and the responsibilities of the stakeholders towards them (Stough and Kang 2015). When it comes to children, Children with special needs seem expelled from disaster risk reduction (DRR) efforts thus leading to enhanced problems related to physical, psychological, and educational growth and development starting from the early stage of their lives (Ronah et al. 2015). According to Nobuyuki et al. (2006), the mortality has been observed among females, elderly and children in Sri Lanka due to the Asian Tsunami 2004. Moreover, related issues and obstacles of how to facilitate and help people with disabilities during an emergency disastrous situation, how to deal and support to each other, and help policy and emergency planners, have been disregarded and unnoticed (Alexander et al. 2012). The content of literature and research on the aforementioned context is trivial compared to the need and demand (Parr 1987) and latest efforts to reiterate it have been neither numerous nor copious (Fjord and Manderson 2009).

The ultimate goal of this project is to ameliorate and promote efforts and plans for an enhanced preparedness delivered by government and non-government institutions related to people with special needs during every stage of a disaster such as before, during and after the disaster. The aforementioned objective was polished through observing the severity and the occurrence of coastal disasters, livelihood of the people with special needs, emergency needs, government cooperation and other relevant community conditions.

The research team collected and assessed data related to ongoing issues of people with disabilities and their domiciliary environment when a disastrous situation strikes. Withal, the research team also assembled data from key government officers such as "Grama Niladhari" and divisional secretariats, also comprising emergency planners, multi service providers, and other respective government and non-government bodies. People were gathered at the regional offices and the discussions were carried out. The aim of this data collection was to congregate people with special needs and the organizations related to disaster management to build up strong relationship and review the issues and matters related to preparedness and emergency evacuation in a more productive and lively manner.

The current study also aims at sharing the findings with the emergency planners, managers and decision makers who directly or indirectly work for the disaster risk reduction and making efforts to include people with special needs to consider as a particularly and admittedly vulnerable group during coastal disasters.

1.1 Disability and Its Implications in Sri Lanka

According to the National Policy on Disability, the legitimate interpretation of disability in Sri Lanka which was stated in the protection of the rights of persons with disabilities which was originated by the Sri Lankan Parliament in 1996 is given below. A "person with disability means any person who, as a result of any deficiency in his physical or mental capabilities, whether congenital or not, is unable by himself to ensure for himself, wholly or partly, the necessities of life" (De Silva et al. 2007). This definition seems comprehensive, including most aspects of medical, social and economic factors binding with people with disabilities. However, there are numerous definitions and implications for persons with disabilities in various acts and curriculums contemporary prevailing in Sri Lanka. Furthermore, people with disabilities and special needs are considered as a vulnerable group in this all above context (Hemingway and Priestley 2006) and still this so-called vulnerability is barely even contemplated or said. Nevertheless, according to the literature available it appears that there is a smidgen of actions related to the disaster risk reduction in the context of disability. According to the census and statistics in Sri Lanka in the year 2001, there are a significant number of people with special needs living in the country including the areas prone to coastal disasters (Ministry of Social Welfare 2003). Table 1 was extracted from the national report of census and statistics and it shows the statistical illustrations of the different disabilities in the areas we concerned for the present study. It is important to note that Table 1 includes only a few districts that we covered our scope of work.

Table 1. The census and statistics of six major near coastal districts in Sri Lanka in the year of 2001, according to type of disability, number and rate per 10,000 population

District and sex	Total		Seeing		Hearing/Speaking		Disability in Hands		Disability in Legs		Other Physical Disability		Mental Disability	
	No	Rate	No	Rate	No	Rate	No	Rate	No	Rate	No	Rate	No	Rate
Total(18 Districts)														
Both	274711	162.9	69096	41	73343	43.5	48131	28.5	90576	53.7	13275	7.9	69026	40.9
Male	158446	189.9	35419	42.5	40584	48.6	31070	37.2	56529	67.8	7344	8.8	37181	44.6
Female	116265	136.4	33677	39.5	32759	38.4	17061	20	34047	40	5931	7	31845	37.4
Colombo														
Both	27430	122.8	5998	26.8	6414	28.7	4854	21.7	9168	41	1219	5.5	8730	39.1
Male	15355	135.4	2997	26.4	3507	30.9	3208	28.3	5655	49.9	698	6.2	4391	38.7
Female	12075	109.8	3001	27.3	2907	26.4	1646	15	3513	31.9	521	4.7	4339	39.4
Kalutara														
Both	18431	173.7	4690	44.2	5043	47.5	3272	30.8	6209	58.5	808	7.6	4809	45.3
Male	10471	201.1	2303	44.2	2757	52.9	2096	40.3	3818	73.3	431	8.3	2605	50
Female	7960	147.4	2387	44.2	2286	42.3	1176	21.8	2391	44.3	377	7	2204	40.8
Galle														
Both	17873	180.4	4722	47.7	5051	51	3051	30.8	5762	58.2	666	6.7	4981	50.3
Male	9796	204.3	2235	46.6	2727	56.9	1888	39.4	3389	70.7	346	7.2	2585	53.9
Female	8077	158	2487	48.7	2324	45.5	1163	22.8	2373	46.4	320	6.3	2396	46.9
Matara														
Both	15255	200.4	3773	49.6	4320	56.7	2535	33.3	4764	62.6	681	8.9	4379	57.5
Male	8486	231	1849	50.3	2373	64.6	1574	42.8	2772	75.4	336	9.1	2343	63.8
Female	6769	171.9	1924	48.9	1947	49.4	961	24.4	1992	50.6	345	8.8	2036	51.7
Hambanthota														
Both	12736	242.4	3682	70.1	3287	62.6	2079	39.6	3745	71.3	628	12	3391	64.5
Male	7181	274.8	1852	70.9	1836	70.3	1272	48.7	2203	84.3	317	12.1	1882	72
Female	5555	210.3	1830	69.3	1451	54.9	807	30.6	1542	58.4	311	11.8	1509	57.1
Ampara														
Both	8462	143.6	2339	39.7	2367	40.2	1289	21.9	2511	42.6	324	5.5	1670	28.3
Male	4973	168.4	1247	42.2	1298	44	839	28.4	1623	55	191	6.5	946	32
Female	3489	118.7	1092	37.1	1069	36.4	450	15.3	888	30.2	133	4.5	724	24.6

2 Methodology and Study Plan

According to the above explained content, the motif of the current study is to find out some of the gaps in the currently prevailing system in Sri Lanka related to people with special needs and their emergency evacuation during a coastal disaster. It was performed with regards to the various hazards occur, disability and its' implications, livelihood of the people with special need, evacuation routes and shelters, different emergency needs and also the level of local government cooperation before and during a disaster.

2.1 Data Collection: Survey with People with Special Needs and Their Relatives, Local Public and Government and Non-Government Agencies

The approach was consisted with telephone calls, open and closed interviews and also an online survey. Basically, the evacuation of people with special needs by the normal people was considered while determining the administrative gaps and existing methodologies. The interviews were done in 6 major near coastal districts. (Colombo, Kaluthara, Galle, Matara, Hambanthota, Ampara). A part of the interviews was collected through telephone survey of the people with special needs and as well as normal public. The other part of the responses was collected through an online questionnaire survey of persons with reference to the evacuation of people with special needs who live in the community. The questionnaire was developed according to the open interviews and fine-tuned in three sandpit events employing various local and foreign professionals in academics. The questionnaires were distributed online and the responses were analyzed.

3 Results and Discussion

The responses were obtained from people with special needs, normal people in being with the perception of evacuation of people with special needs and a qualitative response from professionals were gathered in order to visualize and determine the capacity of the main topic. Half of the responses from the normal people reveals that they have someone with special needs near their vicinity and more than sixty percent of the respondents expressed that they put an extra effort to evacuate that sort of person while risking their own lives. It seemed that the majority of people tends to help the people with special needs in any kind of emergency which can be highly evaluated as a great attitude in the community. The rest has mentioned that basically they do not have the capacity to do so since they are more concerned about their own evacuation. But findings about priorities stated that nearly eighty percent of people tend to prioritize their family evacuation.

3.1 Key Findings on Different Types of Hazards and Severity

From the telephone conversations and open interviews conducted, it was identified that the most common coastal disasters are coastal flood, sea level rise, storms and heavy rains and extreme wind forced- waves. Although Tsunami has once occurred and has brought lot of damage, it is not considered as a frequent disaster. As reported storms have high frequency of occurring. Although Asian Tsunami destroyed almost everything belong to them, that kind of higher devastative Tsunami has low frequency of occurring.

3.2 Key Findings on Early Warning Systems

Whatever to achieve, the early warnings should be reached to the last mile so that people can react to the disasters easily to an extent where it is possible to the maximum. It was found that main early warning systems currently prevailing are SMS alert, Tsunami early warning towers, television/radio, door to door, telephone/internet and police warnings. But the most common system was TV/Radio that was more than 80%. Some people complained that they totally have not received any kind of early warnings. When it comes to the effectiveness, majority of respondents nearly 60% have mentioned that it was approximately fifty percent effective.

3.3 Key Findings on Evacuations and Disabilities

As per the findings majority of people that have evacuated someone with special needs during a coastal disaster reported that the incident was with majorly pregnant woman. Nevertheless, there were significant amount of people with waling disability as well. Few were with other disabilities. Almost everyone in this category mentioned that they faced difficulties in providing mobility assistance like transportation. The other considerable fact was that a mentioned above large number of people have not received the early warnings before the Asian Tsunami 2004 and even for the disasters occurred later.

3.4 Key Findings on Trainings/Drills

According to the survey, approximately over 85% of normal people have undergone some sort of a training drill after the Asian Tsunami 2004 but not before. The drills have been composed with self-evacuation and as well evacuation of people with special needs. As per the telephone discussions and open interviews conducted with the people with special needs it is reported that they rarely had training or any kind of drills regarding this. This fact might reveal the low attention towards those people in disasters as the government and administrations mostly believe that the people with special needs are never left behind by the normal people. Suggestions for having a drill affirmed that there were few major opinions such as swimming training, first aid training, trainings on specific ways of carrying for safety of people with special needs and trainings on finding a shelter. There were lot of individual various and dynamic opinions as it depends on the mentality and the knowledge of the respondent. However, a majority of people over 80% mentioned that the training drills were held but they have not been informed it prior and they could not participate. So, the proper coordination and lack of responsibilities of the administrations were identified. Some of the respondents pointed out that the drills are not regularly or periodically happening but it is just only after a certain disaster.

3.5 Key Findings on Shelter

According to the respondents, it appears that nearly 60% of people know about all the shelters that they might transfer during a disaster. As they told, after the Tsunami 2004 the sign boards and maps of evacuation paths and shortest distances to the evacuation shelters were established but not maintained. Nowadays they are hardly seen. Somehow over 50% of respondents know about their specific evacuation shelters. What that means is in almost every administrative division (Grama Niladhari divisions) disaster management centre has specified site specific evacuation shelters and routes. As per the findings, over 70% of people go to the nearest temples because traditionally the temples have been constructed at places where it has a higher elevation related to the particular area. Approximately 20% of people use schools as their evacuation shelters while the rest use some other specified location or some people do not have that kind of place due to various reasons. It is important to note that the findings about the facilities of those shelters strongly disclosed that the places do not have proper utilities and facilities. Nearly 90% of respondents firmly mentioned about the lack of facilities and utilities for the people who has special needs. The major facts they addressed are mentioned below. As per the findings, there are no specifically built in evacuation shelters for even normal people instead of people with special needs rather than a temple or public school. Since they are not built or constructed upon that purpose the facilities might be inadequate obviously. Majority of the people mentioned about the lack of space, sanitary facilities, gender bias irregularities and more often ethnic bias complications are also there. It is a very paramount discussion to bring forward that according to the respondents there have been a tendency for sexual harassments and the incidents have been reported several times. Because of the difficulties of the people with special needs they more often face these kinds of situations. The government and respective administrations should attempt this case immediately and at the same time the attitudes of the people should be questioned.

3.6 Key Findings Related to Emergency Needs

According to the telephone conversations and the open interviews conducted the emergency needs of those people who have special needs were identified for a certain extent. Self-care materials and equipment, dependency on utility services (water, electricity, telephone etc.), removing household debris and evacuation, transport, evacuation assistance, resupply of very essential goods, asking for help and assistance, necessity of a guide dog or self-evacuation plan and local government cooperation. The respondents stated that the most important emergency need is transport and mobility. At the same time, they were asked whether they are given or provided those emergency needs at the time of a disaster by the government or any sort of responsible authorities. The respondent's mentioned that they are only given food and first aid more often and even that provisions are not equalized among the community because of several socio cultural and ethnic irregularities. Eventually, suggestions were asked about the emergency needs. Majority of people mentioned that the needs of infants and pregnant women are highly conspicuous.

3.7 Key Findings Related to Local Government Cooperation and the Improvement

Over 80% of people stated firmly that the cooperation of the local government is not sufficient due to several reasons. The major reasons that can be emphasised are as follows. Lack of household preparedness of the government administrations, lack of proper statistics about the people with special needs, funding, lack of pre-planned evacuation procedures conducted by the government lack of information about the disasters and new technology. There were several other opinions depending on the various perspectives. In order to improve the local system few improvement suggestions could be highlighted according to the respondents. Majority of people have suggested that it is important to gain understanding of the present hazards and some sort of technical and general awareness is needed. It is a learnt lesson that when Tsunami 2004 occurred, none of the people in coastal areas knew that this was a Tsunami and the severity of the oncoming disaster. Furthermore, estimating the capacity of the people who have special needs was also considered important by the majority of respondents. Since the level of need and assistance differ depending on the capacity of the people it would be a proper procedure to identify specific people for specific need and assistance. Creating a personal care network consisting of at least 3 people for each site was also proposed by significant number of respondents. So that special care can be addressed for the people who need help. Moreover, wearing a label or bracelet in order to identify those type people was another major suggestion since during those kinds of an event people might understand that the particular person with the label or bracelet needs help and they can attend. All above suggestions should come with proper planning and preparedness.

On the other hand, there is a school of thought that is composed with government officers and volunteers, who has devoted their lives during these types of coastal disasters to safeguard and protect the lives of the community, pointedly people with special needs. Despite of the ardent followers who think that an extra attention towards people with special needs is not an imperative task during a coastal or any disaster, it can be seen that a vigorous force is developing through the social media and academic research supporting the aforementioned issues of the people with special needs.

4 Conclusions

It is evident from the findings that there is an evoking interest towards the establishment of proper policies and frameworks for the people with special needs by extending the current disaster management frameworks by cooperating existing context of people with special needs with regards to the coastal disasters. The cooperation and association of agencies and institutions working for people with special are still substantive questions when integrating people with special needs into all the stages of a disaster such as planning and preparedness, response and recovery. It is increasingly necessary to establish adequate subsidies to cater planning and preparedness, response and recovery for persons with special needs. Moreover, an island wide disaster planning programme with proper statistics of people with special needs and equally caring for those people without any boundaries is a must.

Acknowledgements. This research was conducted in collaboration with the University of Peradeniya, Faculty of Engineering and the financial aid from CABARET (Capacity Building in Asia for Resilience Education), a project of *the European Union's Erasmus + programme – Key action 2 – Capacity building in the field of higher education.*

References

Alexander D, Gaillard JC, Wisner B (2012) Disability and disaster. The Routledge handbook of hazards and disaster risk reduction. Routledge, London/New York, pp 413–423

De Silva WI, Amarabandu WP, Gunasekara HR (2007) Disability amongst the elderly in Sri Lanka: research studies series, Number 1. Institute for Health Policy, Colombo: Department of Census and Statistics, Charachteristics of the disabled Persons, Census of Population and housing 2001, Sri Lanka

Fjord L, Manderson L (2009) Anthropological perspectives on disasters and disability: an introduction. Hum Organ 68(1):64–72

Hemingway L, Priestley M (2006) Natural hazards, human vulnerability and disabling societies: a disaster for disabled people? Rev. Disabil. Stud. 2(3):57–67

Ministry of Social Welfare (2003) National Policy on Disability for Sri Lanka

Nishikiori N, Abe T, Costa DG, Dharmaratne SD, Kunii O, Moji K (2006) Who died as a result of the Tsunami? – risk factors of mortality among internally displaced persons in Sri Lanka: a retrospective cohort analysis. BMC Public Health 6(1):73

Parr AR (1987) Disasters and disabled persons: an examination of the safety needs of a neglected minority. Disasters 11(2):148–159

Ronoh S, Gaillard JC, Marlowe J (2015) Children with disabilities and disaster risk reduction: a review. Int J Disaster Risk Sci 6(1):38–48

Report of the Online Forum on Disabled and other Vulnerable People in Natural Disasters (2006) World Bank Disability & Development Team (Human Development Network), World Bank Institute's Multimedia Center, Center for International Rehabilitation, Leonard Cheshire Centre of Conflict Recovery, University College London

Stough LM, Kang D (2015) The Sendai framework for disaster risk reduction and persons with disabilities. Int J Disaster Risk Sci 6(2):140–149

Community Environment and the Ecosystem

Wetland is identified as an ecosystem service that helps to control flooding and erosion, replenish the ground water table and enhance biodiversity (Niering 1970). Several early research done on wetlands and flooding demonstrated the ability of wetlands to store water (Cheng et al. 2014; Brody et al. 2008, 2007). Wetlands not only provide the ecological infrastructure for watershed systems but also offer natural flood mitigation by maintaining a properly functioning water cycle (Brody et al. 2008). Lack of knowledge and awareness on the importance of wetlands has resulted in their significant reduction.

While it is likely that the rate of flooding will increase due to large developments and impervious surfaces, results of research prove that wetland alteration in a small-scale has a greater impact on flooding in the long run (Brody et al. 2007). The results of the research conducted by Brody et al. (2008) confirm that "the alteration of naturally occurring wetlands is the most important built environment indicator of flood damage" (p. 13).

2.3 Landscape Architectural Approach

It is now evident that due to development of cities, there is a great increase in the impacts on the natural environment and to collectively achieve both economic growth and environmental protection, sustainable management through landscape needs to be implemented. This includes not only addressing water quantity and flow issues but integrating society and enhancing the character of the area while managing floods (Cheng et al. 2014). At present, there is a challenge to move away from traditional approaches toward a more sustainable approach. For this purpose, "landscape infrastructure is being explored as a concept that expands the traditional planning and design strategies towards the realization of a multifunctional system" (Bacchin et al. 2014, p. 2).

It is also evident that existing drainage infrastructure, built to manage urban runoff, will become obsolete in the near future due to climate and urbanization uncertainties (Bacchin et al. 2014). Hence, water management systems need to be designed to be adaptable to change and possibly with a multi purpose use. Furthermore, in order to move away from the common notion that cities and floods cannot co-exist, a shift in perception and creative design and planning of cities is required. As an example, open spaces become multifunctional to convey and store flood water during rainfall events (Turner 2016; Bacchin et al. 2014).

2.4 Multifunctional Landscapes

As the term implies, multifunctionality is the idea that a single space can have multiple uses in order to benefit both people and their environment (Macfarlane 2007). In dense urbanized environments, it is not always easy to find suitable areas for new stormwater infrastructure. In such situations a more effective option is the use of multifunctional landscapes where city revitalization and flood control can go hand in hand (Rezende et al. 2011; Miguez et al. 2007). Hence, a multipurpose landscape not only manages the quantity of water but also focuses more on the social integration and aesthetic quality of the landscape (Wong and Eadie 2000). This forms a parallel argument to that of Van Herk et al. (2013) who state that we should move from acting against water, to living with it. Further research provides a direct connection to multifunctionality of space as

Liao (2012) states, "the theory of urban resilience to floods challenges the conventional wisdom that cities cannot live without flood control" (p. 01).

Among several scholarly interpretations of multifunctional landscapes, the basic idea is that a given area should fulfil different functions and objectives. It is to treat the problem at its source by making use of the urban landscape and recovering natural flow conditions while enhancing urban space through an integrated design of built and natural environment (Miguez et al. 2007).

As Lawson (2014) states "Most of the time we are not under flood conditions - so why not design the landscape to be multifunctional and accrue benefits?" (p. 51). Further, Kato and Ahern (2009) argue that multifunctional landscapes including ecological greenways and mixed use urban developments, which provide ecological, recreational, cultural, aesthetic or other purposes are "an important and necessary component of sustainable landscapes" (p. 1). However, according to a study done by Cesar et al. (2007) on the use of public squares as flood control measures, one major challenge that had to be overcome was to develop public acceptance and convince the local community of its importance.

2.5 Principles of Urban Flood Management for Multifunctionality

Several principles for urban flood management have emerged through studies and implementation of ideas. They take into consideration the theories of urban resilience and multifunctionality as well as the enhancement of the landscape.

2.5.1 Best Management Practices

Best Management Practices (BMP) is a flood management principle associated with the concept of green infrastructure. BMP include both structural strategies and non-structural strategies to manage stormwater runoff (Cheng et al. 2014). They focus on providing effective sustainable solutions for water management while ensuring better community appreciation. The structural strategies include;

- Rain Gardens
- Vegetated Swales
- Pervious pavement
- Filtration-Storage system
- Detention Basin

2.5.2 East Asian Water Management Principles

In the East Asian context, the research done by Luo et al. (2015) provides comprehensive details on the Japanese and Chinese flood management systems. In Singapore, for the purpose of water management, they follow a holistic and strategic plan which follows the vision of converting engineered concrete drains, canals and reservoirs into recreational (multifunctional) channels, and manage them to supply water and mitigate floods (Hua and Wen 2011). Further, many countries such as Singapore and China, address the need for outdoor spaces for social interaction through water front development. This is carried out in a multifunctional manner where these developments also help the city adapt to flood situations while creating visually pleasing landscapes.

- Integrated river development
- Green dams and embankments
- Social integration
- Urban regeneration

2.5.3 Blue-Green Cities

Constructed water channels in cities carry away urban rainfall, which nature can do more aesthetically and at no cost. Dekay and O'Brien (2001) express, "Imagine a 'blue-way network' where streams meander through our neighbourhoods, carrying away runoff and slowing its flow, providing rich natural habitat…" (p. 22). This principle, known as the concept of 'Blue-Green Cities', focuses mainly on recreating the natural water cycle and contributing to the aesthetic development and amenity of the city. It creates a bond between water management and green infrastructure (Turner 2016; O'Donnell 2016). Blue-Green infrastructure aims to reduce the amount of water that is directed towards artificial drainage systems through natural means such as infiltration, interception, transpiration and temporary or long-term storage. Hence working with nature to control water at source (Lawson 2014).

- Multifunctional spaces
- Stormwater retention/detention
- Living with floods

2.6 Theoretical Framework

As a summary of the literature review, to determine the degree of multifunctionality the following framework has been identified to analyze the selected case study (Fig. 1).

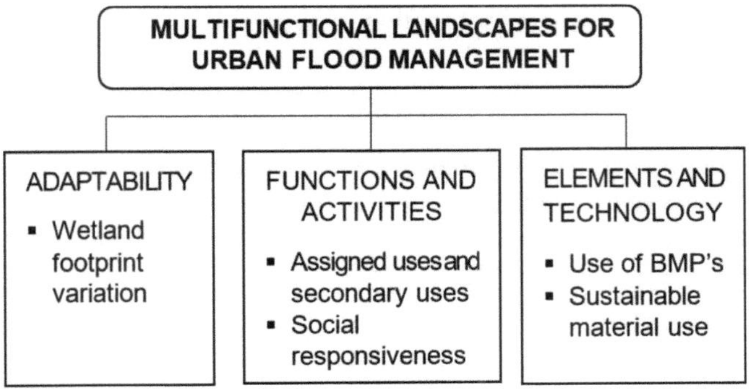

Fig. 1. Theoretical framework

3 Methodology

The study adopted a mixed method which is twofold; a literature survey and case study. Concepts and arguments related to the integration of multifunctionality for flood management were identified through a literature survey. The factors which assist in determining the degree of multifunctionality of flood responsive landscapes are summarized and applied to analyze the case study. The public open space developments in the Kotte wetland around the Diyawanna Oya, Sri Jayawardenepura Kotte in lieu of flood mitigation and landscape development were identified for further study.

Data was collected through existing literature, mapping, on-site observations, documentation and interviews. Both quantitative and qualitative methods are used to analyze the data. However, the effectiveness of multifunctional landscapes is discussed qualitatively.

3.1 Limitations

The analysis only focuses on analyzing the multifunctionality of the landscape. The study does not include a quantitative analysis of flood such as water quantity and rate of runoff and does not discuss in detail the ecological details such as the diversity of flora and fauna, and water quality. The social survey was also limited to 30 respondents.

4 Case Study

The city of Colombo situated in the wet zone of Sri Lanka and experiences rain year-round. Flooding in Colombo has worsened in recent years primarily due to the city's inability to absorb water as the natural flood storage capacity has been reduced by inadequate planning and wetland encroachment. The most recent major floods in the year 1992 recorded the highest ever rainfall (493 mm) followed by the years 2010 and 2016.

As a result of several studies, five wetlands including the Kotte Wetland and a "green belt" around the Diyawanna water body were established as flood detention areas in Colombo (CEA and Euroconsult 1995). The recent developments in the green belt aimed to preserve the flood detention area through sustainable and social landscape developments, which are considered significant for flood mitigation in Colombo (Fig. 2).

4.1 Kotte Wetland

Colombo is historically a large area of wetland and natural landscapes. As a result of imbalanced development of Colombo and mismanagement of water related landscapes, these wetlands have now depleted. Diyawanna Oya and the Kotte wetlands are some of the last remaining extensive areas of wetland in Colombo having the capacity to mitigate floods. The green belt of wetlands around the Oya act as temporary storage basins for flood water. According to the present land use, the area of wetland has reduced to approximately 230 Ha. Hence, for the purpose of conserving the remaining

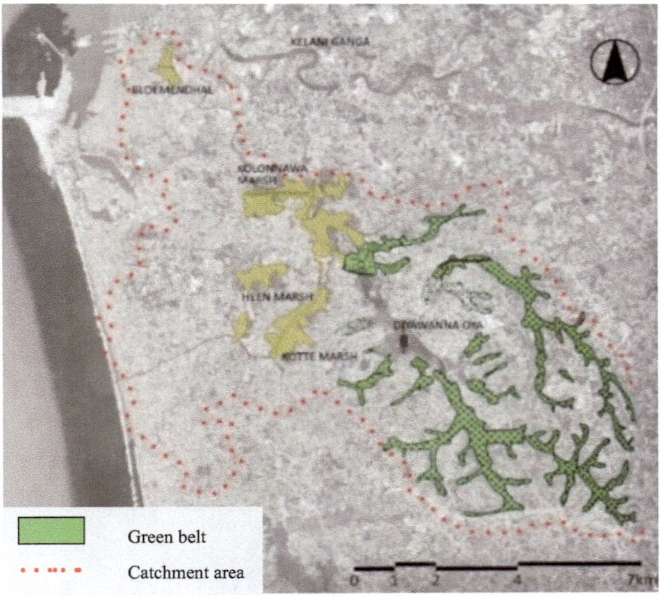

Fig. 2. Colombo flood detention areas (Source: CEA and Euroconsult 1995)

wetlands and flood detention areas, the local authorities have taken measures to prevent further landfill and encroachment. This has been implemented through multifunctional landscape developments in the area (Fig. 3).

Diyatha Uyana

The present location of Diyatha Uyana has undergone several landscape changes in the past. It was initially a natural wetland area, then infilled to construct a golf course. In 2010, this area was developed as a public recreational space while serving purpose of a flood detention area. It includes a waterfront development, recreational park and sales area, and acts as the center for local horticultural sales.

Diyawanna Oya Linear Park

With the development of the city for improved flood retention due to the flood of 2010, the waterfront area around Diyawanna Oya facing the parliament was developed as a linear park with jogging tracks while increasing the retention capacity of the Oya. This was developed to serve the purpose of flood retention, recreation and health benefits for the community.

Beddagana Wetland Park

This park is located close to the parliament complex, bordering the Diyawanna Oya. It is a wetland protected within a wildlife sanctuary and classified as a Wetland Protection Zone due to its high bio-diversity and bird habitats, home to many endemic birds and other faunal species. This area was developed as a wetland park, enhancing the natural environment and bio diversity through recreational and educational facilities, while maintaining and preserving it for flood retention.

Fig. 3. Map of study area and selected landscape developments

5 Data Presentation and Analysis

Landscape Architectural solutions for flood management could have subjective outcomes. Hence, determining the multifunctionality of the selected parks takes into consideration these variations and provides a qualitative outcome.

5.1 Wetland Footprint Variation

Sri Jayawardenepura Kotte was home to more than a 1000 Ha of wetland before the administrative capital was shifted here. The dredging of the wetlands for the construction of the parliament complex and surrounding lake in the years 1979–1983 resulted in reclamation of about 272 Ha of natural flood retention area. Since then this area has seen a rapid decrease in wetlands (CEA and Euroconsult 1995) (Fig. 4 and Table 1).

It is evident that within a period of 17 years, the amount of marshes and agricultural fields which contribute to natural flood retention have decreased by almost half. Although there is a significant increase in the area of water bodies, due to the flood management and development activities that took place after 2010, this increase does not adequately compensate for the loss of wetlands as evident through the recent flood

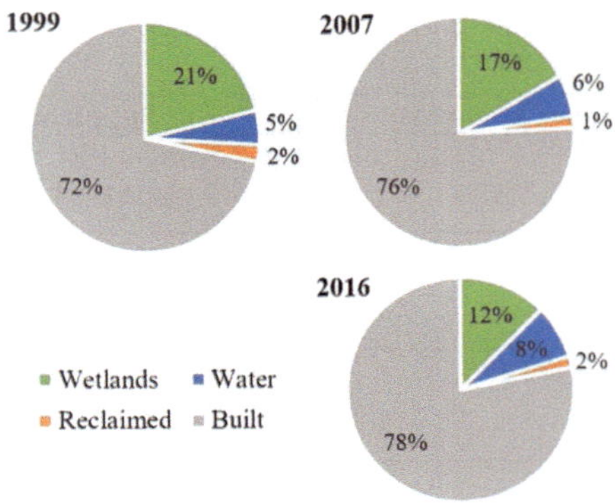

Fig. 4. Graphical illustration of land use

Table 1. Land use variation in the study area

	1999 (Ha)	%	2007 (Ha)	%	2016 (Ha)	%
Wetlands	391	21	314	16.9	233	12.5
Water bodies	91	4.9	107	5.8	145	7.8
Reclaimed land	44	2.4	31	1.6	30	1.6
Built area	1334	71.7	1408	75.7	1451	78.1
Total	1860	100	1860	100	1860	100

event of 2016. The built area as expected shows an increase. However, the rate of increasing built area has decreased.

Although many steps were taken during the last few years to conserve what is left of the natural wetlands, for the many benefits we gain from this ecosystem and also to manage the increasing flood risk, the loss of wetlands have not decreased. The developments that took place have somewhat succeeded in preserving the flood detention in the public open spaces. However, a larger area has been dredged without preserving what is left of the wetland or creating man-made wetlands to compensate for the lost wetland area. Hence, there is an imbalance between open water and marsh wetlands when considering their flood retention capacity.

5.2 Functions and Activities

This factor will be discussed separately for each park.

Diyatha Uyana

Table 2 indicate the activities in this public open space. Most functions in this public space have been designed to incorporate supplementary functions. This is a good application of multifunctional landscapes (Miguez et al. 2007). Most of these secondary functions cater to social requirements, hence, it is evident that the local community has been benefitted through this development.

Table 2. Functions at Diyatha Uyana

Main function / Elements	Supplementary functions
Water retention lake	▪ Rowing track ▪ Sea plane track
Retention ponds and constructed wetlands	▪ Landscape feature
Jogging track	▪ Scenic walkway ▪ Street art
Water front seating and lawn	▪ Inundation space for major floods ▪ Children's play area
Vehicle parking	▪ Public performance space
Children's play space	
Food outlets and seating	
Weekly sales outlets	▪ Social gathering space ▪ Seating area
Boat ride	

Diyawanna Oya Linear Park

This park was developed on the idea of increasing the water retention capacity of the lake and a water front development to provide recreation and health benefits to the community (Table 3).

Table 3. Functions at the Linear Park

Main function / Elements	Supplementary functions
Jogging track along Diyawanna Oya	▪ Inundation area for high water level of the lake
Waterfront seating	▪ Flood inundation area / Water detention
Parking	▪ Flood inundation area / Water detention
Open play space	▪ Exercising space ▪ Activities related to paddy farming
Food vendors	
Paddy fields	▪ Water detention ▪ Natural landscape
Jogging/cycle track along paddy fields	
Exercising space	
Dry weather children's play ground	▪ Flood water detention space
Canal for excess runoff drainage	▪ Landscape feature

Beddagana Wetland Park

The Beddagana wetland park functions more as a nature park for observation and education and does not include a wide variety of activities (Table 4).

Table 4. Functions at Beddagana Wetland Park

Main function / Elements	Supplementary functions
Jogging path	
Boardwalk	
Bird watching	
Dry weather play ground	▪ Flood water detention space ▪ Bird watching
Outdoor learning	▪ Outdoor seating
Nature trail	▪ Scenic walkway
Open viewing decks	
Vehicle park	
Canal as physical barrier	▪ Landscape feature

The main activities that have been integrated into the design concerns involvement of the public in effectively conserving the wetland. This ensures that the natural wetland is protected through multifunctionality combining both the hydrological/ecological benefits of wetland and social awareness and appreciation.

Social Responsiveness

The social responsiveness to the public landscapes of Diyatha Uyana, Diyawanna Oya linear park and Beddagana wetland park were studied through a questionnaire mainly by the frequency and reasons of visit, and overall perception of the landscape (Table 5).

Table 5. Frequency of visit by respondents

Public landscapes	No of respondents		
	Frequently	Occasionally	Total
Diyatha Uyana	8	22	30
Diyawanna Oya linear park	8	14	22
Beddagana wetland park	2	13	15

Diyatha Uyana is a park with diverse activities that cater to a wider range of users and it is clearly the more popular public space. It is evident that these multifunctional landscapes have been accepted by the community as 87% of the respondents stated that they prefer spending time in public open spaces (Fig. 5).

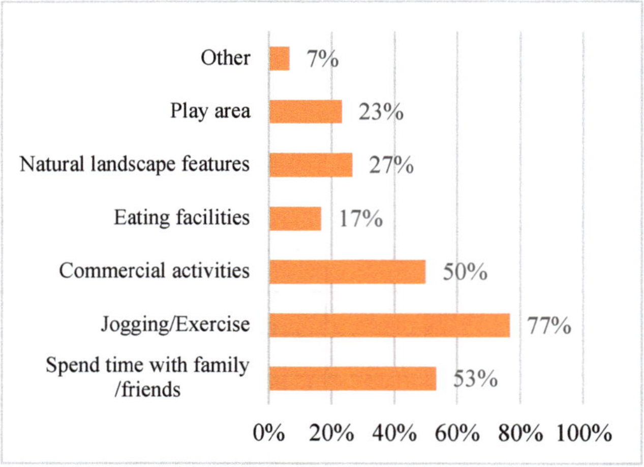

Fig. 5. Reason for visiting public open spaces

5.3 Landscape Elements and Technology

This factor will also be discussed separately for each park.

Diyatha Uyana

The landscape elements and material used in the design complement its functions in most instances and contribute to create flood resilience (Table 6).

Table 6. Materials and landscape elements used at Diyatha Uyana

Landscape elements / uses	Materials / technology used
Walking/jogging track	Sand paved, tar bitumen, concrete paving stone At a higher elevation to water level of the lake
Seating area	Maintained lawn Closer to the water level of the lake
Open viewing deck	Timber with gaps, railway sleepers
Retention ponds	Grass slope (non-vegetated)
Constructed wetlands	Curb stone, grass slope
Dredged lake	Grass, steep slope
Parking and commercial area	Concrete paving
Children's play area	Sand

There are three main walkways designed at different elevations (Fig. 6). The surrounding lawn and seating area along the edge of the Oya is at a lower level -

approximately 1 m above the mean water level. In the event of a high precipitation, this area will get inundated creating a new high-water level (Fig. 7). It is similar to the use of a green dam/embankment for natural flood control (Luo et al. 2015). This multifunctional use of the open space enables floods to occur without causing great inconvenience in this particular area.

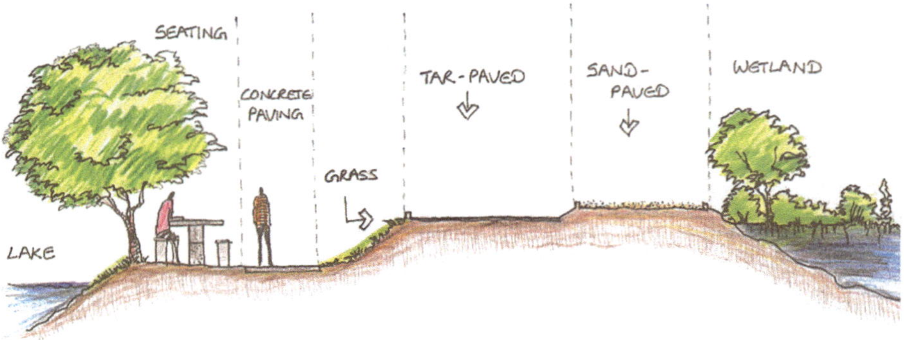

Fig. 6. Different levels of walkways and multifunctional embankments

Fig. 7. Seating area inundated during the flood of 2016

Fig. 8. Retention pond surrounded by well-maintained lawn

Most materials used for walkways are not permeable and will further contribute to the increasing impervious surfaces in the city, which will cause further surface runoff. This water will potentially runoff into the lake or the constructed wetlands and contribute to unnecessary increase in the water level of the Diyawanna Oya.

The adjacent part of the Diyawanna lake has been dredged to increase the water retention capacity but there are also problems of silting. The retention ponds and constructed wetlands directly contribute to flood management (Fig. 8). However, they are regularly maintained thereby, disturbing ecological succession.

Fig. 9. Main element of the design – the Jogging track

Fig. 10. Stormwater canal integrated into the landscape

Diyawanna Oya Linear Park

Most activities and elements incorporated in this development are designed to be multifunctional in events of high precipitation (Table 7).

Table 7. Materials and landscape elements used at Diyawanna Linear Park

Landscape elements / uses	Materials / technology used
Walking/jogging track	Sand paved, compacted earth Intermediate level between water level and road
Seating area	Maintained lawn
Retention ponds	Grass, steep slope
Carpark	Concrete paving
Open play space	Compacted earth
Dry weather children's play area	Sand, natural ground
Canals	Grass slopes
Drains for runoff from roads	Concrete Direct water to nearest waterbody

Figure 9 shows the main element in this area, the jogging track. Most part of it including the carpark has been infilled. They are at a lower level than the main road and act as a high-water level for the lake in the event of a flood. In this case, the main road acts as the embankment for the city, and the jogging track carpark and most seating areas become part of the detention basin.

The paddy fields in the vicinity act as natural flood detention areas. However, due to the significant absence of natural or constructed wetland or any other BMPs, it is possible that excess water will continuously flow downstream and not be detained at source. The dry weather playground incorporates multifunctionality. It is surrounded on one side by a stormwater canal and on the other by an existing retention pond (Fig. 10). Hence, in the instance of high precipitation, this playground will be inundated converting it to a flood detention landscape. The blue green principles are implemented in the design through the use of open drains leading to natural water bodies.

Beddagana Wetland Park

The Beddagana park focuses on conserving the natural wetland to maintain flood detention capacity (Table 8).

Table 8. Materials and landscape elements used at Beddagana Wetland Park

Landscape elements / uses	Materials / technology used
Access walkway/jogging track	Compacted earth Closer to wetland level, lower level than the main road
Carpark	Concrete paving
Raised walkway	Timber with gaps Minimum impact on wetland
Dry weather play ground	Grass lawn Maintains natural edge
Bird watching tower	Steel columns Minimum impact on wetland
Bund roads/ internal pathways	Compacted earth, gravel
Open viewing decks	Timber with gaps

This wetland is subject to periodic flooding lasting for 3–7 days (Uni Consultancy Services 2011). Hence it has been designed without obstructing the existing drainage paths and has ensured the least impact to the wetland so that the flood retention and detention capacity is not compromised. The jogging track, internal nature trails and existing bund roads which have been integrated into the design are of compacted earth or are gravel-paved, thereby encouraging ground water absorption. The raised timber walkways ensure minimum impact on the natural landscape. Similar treatment has also been given to the other built structures.

The dry weather playground as well as most part of the wetland acts as a detention basin for flood water at source. However, the area for parking has taken up a significant area of wetland. Although they are essential for a public space, BMPs such as permeable paving and vegetated swales could have been introduced.

6 Conclusion

Based on the factors to assess the multifunctionality of the landscape developments in the Kotte wetland and Diyawanna Oya, the research attempted to determine if the three selected parks were multifunctional in terms of its ability to resist urban flooding.

All three landscape developments made use of multifunctionality in their designs, incorporating principles of urban flood management, integrating multifunctional spaces rather than single-use spaces and multifunctional elements. The use of technology similar to embankments is an effective method to create resilient landscapes. In the Diyatha Uyana and Linear park, there is a concern that the dredging of large areas for lakes will not detain the expected amount of water. There is a possibility of downstream flow due to the loss of natural vegetation and wetland on the edges of the water bodies. Further, these developments have been implemented for a 10-year flood return period and will need to be constantly improved and maintained. Hence, they may not be effective in the case of a 50-year flood but resilient to minor floods. The Beddagana park, however, is a good solution conserving valuable wetland that could be encroached in the future. This landscape design effectively integrates the natural hydrological functions of the area to ensure an effective and flood resilient landscape design.

By analyzing the social responsiveness for multifunctional open space developments, it is evident that people interact with these public open spaces and that it addresses public recreation, the health and well-being. However, the community needs to be given awareness on the objectives of such developments in order to have a better a sense of ownership of public landscapes. Taking into consideration an overall perspective, these landscapes effectively address the multifunctional use of the landscape for flood mitigation as well as social integration.

The landscape developments analyzed in the study area all aim to conserve the natural flood retention capacity, but in most cases have focused on increasing the water retention capacity of the Diyawanna Oya. The use of constructed vegetated wetlands would have been a more effective water management system for flood detention. Although an effort has been taken to maintain the water retention capacity of the natural environment, it has not reached a satisfactory level for flood management due to the continuous loss of natural wetland as proved by the findings.

Further, no planning solution has been given for the inundation of residential areas caused by urbanization and development of property. To mitigate this, multifunctional use and BMPs such as vegetated swales, rain gardens and filtration-storage systems to detain water at source should be further encouraged in private developments. This will not only help in flood management but will integrate the community in the process (Fig. 11).

With further studies in this field, future developments for sustainable flood management by means of multifunctional landscapes will have the potential to successfully manage any degree of floods, thereby establishing flood resilient cities.

Fig. 11. Use of BMP's in residential areas to integrate the community in flood mitigation

References

Bacchin TK et al (2014) Green-blue multifunctional infrastructure: An urban landscape system design new approach. S.N., Sarawak, pp 1–8

Brody SD, Highfield WE, Ryu HC, Spanel-Weber L (2007) Examining the relationship between wetland alteration and watershed flooding in Texas and Florida. Nat Hazards 40(2):413–428

Brody SD et al (2008) Identifying the impact of the built environment on flood damage in Texas. Disasters 32(1):1–18

CEA and Euroconsult (1995) Wetland conservation project - Colombo flood detention areas. CEA, Colombo

Cesar F et al (2007) Comparison of different multifunctional landscapes approaches for flood control in developing countries. NOVATECH, pp 83–90

Cheng C, Brabec EA, Yang Y-CE, Ryan RL (2014) Rethinking stormwater management in a changin world: effects of detention for flooding hazard mitigation under climate change scenarios in the Charles river watershed. In: Li M (ed) Landscape research record, pp 214–228

Dekay M, O'Brien M (2001) Gray city, green city. In: Forum for applied research and public policy, pp 19–27

Hua D, Wen LY (2011) Managing sotrmwater in urbanised Singapore for flood control. PUB, Tokyo

Jha AK, Bloch R, Lamond J (2012) Cities and flooding: a guide to integrated urban flood risk management for the 21st century. The World Bank, Washington, DC

Kato S, Ahern J (2009) Multifunctional landscapes as a basis for sustainable landscape devlopment. J Jpn Inst Landscape Architect 72(5):799–804

Lawson E (2014) Managing urban flood risk: The blue-green appraoch. http://www.bluegreencities.ac.uk/bluegreencities/documents/publicsciencelecture-dec2014-bgc.pdf

Leopold LB (1968) Hydrology for urban planning: a guidebook on the hydrologic effects of urban land use. USGS Circular

Liao K (2012) A theory on urban resilience to floods—a basis for alternative planning practices. Ecol Soc 17(4). http://www.jstor.org/stable/26269244

Luo P, He B, Takara K, Xiong YE, Nover D, Duan W, Fukushi K (2015) Historical assessment of Chinese and Japanese flood management policies and implications for managing future floods. Environ Sci Policy 48:265–277. https://doi.org/10.1016/j.envsci.2014.12.015

Macfarlane R (2007) Multifunctional landscapes: conceptual and planning issues for the countryside. In: Benson JF, Roe M (eds) Landscape and sustainability, 2nd edn. Taylor & Francis Group, New York, pp 138–166

Miguez MG, Mascarenhas FC, Magalhaes LP (2007) Multifunctional landscapes for urban flood control in developing countries. Int J Sustain Dev Plan 2(2):153–166

Niering WA (1970) The ecology of wetlands in urban areas. In: Dansereau P (ed) Challenge for survival: land, air, & water for man in Megalopolis. Columbia University Press, New York

O'Donnell E (2016) University of Nottingham blogs/blue green cities. http://www.blogs.nottingham.ac.uk

Rezende OM, Carneiro PR, Miguez MG (2011) Sustainable stormwater management: use of multifunctional landscapes in urban drainage for flood control. S.N., Porto Alegre, pp 1–8

Turner TF (2016) How landscape change causes flooding in towns and cities. http://www.landscapearchitecture.org.uk

Uni Consultancy Services (2011) Environmental screening report for establishment of Beddagana and Kotte ramparts parks

Van Herk S, Rijke J, Zevenbergen C, Ashley R (2013) Understanding the transition to integrated flood risk management in the Netherlands. Environ Innov Soc Trans 15:84–100

Wong THF, Eadie ML (2000) Water sensitive urban design - a paradigm shift in urban design. In: 10th world water congress

Zevenbergen C et al (2010) Urban flood management. Taylor & Francis Group, Florida

Zevenbergen C, Veerbeek W, Gersonius B, Ven Herk S (2008) Challenges in urban flood management: travelling across spatial and temporal scales. J Flood Risk Manag 1:81–88

Single and Binary Adsorption of Heavy Metal Ions from Aqueous Solutions Using Coirpith as the Adsorbent

E. N. Ranasinghe and B. M. W. P. K. Amarasinghe(✉)

Department of Chemical and Process Engineering, University of Moratuwa, Moratuwa, Sri Lanka
padma@cheng.mrt.ac.lk

Abstract. The adsorption of nickel and zinc ions onto coir pith was studied using batch experiments at 30 °C and 40 °C. Adsorption kinetics and equilibrium isotherms were obtained for both single component adsorption and binary adsorption. The kinetics of single system and binary system adsorption of nickel and zinc was studied using Lagergren pseudo first order model, pseudo second order model and the intraparticle diffusion model. The experimental data showed all of the systems could be described using pseudo second order model. In the single system the coir pith had a higher capacity for zinc than nickel. In both the nickel and zinc single systems the adsorption decreased when temperature increased from 30 °C to 40 °C. In the binary system at 30 °C the coir pith had a larger capacity for zinc adsorption then nickel but at 40 °C coir pith had a larger capacity for nickel than zinc. Equilibrium data were satisfactorily fitted to Langmuir and Freundlich isotherms. When comparing the adsorption capacity, q_{max}, the experimental data showed the binary system had a lesser capacity (q_{max} 13.6 mg g^{-1}) than the single systems nickel (q_{max} 27 mg g^{-1}) and zinc (q_{max} 22.7 mg g^{-1}) under similar conditions. This suggests the competitive nature of the systems decreases the overall adsorption.

Keywords: Binary adsorption · Heavy metal · Nickel · Zinc · Coir pith

1 Introduction

Heavy metals, such as zinc, cadmium, nickel and lead occur naturally in very low concentrations in both soil and water as dissolved metals ions and in their native forms. However due to increased industrial and agricultural activity and poor management of waste produced there has been an increase in heavy metals introduced into the environment. This can lead to many harmful effects to aquatic environments, plants and human health as in many developing countries such as India and Sri Lanka. In reviews by Wijayawardhana et al. (2016) and (Adikaram and Pitawala 2017), it was found that the reported levels of heavy metals found in Sri Lankan soil were within the safe levels for daily activity. However as heavy metals do not degrade overtime they tend to build up in water and soil, especially in coastal regions.

The toxicity of the heavy metals depend both on the chemistry of the metal and the concentration in water and soil. Nickel and zinc are harmless in low concentrations but

R. Dissanayake and P. Mendis (Eds.): ICSBE 2018, LNCE 44, pp. 75–90, 2020.
https://doi.org/10.1007/978-981-13-9749-3_7

the presence of these chemicals in fertilizers, pesticides and galvanized items have lead to higher concentrations in natural water sources such as rivers and nearby wells. One method of removing these heavy metals which is appealing due to its easy of application is adsorption, where the pollutant is adsorbed onto the adsorbent and then filtered out with the solid adsorbent. The efficacy of the adsorbent depends on the surface chemistry and structure of the adsorbent. Ideally the adsorbent should be both cheap and readily available. A cheap alternative is to use agricultural waste such as rice husks, vegetable peel, straw or coir pith as bioadsorbents. In a study by Amarasinghe (2007) coir pith and tea waste was shown to have a higher capacity for metals such as lead compared to activated carbon.

In a review by Nassrulla et al. (2016) it was shown that lignin can be a good nonspecific adsorbent to remove heavy metal pollutants from waste water. (Ho and McKay 1999) showed peat was an effective adsorbent in the removal of nickel and copper ions from aqueous solutions. In a study by Amarasinghe (2011) showed that coir pith can be used to remove heavy metals lead and cadmium from waste water. Based on previous research and the low cost and availability, for this study coir pith was chosen as the adsorbent. Coir pith is made up of mostly lignin and cellulose, which are organic polymers which contain many polar surface functional groups such as esters, ethers and carbonyl groups.

Nickel and zinc are common heavy metals used in the electroplating industry as well as being present in many fertilizers and pesticides used in agriculture. Nickel and zinc often occur together both naturally and in industrial applications. Zinc and nickel have similar atomic mass, ionic size and density. Studies of the competitive adsorption of similarly sized metal ions by (Sdiri et al. 2014) which compared zinc and copper and Shaheen et al. (2012) which compared copper, nickel and zinc showed that small differences in ionic size can result in large differences in adsorption capacity and preference.

This study will focus on the adsorption of nickel and zinc in both single system and binary system environments. The adsorption isotherm will be studied at 30 °C and the kinetics of adsorption will be studied at 30 °C and 40 °C. From the experimental analysis the change in the kinetics of the adsorption at the two different temperatures as well as the mechanism of adsorption of nickel and zinc will be deduced.

2 Methods and Materials

2.1 Preparation of the Adsorbent

Coir pith was obtained from Sampath Coir Mill, Singakkuliya, Sri Lanka and dried in the oven using stainless steel trays at 60 °C for 12 h. It was sieved using a sieve shaker to obtain particles with a size distribution between 335 μm and 1 mm. The coir pith was washed with distilled water and then dried in the oven at 60 °C for 12 h.

2.2 Characterization of the Adsorbent

To determine the surface functional groups, Fourier transform infrared spectrometer (FTIR) analyses were performed. The coir pith was imaged using scanning electron microscopy (SEM) at 500× magnification.

2.3 Batch Adsorption Test

The adsorbate stock solution of 1000 mg L^{-1} Ni(II) and Zn(II) were prepared from analytical grade nickel nitrate $Ni(NO_3)_2.6H_2O$ and zinc nitrate $Zn(NO_3)_2.6H_2O$ (supplied by Sigma-Aldrich chemicals, Germany) respectively in distilled water.

2.3.1 Adsorption Isotherm

For the isotherm study, 500 ml solutions of Ni, Zn or a mixture of both in a 1:1 ratio with initial concentrations of 10 mg L^{-1}, 20 mg L^{-1}, 50 mg L^{-1}, 100 mg L^{-1} and 200 mg L^{-1} were made by diluting stock solutions of 1000 mg L^{-1} using distilled water. 0.75 g of the coir pith adsorbent was added to the solution and stirred using an agitator system at a speed of 135 rpm at 30 °C for 120 min. The sample concentrations were determined using inductively coupled plasma mass spectrometry (ICP-MS).

2.3.2 Adsorption Kinetics

For the study of kinetics over different temperatures 500 ml solutions of nickel or zinc or a mixture of both in a 1:1 ratio with total initial concentrations of 50 mg L^{-1}. 0.75 g of the coir pith adsorbent was added to the solution and stirred constantly. Samples were taken out at times of 1 min, 2 min, 5 min, 10 min, 30 min, and 120 min. The experiments were carried out at 30 °C and 40 °C. The concentrations of the samples were determined using ICP-MS.

2.4 Analytical Models

2.4.1 Isotherm Models

Several adsorption isotherm models have been developed, and are widely used, to describe adsorption monolayer formation.

Langnuir isotherm model assumes a complete monolayer coverage of the solid adsorbent (Langmuir 1916) and is given by Eq. 1, where q_e is the equilibrium adsorption capacity, q_{max} is the maximum adsorption capacity corresponding to complete coverage of available sites, C_e the equilibrium concentration of metal ions in solution, k_L is the Langmuir isotherm constant (L/mg).

$$q_e = \frac{q_{max} k_L C_e}{1 + k_L C_e} \tag{1}$$

Equation 1 can be rearranged to the linear form expressed in Eq. 2.

$$\frac{C_e}{q_e} = \frac{1}{k_L q_{max}} + \frac{C_e}{q_{max}} \tag{2}$$

The Freundlich isotherm is an empirical equation which takes into account heterogeneous adsorption (Freundlich 1906a, b). The Freundlich isotherm equation is described in Eq. 4 where q_e is the equilibrium adsorption capacity, C_e is the equilibrium concentration, k_F is the Freundlich constant which is the relative adsorption capacity of the adsorbent, n is the Freundlich constant which indicates adsorption intensity.

$$q_e = k_F \cdot C_e^{\frac{1}{n}} \tag{3}$$

Equation 3 can be rearranged to the linear form expressed in Eq. 4. The isotherm can be obtained by plotting log q_e against log C_e.

$$log q_e = log k_F + \frac{1}{n} log C_e \tag{4}$$

Extended Langmuir model is used to analyse the isotherm of the binary system. This model takes into consideration the competitive forces that occur during adsorption (Singh et al. 2017). $q_{e,Ni}$ is the equilibrium adsorption capacity of nickel at equilibrium, $C_{e,Ni}$ is the equilibrium concentration of nickel, $k_{L,Ni}$ is the Langmuir constant for nickel, $q_{e,Zn}$ is equilibrium adsorption capacity of zinc at equilibrium, $C_{e,Zn}$ is the equilibrium concentration of zinc, $k_{L,Zn}$ is the Langmuir constant for zinc, and q_{max} is the maximum capacity for the binary system.

$$q_{e,Ni} = \frac{(q_{max} K_{L,Ni} C_{e,Ni})}{(1 + K_{L,Ni} C_{e,Ni} + K_{L,Zn} C_{e,Zn})} \tag{5}$$

$$q_{e,Zn} = \frac{(q_{max} K_{L,Zn} C_{e,Zn})}{(1 + K_{L,Ni} C_{e,Ni} + K_{L,Zn} C_{e,Zn})} \tag{6}$$

2.4.2 Kinetic Models and Thermodynamic Analysis

Lagergren pseudo-first model has been used by many researches to study adsorption where the sorbate interaction was not significant, and usually modelled only the initial adsorption accurately (Kumar et al. 2005). The pseudo-second order model was shown to be accurate in modelling the whole adsorption process.

Lagergren pseudo first order which was proposed by Lageregen (1898), describes physisorption and can be derived from the integrated rate law (Eq. 8) where k_1 is the rate constant of first order sorption, q_e is the amount of solute adsorbed at equilibrium (mg g^{-1}), q_t is amount of solute adsorbed on the surface of the sorbent at any time t (mg g^{-1}).

$$\frac{dq_t}{dt} = k_1(q_e - q_t) \tag{7}$$

Equation 7 can be rearranged to a linearized form which is more commonly used. Log(q_e) and k_1 can be determined from the intercept and slope of the plot of log ($q_e - q_t$) against t.

$$\log(q_e - q_t) = \log(q_e) - \frac{k_1}{2.303}t \tag{8}$$

Lagergren pseudo second order model, which was proposed by Ho (1995) can be used to study processes where the adsorption reaction is the rate determining step. The solute and the adsorbent undergo a chemical reaction i.e. chemisorptions takes place. Lagergren pseudo second order can be derived from the integrated rate law (Eq. 10) where k_2 is the rate constant of adsorption (g mg^{-1} min^{-1}), q_e the amount of divalent metal ions adsorbed at equilibrium (mg g^{-1}), and q_t is the amount of divalent metal ions on the surface of the adsorbent at any time, t (mg g^{-1}).

$$\frac{dq_t}{dt} = k_2(q_e - q_t)^2 \tag{9}$$

The integrated form of Eq. 9 can be rearranged to its linear for Eq. 10. The values of k_2 and q_e can be determined from the intercept and slope from the plot of $\frac{t}{q_t}$ against t.

$$\frac{t}{q_t} = \frac{1}{k_2 q_e^2} + \frac{1}{q_e}t \tag{10}$$

Intraparticle diffusion model (Cheung et al. 2007) was used to analyse the adsorption and involves the transportation of the solute from the aqueous phase to the solid surface and then a diffusion of the solute into the pores of the solid. This equation is described in Eq. 11 where q_t is the amount adsorbed (mg g^{-1}) at time t (min). The rate constant k_{id} is directly evaluated from the slope of the regression line. The intercept c provides information of the thickness of the boundary layer, the larger the intercept the larger the boundary effect.

$$q_t = k_{id} \cdot t^{\frac{1}{2}} + c \tag{11}$$

The thermodynamic potential of the adsorption systems at two different temperatures were determined using the value of K_d calculated using Eq. 12 and then substituted into Eq. 13 to calculate ΔG (J mol^{-1}) which is the change in free energy, R is the gas constant (8.314 J mol^{-1} K^{-1}) and T is the temperature (K) (Sharifipour et al. 2015)

$$K_d = \frac{q_e}{C_e} \tag{12}$$

$$\Delta G = -RTlnK_d \tag{13}$$

3 Results and Discussion

3.1 Characterization of Adsorbent

The surface morphology was examined by scanning electron microscopy (SEM) is shown in Fig. 1. Coir pith contains pores of various sizes with diameters between 25 to 50 μm. The low density and porous structure of coir pith allows a large surface of the adsorption of heavy metals.

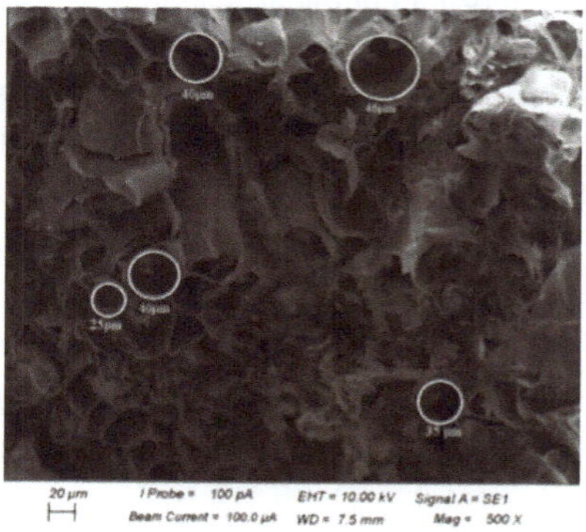

Fig. 1. SEM image of coir pith

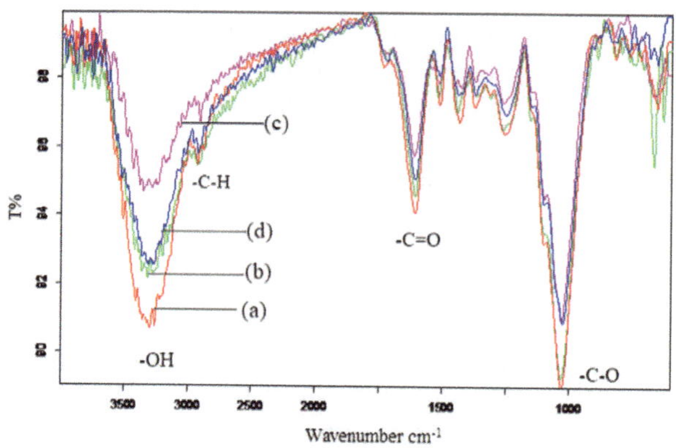

Fig. 2. FTIR spectra of (a) coir pith (b) Ni(II) loaded coir pith (c) Zn(II) loaded coir pith (d) Ni (II) and Zn(II) loaded coir pith

Fourier transform infrared spectra (FTIR) was used to measure the change in vibrational frequency in the functional groups of the adsorbent. The adsorbent was scanned in the wavelength range of 4000–8000 cm^{-1}. Figure 2 shows the FTIR spectrum of (a) coir pith before adsorbent, (b) the spectrum of coir pith after nickel adsorption, (c) the spectrum of coir pith after zinc adsorption, (d) the spectrum of coir pith after nickel and zinc binary adsorption. The FTIR analysis shows the presence of surface functional groups at their corresponding wavelengths; -OH (3000–3500 cm^{-1}), C-H (2900 cm^{-1}), C = O (1600 cm^{-1}), and C-O (1200–100 cm^{-1}). These peaks correspond to the polar groups of the lignin and cellulose of which coir pith is made of. The intensity of the OH peak decreases when adsorption takes place, suggesting that during adsorption the OH bond was broken and the metal ion chemically bonds to the functional group via the oxygen atom. The most significant decrease was after zinc adsorption suggesting that coir pith has a greater capacity for zinc than nickel. The amount of adsorption in the binary system is comparable to that of the nickel single system, and therefore less than the average of the nickel and zinc single systems. This suggests that due to the competitive nature of the binary system, adsorption is suppressed.

3.2 Adsorption Isotherms for Ni and Zn

The adsorption isotherm of single system nickel, single system zinc and total binary system was obtained by plotting adsorption capacity at equilibrium, q_e against equilibrium concentration, C_e as shown in Fig. 3. The adsorption isotherm of nickel shows a mostly linear relationship between q_e and C_e. The isotherm of zinc shows, q_e increases as the solution concentration increases. The shape of the zinc isotherm suggests there is an exponential relationship between q_e and C_e. The difference in the isotherms is possibly because the nickel ion is smaller than zinc which therefore allows the adsorbent to absorb more nickel in its pores than zinc. At higher initial concentrations the adsorption in the binary system is much less than that of the single systems. The competitive nature of the binary system results in less adsorption and possibly less diffusion into the cavities of the adsorbent.

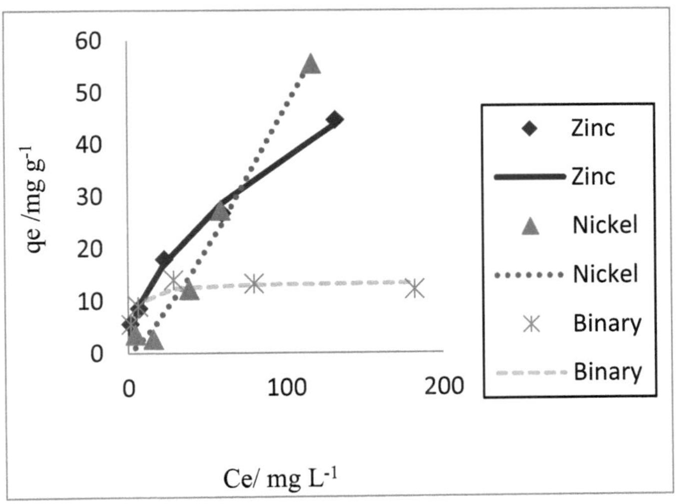

Fig. 3. Adsorption isotherms of single system nickel and zinc with trend lines deduced by fitting the data into the Freundlich model and the adsorption isotherm of the binary isotherm with the trend line deduced by fitting the data into the Langmuir isotherm model.

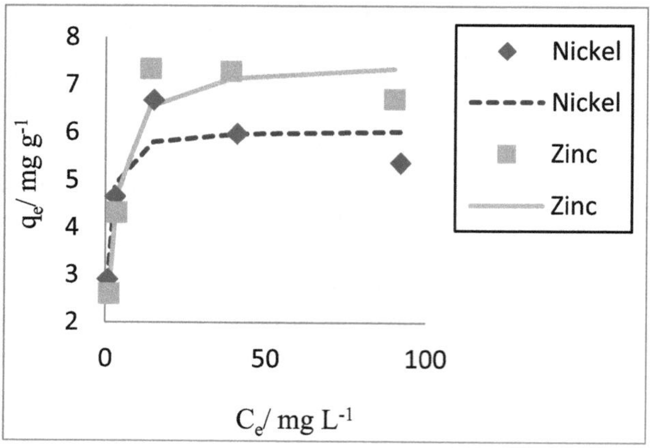

Fig. 4. Adsorption isotherms of the nickel and zinc components of the binary adsorption system. The regression lines were drawn by fitting the data into the Langmuir model.

The adsorption isotherm of the nickel and zinc components of the binary system shown in Fig. 4 is a type 1 isotherm where there is a formation of a monolayer followed by desorption at higher concentrations.

Table 1. Isotherm parameters for the single system and binary adsorption of nickel and zinc solutions onto 0.75 g of coir pith at 30 °C and 40 °C

Isotherm model	Single system		Binary system		
Parameter	Ni	Zn	Ni	Zn	Total
Langmuir					
q_{max}/mg g^{-1}	27	22.7			13.5
k_L/L mg^{-1}	0.03	0.14			0.38
R^2	0.82	0.83			0.90
Freundlich					
k_F	0.10	3.17			6.95
1/n	1.32	0.53			0.13
R^2	0.98	0.97			0.57
Extended Langmuir					
q_{max}/mg g^{-1}			13.6		
k_L/L mg^{-1}			0.304	0.345	
R^2			0.50	0.54	

The data from the nickel single system, zinc single system and binary adsorption at 30 °C was fitted to the Lagmuir, and Freundlich isotherm models. The Langmuir parameters q_{max} and k_L were calculated using Eq. 2. The Freundlich parameters k_F and 1/n were calculated using Eq. 4 and the results are shown in Table 1. The maximum adsorption capacity corresponding to complete monolayer coverage and saturation of all available adsorption sites, q_{max}, of nickel was 27 mg g^{-1}, and q_{max} of zinc was 22.7 mg g^{-1}. The Langmuir constant, k_L, which is related to the free energy of adsorption is lower for nickel than for zinc being 0.03 L mg^{-1} and 0.14 L mg^{-1} respectively. This suggests that coir pith has a greater capacity for nickel than for zinc.

The Freundlich parameter, k_F, which describes relative adsorption capacity, shows that that of zinc is greater than that of nickel being 3.17 and 0.1 respectively. This suggests that zinc has a greater affinity for the adsorbent than nickel. The Freundlich parameter 1/n which is the adsorption intensity, is greater for nickel than for zinc being 1.32 and 0.53 respectively. As both values are greater than 0 the processes are confirmed to be chemisorptions. As 1/n of zinc is closer to 0, the curvature of the isotherm is greater and the process can be interpreted as being more heterogeneous.

In order to calculate q_{max} using the Freundlich model, the experiment needs to be carried out using a fixed initial concentration and varied dosages of adsorbent and then analysed using Eq. 14 suggested by Halsey (1952). The Freundlich expression is an exponential equation and therefore assumes that as the solution concentration increases, the concentration of the solute on the adsorbent increases. The value of k_F is larger for zinc but the value for 1/n is larger for nickel. At a lower initial concentrations C_o the q_{max} of zinc would be higher than nickel while a larger C_o would show a larger q_{max} for nickel.

$$k_F = \frac{q_{max}}{C_o^{\frac{1}{n}}} \tag{14}$$

The data from the total binary system at 30 °C was fitted to the Langmuir and Freundlich isotherm models while the nickel and zinc components of the binary adsorption were analysed using the extended Langmuir isotherm model. When fit into the Langmuir model, it fit well with a R^2 value of 0.90 while the q_{max} value is 13.5 mg g^{-1}. When the binary data was fit into the Freundlich model, the R^2 value was very low at 0.57 which means that the model is not suitable for this system. As the binary system fit the Langmuir model best, the adsorption in the binary system most likely forms a monolayer with very little diffusion into the adsorbent. The extended Langmuir isotherm model calculates the parameters pertaining to nickel and zinc simultaneously. The q_{max} for the whole system which includes both nickel and zinc adsorption was 13.6 mg g^{-1}, which is very similar to the Langmuir. The R^2 value is 0.5 and 0.54 for nickel and zinc respectively. This low correlation coefficient suggests that the extended Langmuir model is unsuitable for this system. Since the Langmuir model shows a better fit for the data and is simpler to model it would be the better model to use to analyse this system.

3.3 Kinetic Study of Adsorption of Nickel and Zinc Single System and Binary Nickel and Zinc Adsorption

Both single system and binary adsorption of nickel and zinc were studied at 30 °C and 40 °C by plotting the mass of metal ion adsorbed onto the adsorbent at time t, q_t, was against time. The data from the adsorption of single systems and binary system nickel

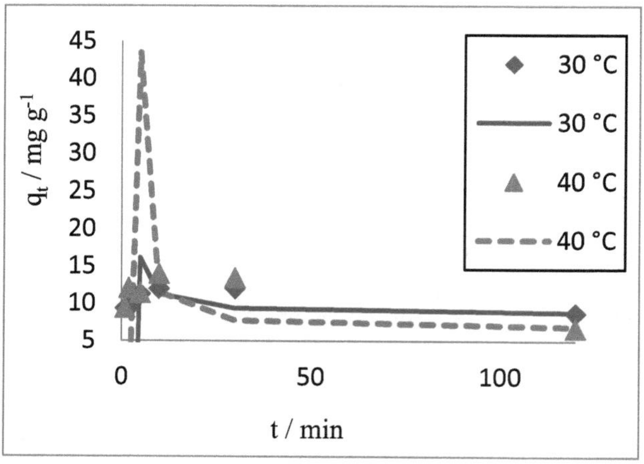

Fig. 5. Kinetics study of nickel solution of initial concentration 50 mg L^{-1} onto 0.75 of coir pith at 30 °C and 40 °C. The regression lines were drawn by fitting the data to Lagergren pseudo second order model.

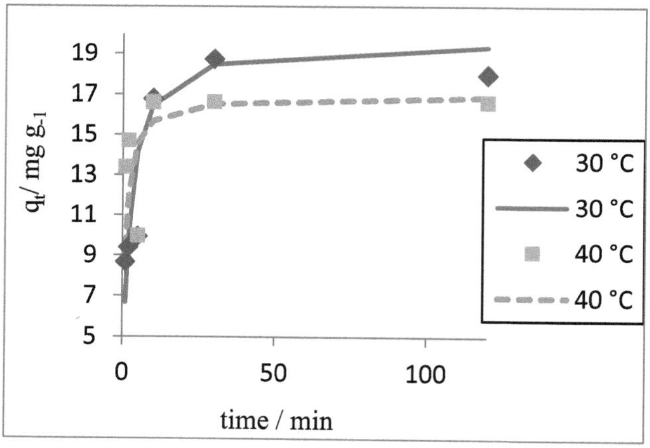

Fig. 6. Kinetic study of zinc solution of initial concentration 50 mg L^{-1} at 30 °C and 40 °C. The regression lines were drawn by fitting the data to the Lagergren pseudo second order model.

and zinc at 30 °C and 40 °C were analysed using Lagergren first order, Lagergren second order models and intraparticle diffusion model to study the mechanism of adsorption. The kinetic parameters $q_{e,1}$ and k_1 were calculated using Eq. 8, $q_{e,2}$ and k_2 were calculated using Eq. 10. The diffusion parameters kid and c were calculated by fitting the data into the intraparticle diffusion model described in Eq. 11. The calculated parameters for single and binary systems were summarised in Tables 2 and 3 respectively. Figure 5 shows the qt against time for the adsorption of nickel at 30 °C

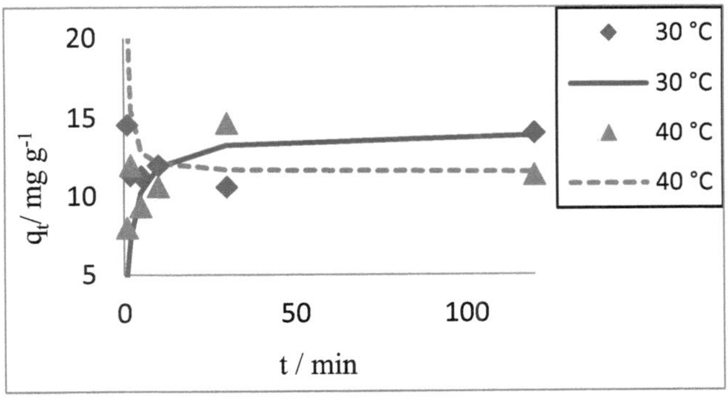

Fig. 7. Kinetic study of binary zinc and nickel solution of initial concentration 50 mg L^{-1} at 30 °C and 40 °C. The regression lines were drawn by fitting the data to Lagergren pseudo second order model

Table 2. Kinetic parameters for the adsorption of single system nickel and single system zinc at 30 °C and 40 °C from a solution of initial concentration 50 mg L^{-1} onto 0.75 g of coir pith.

Kinetic model	Single system			
Parameter	Ni		Zn	
	30 °C	40 °C	30 °C	40 °C
Lagergren pseudo first order				
$q_{e,1}$/mg g^{-1}	–	–	13.19	5.16
k_1/min^{-1}	–	–	0.096	0.105
R^2	–	–	0.91	0.57
Lagergren pseudo second order				
$q_{e,2}$/mg g^{-1}	8.62	6.62	18.52	16.98
k_2/g mg^{-1} min^{-1}	−0.052	−0.036	0.026	0.082
R^2	0.99	0.99	0.98	0.99
qe experimental	8.67	6.69	17.95	16.63
Intraparticle diffusion				
k_{id}/mg g^{-1} $min^{-1/2}$	−0.09	−0.39	2.44	0.83
c/mg g^{-1}	10.78	12.72	6.18	12.04
R^2	0.06	0.29	0.86	0.29

and 40 °C. This shows that as temperature of the solution increases, the rate as well as the amount of metal ion adsorbed increases initially. This is possibly due to the expansion of the solid adsorbent and increase in kinetic energy as the temperature increases.

Table 3. Kinetic parameters for the adsorption of binary system nickel and zinc at 30 °C and 40 °C from a solution of initial concentration 50 mg L^{-1} onto 0.75 g of coir pith.

Kinetic model	Binary system					
Parameter	Ni		Zn		Total	
	30 °C	40 °C	30 °C	40 °C	30 °C	40 °C
Lagergren pseudo first order						
$q_{e,1}$/mg g^{-1}	0.51	2.66	1.19	–	1.75	1.72
k_1/min^{-1}	−0.014	0.04	−0.0299	–	−0.023	0.023
R^2	0.01	0.69	0.54	–	0.34	0.21
Lagergren pseudo second order						
$q_{e,2}$/mg g^{-1}	6.67	7.44	7.35	3.98	14.08	11.12
k_2/gmg^{-1} min^{-1}	0.234	0.059	0.04	−0.077	0.038	0.15
R^2	0.99	0.99	0.99	0.98	0.98	0.96
$q_e$experimental	6.67	7.34	7.33	4.01	14	11.35
Intraparticle diffusion						
k_{id}/mg g^{-1} $min^{-1/2}$	−0.001	0.274	0.103	−0.026	0.102	0.248
c/mg g^{-1}	6.42	4.55	5.45	5.44	11.87	9.997
R^2	0	0.68	0.15	0	0.06	0.16

However, as the system reaches equilibrium desorption occurs. Figure 6 which shows qt against time for the adsorption of zinc shows that the rate increases with temperature but the amount of zinc adsorbed decreases when the temperature increases from 30 °C to 40 °C. Figure 7 shows the amount of metal ion adsorbed, qt, against time for the binary system of nickel and zinc. The overall adsorption decreases as the temperature increases from 30 °C to 40 °C. For the binary adsorption at 40 °C there is initially more adsorption than at 30 °C but there is a significant amount of desorption that occurs as it reaches equilibrium. As shown in Table 3, the equilibrium adsorption capacity, qe, of nickel in the binary system increases from 6.67 mg g^{-1} to 7.34 mg g^{-1}while qe of zinc decreases from 7.33 mg g^{-1} to 4.01 mg g^{-1}. The data of the single system adsorption of nickel was fit into the Lagergren pseudo first order model. Due to the equilibrium adsorption capacity, qe, being much smaller than preceding qt values it was not possible to use this model. The adsorption data for zinc adsorption at 30 °C fit well into the Lagergren pseudo first order model with a R2 of 0.91. However the calculated qe,1 value 13.19 mg g^{-1} was much smaller than the experimentally obtained value of qe 18.52 mg g^{-1}. The adsorption capacity decreases as temperature increases from 13.19 to 5.16 mg g^{-1}, this suggests that due to the increased kinetic energy of the system, it overcomes the attraction between zinc and the adsorbent leading to more desorption. The same adsorption data was analysed using the Lagergren pseudo second order model. The calculated qe,2 values for both nickel and zinc at 30 °C and 40 °C, more closely matched the experimentally obtained values. The correlation coefficient, R2, for all four systems were between 0.98 and 0.99. The values of qe,2 for nickel shows the adsorption capacity decreased as temperature increased and reduced from 8.62 mg g^{-1} to 6.62 mg g^{-1}. Similarly for zinc, the capacity decreases from 18.52 mg g^{-1} to 16.98 mg g^{-1}. This is likely due to the increase in kinetic energy of the system at higher temperatures promoting desorption. Overall, the adsorption of nickel was about half that of zinc, this is likely because nickel ion is smaller and with a higher charge density than the zinc ion, so it would prefer the more polar environment of water whereas zinc would more easily bind to the polar coir pith functional groups. Therefore, the amount of zinc adsorbed is greater than nickel under similar conditions. The kinetic data for the single system adsorption was fit into the intraparticle diffusion model as shown in Table 2. The R2 values were all very low except for the zinc adsorption at 30 °C. This indicates that the intraparticle diffusion was not a rate determining step for the nickel adsorption or zinc adsorption at 40 °C. As the isotherm data also suggests the adsorption is primarily dependent on the solution concentration, it is likely that the rate determining step in nickel adsorption is the chemisorptions to the surface. The intraparticle diffusion in the zinc system at 30 °C was slow enough that it affects the rate.

Table 4. Thermodynamic parameters for the single and binary systems of nickel and zinc at 30 °C and 40 °C

Parameters		Single		Binary		
		Ni	Zn	Ni	Zn	Total
K_d/L g^{-1}	30 °C	0.234	0.780	0.444	0.524	0.483
	40 °C	0.167	0.665	0.524	0.211	0.343
ΔG/J mol^{-1}	30 °C	3672	625	2043	1629	1835
	40 °C	4674	1062	1683	4056	2782

Based on the FTIR spectra and kinetic data, it can be deduced that adsorption involves a chemical reaction at the –OH groups of the functional groups on the coir pith surface. As the adsorption in the binary system is competitive it can be presumed that nickel and zinc bind to the surface at the same sites. The mechanism of adsorption is likely a complexation of the metal ion and the surface functional groups as illustrated in Fig. 8 where M can represent either nickel or zinc ions.

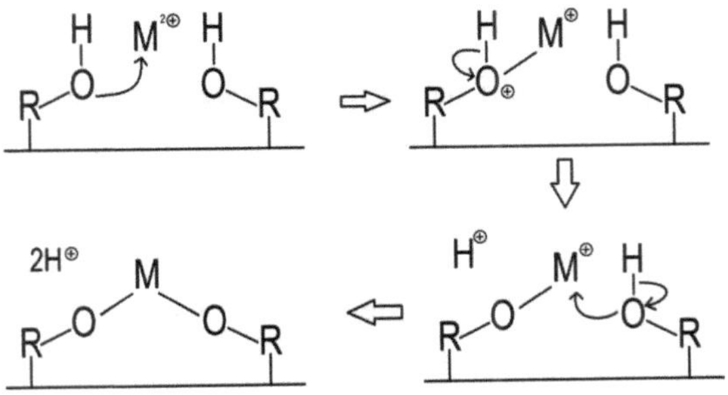

Fig. 8. Mechanism of adsorption of the metal ion onto the coir pith surface

The equilibrium constant K_d, was calculated using Eq. 12 and the Gibbs free energy, ΔG, which was calculated using Eq. 13, was summarised in Table 4. ΔG for nickel becomes more positive from 3672 J mol^{-1} K^{-1} to 4674 J mol^{-1} K^{-1} as the temperature increases meaning the reaction becomes less spontaneous. ΔG increases from 625 J mol^{-1} K^{-1} to 1061 J mol^{-1} K^{-1} for the zinc adsorption. When comparing the ΔG of the adsorption of zinc and nickel, the zinc adsorption was shown to be more spontaneous. The total adsorption in the binary system at 30 °C is 14 mg g^{-1} which is larger than the average adsorption of the single systems at 30 °C. The percentage of nickel adsorption is greater than in the single system while the zinc adsorption is less than the adsorption in the single system. The total binary adsorption at 40 °C is 11.35 mg g^{-1}, which is about the average of the adsorption in the nickel and zinc single system under similar conditions. However the relative proportions of nickel adsorbed was greater than zinc. This suggests that the adsorptions of nickel and zinc are not independent of each other and the presence of competitive ions suppresses adsorption of zinc but encourages adsorption of nickel. The binary adsorption data was studied using the same single system models. When the binary adsorption data was fit into the Lagergren pseudo first order model, the calculated q_e values for the total binary system at 30 °C and 40 °C was 1.75 mg g^{-1} and 1.72 mg g^{-1} respectfully varied greatly from the experimentally obtained q_e value, 14 mg g^{-1} and 11.12 mg g^{-1}. The correlation coefficients, R^2, of the data, fit to the pseudo first order model for nickel and zinc where 0.34 and 0.21 at 30 °C and 40 °C. This shows that Lagergren pseudo first order model is not suitable to analyse the binary adsorption of nickel and zinc. The data was also fit to Largergren pseudo second order model. The q_e value calculated using this model at 30 °C and 40 °C were 14.01 mg g^{-1} and 11.12 mg g^{-1} which were very similar to the experimentally obtained values 14 mg g^{-1} and 11.35 mg g^{-1}. The R^2 values are 0.98 and 0.96 at 30 °C and 40 °C. This shows the adsorption occurring is primarily chemisorptions and that this model can be used to analyse the binary adsorption of nickel and zinc onto coir pith. Similar to the single system adsorption, the adsorption of nickel was less than the adsorption of zinc at 30 °C. When the temperature was increased to 40 °C the adsorption of nickel increases from 6.67 mg g^{-1} to 7.35 mg $g^{-1}$and the adsorption of zinc decreased from 7.34 mg g^{-1} to 3.98 mg g^{-1}. Unlike in the single system, there is a greater decrease in the amount of zinc adsorbed at 40 °C compared to zinc adsorbed at 30 °C. While there is more zinc adsorption than nickel adsorption at 30 °C the difference between the two amounts is much less than in the single system and at 40 °C there is also more nickel adsorption than zinc adsorption. The binary adsorption data was fit to the intraparticle diffusion models. However, the R^2 values were very small, making it very unlikely that the model is suitable to be used to analyse this data. This, along with the isotherm data, suggests that there is little to no diffusion of the metal ions into the pores of the coir pith in the binary system. The ΔG value of the total binary adsorption at 30 °C was 1835 J mol^{-1} K^{-1} and increases to 2782 J mol^{-1} K^{-1} as the temperature increases to 40 °C. Therefore the adsorption gets less spontaneous as temperature increases.

4 Conclusion

The data from both the single systems were fit into the Langmuir, and Freundlich isotherm models. Nickel and zinc single systems fit both models satisfactorily. However, they fit the Freundlich model the better, suggesting the adsorption surface is heterogeneous. The isotherm data from the binary system was fit to the Langmuir, Freundlich and extended Langmuir models. However they only fit only the Langmuir model satisfactorily, suggesting a homogenous adsorption surface. The kinetic data for single and binary system adsorption fit Lagergren pseudo second order model best, suggesting the adsorption is mostly chemisorption. Single system zinc adsorption and single system nickel adsorption decreased as the temperature did. In the binary system, the adsorption capacity was less than the average of the single system under similar conditions, where nickel percentage of nickel adsorption is more compared to the single system, suggesting the competitive nature of the binary system promotes the adsorption of nickel while it demotes zinc adsorption.

Acknowledgements. The authors wish to acknowledge the financial support from the University of Moratuwa, Senate Research Committee for the grant SRC/ST/2017/31.

References

Adikaram AMNM, Pitawala HMTGA (2017) An overview of heavy metal contamination in coastal sediments of Sri Lanka. JTFE 7(2):1–9

Amarasinghe BMWPK (2007) Comparison of adsorption characteristics of waste biomass materials for the removal of Pb ions from industrial effluents. Eng: J Inst Eng Sri Lanka 40 (4):167–174

Amarasinghe BMWPK (2011) Lead and cadmium removal from aqueous medium using coir pith as adsorbent: batch fixed bed column studies. JTFE 1(1):36–47

Cheung WH, Szeto YS, McKay G (2007) Intraparticle diffusion processes during acid dye adsorption onto chitosan. BioresourTechnol 98:2897–2904

Freundlich H (1906a) Über die adsorption in lösungen (Adsorption in solution). J Chem Phys 57:384–470

Freundlich HMF (1906b) Über die adsorption in lösungen. Z Phys Chem 57:385–470

Halsey GD (1952) The role of surface heterogeneity. Adv Catal 4:259–269

Ho YS (1995) Adsorption of heavy metals from waste streams by peat. Ph.D. thesis, University of Birmingham, Birmingham, UK

Ho YS, McKay G (1999) Competitive sorption of copper and nickel ions from aqueous solution using peat. Adsorption 5:409–417

Kumar KV, Ramamurthi V, Sivanesan S (2005) Modeling the mechanism involved during the sorption of methylene blue onto fly ash. J Colloid Interface Sci 284(1):14–21

Lagergren S (1898) About the theory of so-called adsorption of soluble substances. Kungliga Svenska Vetenskapsakademiens. Handlingar 244:1–39

Langmuir (1916) The constitution and fundamental properties of solids andliquids. J Am Chem Soc 38:2221–2295

Nasrullah A, Bhat AH, Isa MH (2016) Lignin: a sustainable biosorbent for heavy metal adsorption from wastewater, a review. In: AIP conference proceedings

Sdiri AT, Higashi T, Jamoussi F (2014) Adsorption of copper and zinc onto natural clay in single and binary systems. Int J Environ Sci Technol 11:1081–1092

Shaheen SM, Derbalah AS, Moghanm FS (2012) Removal of heavy metals from aqueous solution by zeolite in competitive sorption system. IJESD 3(4):362–367

Sharifipour F, Hojati S, Landi A, Faz Cano A (2015) Kinetics and thermodynamics of lead adsorption from aqueous solutions onto Iranian Sepiolite and Zeolite. Int J Environ Res 9 (3):1001–1010

Singhe N, Balomajumder C (2017) Equilibrium isotherm and kinetic studies for the simulataneous removal of phenol and cyanide by use of S. Odorifera (MTCC 5700) immobilized on coconut shell activated carbon. Appl Water Sci 7:3241–3255

Wijayawardhana D, Herath V, Weerasinghe A (2016) Heavy metal pollution in sri lanka with special reference to agriculture: a review of current research evidence. Rajarata Univ J 4 (1):52–66

Sustainable Autarky of Food-Energy-Water (Safe-Water)

N. Khandan(✉), D. Tchinda, S. M. Henkanatte-Gedera, I. S. A. Abeysiriwardana-Arachchige, H. M. K. Delanka-Pedige, S. P. Munasinghe-Arachchige, and Y. Zhang

New Mexico State University,
PO Box 30001 MSC 3CE, Las Cruces, NM 88003, USA
nkhandan@nmsu.edu

Abstract. Urban wastewaters (UWW) are now being recognized as a resource, rich in nutrients and energy, rather than a waste stream that has to be treated and disposed of at the input of considerable energy and associated environmental emissions. This paper presents an algal-based approach for treating UWWs to the mandated discharge levels in a single step and at the same time, recover net energy, nutrients, and water from UWWs for use in the food-energy-water sectors. The proposed approach entails cultivation of energy-rich algal biomass in primary-settled UWW, followed by hydrothermal liquefaction of the resulting algal biomass to recover its energy- and nutrient-contents. Results from a pilot scale field demonstration project are presented to demonstrate the feasibility of this approach. The pilot scale study conducted at a local wastewater treatment plant confirmed that the discharge standards for biochemical oxygen demand (BOD), nitrogen, and phosphorous could be met with a batch process time of 3 days. Recovery experiments showed that more than 20% of the ammoniacal nitrogen (N-NH_3) and more than 90% of the phosphorous (P) in the primary-settled wastewater could be recovered as struvite by the proposed process.

Keywords: Algal wastewater treatment · Energy recovery · Nutrient recovery · Mixotrophic algal growth · Pilot scale demonstration

1 Introduction

With nearly 70% of the world population already concentrated in towns and cities, and with continuing trends in population growth and sprawl of metropolitan areas, provision of the basic human needs of food, energy, and water to the inhabitants is seen as a major challenge. Being net consumers of food, energy, and water, current urban infrastructure that has continued to rely on obsolete technologies to provide these basic needs is not sustainable. Development of a modern infrastructure that can recover and recycle energy and materials within the food-energy-water (F-E-W) sectors at the urban-agriculture interface could be an option to improve the sustainability of both urban and agricultural systems.

In this paper, we report on algal-based, **p**hotosynthetically **o**xygenated **w**aste-to-**e**nergy **r**ecovery (POWER) system for recovering energy, water, and nutrients from

R. Dissanayake and P. Mendis (Eds.): ICSBE 2018, LNCE 44, pp. 91–98, 2020.
https://doi.org/10.1007/978-981-13-9749-3_8

urban wastewaters (UWW) that can be utilized in the food production in the agriculture sector. The POWER system affords waste treatment, net energy production, and recovery of irrigation-quality water and high-purity crop fertilizers. The strain used in the POWER system is an extremophile, *Galdieria sulphuraria*, capable of mixotrophic growth. Optimal growth conditions for *Galdieria sulphuraria* are a temperature range of 45–55 °C and a pH range of 1–4. Mixotrophic metabolism of *Galdieria sulphuraria* enables single-step removal of carbon (BOD) and nutrients (nitrogen, N and phosphates, P) from UWWs. This mixotrophic algal system is capable of incorporating all the C, N, and P in UWW into energy- and nutrient-rich biomass without any energy input and without any loss of C as CO_2 as in the current activated sludge systems. The extreme culture conditions of above-ambient temperature and acidic pH levels in the POWER system contribute to simultaneous deactivation of pathogens and viruses. Additionally, solar energy captured via photosynthesis is incorporated into the biomass.

The energy- and nutrient-rich biomass generated in the POWER system is then hydrothermally processed to yield biocrude, biochar, and an aqueous phase with solubilized N and P in concentrated form as by-products for recovery and use as fertilizers. This paper includes details of the POWER system and results from a pilot scale study conducted under field conditions at an urban wastewater treatment plant.

2 POWER System

The first step in the POWER system entails cultivation of *Galdieria sulphuraria*, 5587.1 (GS) in primary-settled UWW. Due to its metabolic versatility, GS is capable of removing dissolved organic carbon (measured as biochemical oxygen demand, BOD), and nutrients (quantified as ammoniacal nitrogen and phosphates) in primary-settled wastewater in a single reactor, without any external energy demand. The algal biomass cultivated in this step incorporates most of the organic carbon, nitrogen, and phosphates contained in the wastewater. In addition, this biomass embodies most of the energy content of the wastewater captured by the heterotrophic mode as well as solar energy captured by photosynthesis in the autotrophic mode.

Subsequent steps in this system entail harvesting of the energy-rich biomass cultivated in the first step followed by hydrothermal liquefaction (HTL) at moderate temperatures (<350 ºC) and pressures (<20 MPa). In the HTL process, lipid-, protein-, and carbohydrate-contents of the algal biomass are converted to four product streams: an energy-rich oil phase; an N-rich aqueous phase; a P-rich solid residue; and waste off-gases. These products of HTL afford efficient recovery of the energy- and nutrient-contents of the algal biomass in the third step.

The oil-phase in the aqueous phase is separated by gravity settling and processed further to yield biocrude. The P-rich solid phase, referred to as biochar, is separated from the product mix by filtration. Particulate form of phosphorous in the biochar is then extracted by leaching under basic conditions. The biochar leachate and the oil-extracted aqueous phase are then mixed together and dosed with magnesium chloride to precipitate struvite crystals for use as crop fertilizers.

2.1 Pilot Scale System

A pilot scale version of the POWER system fed with primary-settled wastewater has been deployed at a local wastewater treatment plant. The bioreactors of the POWER system were seeded with *G. sulphuraria* CCMEE 5587.1 obtained from Culture Collection of Microorganisms from Extreme Environments (University of Oregon). Since the optimal pH for *G. sulphuraria* is 1 to 4, the initial pH of the culture volume in the bioreactors was adjusted to 2.5 by adding 10N of H_2SO_4.

The bioreactors are of enclosed design, fabricated of clear polyethylene and fitted with a motor-driven paddlewheel to enable circulation and mixing of its contents. The active volume of the bioreactors was 700L and the culture depth was maintained at 20 cm. The headspace of the bioreactor was continuously fed with 2% carbon dioxide. The bioreactors were fed with primary effluent and evaluated first in batch mode and then in fed-batch mode. Samples were collected daily and analysed for ammoniacal nitrogen (NH_3-N) and phosphates (PO_4); and every two days, for BOD_5 (Fig. 1).

Fig. 1. Schematic of the pilot scale system and field installation at a wastewater treatment plant.

3 Results

3.1 Removal of BOD, N, and P

Results from tests on two identical bioreactors of the POWER system (R1 and R2) conducted over a period of 120 days in 25 fed-batch cycles are summarized in Fig. 2. As can be noted from Fig. 2, the bioreactors were able to reduce BOD_5, NH_3-N, and PO_4 to near-discharge standards in 2 days, irrespective of the variations in the initial concentrations and ambient conditions. The high degree of reproducibility of the system is also illustrated in Fig. 2.

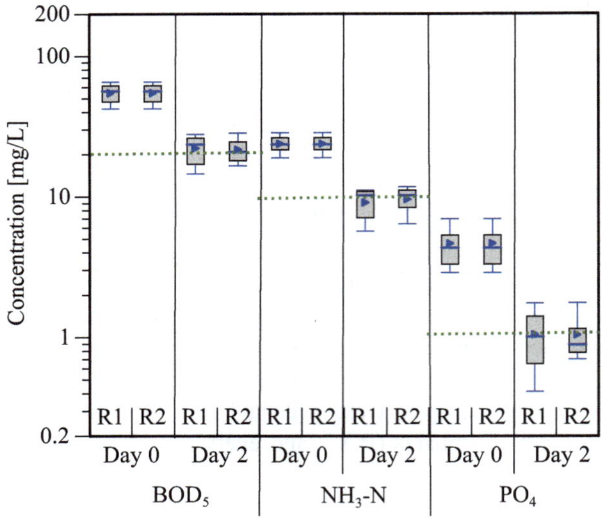

Fig. 2. Reduction of BOD, NH_3-N, and PO_4 in 2 days in 25 fed-batch cycles run over 120 days. Dashed line shows the respective discharge standards.

3.2 Removal of Pathogens

Fecal coliform count in the primary effluent was reduced instantly in the bioreactors to non-detectable level (<1 CFU/mL). Total coliform count was reduced to non-detectable levels (<1 CFU/mL) within one day of treatment. Total bacteria population was reduced by 1.68 logs in the algal system at the end of treatment. Enterococcus faecalis and Escherichia coli copy numbers were reduced to non-detectable levels (<2 copies/mL) resulting in 3.85 and 5.44 log reductions, respectively (Delanka-Pedige 2018). Of special note, *Pseudomonas aeruginosa*, Salmonella *enterica* and *Staphylococcus aureus* were not detected in the primary effluent.

Based on a mechanistic study undertaken to discern the key factors responsible for the high bacterial inactivation noted in this algal system, it was concluded that the inactivation could be attributed primarily to the low pH conditions maintained in the reactor. Additionally, the synergistic effect of high dissolved oxygen levels (5.1 to 7.8 mg/L; average = 6.2 ± 0.53 mg/L) during sunlight hours with the potential to produce reactive oxygen species could also have contributed to bacterial inactivation. Even though the enclosed reactor design enabled above-ambient culture temperatures during the test period (27 to 46 °C; average = 34.9 ± 4.3 °C), temperatures below 45 °C were found to be insignificant in inactivating bacteria (Munasinghe-Arachchige 2018).

3.3 Recovery of Energy

A core technology embedded in the POWER system is hydrothermal liquefaction of the algal biomass that enables recovery of energy and nutrients for reuse. The yields of the four products of HTL (biocrude, biochar, aqueous phase, and off-gases) are functions of the HTL process temperature. Given the energy content of the wastewater (6.3–7.6 kJ/L) the net energy recovery from the integrated process is dependent primarily on the energy input to the HTL process to maintain the HTL process temperature on one hand, and the resulting yield of biocrude, on the other.

Simulation models and experimental results were combined to estimate the energy input for heating the feed to the HTL process temperature, energy that could be recovered from the biocrude and biochar, and the net energy yield at various HTL process temperatures. Results of this analysis are summarized in Fig. 3 that show that the optimal HTL process temperature is 180 °C, if energy could be extracted from biochar as well. However, technologies for efficient recovery of energy from biochar are not common. With biocrude as the sole energy product, 300 °C is seen as the optimal HTL process temperature. Considering the energy content of UWW (6.3 to 7.6 kJ/L), the net energy output estimated here including biochar is 4.5 kJ/L, which translates to energy recovery efficiency of 60%.

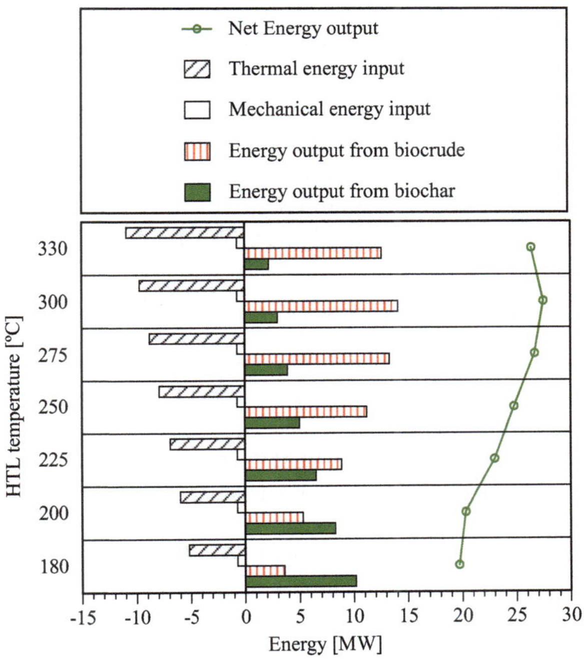

Fig. 3. Energy input to, output from, and net energy from hydrothermal liquefaction process as a function of process temperatures.

3.4 Recovery of Nutrients

The efficiency of recovery of nutrients from the algal biomass is a function of the HTL process temperature. Concentrations of total N and total P in the aqueous resulting from hydrothermal liquefaction of the warm, cold, and mixed cultures of GS cultivated in primary settled wastewater were measured These results indicated that typical total N averaged 1,450 mg/L and typical total P averaged 70 mg/L. These high concentrations enable efficient recovery of N and P for beneficial use. Available P in the biochar samples was typically less than 50 mg P/kg. However, total phosphorous in the biochar exceeded 50,000 mg/kg (or more than 5% by weight).

It was hypothesized that the non-available P could be extracted from the biochar by leaching under acidic or basic conditions. Experiments were conducted to assess the leaching efficiency at pH = 7, pH = 1.5, and pH = 12 to establish the optimum conditions to extract P into the available form. Based on the results of the leaching experiments it was concluded that pH > 12 would be optimal for complete extraction of P from biochar.

Magnesium in the form of magnesium chloride was selected to recover N and P as struvite. Struvite was rapidly precipitated upon addition of magnesium chloride to the mixture of the biocrude-extracted aqueous phase and the biochar-leachate. Chemical analysis of the algal struvite precipitated in this manner indicated that its composition was close to that of pure struvite.

A mass balance was constructed across the entire system to assess the efficiency of recovery of nutrients by the POWER system. From the results of a typical overall mass balance summary shown in Fig. 4, N-recovery as struvite can be seen as 16% and P-recovery as 89%. This analysis showed that the supernatant from the struvite precipitation was rich in N; about 71% of the N in the primary-settled waste remained unrecovered. One option for utilizing this resource is to recycle it to the algal cultivation step to boost biomass production. Future research needs to be directed towards developing technologies for recovering N from this by-product stream.

3.5 POWER System in F-E-W Sectors

Results of long-term evaluation of the POWER system suggests that the system has the potential to render wastewater treatment a sustainable operation. It can conserve the energy consumed in the traditional aerobic wastewater treatment practice and the energy consumed in the manufacture of fertilizers. It can minimize environmental emissions associated with the current wastewater treatment practice and recover valuable resources for reuse and thus conserve the limited natural resources. Figure 5 shows a conceptual system for applying the POWER system at the urban-agriculture interface. By recovering irrigation-quality water, net energy, and fertilizers from the wastewaters from urban and agricultural sources for reuse overall sustainability could be improved.

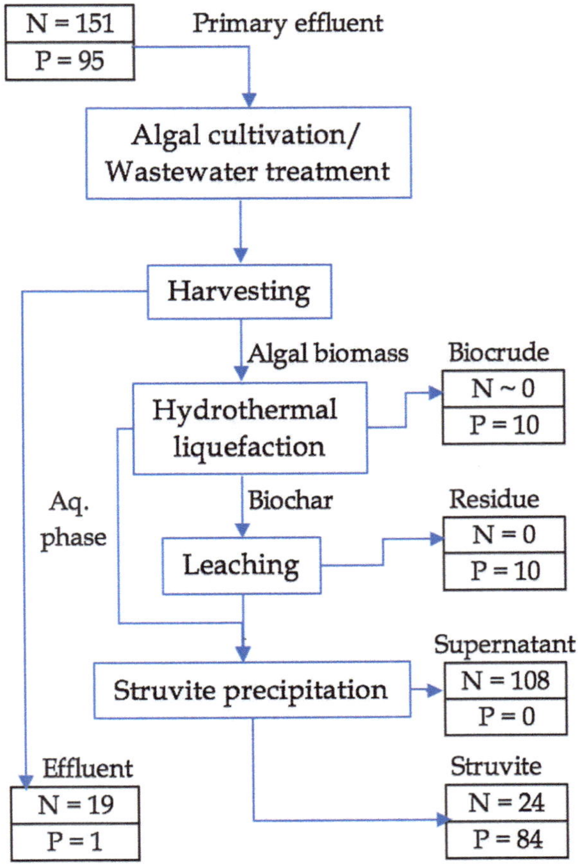

Fig. 4. Overall mass balance on ammoniacal nitrogen (N, kg/d) and phosphates (P, kg/day) for a flow rate of 1 MGD.

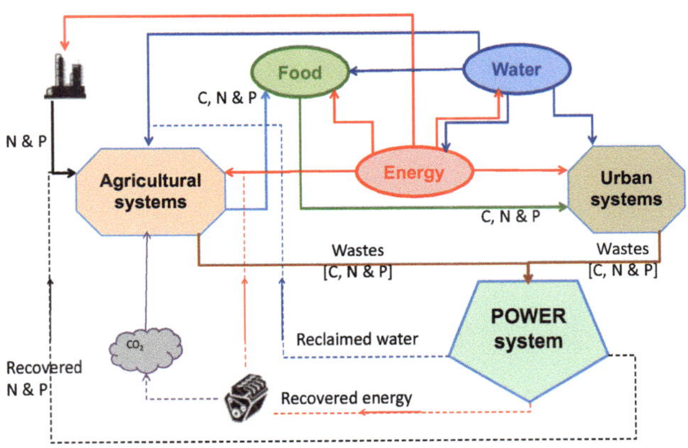

Fig. 5. Application of POWER system at the urban-agriculture interface to recover irrigation-quality water, net energy, and fertilizers.

4 Conclusions

Feasibility of the POWER system in treating primary-settled urban wastewater and recovering net energy and nutrients from it was demonstrated at pilot scale. Long-term, stable operation of the system in a reproducible manner has been demonstrated. The algal cultivation-hydrothermal liquefaction-struvite precipitation pathway embedded in the POWER system has the potential to afford sustainable wastewater treatment. A conceptual scheme to apply the POWER system at the urban-agriculture interface is proposed to recover and recycle resources within the food-water-energy sectors.

Acknowledgements. This study was supported in part by the National Science Foundation Engineering Research Center for Reinventing the Nation's Urban Water Infrastructure (ReNU-WIt), award EEC 1028968; the National Science Foundation Energize New Mexico EPSCoR award IIA-1301346; College of Engineering at New Mexico State University, and the Ed & Harold Foreman Endowed Chair.

References

Henkanatte-Gedera SM et al (2017) Removal of dissolved organic carbon and nutrients from urban wastewaters by Galdieria sulphuraria: laboratory to field scale demonstration. Algal Res 24:450–456

Selvaratnam T et al (2014) Evaluation of a thermo-tolerant acidophilic alga, Galdieria sulphuraria, for nutrient removal from urban wastewaters. Bioresour Technol 156:395–399

Delanka-Pedige HMK (2018) Algal-based wastewater treatment: assessment of pathogen removal and bacterial community profile of effluent. MS thesis, New Mexico State University, Las Cruces

Munasinghe-Arachchige SP (2018) Algal-based wastewater treatment: factors influencing bacterial inactivation and potential for coliform regrowth, MS thesis, New Mexico State University, Las Cruces

Selvaratnam T et al (2015) Feasibility of algal systems for sustainable wastewater treatment. Renew Energy 82:71–76

Henkanatte-Gedera SM et al (2015) Algal-based, single-step treatment of urban wastewaters. Biores Technol 189:273–278

Study of Evidences on Chronic Kidney Disease Due to Unknown Etiology Based on Environmental, Social, Economic and Health Patterns of Selected Population

A Case Study Done at Wilgamuwa Area

S. M. Rajapakshe[1(✉)], S. K. Weragoda[1], T. Kawakami[2], and W. B. M. L. I. Weerasekara[1]

[1] Water Safety Plan Advisory Unit, National Water Supply and Drainage Board, Katugastota, Sri Lanka
sardhaman@gmail.com
[2] Toyama Prefectural University, Imizu, Japan

Abstract. Chronic Kidney Disease (CKD) is becoming a major rising public health problem in Sri Lanka. In addition, CKD has a very complicated interrelationship with other diseases. Accordingly, a community-based study was conducted to identify the evidences of the prevalence of CKD and correlate the relationship between drinking water and causes for CKD. Further, a study was conducted to identify the impact on lifestyle due to CKD issue. The study particularly focused on the Wilgamuwa Divisional Secretariat area where the burden of CKD is pronounced and the underlying cause is still unknown. The sample of the study was selected representing high and low prevalence of CKD which included both patients and non-patients. A questionnaire survey was conducted to gather data, mainly focused on areas like drinking water sources, social, environmental, health, economic, genetic and cultivation practices. Drinking water sources from the dug wells and deep tube wells used by both patients and non-patients was analysed for physical and chemical parameters including heavy metals. Water quality parameters like pH, conductivity, turbidity, oxidation- reduction potential and dissolved Oxygen were analysed onsite. Heavy metals were analysed from Inductively Coupled Plasma Mass Spectroscopy (ICP-MS) and anions was analysed from Ion Chromatography (IC). Locations of water sources and patients and non-patients were traced by GPS and a groundwater quality and patient's distribution map was prepared using GIS and a study of the patient distribution and causative factors interrelating the water quality was conducted. Based on all the water quality parameters and questionnaire survey results in the study, some of the dug wells in the area were identified as doubtful whereas some were suitable for drinking water. It can be concluded that there is no relationship with Hardiness and Fluoride with CKDU and most of the water quality parameters did not exceed the SLS drinking water standards SLS 614:2013.

R. Dissanayake and P. Mendis (Eds.): ICSBE 2018, LNCE 44, pp. 99–106, 2020.
https://doi.org/10.1007/978-981-13-9749-3_9

1 Introduction

Chronic Kidney Disease (CKD) has become a serious health issue and is increasing at an alarming rate today in Sri Lanka. It is a major health care problem, especially in the dry zone. It is observed that CKD is caused by multiple factors, linked in combination.

Over one thousand people were reported as dead due to CKD and more than 60,000 patients have registered at renal clinics of several government hospitals in the dry zone of the island (Weragoda et al. 2013).

CKD is most prevalent among male farmers in the 40 to 60 age group engaged in rice cultivation, although lately, it has also been detected among women, children, and even cattle (Centre for Science and Environment 2012). There is scientific consensus that this fatal disease in Sri Lanka is not related to known causes commonly identified with kidney disease, such as, diabetes, hypertension, obesity, or other factors such as snake bites (Ministry of Healthcare and Nutrition, Epidemiology Unit, Sri Lanka and World Health Organization 2009). After year 2000, there was a rapid increment of the CKD patients in Sri Lanka and from those, majority were in the dry zone, especially Anuradhapura, Polonnaruwa, Kurunegala and Matale districts. There was a big impact of the disease on social, economic, political, and environmental backgrounds of the country. Identifying causal factors is very essential to answering this problem. So, number of researches are conducted based on different disciplines. Groundwater is the main drinking water resource and more than 90% of the drinking water requirements for the rural communities are obtained by shallow and deep wells.

The selected study area of Wilgamuwa is situated in the Matale District in the Central Province of Sri Lanka. Topographically, the western part of the study area is hilly while the eastern part is flat and undulating. The Mahaweli Ganga is the main surface water body of the region that flows along the direction from south to north. In addition, irrigation tanks, channels, and minor streams are located in the study area (Fig. 2).

Most of the irrigation network of the area is based on water diverted from the Mahaweli River. Geologically, Migmatitic gneiss is the main rock types in this area and the overburden is mainly composed of a weathered profile. Annual precipitation of the area is around 1750 mm and the study area belongs to the intermediate climatic zone.

The main recharging source of both shallow and deep groundwater of the area is from local rainfall. However, in some areas groundwater sources recharge from local rainfall as well as irrigated water. So, contaminated groundwater in the shallow dug wells in the areas may be used by the people of the villages (Figs. 1 and 3).

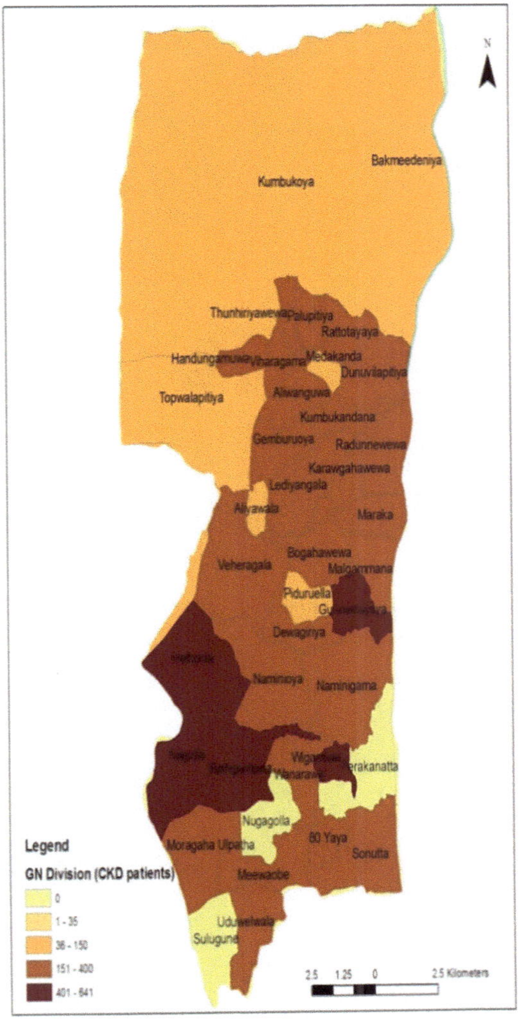

Fig. 1. CKDu patient distribution map Wilgamuwa GS division

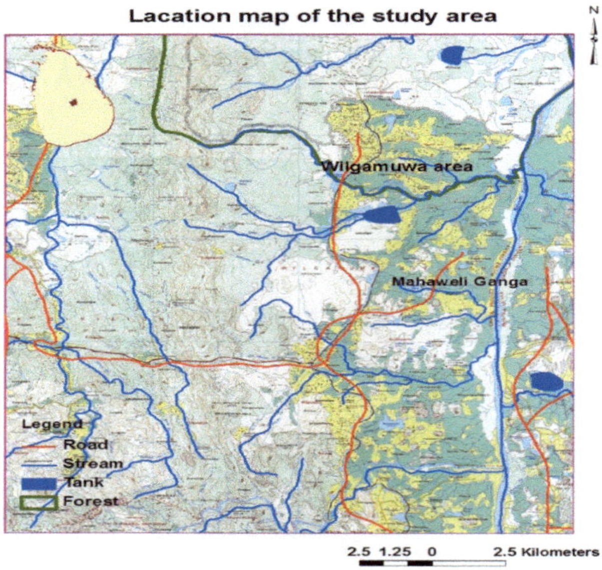

Fig. 2. Location map of the study area

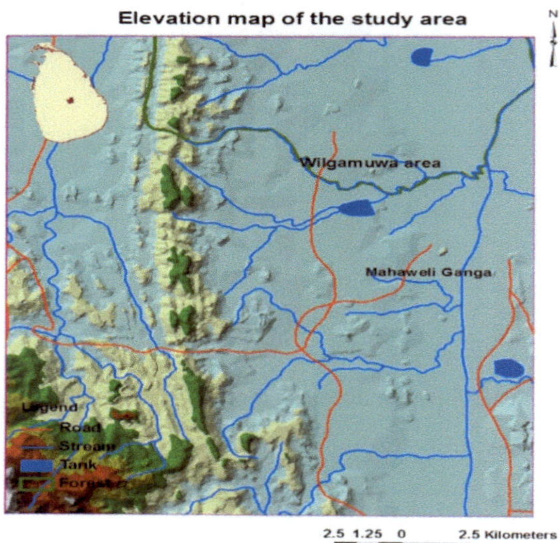

Fig. 3. Elevation map of the study area

2 Materials and Methods

This community-based study was conducted to identify the causes for the prevalence of CKDu and to evaluate the relationship between drinking water quality and causes for CKD. The stratified sampling collection method was applied and samples collected from the selected 11 Grama Niladhari Divisions (GND) representing high and low prevalence of CKDu which included both patients and non-patients.

A questionnaire survey was conducted to gather data, mainly focused on areas like drinking water sources, social, environmental, health, clinical, economic, genetic and occupational practices. Chemical (Cd, As, Pb, F, Mg, Mn, Ca, Hardness) and physical parameters (onsite analysis was conducted for pH, conductivity, turbidity, oxidation-reduction potential (ORP) and dissolved oxygen) were analysed for drinking water sources from the dug wells and deep tube wells used by both patients and non-patients.

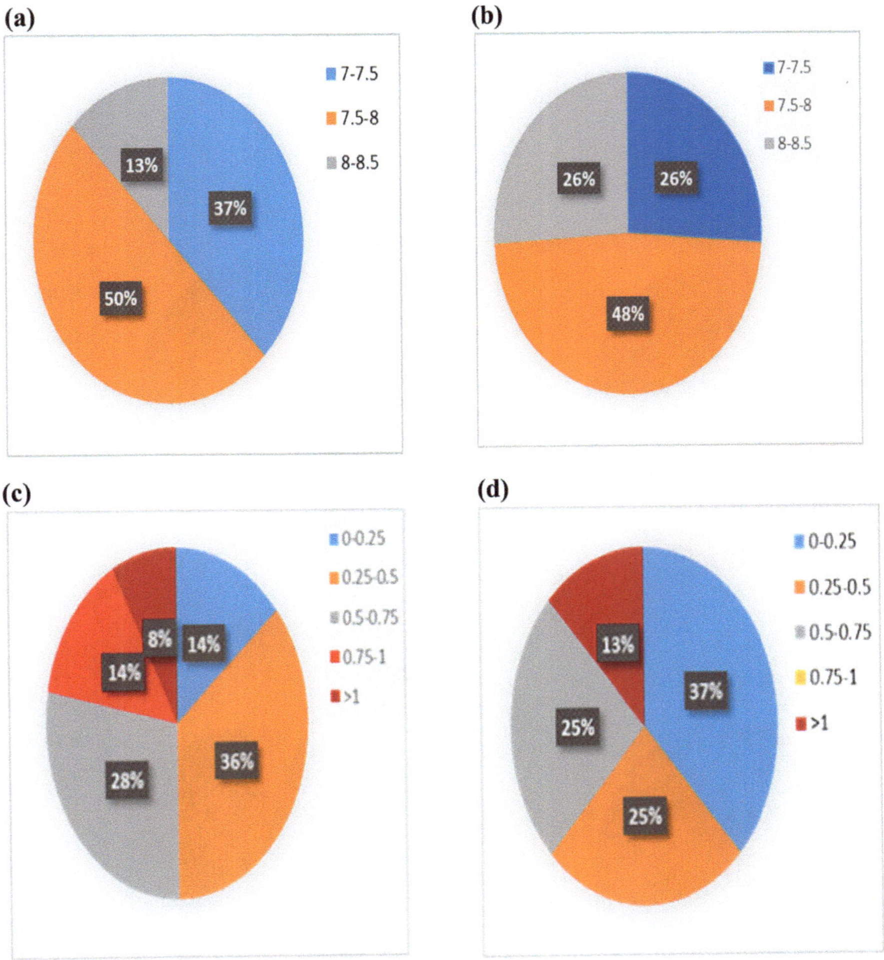

Fig. 4. (a) pH variation of patients (b) pH variation of non-patients (c) Fluoride variation of patients (d) Fluoride variation of non-patients (e) EC variation of patients (f) EC variation of non-patients (g) Calcium variation of patients (h) Calcium variation of non-patients (i) Arsenic variation of patients (j) Arsenic variation of patients (k) Maganese variation of patients (l) Maganese variation of non-patients

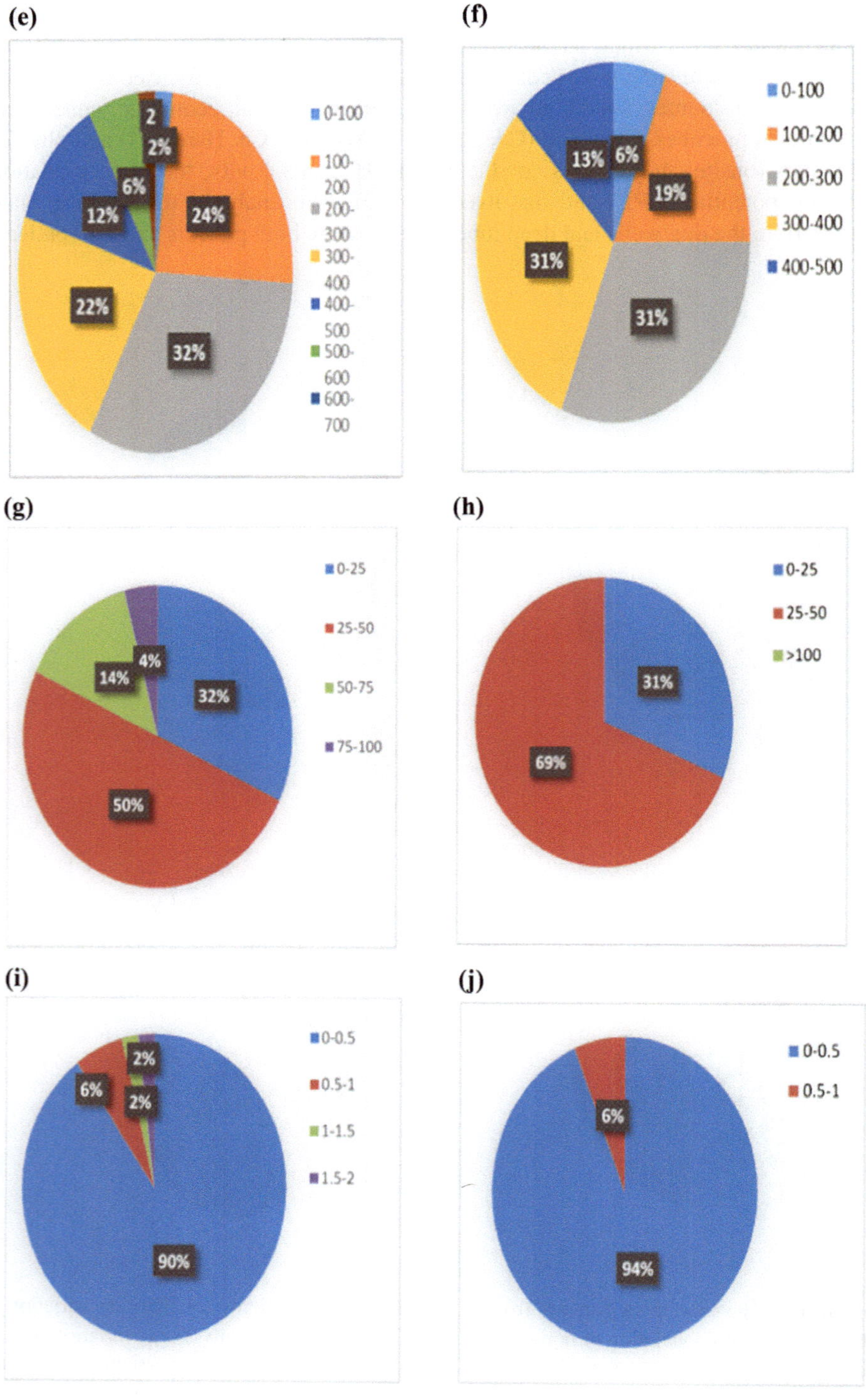

Fig. 4. (*continued*)

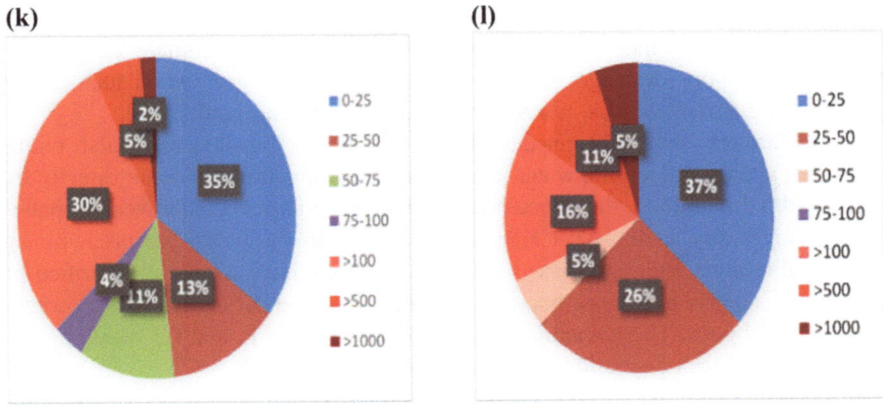

Fig. 4. (*continued*)

Heavy metals were analysed from Inductively Coupled Plasma Mass Spectroscopy (ICP-MS).

Locations of water sources, patient and non-patient, were traced by GPS and groundwater quality and patient distribution map using GIS was prepared and the patient distribution and causative factors interrelating the water quality was studied. Descriptive statistical results are shown in graphical form comparing the chemical parameters with recommended levels stipulated by Sri Lankan standards (Fig. 4).

3 Results

Based on all the water quality parameters and the questionnaire survey results, some of the dug wells were identified as doubtful whereas some were suitable for drinking water and also, significant difference was not observed in CKD prevalent and non-prevalent areas.

Many of those affected by CKDu are young to middle aged male farmers who work on rice farms (paddy fields) and other agricultural commodity cultivations, where they are presumably exposed to unusually high levels of environmental hazards compared to their non-farming peers.

In both patient and non-patients, high Fluoride, Manganese and Magnesium concentration were observed while Calcium concentration is within the limits.

Male farmers 25–65 years of age are the most affected and die of this disease in greatest numbers; they also frequently experience unprotected exposure to agro-chemicals and other toxins in their harsh work environment, and also engage in high-risk behaviour in comparison to women.

On top of regularly getting dehydrated while at work and drinking less water, one of the plausible reasons male farmers are more exposed to potential toxins is that many have disregarded the age-old habit of taking clean water from home to the farmlands. They have adapted an easier and a dangerous habit of drinking polluted water directly from nearby streams and from the paddy fields while, women and children continue to drink water from relatively protected shallow wells and deep tube wells near their homes. Those affected most are male farmers who work in harsh climatic conditions, in flat lands with poor drainage; they do not use protective gear when handling and spraying toxic agrochemicals, nor while working in farming land.

These farmers frequently get dehydrated during work in hot, and harsh climatic conditions, and drink too little water (and some drink contaminated water directly from paddy fields). In addition, the affected areas experience unpredictable weather patterns with prolonged dry spells. All CKDu-affected areas in the world are predominantly agricultural regions with agro-based economies; do not have a centrally purified pipe borne water supply, have less than optimal access to medical care, and are economically relatively poor. These chemicals included Cd, As, Pb, F, and Ca which previous researches have been suggested as causative factors for CKD.

It was seen that in the GND divisions where there is high prevalence of CKD, most of the agricultural practices and drinking water sources were unsatisfactory. Some of the dug wells were contaminated with heavy metals. But still, the water quality parameters of some patients' locations were satisfactory as a drinking water source. Therefore, a further detailed study on all other factors is needed.

4 Discussion

It was seen that in the GND divisions where there is high prevalence of CKD, most of the agricultural practices and drinking water sources were unsatisfactory. Some of the dug wells were contaminated with heavy metals. But still, the water quality parameters of some patients' locations were satisfactory as a drinking water source. Therefore, further detailed study on all other factors is needed.

5 Conclusion

According to the samples analysed, it can be concluded that most of the water quality parameters do not exceed the SLS drinking water standards. But considerably, Manganese, Fluoride, Magnesium exceeds the SLS drinking water standards in both patient and non-patients.

The aetiology of CKDu in of Sri Lanka is probably multifactorial, involving one or more environmental factors and a possible genetic predisposition in vulnerable populations.

References

Centre for Science and Environment (2012) Environmental contamination and its association with Chronic Kidney Disease of Unknown Aetiology in North Central Region of Sri Lanka

Ministry of Healthcare and Nutrition, Epidemiology Unit, Sri Lanka and World Health Organization (2009) Chronic kidney disease of unknown aetiology (CKDu). National Research Programme for the Chronic Kidney Disease of Unknown Aetiology (CKDu) in Sri Lanka

Weragoda S, Kawakami T, Motoyama A, Kodithukakku S (2013) Is groundwater in Dry Zone, Sri Lanka safe to drink? Sri Lanka Collaborative Research

Algal Wastewater Treatment: Study of Polyculture in Fed-Batch Mode in a Winter Climate

I. S. A. Abeysiriwardana-Arachchige, W. Chapman, and N. Nirmalakhandan(✉)

Civil Engineering Department, New Mexico State University, Las Cruces, NM 88003, USA
nkhandan@nmsu.edu

Abstract. Since the 1950s, reactors using algal-bacterial polycultures have been engineered by researchers to treat wastewater. The symbiotic relationship between algae and bacteria accommodates energy-efficient wastewater treatment by eliminating artificial aeration. Here, we investigated the applicability of a novel algal-based system using two polycultures to treat primary-settled municipal wastewater under field conditions in a winter climate (average daily temperature: 4–16.5 °C). The two polycultures were cultured in two closed 700L bioreactors deployed at a local wastewater treatment plant in Las Cruces, NM, United States. The Ammoniacal nitrogen and phosphate concentrations and OD750 of the two reactors were analyzed daily. The BOD_5 in the reactors were analyzed once in two days. The samples from the two reactors were analyzed over 44 days from January to March in 2018. The culture in reactor 1 attained average ammoniacal nitrogen, phosphate, and BOD_5 removals of 99.3%, 70.1%, and 70.0% respectively in a single step within five days. The corresponding efficiencies for reactor 2 were 44.3%, 47.5%, and 50.7%. The average batch processing times for reactor 1 and reactor 2 were 1.02 and 4.02 days respectively. This paper confirms the ability of both the polycultures to treat primary-settled wastewater in a single step.

Keywords: Algal wastewater treatment · Phosphorus removal · Nitrogen removal · Polyculture

1 Introduction

Wastewater treatment utilities have continued to rely on energy-intensive and unsustainable technologies for removing nutrients (nitrogen and phosphorus) from wastewaters to the required discharge standards. Researchers have been looking for alternative wastewater treatment technologies to conserve energy and recover nutrients from wastewaters to render the process sustainable and cost-effective [1]. On realizing the capability of algae to uptake nutrients while growing photosynthetically, several studies have evaluated the feasibility of using algae in wastewater treatment systems. Since the premise to use an algae-bacteria mixed polyculture for wastewater treatment was tested in 1950s [2], many studies have been conducted to develop engineered

R. Dissanayake and P. Mendis (Eds.): ICSBE 2018, LNCE 44, pp. 107–115, 2020.
https://doi.org/10.1007/978-981-13-9749-3_10

systems with algae-bacteria cultures to treat industrial and domestic wastewaters [3–6]. The symbiotic relationship between algae and bacteria has made energy efficient wastewater treatment possible in solar-powered photobioreactors [1].

Microalgae and bacteria have been shown to coexist in culture while mutually benefiting each other. With the presence of sunlight, microalgae photosynthetically produce oxygen which is used by aerobic bacteria to biodegrade organic matter. The algae then consume the carbon dioxide released by bacterial respiration whereby eliminating artificial aeration [1, 6]. More recently, the ability of algal species such as *Galdieria sulphuraria* [7, 8] to mixotrophically metabolize organic carbon and nutrients in a single step to meet the respective discharge standards has been recognized. Since most of the ammoniacal nitrogen and phosphorus in wastewater accumulate in algal cells [2, 6], such a mixotrophic system can achieve efficient and economical wastewater treatment, while generating biomass that can be processed downstream to recover the nutrients as crop fertilizers [6] or recover energy in the form of biocrude [9, 10].

An algae-based, Photosynthetically Oxygenated Waste to Energy Recovery (POWER) system has been developed at New Mexico State University (NMSU) for energy-efficient wastewater treatment and resource recovery. An axenic culture capable of mixotrophic metabolism, *G. sulphuraria,* used in this system has demonstrated superior carbon and nutrient removal where the culture temperatures are near its optimal range of 27–46 °C [7, 8]. Since the temperature drops in winter, *G. sulphuraria* loses its treatment efficiency. Hence, finding an adequate culture replacement for winter had been a long-standing quest. The studies that have been done so far on urban wastewater (UWW) treatment with microalgae in winter climate are very limited. Therefore, a polyculture that could sustain the cold weather was selected assuming it could treat urban wastewater as efficient as by *G. sulphuraria* during summer.

This study evaluated the feasibility of two polycultures to accommodate winter treatment in the POWER system. The two polycultures were assumed to consist of two different algal-bacterial compositions. The specific objective of the study was to investigate the removal efficiency of dissolved organic carbon and nutrients by the two polycultures in a winter climate.

2 Materials and Methods

2.1 Microalgae Strain and Cultivation Methods

Polycultures obtained from the Fabian Garcia Research Facility (New Mexico State University) were grown in two pilot scale photobioreactors deployed at a local wastewater treatment facility, Las Cruces, NM, USA. The bioreactor was fed with primary-settled wastewater. Unlike previous studies on the cultivation of *G. sulphuraria*, the addition of growth mediums of any kind or enrichment of headspace with carbon dioxide was not done during this study. The initial pH in the reactors was not chemically adjusted.

2.2 Equipment Used and Fed-Batch Cultivation

The experiments were conducted in two identical enclosed photobioreactors (R1 and R2) fabricated from a thin clear polymer allowing for light penetration and distribution. The enclosed design enabled the prevention of evaporation and contamination, and conserved heat inside the reactor. The culture depth was maintained at 20 cm and circulation and mixing in the reactor were controlled by a motor-driven paddle wheel.

The reactor was operated in a fed-batch mode, with a working volume of 700L, which comprised of 400L of primary effluent and 300L of culture. Each fed-batch cycle was terminated when the reactor reached the stipulated discharge standards for all the three primary pollutants: ammoniacal nitrogen - 10 mg/L; phosphate - 1 mg/L; and BOD_5 - 30 mg/L. The paddle wheel was then switched off for 24 h to settle the biomass, and 400L of the resulting supernatant was discharged and replenished with a fresh batch of 400L of primary-settled wastewater. These fed-batch experiments were conducted continuously for 44 days from January through March in 2018.

Instead of direct measurement of the dry density, the biomass concentration in the reactor was analyzed daily in terms of optical density at 750 nm wavelength (OD 750) with a HACH DR 6000 spectrophotometer. Ammoniacal nitrogen and phosphate concentrations were also analyzed daily using the same equipment following the salicylate method 8155 and the phosver3 method 8048 respectively. Dissolved BOD_5 was measured every two days with the standard method 5210 B. Furthermore, the temperatures in each reactor and in the field were recorded at an interval of 15 min using Onset HOBO data loggers (UA-002-64).

2.3 Summary of Operation Conditions

The temporal variation of daily average temperatures is illustrated in Fig. 1. The average daily temperature in the reactor and in the atmosphere ranged between 8–21 °C and 4–16.5 °C respectively. The concentrations of ammoniacal nitrogen, phosphate, and BOD_5 in the primary effluent fed to the bioreactors ranged between 19.8–40.3 mg/L, 0.6–2.1 mg/L, and 39.5–70 mg/L respectively.

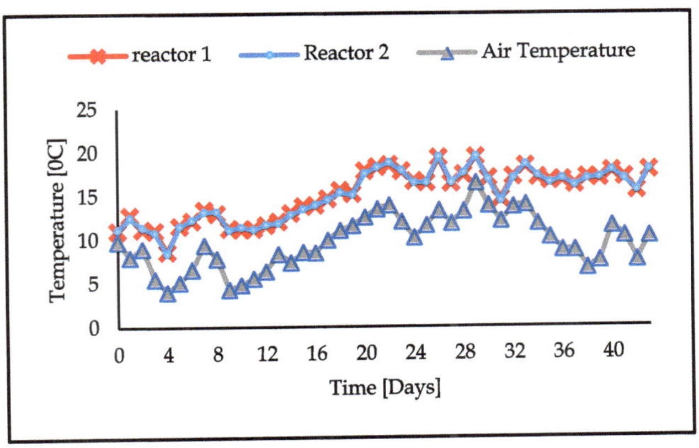

Fig. 1. Temporal variation of average daily temperature

3 Results and Discussion

3.1 Growth, PH, and Reactor Stability

The temporal variation of biomass growth is illustrated in Fig. 2. The two reactors R1 and R2 maintained growth rates of 0.016 ± 0.001 and 0.014 ± 0.001 OD 750 nm day^{-1} respectively. The hypothesis test (H_0: $ß_1 = ß_2$ vs. H_a: $ß1 \neq ß2$) to check equality of the two slopes concluded that there is no statistically significant difference between growth rates at a significance level of 0.05 ($P \simeq 0.88$). However, both the reactors in the current study seemed to experience a lag phase compared to the previous results on the POWER system [11] with *G. sulphuraria* during summer time.

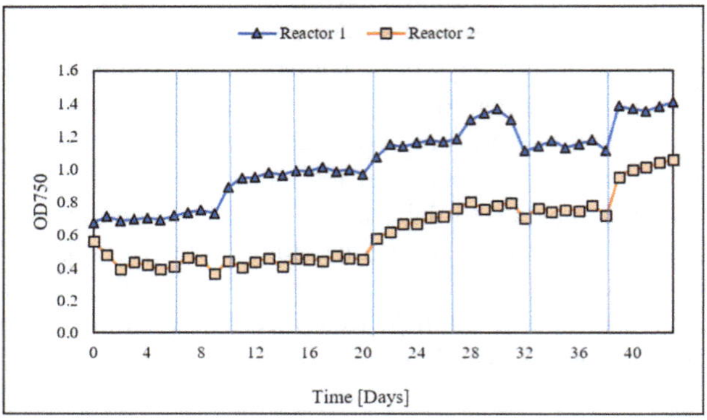

Fig. 2. Variation of optical density at 750 nm of the reactors with time. Vertical lines indicate cycle separation.

As the cultures used in this experiment endured the cold climate, stable operation of both reactors was possible. Unlike in the case of *G. sulphuraria,* the photobioreactors were neither supplemented with CO_2 nor acidified. The pH variation in the two setups during the study period is shown in Fig. 3. As can be seen, the pH of reactor 2 was slightly higher than that of reactor 1; however, the overall pH increased in both reactors as the treatment cycle progressed. As explained by the following equation, the increase of pH can be ascribed to the uptake of photosynthetic carbon by microalgae from the surrounding which promotes the release of hydroxyl ions to the medium [12, 13].

$$HCO_3^- \leftrightarrow CO_2 + OH^- \tag{1}$$

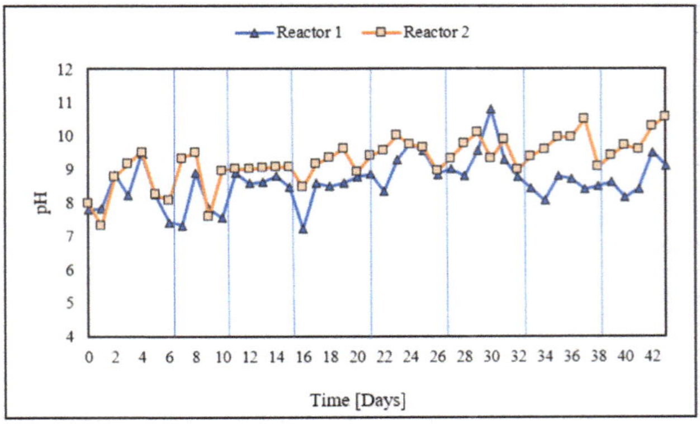

Fig. 3. Variation of pH in the reactors with time. Vertical lines indicate cycle separation.

3.2 Nutrient Removal

The temporal variation of the ammoniacal nitrogen profile during the study period is illustrated in Fig. 4. As seen in the plot, both reactors achieved the discharge standard of 10 mg/L at the end of the operation cycle. For the 8 cycles, the average volumetric nitrogen removal in reactor 1 was 5.34 ± 2.97 mg L^{-1} day^{-1} with 99.3% removal. The corresponding parameters for reactor 2 were 1.78 ± 1.03 mg L^{-1} day^{-1} and 44.3% respectively.

The temporal profile of phosphate concentration in the two reactors along with the allowable discharge limit (1 mg/L) is shown in Fig. 4. The average volumetric phosphate removal considering 8 operation cycles for reactor 1 was 0.11 ± 0.09 mg L^{-1} day^{-1} with 70.1% removal. The corresponding values for reactor 2 were 0.07 ± 0.06 mg L^{-1} day^{-1} and 47.5% respectively. The lower removal rates could be due to the lower average phosphate concentration observed during winter (1.26 mg/L) compared to that in summer (3.74 mg/L). Furthermore, the instantaneous phosphate removal rates on the initial day immediately after feeding primary effluent to reactor 1 and reactor 2 were as high as 0.92 ± 0.54 mg L^{-1} day^{-1}, 0.80 ± 0.66 mg L^{-1} day^{-1} respectively. The phosphate concentration of primary wastewater dropped instantaneously by 67.8% and 52.6% on average after been mixed with the preadapted cultures in R1 and R2. This drop was calculated considering the difference of phosphate concentrations in the primary-settled wastewater and in the reactor on day 0. Apart from the dilution effect, this pronounced drop of phosphorus concentration can also be attributed to adsorption of phosphorus to the surface of cells and the surface of bioreactors [12]. Furthermore, the subsequent rise of phosphorus concentration in the reactors could partially be due to the release of phosphorus by ruptured cells as described by [12]. This phenomenon is common with old cultures, therefore, the subsequent rise of phosphorus concentrations in reactor one during the last two cycles can be explained [14] (Fig. 5).

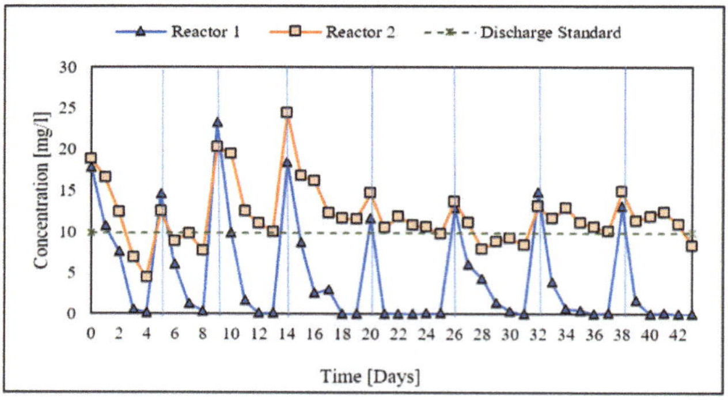

Fig. 4. Variation of NH_4-N concentration with time. Vertical lines indicate cycle separation.

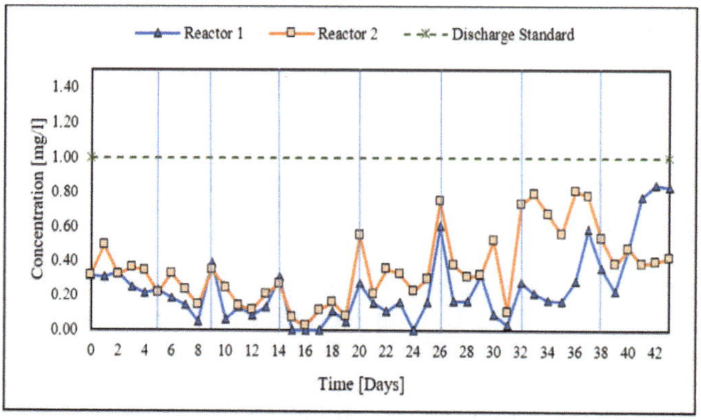

Fig. 5. Variation of phosphate concentration with time. Vertical lines indicate cycle separation

3.3 Carbon Removal

The temporal profile of organic carbon during the study period is illustrated in Fig. 6. As can be noted, both reactors achieved the permissible discharge level by the end of the cycle time. Considering the 8 operation cycles, the average volumetric BOD removal for reactor 1 was 12.93 ± 6.53 mg L^{-1} day^{-1} with 70.0% removal; the corresponding values for reactor 2 were 6.77 ± 3.51 mg L^{-1} day^{-1} and 50.7%.

The required minimum cycle time to achieve all three discharge standards for each cycle is demonstrated in Fig. 7. The average cycle time in reactor 2 (4.02 days) appeared nearly 4-fold higher compared with that in reactor 1 (1.02 days)

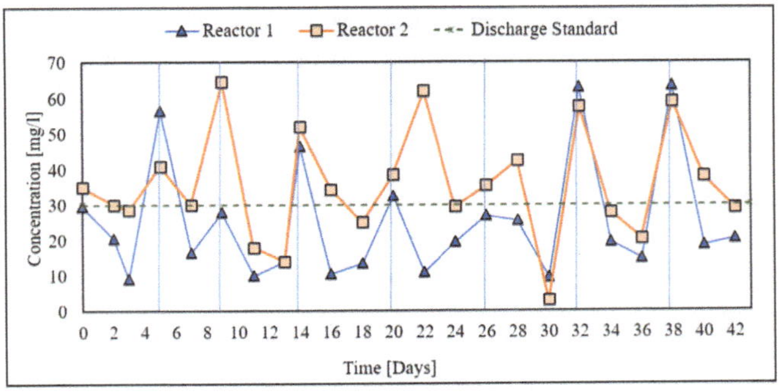

Fig. 6. Variation of BOD concentration with time. Vertical lines indicate cycle separation.

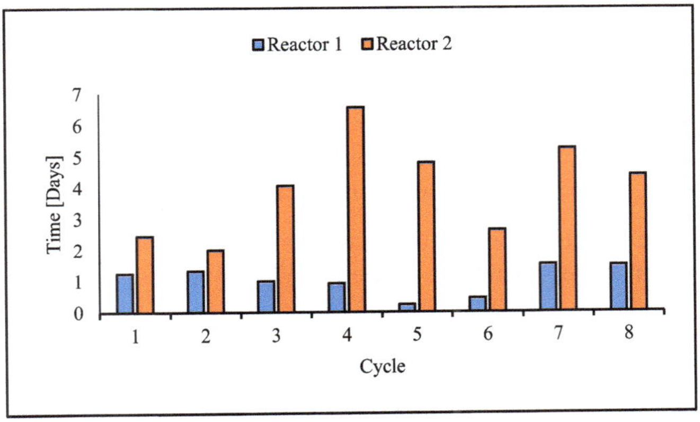

Fig. 7. Required minimum cycle time considering all three discharge standards

4 Comparison of Previous Studies

The availability of literature for studies conducted with algal-bacterial cultures during cold climates (5–15 °C) is limited due to the widely experienced performance limitations of such systems in extreme environments. A study conducted by [15] using cyanobacterial strains isolated from the Arctic and Antarctic environments reported maximum nitrogen and phosphorus removal efficiencies of 2.3 mg $NL^{-1}d^{-1}$ and 0.07–0.15 mg $PL^{-1}d^{-1}$ at 15 °C under laboratory conditions. The maximum nitrogen removal rate reported in the above study is higher compared to the average nitrogen removal rate in reactor 2 (1.78 mg L^{-1} day^{-1}). However, reactor 1 showed a higher average removal rate (5.43 mg L^{-1} day^{-1}) under field conditions (average culture temperature: 8.45–21.25 °C). The average phosphate removal rates were however similar to the highest phosphorus removal rates reported by [15].

[16] reported the average batch processing times for experiments conducted in warmer temperatures (culture temperature: 24–36 °C) with *G. sulphuraria Soos* strain, *G. sulphuraria* 5587.1 and, mixed (a mixture of 5587.1 & Soos) as 4.62 days, 3.87

days and 4.16 days respectively. The batch processing time reported in reactor 2 was similar to those reported in the above study. However, the culture in reactor 1 managed to reach the discharge standards in lesser time (1.02 days) despite the cold climate.

5 Conclusion

The study evaluated the suitability of two polycultures to serve in the POWER system during cold temperatures. A summary of the research findings from January to March is documented below:

- Average removal efficiencies of BOD_5, ammoniacal nitrogen, and phosphate respectively by 70.0%, 99.3% and 70.1% by reactor 1 and 50.7%, 44.3% and 47.5% by reactor 2.
- Both reactors remained stable despite the cold weather. reactor 1 and 2 reached the mandated discharge standards respectively in 1.02 days and 4.02 days.
- The initial phosphate concentration in the POWER system drops rapidly. This could both be due to the dilution effect and surface adsorption.
- Carbon dioxide and nutrients were not supplied to support the growth. Therefore, winter operation of the POWER system with the polycultures can be more economical even compared to the summer operation with *G. sulphuraria*.

Based on the results, a polyculture with a similar composition as in reactor 1 is recommended for winter operations of the POWER system.

References

1. de Godos I, Blanco S, García-Encina PA, Becares E, Muñoz R (2009) Long-term operation of high rate algal ponds for the bioremediation of piggery wastewaters at high loading rates. Bioresour Technol 100:4332–4339. https://doi.org/10.1016/J.BIORTECH.2009.04.016
2. Oswald WJ, Gotaas HB, Golueke CG, Kellen WR, Gloyna EF, Hermann ER (1957) Algae in waste treatment [with discussion]. Sewage Ind Wastes 29:437–457. http://www.jstor.org/stable/25033322
3. Muñoz R, Guieysse B (2006) Algal–bacterial processes for the treatment of hazardous contaminants: a review. Water Res 40:2799–2815. https://doi.org/10.1016/J.WATRES.2006.06.011
4. García J, Mujeriego R, Hernández-Mariné M (2000) High rate algal pond operating strategies for urban wastewater nitrogen removal. J Appl Phycol 12:331–339. https://doi.org/10.1023/A:1008146421368
5. Bordel S, Guieysse B, Muñoz R (2009) Mechanistic model for the reclamation of industrial wastewaters using algal–bacterial photobioreactors. Environ Sci Technol 43:3200–3207. https://doi.org/10.1021/es802156e
6. Su Y, Mennerich A, Urban B (2011) Municipal wastewater treatment and biomass accumulation with a wastewater-born and settleable algal-bacterial culture. Water Res 45:3351–3358. https://doi.org/10.1016/J.WATRES.2011.03.046

7. Selvaratnam T, Henkanatte-Gedera SM, Muppaneni T, Nirmalakhandan N, Deng S, Lammers PJ (2016) Maximizing recovery of energy and nutrients from urban wastewaters. Energy 104:16–23. https://doi.org/10.1016/j.energy.2016.03.102
8. Henkanatte-Gedera SM, Selvaratnam T, Karbakhshravari M, Myint M, Nirmalakhandan N, Van Voorhies W, Lammers PJ (2017) Removal of dissolved organic carbon and nutrients from urban wastewaters by Galdieria sulphuraria: laboratory to field scale demonstration. Algal Res 24:450–456. https://doi.org/10.1016/J.ALGAL.2016.08.001
9. Mehrabadi A, Craggs R, Farid MM (2015) Wastewater treatment high rate algal ponds (WWT HRAP) for low-cost biofuel production. Bioresour Technol 184:202–214. https://doi.org/10.1016/J.BIORTECH.2014.11.004
10. Park JBK, Craggs RJ, Shilton AN (2013) Enhancing biomass energy yield from pilot-scale high rate algal ponds with recycling. Water Res 47:4422–4432. https://doi.org/10.1016/J.WATRES.2013.04.001
11. Tchinda D, Henkanatte-Gedera SM, Abeysiriwardana-Arachchige ISA, Delanka-Pedige HMK, Munasinghe-Arachchige SP, Zhang Y, Nirmalakhandan N (2019) Single-step treatment of primary effluent by Galdieria sulphuraria: removal of biochemical oxygen demand, nutrients, and pathogens. Algal Res 42:101578. https://doi.org/10.1016/J.ALGAL.2019.101578
12. Martínez M, Sánchez S, Jiménez F, El Yousfi L, Muñoz L (2000) Nitrogen and Phosphorus removal from urban wastewater by the microalga Scenedesmus obliquus. Bioresour Technol 73:263–272. https://doi.org/10.1016/S0960-8524(99)00121-2
13. Axelsson L (1988) Changes in pH as a measure of photosynthesis by marine macroalgae. Mar Biol 294:287–294. https://doi.org/10.1007/BF00391314
14. Droop M (1975) The nutrient status of algal cells in batch culture. J Mar Biol Assoc UK55 (3):541–555. https://doi.org/10.1017/S0025315400017240
15. Chevalier P, Proulx D, Lessard P, Vincent WF, de la Noüe J (2000) Nitrogen and phosphorus removal by high latitude mat-forming cyanobacteria for potential use in tertiary wastewater treatment. J Appl Phycol 12:105–112. https://doi.org/10.1023/A:1008168128654
16. Henkanatte-Gedara S (2018) Urban wastewater treatment by Galdieria sulphuraria: single step process for dissolved organics and nutrient removal. New Mexico State University

Comparative Study of Spatial and Temporal Variation of Drought Using Remotely Sensed Data - A Case Study for Kirindi Oya Basin

S. U. Hendawitharana[1(✉)], M. K. D. D. Priyasad[2], and R. L. H. L. Rajapakse[1]

[1] Department of Civil Engineering, University of Moratuwa, Katubedda, Sri Lanka
sahaniuththara@gmail.com
[2] Department of Computer Science and Engineering, University of Moratuwa, Katubedda, Sri Lanka

Abstract. Droughts have turned out to be a standout amongst the most discussed natural hazards in the recent past due to its adverse impacts on humankind. The effects of drought differ from region to region and time to time. Unlike, the other cataclysmic events, a drought is a mind-boggling phenomenon due to its longer propagation time and the effects being long-lasting. This study focuses on analysing the spatial and temporal variations of the drought in the Kirindi Oya basin which lies in the intermediate, dry and arid climatological zones. Daily remotely sensed precipitation data for a span of 37 years from 1979–2017 has been used with a spatial resolution of 0.1°. Several drought indices, namely Standardized Precipitation Index (SPI), Deciles, Percent of Normal, Rainfall Anomaly Index and Z index have been considered. A comparative analysis of the applicability of each drought index along with the validation by ground reality is presented.

The results of the study imply the importance of conducting a study at the sub-basin scale for the river basins lying in several vastly different climatological regions. The Kirindi Oya basin that is considered, lies in three different climatological regions, where the upper Kirindi Oya subbasin is in the intermediate zone, the lower Kirindi Oya in the arid zone and the rest in the dry zone. The results indicate that the effects of drought can vary even within a river basin and a sub basin scale analysis of drought gives a comparatively accurate interpretation of the variation of drought within the basin. Furthermore, the recent findings confirms that the drought prediction at a sub basin scale is more effective for river basins which lie in several climate zones.

Keywords: Drought indices · Remotely sensed data · Spatial and temporal analysis

1 Introduction

Drought can be identified as a natural phenomenon which propagates gradually and unlike other natural disasters, its effects can last for an extended time (Wilhite 2000; Shahabfar and Eitzinger 2013). The unavailability of a solitary definition for droughts

R. Dissanayake and P. Mendis (Eds.): ICSBE 2018, LNCE 44, pp. 116–130, 2020.
https://doi.org/10.1007/978-981-13-9749-3_11

has made decision making using drought-related studies more complex (Wilhite and Glantz 1985; Wilhite 2000). For the simplicity of analysis, droughts have been categorized into four main types namely; meteorological, hydrological, agricultural and socio-economic (Bayissa et al. 2018). All three former drought types occur following a meteorological drought and an extended meteorological drought can lead to the latter drought type (Zargar et al. 2011; Shah and Mishra 2015).

Spatial and temporal studies of droughts can be done using drought indices. The drought index is a measure of drought and for each drought type, a different set of drought indices have been developed. There are numerous drought indices developed as a measure of different drought types and its appropriateness may differ depending upon the geography, information accessibility and several other factors (Hayes 2017). Drought indices provide a qualitative indication of droughts, where the results would suggest the severity of the drought (Richard and Heim 2002; Svoboda and Fuchs 2016). However, the impacts from the same severity class can vary from region to region and is dependent upon the climate conditions of the region under consideration. Even though precipitation is the main input for most drought indices, it cannot be taken directly as a measure of drought, as the amount of rainfall differs spatially (Svoboda and Fuchs 2016). Thus, the drought indices tend to report the deviation from the mean using different formulae, which sometimes end up in unrealistic interpretations. The areas with very high annual rainfall may be considered as undergoing a severe drought when the rainfall is reduced, even though the adverse impacts of a drought are not significantly observed by the residents in that area. The complexity of the drought phenomenon, the absence of a unique definition and data inaccessibility for a longer duration may lead to variations in results given by different drought indices for the same location and time period. Hence, it is a requirement to identify a suitable drought index for a given area (Bayissa et al. 2018). Some indices such as Standardized Precipitation Index (SPI) and deciles give indications of several drought types, namely; meteorological, hydrological and agricultural and this is due to the interdependency of these droughts over the temporal scale (Zargar et al. 2011).

This study mainly focuses on two main aspects of drought evaluation. The first aspect is identifying the applicability of drought studies in larger resolutions, especially at sub-basin levels where the most of the empirical studies focus on an entire basin, a large district or country and evaluating the usefulness of such a study in a country with high ecological diversity. The other aspect is finding the most suitable drought evaluation methodologies (indices) to represent the drought patterns and behaviour in a selected area. Most of the existing literature is based on larger extents, assuming that the climatological patterns are approximately similar. However, the validity of such methods for a country like Sri Lanka which has a higher variation in climatological zones is questionable and the majority of the rivers originate from the central highlands and reach the ocean at different destinations all around the country traversing over several such zones.

One of the major challenges for a high spatial resolution drought study is the availability of data. Precipitation and temperature are the most prominent parameters in drought study. Temperature over a large extent of the area remains nearly constant while rainfall patterns change rapidly. This difference cannot be easily tracked by traditional methodologies since the rainfall gauge stations are sparsely distributed. Therefore, moving towards high resolution remotely sensed data sources is important.

The SPI is one of the most common drought indices used around the world, including Sri Lanka. Other than that, many other drought indices have been considered for research studies in different regions of the country.

Considering the drought studies conducted, it can be identified that most of the studies on drought analysis have been conducted using gauge station data. Main issues associated with gauge station data are the significant distances between the gauge stations, high cost and the data discontinuity (Villarini et al. 2008). However, use of remotely sensed data is becoming popular and the data sources selected by the researchers include NOAA, TRMM, GMP, PERSIANN, MSWEP (Johnson et al. 1993; Thenkabail et al. 2004; Beck et al. 2017). The spatial and temporal resolution of the available data is the main concern, since even though some regions of the world have a complete set of satellite-based gridded data with high spatial accuracy, there are limited data sources for Sri Lanka which have both the required temporal and spatial resolutions. The data required for a drought study is high in comparison to other studies and it requires a minimum of 30 years of continuous data in order to calculate the available drought indices with sufficient accuracy (Juliani and Okawa 2017). Considering the inaccessibility of continuous data, drought-related studies using remotely sensed data have received considerable attention and this study uses remotely sensed gridded precipitation data with a spatial resolution of 0.1°.

The next challenge is determining the most suitable method for qualitative analysis of drought impact and its severity. The SPI is one of the most common drought indices used around the world including Sri Lanka. Other than that, many other drought indices have been considered for research studies in different regions of the country.

Objectives of this research are,

- Evaluating the applicability and suitability of sub-basin scaled drought study over basin-scaled drought study
- Evaluating different quantitative drought measures to identify the most suitable measure

This paper critically examines the notion that drought-related studies can be more accurate when they are conducted at the sub-basin scale using the Kirindi Oya as a case study.

This study uses five selected drought indices with the intention of performing a critical analysis of the applicability of them in the selected basin.

2 Study Area

Sri Lanka is a tropical island nation situated towards the North of equator in the Indian Ocean. The island comprises of four main climatological regions; wet zone, dry zone, intermediate zone and arid zone. Towards the southern central part of the island, a hilly territory is observed while the coastal belt has a nearly flat terrain. These central hill lands have a significant impact on the country's climate. The island has an intricate system of waterways which take water from the inland to the ocean.

Considering the river basins of the island, they are either totally in one climate zone or span over more than one climate zone. Hence, studying droughts in a river basin should consider the fact that there can be an influence from the climate zone for drought in that area. In order to study the spatial and temporal variation of meteorological drought, the authors have selected the Kirindi Oya basin, which spans in three climate zones; intermediate zone, dry zone and arid zone while a major part lies in dry and arid zones. According to the agro-ecological region map of Sri Lanka, the Kirindi Oya spans over the zones; IM2, IU3, I(L1-L2), I(U2-U3), DL1 and DL5. Figure 1 indicates the location of the Kirindi Oya basin in Sri Lanka and the agro-ecological zoning for Kirindi Oya basin is shown in Fig. 2.

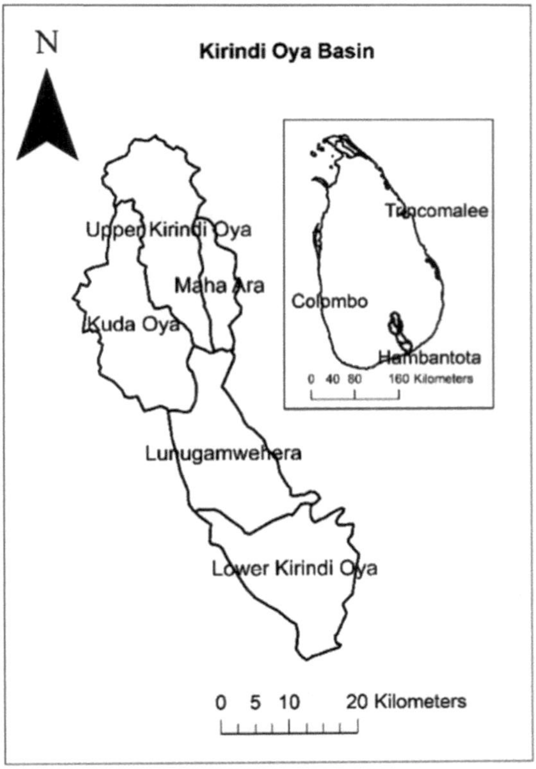

Fig. 1. Kirindi Oya basin

The Kirindi Oya starts from Bandarawela and reaches the Indian Ocean at Bundala. The basin is of a land extent of 1176 km^2 and is situated in the Badulla, Moneragala and Hambantota districts. High temperatures and low relative humidity leading to higher evaporation and comparatively low mean annual rainfall are common features of the area. A bimodal rainfall pattern can be observed in the area with a major proportion of rainfall being contributed by the North-Eastern Monsoon (September–February) which goes in line with the Maha season.

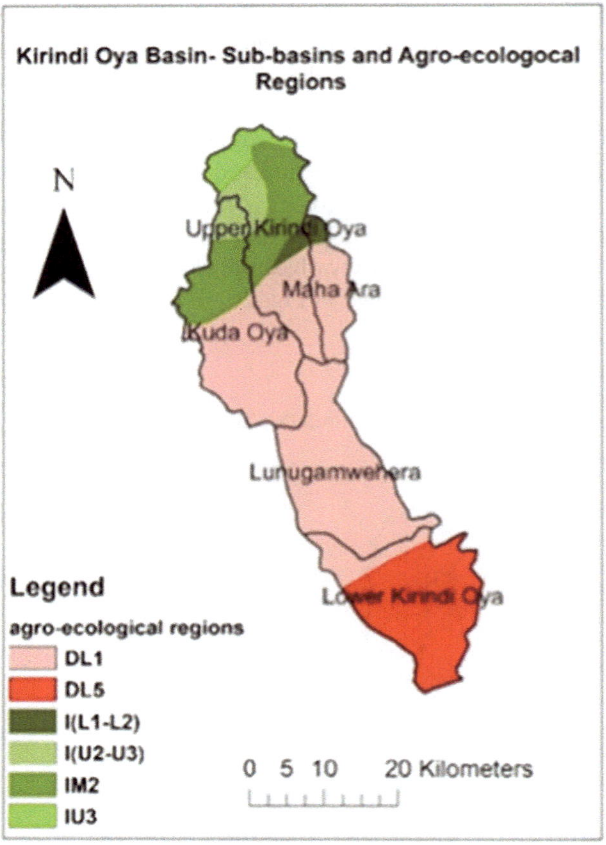

Fig. 2. Agro-ecological regions - Kirindi Oya

3 Data and Methodology

Depending on the drought index being used, the past data requirement for drought analysis varies. The SPI is the current drought index used in Sri Lanka which requires a minimum of 30 years data (Juliani and Okawa 2017).

Obtaining continuous rainfall data for 30 years is challenging for some areas of the country due to data discontinuity and other management issues. Remotely sensed precipitation data with a spatial resolution of 0.1° can be identified as an effective solution.

For this analysis, daily gridded precipitation data obtained from Multi-Source Weighted-Ensemble Precipitation (MSWEP) was used for a temporal duration of 37 years (1979–2017).

Detection and real-time monitoring of droughts can be done by using drought indices which enables us to identify the drought duration and drought severity. However, the definitions of drought indices differ from one another and the drought measure is a qualitative one. Hence, a comparison of different drought indices is complex.

For this study, five identified drought indices, namely Standardized Precipitation Index (SPI), Deciles (DI), Percent of Normal (PN), Rainfall Anomaly Index (RAI) and Z index, were calculated for each sub-basin in the Kirindi Oya basin.

3.1 Standardized Precipitation Index (SPI)

In calculating the SPI index, the rainfall received in a particular area is fitted to a corresponding gamma distribution and the standard deviation from the long-term mean is considered. This measure indicates the deviation from the historical mean rainfall of a particular area and hence can be considered as a measure of drought. The SPI is calculated for several durations; 1 month, 3 months, 6 months, 9 months, 12 months, 18 months, 24 months and 48 months. However, the SPI values corresponding to a given month can differ from each other, when different durations are considered.

Hence, these different durations can be used when analysing diverse drought scenarios. The 1-month SPI would provide an indication of short-term soil moisture conditions and crop moisture levels and is useful in decision- making during crop growing seasons. The 3-month SPI reflects the seasonal rainfall estimates and meteorological drought. Indications of agricultural drought are given by 9-month SPI while hydrological drought is reflected by 12-month SPI (Senay et al. 2015). The SPI is calculated using Eqs. 1 and 2.

$$\mathrm{g(x)} = \frac{1}{\beta^{\alpha}\tau(\alpha)} x^{\alpha-1} e^{\frac{-x}{\beta}},\ x > 0 \tag{1}$$

$$A = \ln(x) - \frac{\sum \ln(x)}{n} \tag{2}$$

Calculated SPI values for each month are categorized into regions as indicated in Table 1 (Morid et al. 2006).

Table 1. SPI impact levels

SPI value	Impact level
2.0 or more	Extremely wet
1.5 to 1.99	Very wet
1.0 to 1.49	Moderately wet
0.99 to −0.99	Near wet
−1.0 to −1.49	Moderately dry
−1.5 to −1.99	Severely dry
−2.0 or less	Extremely dry

The table is then converted to a common scale with the intention of analysing the results. All the other drought indices are also converted to this common scale for the ease of comparison.

The drought period is indicated by the time period between the first month with below −1.0 index value and first month after that month which has an index value greater than or equal to 0.

3.2 Deciles (DI)

This method is one of the simplest methods used to calculate the impact of drought where the precipitation data is arranged in decreasing order and the cumulative frequency distribution is obtained. Then, it is split into ten parts using equal probability distribution and each part is called a decile. According to the original work, precipitations that are below the 5th decile is considered as drought (Sivakumar et al. 2010). The original categorization is converted to the above-mentioned scale as in Table 2.

3.3 Percent of Normal (PN)

In this method, precipitation on a given month (P_i) is expressed as a percentage to the long-term median/normal rainfall (P_n) of a particular area. Precipitations with PN value below 80% is identified as droughts which are calculated as indicated in Eq. 3.

$$PN = \frac{P_i}{P_n} * 100\% \quad (3)$$

However, since this method does not consider a time window, actual drought can be identified by consecutive months with PN values in drought range. Table 2 indicates the impact levels for the PN values.

Table 2. Impact levels

	SPI	RAI	PN	Z Index	DI
No drought	>0.99	>−0.3	>80	−0.25 to +0.25	≥5
Weak drought	0.99 to −0.99	−0.3 to −1.2	70–80	−0.52 to −0.25	4
Moderate drought	−1.0 to −1.49	−1.2 to −2.1	55–70	−0.84 to −0.52	3
Severe drought	−1.5 to −1.99	−2.1 to −3.0	40–55	−1.28 to −0.84	2
Extreme drought	2 or less	<−3.0	<40	>−1.28	1

3.4 Rainfall Anomaly Index (RAI)

This index is used to calculate meteorological drought events and considered independent of time and space. This is calculated as in Eq. 4, where Pn and m refer to mean precipitation of a 30 years span and mean of 10 greatest rainfalls respectively. The results are interpreted in Table 2.

$$RAI = \pm\left(\frac{P - P_n}{m - P}\right) \quad (4)$$

3.5 Z Index

Z index is calculated using Eq. 5 where P_n and δ refer to the mean precipitation and the standard deviation of the region.

$$Z\ Score = \frac{P - P_n}{\delta} \tag{5}$$

Two approaches were taken in calculating the indices, where the first approach is to consider the mean precipitation of each sub-basin in an index calculation. The other approach was to consider the mean value for the basin as a whole. The results obtained from the two approaches for five drought indices were compared and analysed.

4 Analysis of Results

In this study, spatial and temporal variations of drought within the Kirindi Oya basin is analysed by using remotely sensed precipitation.

4.1 Rainfall Variation Within the Basin

First, the rainfall variation within the basin is analysed to ensure that our main hypothesis – more accurate results can be obtained by sub-basin scaled analysis remains valid for this basin. Since the precipitation distributions are significantly different among basins, it can be argued that drought patterns and severity may show different distributions.

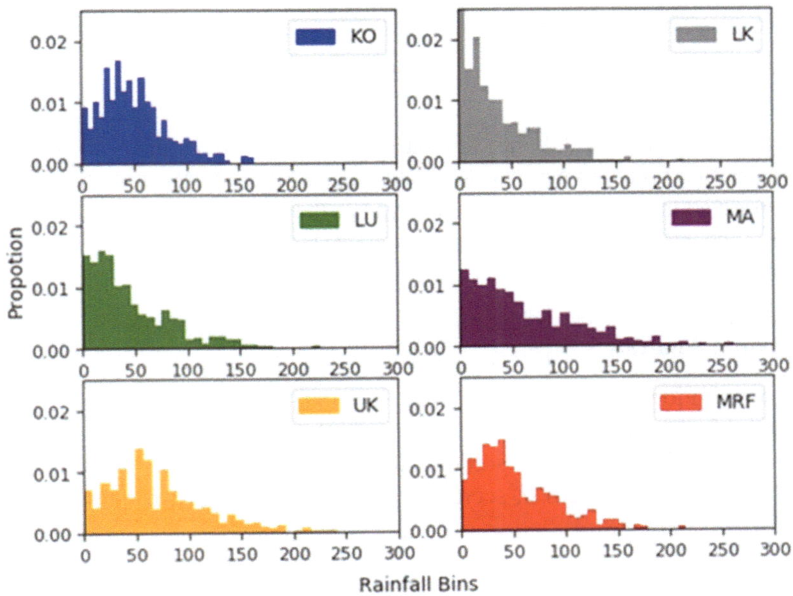

Fig. 3. Rainfall distribution histogram

According to Fig. 3, Lower Kirindi Oya and Lunugamwehera sub-basins have undergone more no-rainfall days compared to others. Moreover, they indicate a lower frequency of high-rainfall days compared to the others. The highest proportion of rainfall obtained for a given sub-basin differs from each other. The subbasins closer to each other show a slight similarity while the subbasins away from each other show a significant variation.

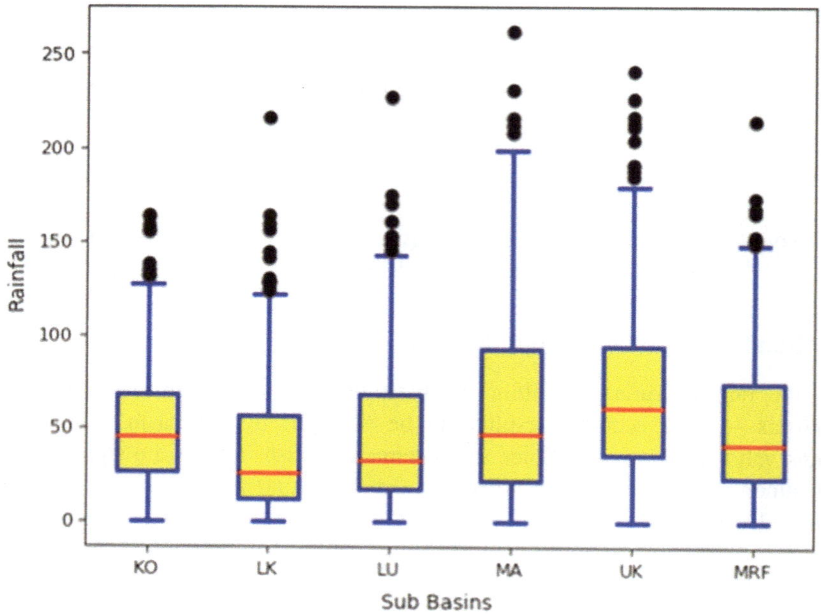

Fig. 4. Distribution of rainfall

According to Fig. 4, all the sub basins have a median of ~50 mm rainfall. However, outlier points (precipitation greater than 1.5IQR, extreme rainfall conditions) can be observed in each sub-basin. Kuda Oya has the least number of outlier points and they are concentrated around 150 mm whereas Upper Kirindi Oya and Maha Ara which shows extreme rainfall events ~250 mm.

4.2 Spatial Analysis

Spatial distribution of drought categories is indicated in Fig. 5. The frequency of occurrence of each drought type identified by each drought index is shown. From the total of 456 months considered, ~200 months show a no drought condition according to the majority of the indices. Z score has identified the lowest number of no drought months.

The drought indices, PN, RAI, Z score and SPI1 have identified a significant variation in occurrence of extreme drought events within the basin and Lower Lunugamwehera sub basin shows the highest frequency.

The indices, DI, PN, RAI, Z score and SPI 1 have been calculated for 1-month time scale. However, the drought identification shows that SPI 1 has identified remarkably low number of extreme, sever and moderate drought events compared to the other indices. In contrast, SPI for all the time scales considered have identified higher number of weak drought events.

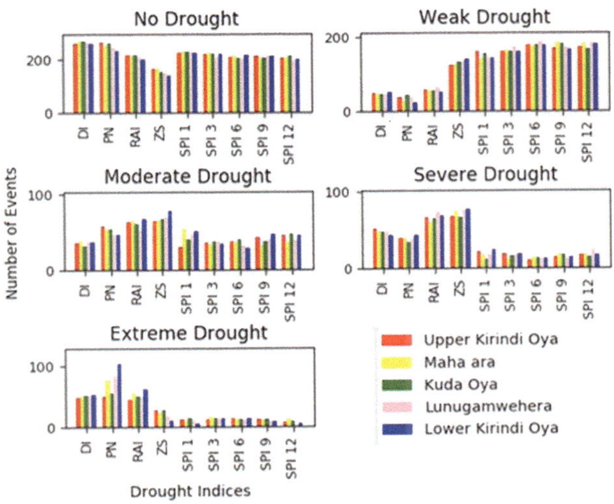

Fig. 5. Drought variation in basins

It can be seen that a significant difference in the number of drought events of drought classes among DI, PN, RAI, ZS which do not use time windows and SPI which uses time windows respectively. Figures, 6, 7 and 8 represent temporal drought variation of three main basins.

PN has identified more drought events in all the three basins compared to the other indices due to calculating drought in one-month intervals. This gives an inaccurate result where 'no precipitation' events in between heavy rain events are also identified as droughts. Therefore, it is suitable to use a windowing technique which can get an average over windowing period and then quantify drought which results in an accurate estimation. For this basin under consideration, SPI can be identified as the most suitable one.

Confusion matrices in Figs. 9 and 10 explain the correlation of drought events among sub-basins. Considering the drought events that occurred for the period under consideration, for the spatial variation of the subbasins, the matrices provide a clear picture of the drought occurrence.

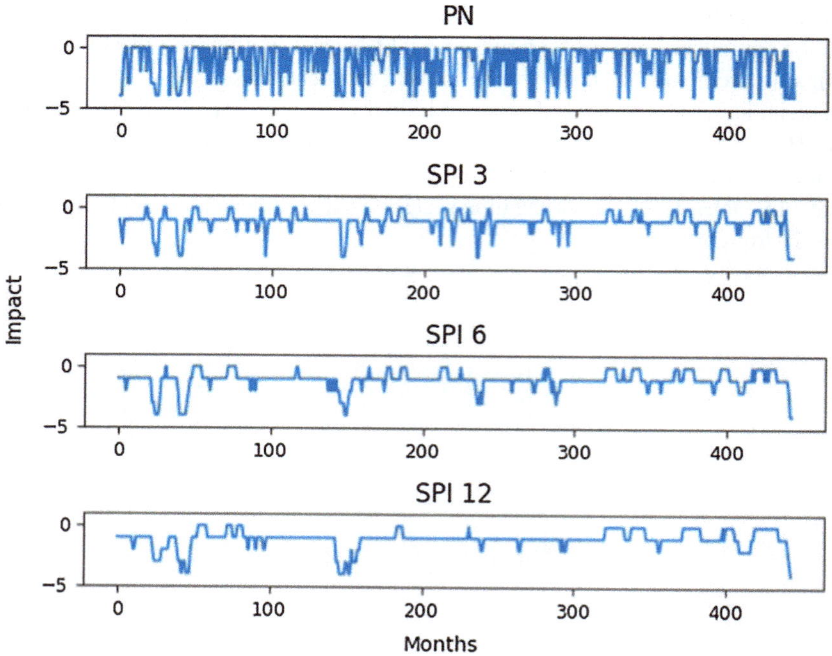

Fig. 6. Temporal drought variation in Kirindi Oya

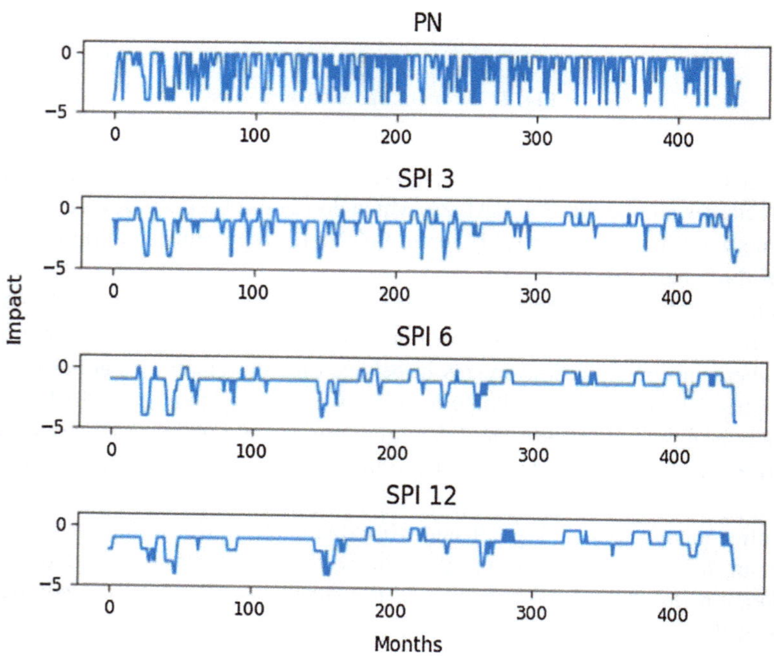

Fig. 7. Temporal drought variation in Lunugamwehera

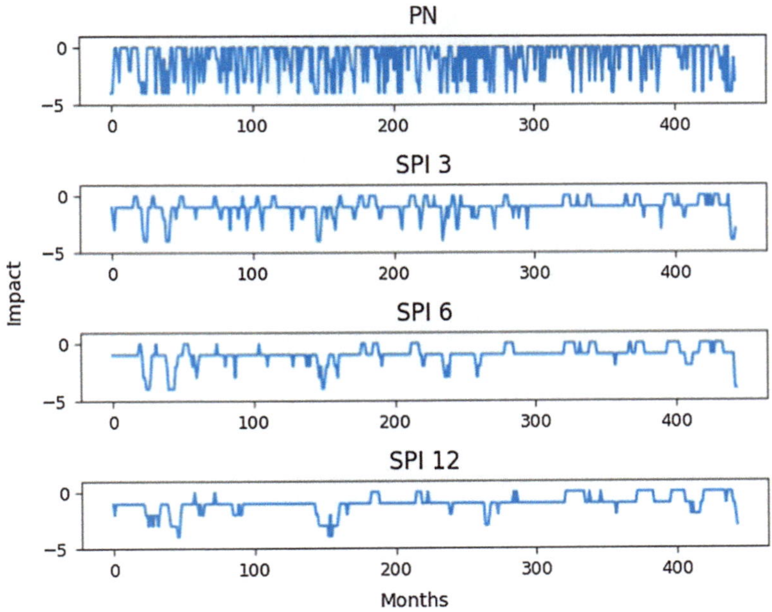

Fig. 8. Temporal drought variation in Lower Kirindi Oya

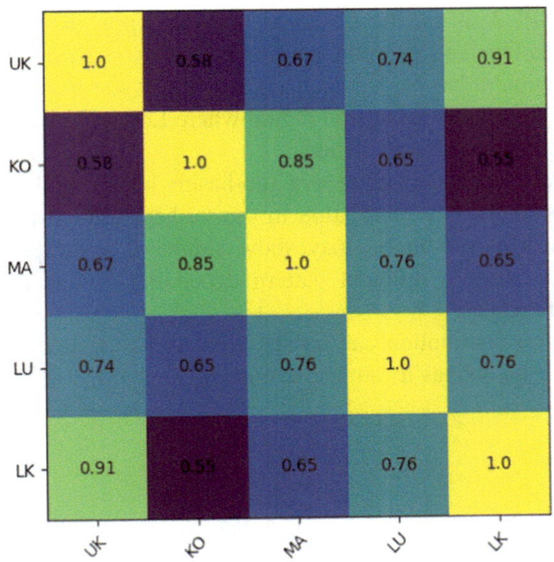

Fig. 9. Correlation matrix - SPI 3

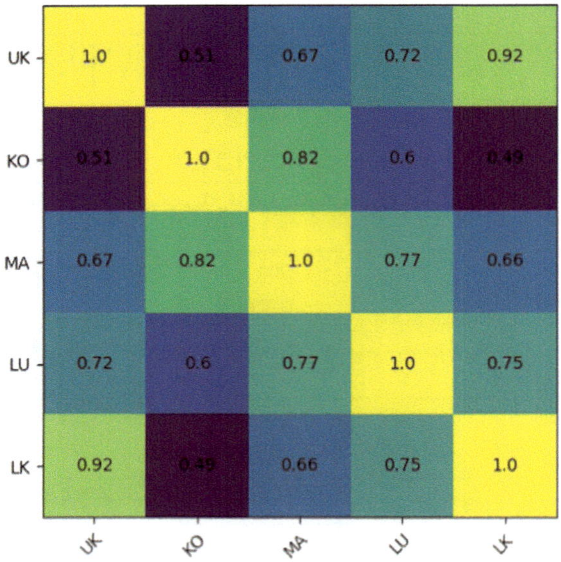

Fig. 10. Correlation matrix - PN

Comparing SPI 3 and PN matrices, it can be observed that PN has identified a higher variation of drought among the sub-basins while SPI 3 shows a comparatively low variation.

Considering SPI 3, it can be observed that the Upper Kirindi Oya has a high correlation to Kuda Oya and very low correlation to Lower Kirindi Oya. According to the spatial distribution of basins, Kuda Oya is the nearest sub-basin to the Upper kirindi Oya and is in the same agro-ecological region where Lower Kirindi Oya is the farthest and in a different agro-ecological region.

Both Maha Ara and Lunugamwehera sub-basins have a high correlation to their nearby sub-basins and a low correlation to the farthest sub-basins. Hence, it can be concluded that nearby sub-basins may show similar drought patterns whereas the farthest sub-basins show a different pattern depending on the agro-ecological and climate region differences.

Same correlation distribution can be identified in PN drought index with slightly low correlation for all subbasins compared to SPI-3.

5 Conclusion

The drought identification using the drought indices considered in this study shows that a higher number of drought events including all the drought events identified by SPI have been identified by PN, DI, RAI and Z score. However, they have also identified a significant difference in rainfall in one month which falls amidst a drought event as the cease of drought which is impractical. Hence, drought predictions at the Kirindi Oya basin can be effectively done using drought indices which use windowing technique; hence, SPI.

Drought events in the Upper Kirindi Oya basin has significant differences during some drought events while the drought variation in the nearby sub-basins being nearly similar is the main observation of this study. Hence, the findings of this study help understand that the effects of drought can vary even within a river basin and, that sub-basin-scale analysis of drought gives a comparatively accurate interpretation of variation of drought within the basin. Furthermore, the present findings confirm that the drought prediction at a sub-basin scale is more effective for river basins which lie in several climatological regions.

Acknowledgement. The authors would like to acknowledge that the research project was financially supported by the Senate Research Committee Grant (SRC/LT/2018/25), University of Moratuwa.

References

Bayissa Y et al (2018) Comparison of performance of six drought indices in characterizing historical drought for the Upper Blue Nile basin. Geosciences 8(3):81

Beck HE et al (2017) MSWEP: 3-hourly 0.25 global gridded precipitation (1979–2015) by merging gauge, satellite and reanalysis data. Hydrol Earth Syst Sci 21:589–615

Hayes MJ (2017) Drought indices. The intermountain west climate summary

Johnson GE, Achutuni VR, Thiruvengadachari S, Kogan F (1993) The role of NOAA satellite data in drought early warning and monitoring: selected case studies. Kluwer Academic Publishers

Juliani BHT, Okawa CMP (2017) Application of a standardized precipitation index for meteorological drought analysis of the semi-arid climate influence in Minas Gerais. Brazil. Hydrology 4(2):26

Morid S, Smakhtin V, Moghaddash M (2006) Comparison of seven meteorological indices for drought monitoring in Iran. Int J Climatol 26:971–985

Richard R, Heim J (2002) A review of twentieth-century drought indices used in the United States. Am Meteorol Soc 83(8):1149–1166

Senay G et al (2015) Drought Monitoring and Assessment: Remote Sensing and Modelling Approaches for the Famine Early Warning Systems Network. In: Drought Monitoring and Assessment, pp 233–262. USGS Staff

Shahabfar A, Eitzinger J (2013) Spatio-temporal analysis of drought in semi-arid regions by using meteorological drought indices. Atmosphere 4:94–112

Shah RD, Mishra V (2015) Development of and experimental near-real-time drought monitor for India. J Hydrometeorol 16:327–345

Sivakumar MVK et al (2010) Drought risk and meteorological droughts

Svoboda M, Fuchs BA (2016) Handbook of drought indicators and indices. Integrated drought management programme (IDMP), integrated drought management tools and guidelines series 2. World Meteorological Organization (WMO) and Global Water Partnership, Geneva

Thenkabail PS, Gamage MSD, Smakhtin VU (2004) The use of remote sensing data for drought assessment and monitoring in Southwest Asia. International Water Management Institute, Research Report 85

Villarini G, Mandapaka PV, Krajewski WF, Moore RJ (2008) Rainfall and sampling uncertainties: a rain gauge perspective. Adv Earth Space Sci 113(D11). https://agupubs.onlinelibrary.wiley.com/doi/full/10.1029/2007JD009214

Wilhite DA (2000) Drought as a natural hazard: concepts and definitions. In: Drought: a global assessment. Drought Mitigation Center Faculty Publications, pp 3–18

Wilhite DA, Glantz MH (1985) Understanding: the drought phenomenon: the role of definitions. Water Int 10(3):111–120

Zargar A, Sadiq R, Naser B, Khan FI (2011) A review of drought indices. Environ Rev 19:333–349

Potential for Formation of Trihalomethane in Diverted and Non-diverted Areas of Mahaweli River in Sri Lanka

H. R. L. C. Bandara(✉), W. B. M. L. I. Weerasekara, and S. K. Weragoda

Water Safety Plan Advisory Unit, National Water Supply and Drainage Board, Kandy, Sri Lanka
lbandaraocu@gmail.com

Abstract. Under the Mahaweli Development project the Mahaweli River was diverted to Rajarata area at polgolla barrage in 1976. Along the river concentration of the dissolved organic (DO), inorganic and suspended solids are gradually increased. In drinking Water Treatment Plants (WTPs) can be removed mainly suspended solid and some part of the dissolved component. As a result of remaining DO in filtered water disinfection by products (DBPs) formed such as Trihalomethane (THM) which is carcinogenic. This study was conducted to investigate THM formation potential in Mahaweli Diverted Areas (DA) and Non-Diverted Areas (NDA). Water samples were collected from following WTPs; Ukuwela, Matale, Dambulla, Bakamuna, Minneriya, Medirigiriya and 2- Ela as a DA and Meewathura, Greater Kandy, Pathadumbara, Mahiyanganaya, Dehiattakandiya, Gallella, Kantale and Mutur as a NDA. The raw water source of Ukuwela and Matale is Sudu Ganga and others are Ibbankatuwa Lake, Yoda Ela, Minneriya Lake, Kaudulla Lake and Parakrama Samudra respectively. Four THM species, Trichloromethane, Bromodichloromethane, Dibromochloromethane and Tribromomethane were measured using Gas Chromatography (GC) system and finally Total THM (TTHM) were calculated. The highest TTHM concentration was reported in 2- Ela (84.67 ± 35.17) μg/L and Mutur (83.85) μg/L which is exceeded the United State Environmental Protection Agency (USEPA) maximum contaminant level of 80 μg/L. Lowest concentration was reported from Matale 8.16 μg/L. According to the Kruskal-Wallis Test the median TTHM concentration of DA (22.73 μg/L) was not significantly different ($p > 0.05$) from the NDA (25.36 μg/L). The mean Conductivity (μS/cm), pH, turbidity (NTU) and alkalinity (mg/l) of the DA were 224.25 ± 65.95, 7.48 ± 0.36, 3.69 ± 1.91 and 87.5 ± 28.72 respectively and in NDA; 139.52 ± 62.39, 7.31 ± 0.37, 12.71 ± 12.13 and 56.25 ± 28.99 respectively. According to the statistical analysis there was no significant difference ($p > 0.05$) between the measured water quality parameters of DA and NDA.

Keywords: Diverted area · Mahaweli river · Non-Diverted Area · Trihalomethane · Water treatment plant

R. Dissanayake and P. Mendis (Eds.): ICSBE 2018, LNCE 44, pp. 131–136, 2020.
https://doi.org/10.1007/978-981-13-9749-3_12

1 Introduction

The Mahaweli River is the longest river in Sri Lanka which starts in the Central Mountains of the country. The mean annual runoff of the river is 7.2 million acre feet. (Mahaweli Development Board 1970). The United Nations Development Programme (UNDP) and the Food and Agriculture Organization (FAO) formulated the master plan for the development of the Mahaweli river basin in 1960. (Hewavisenthi 2009). In 1976, the Mahaweli River was diverted to the Rajarata area at the Polgolla barrage. Researchers have found that most patients having Chronic Kidney Diseases due to unknown aetiology (CKDu) were reported in the Mahaweli Diverted area. (Bandara et al. 2010). Along the river, the concentration of the dissolved organic (DO), inorganic and suspended solids is gradually increasing. In drinking Water Treatment Plants (WTPs) suspended solid can be removed mainly and some part of the dissolved components.

Most of the water treatment plants in Sri Lanka have been using chlorine as a disinfectant. Chlorination can help to reduce risk of water borne diseases such as typhoid, cholera and hepatitis (Morris and Levine 1995). As a result of chlorination in water treatment plants, disinfection by-products (DBPs) can be formed. (Weragoda 2005). The formation of these compounds is a function of precursor concentration, contact time, chlorine dose, and pH (APHA 1995). Trihalomethanes (THMs) can be considered as one of most important group of DBPs. Trihalomethanes (THMs) are halogen-substituted single-carbon compounds with the general formula CHX_3, where X represents a halogen, which may be fluorine, chlorine, bromine, or iodine, or combinations thereof. The THMs most commonly present in drinking-water are Trichloromethane ($CHCl_3$), Bromodichloromethane ($CHCl_2Br$), Dibromochloromethane ($CHBr_2Cl$) and Tribromomethane ($CHBr_3$). This study was conducted to investigate THM formation potential in the Mahaweli Diverted Areas (DA) and Non-Diverted Areas (NDA).

2 Methodology

2.1 Sampling Location

Water samples were collected from the following WTPs; Ukuwela, Matale, Dambulla, Bakamuna, Minneriya, Medirigiriya and 2- Ela as DAs and Meewathura, Greater Kandy, Pathadumbara, Mahiyanganaya, Dehiattakandiya, Gallella, Kantale and Mutur as NDAs. The raw water source of Ukuwela and Matale is Sudu Ganga and others are Ibbankatuwa Lake, Yoda Ela, Minneriya Lake, Kaudulla Lake and Parakrama Samudra respectively.

2.2 Analytical Method

Water samples were collected to the pre washed clear glass bottles. Samples were preserved by using ascorbic acid. Recommended preservation dose of ascorbic acid was 25 mg for 40 ml of sample. Then, samples were transported to the laboratory by maintaining temperatures below 4 °C. Four THM species, Trichloromethane ($CHCl_3$), Bromodichloromethane ($CHCl_2Br$), Dibromochloromethane ($CHBr_2Cl$) and Tribromomethane ($CHBr_3$) were measured by using Thermo scientific trace 1300 computerized gas chromatograph system with electron capture detector (GC-ECD) coupled with Tri Plus RSH auto sampler, (Thermo Fisher scientific Inc., USA). GC-ECD capillary column used to separate THMs (TRACE TR-5 GC) having 15 m × 0.25 mm × 0.25 μm film thickness. Samples were injected through auto sampler (Sampler consist heating, incubation agitation oven and standard X-length rail for headspace injection) in to split/split less injector. GC-ECD system was configured according to headspace THMs determination method proposed by (Kuivinen and Johnsson 1999). Finally, Total THM (TTHM) was calculated, which is a cumulative value of the four THM species. Conductivity, pH and turbidity were measured by using portable water quality testing equipment. Alkalinity was measured by a titrimetric method.

3 Results and Discussion

3.1 Diverted Area

Previous studies indicate that there was adverse water quality effect in Mahaweli River water after the diversion, especially related to the Chronic Kidney Diseases due to Unknown aetiology. (Bandara et al. 2010). One of the main targets of this study was to identify the THM formation potential in Mahaweli DA (Table 1).

Table 1. Reported THM concentration (μg/L) in WTP in Mahaweli DA

Location	$CHCl_3$ ± SD	$CHCl_2Br$ ± SD	$CHBr_2Cl$ ± SD	$CHBr_3$ ± SD
Ukuwela	17.6	4.53	0.6	ND
Matale	5.445	1.93	0.785	ND
Dambulla	13.78	2.99	0.37	ND
Bakamuna	10.58 ± 0.21	5.78 ± 1.01	2.13 ± 0.61	ND
Minneriya	16.23 ± 2.43	7.97 ± 0.62	2.18 ± 0.07	ND
Medirigiriya	13.74 ± 3.33	4.83 ± 0.34	1.23 ± 0.36	ND
2-Ella	67.94 ± 29.45	14.59 ± 4.81	2.14 ± 0.90	ND

Lower concentration of THM values were reported in the Mahaweli DAs compared to the NDAs. The following graph shows the variation of TTHM in Mahaweli DAs. Highest TTHM values were reported in 2-Ella WTP which extracts source water from the Parakrama Samudra. Water flow was divided into two directions from the Elahera anicut. One flowing toward the Minneriya and Kaudulla tank along the Elehera Minneriya Yoda Ella and the other one flowing toward the Parakrama Samudra with the combination of the Kalu Ganga. Parakrama samudra can be more polluted from agricultural runoff coming from both the Kalu Ganga and the surrounding area (Fig. 1).

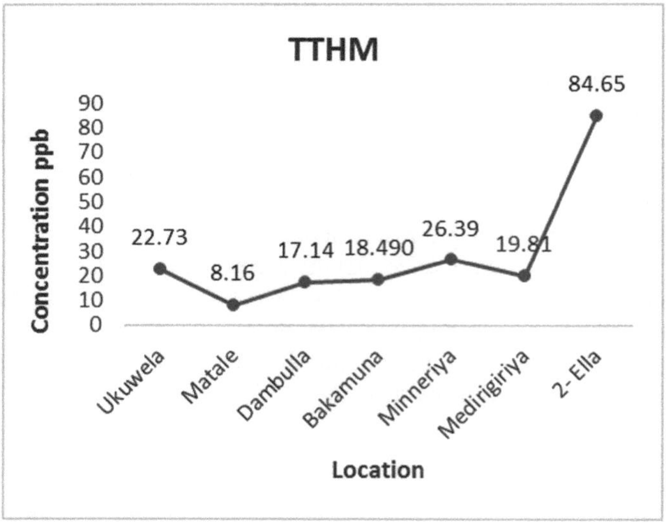

Fig. 1. Variation of TTHM concentration in Mahaweli Diverted area

3.2 Non Diverted Area

The Mahaweli river starts its journey in the wet zone of Sri Lanka and flows through the dry zone. It carries most organic and inorganic pollutants from the urban and agricultural areas. Organic level gradually increases when it comes downstream. Therefore, THM formation potential also high in those areas. The following table shows the recorded THM values in WTP of Mahaweli NDAs (Table 2).

Table 2. Reported THM concentration (μg/L) in WTP in Mahaweli NDA

Location	$CHCl_3$ ± SD	$CHCl_2Br$ ± SD	$CHBr_2Cl$ ± SD
Meewathura	7.58 ± 0.71	1.55 ± 0.10	0.70 ± 0.10
Greater Kandy	15.98 ± 8.18	4.47 ± 1.78	0.76 ± 0.58
Pathadumbara	27.64 ± 12.4	3.99 ± 1.29	0.40 ± 0.08
Mahiyanganaya	23.14 ± 1.29	6.61 ± 0.33	1.70 ± 0.12
Dehiaththakandiya	15.46 ± 8.46	5.58 ± 2.02	1.37 ± 0.45
Gallella	41.89 ± 18.1	9.72 ± 4.71	1.60 ± 0.91
Kantale	20.67	7.07	1.78
Mutur	63.87	16.76	3.22

The highest concentration of Trichloromethane, Bromodichloromethane and Dibromochloromethane ($CHBr_2Cl$) were recorded in the Mutur WTP and the lowest values were recorded in Meewathura WTP. The Meewathura WTP is located in an upstream area. Therefore, water contains less amounts of organic matter. Electrical conductivity and turbidity value was 60.0 μs/cm and 3.23 NTU respectively.

Mutur is the final WTP which extracts source water from the Mahaweli River before entering the sea. Source water flows through many of the agricultural areas and contains considerable levels of organics. Electrical conductivity and turbidity value was 210.0 µs/cm and 17.7 NTU respectively in Mutur WTP source water.

Tribromomethane was not reported in any WTP along the Mahaweli River. Tribromomethane forms when water contaminates with Bromide. Bromide reacts with applied disinfectants and forms bromine, which oxidizes organic matter and finally creates Tribromomethane. (Singer 1994; McTigue *et al.* 2014). Water can be polluted from the bromide due to anthropogenic activity such as sewage or industrial effluent discharges and landfill leachate. (Siddiqui *et al.* 1993; Boyer *et al.* 2013).

The following graph show the variation of TTHM (cumulative value of four THM species) along the river. TTHM values gradually increased along the river. But there were a few reductions at the Dehiaththakandiya and Kantale WTP. TTHM concentration can be largely affected by treatment adequacy and the treatment technology of the WTP. However, TTHM concentration of the all WTP were below the USEPA maximum contamination level of 80 µg/L except at Mutur WTP (Fig. 2).

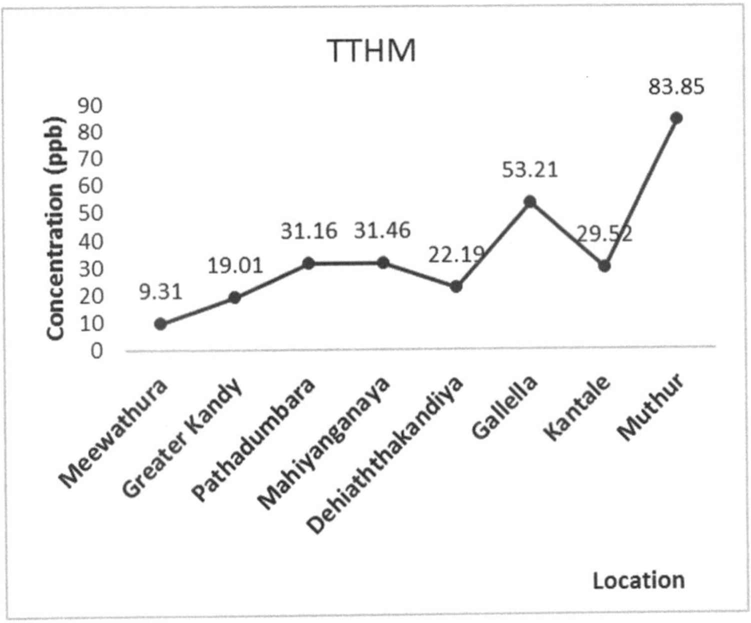

Fig. 2. Variation of TTHM concentration in Mahaweli Non-Diverted area

However, according to the Kruskal-Wallis Test, the median TTHM concentration of DAs (22.73 µg/L) was not significantly different ($p > 0.05$) from the NDAs (25.36 µg/L). The mean Conductivity (µS/cm), pH, turbidity (NTU) and alkalinity (mg/l) of the DAs were 224.25 ± 65.95, 7.48 ± 0.36, 3.69 ± 1.91 and 87.5 ± 28.72 respectively and in the NDAs; 139.52 ± 62.39, 7.31 ± 0.37, 12.71 ± 12.13 and

56.25 ± 28.99 respectively. According to the statistical analysis, there was no significant difference ($p > 0.05$) between the measured water quality parameters of DAs and NDAs.

4 Conclusion

This study was conducted to investigate THM formation potential in the Mahaweli Diverted areas and Mahaweli Non-Diverted areas. Some researchers have found that there was an adverse effect on water quality after the diversion. Large number of CKDu patient were reported in the Mahaweli Das, especially in the North Central province. However, it was found that THM concentration in the Mahaweli DAs and NDAs were below the USEPA maximum contamination level of 80 μg/L except at Mutur WTP. According to the statistical analysis, it was found that there was no significant difference in THM concentration between Mahaweli DAs and NDAs.

References

APHA, AWWA, WEF (1995) Standard methods for the examination of water and wastewater, 19th edn, Washintion D.C

Bandara JMRS, Senevirathna DMAN, Dasanayake DMRSB, Herath V, Bandara JMRP, Abeysekara T, Rajapaksha KH (2008) Chronic renal failure among farm families in cascade irrigation systems in Sri Lanka associated with elevated dietary cadmium levels in rice and freshwater Wsh (Tilapia). Environ Geochem Health 30:465–478

Boyer TH, Ged E, Motz L, Chadik P, Frank K, Hill CP (2013) Impact of sea-level rise on saltwater intrusion and formation of brominated disinfection byproducts. In: Water quality technology conference and exposition (WQTC) conference proceedings. American Water Works Association, Denver

Hewavisenthi ACS (2009) Management of the Mahaweli, a river in Sri Lanka. Water Int 22 (2):98–107

Kuivinen J, Johnsson H (1999) Determination of trihalomethanes and some chlorinated solvents in drinking water by headspace technique with capillary column gas-chromatography. Water Res 33(5):1201–1208

Mahaweli Ganga Development Sri Lanka (1970) Mahaweli Development Board of Sri Lanka

McTigue NE, Cornwell DA, Graf K, Brown R (2014) Occurrence and consequences of increased bromide in drinking water sources. J Am Water Works Assoc 106(11):E492–E508

Morris RD, Levine R (1995) Estimating the incidence of waterborne infectious disease related to drinking water in the United States. Reichard

Siddiqui MS, Amy GL (1993) Factors affecting DBP formation during ozone–bromide reactions. J Am Water Works Assoc 85(1):63–72

Singer PC (1994) Control of disinfection by-products in drinking water. J Environ Eng 120 (4):727–744. https://doi.org/10.1061/(ASCE)0733-9372(1994)120:4(727)

Weragoda SK (2005) THM formation modeling in treated water using the rapid organic characterization technique. http://www.faculty.ait.ac.th/visu/Data/AIT-Thesis/MasterThesis final/Weragoda05.pdf

Studying About the Manganese Phase Diagram When Manganese Is Reacting with Calcium Hypochlorite

S. A. Abeysinghe[1(✉)], S. K. Weragoda[2], and W. B. M. L. I. Weerasekara[2]

[1] Department of Science and Technology, Faculty of Science and Technology, Uva Wellassa University, Badulla, Sri Lanka
samuthianji@gmail.com
[2] National Water Supply and Drainage Board, Kandy, Sri Lanka

Abstract. Oxidation Reduction behaviour is highly contributes to the condition of the water. Oxidation reduction potential [ORP] gives an idea about the Redox behaviour of water while the pH value give an indication about the behaviour of the Hydrogen ion concentration. When Manganese is in water, it can exist either as particulate matter or as dissolved matter and this can be showed in Manganese phase diagrams. Due to the inconvenient consequences, Manganese should be removed from the water if only it has a higher concentration. For this, oxidation can be done by using Calcium hypochlorite and the obtained precipitate can be filtered. The behaviour of different Manganese species fluctuates with different pH values and Eh values. The objective of this study was to determine the effect of pH and Eh while changing the different concentration of Calcium hypochlorite to the blended Manganese solution which is prepared by using $MnSO_4$. Final concentration of Manganese was reduced with lower concentration of Calcium hypochlorite and ORP values were reduced with the increasing pH. To conclude the result, Mn_3O_4, Hausmannite was obtain as the Manganese species according to the Manganese phase diagram.

Keywords: Oxidation reduction potential · Redox behavior · Eh · Manganese phase diagram · Calcium hypochlorite · Hausmannite

1 Introduction

1.1 Effect of Manganese to the Water

Manganese naturally exists in earth's crust and it aggregates to water due to the dissolution of Manganese bearing minerals and rocks such as Manganosite, Manganite, Bixbyte when there is weathering. Until Manganese is in groundwater, Manganese exists as a stable form, but as soon as it is exposed to the outside air it oxidizes and precipitates out naturally (Salem *et al.* 2012).

Manganese generally does not affected human health but it may cause undesirable taste, appearance and straining (Rozainy *et al.* 2015). Presence of Manganese bearing

R. Dissanayake and P. Mendis (Eds.): ICSBE 2018, LNCE 44, pp. 137–146, 2020.
https://doi.org/10.1007/978-981-13-9749-3_13

bacteria in water distribution systems causes bacterial growth in pipes. From the consumer's view, washing machines and boilers can be damaged due to oxidization and precipitation under conditions of excess Manganese (Barlokova and Ilavský 2009).

1.2 Manganese Phase Diagram

The Manganese phase diagram, (Fig. 1) shows the relationship of the redox potential and hydrogen-ion concentration that are relevant with specific dissolved ions and solids at chemical equilibrium in the presence of water. Phase boundary lines are called Stability field boundaries and in their activities, each adjacent species is nearly equal.

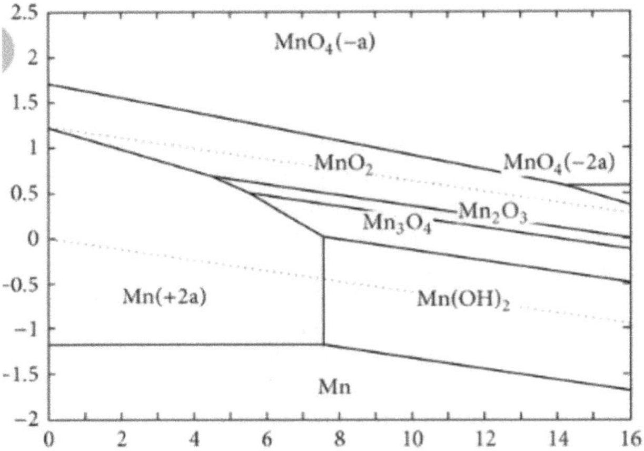

Fig. 1. Manganese phase diagram

pe is accessible as tendency of solution to donate or accept protons while Eh is known as the potential of the solution relative to the Standard Hydrogen Electrode [SHE]. In here SHE is the reference point and when all conditions are constant no reactions occur and the Eh value can be obtained from connecting the electrode containing another solution.

When Eh is higher, the solution is desirable to accept electrons from another species and when Eh is lower, the solution is desirable to donate electrons to the species. Therefore solution is reducing and oxidizing respectively.

1.3 Redox Reactions

Redox reactions are typically happens in a solution by oxidizing or reducing an electron. Hence, these reactions of oxidation or reduction do not exist in nature individually because free electrons never exist in a solution. Releasing of any electron from one species and its gain by another species happens simultaneously (Appelo and Pstma 2005).

$$Ox_1 + Red_2 \rightarrow Red_1 + Ox_2$$

Due to this, pe of the solution cannot be measured practically. Eh can be discovered not only by measuring but also by calculations of the Nernst equation. Therefore, Eh is the integration of both the calculated Eh and the measured Eh (Christensen *et al.* 2000).

$$\text{Eh} = \text{E}_{\text{measured}} + \text{E}_{\text{reference}} \quad (1)$$

Both pe and Eh measure essentially the same and therefore the relationship between them can be written as follows;

$$Eh = \frac{2.303\ RT}{F} pe \quad (2)$$

1.4 Behavior of Manganese in Water

Manganese is the 25th element of the periodic table and the most stable oxidation state is +2. When Manganese reacts with water, $Mn(OH)_2$ is obtained but it can change with the condition of the water.

$$Mn^{2+} + 2H_2O \rightarrow Mn(OH)_2 + H_2$$

1.5 Behavior of the Calcium Hypochlorite in Water

Calcium hypochlorite, which has the commercial name of bleaching powder, is an inorganic compound. It is mostly used in water treatments because of the high efficiency and low cost. It exists as a powder form and it should be dissolved with water to acquire Calcium hypochlorite solution.

$$Ca(OCl)_2 + 2H_2O \rightarrow Ca(OH)_2 + 2HClO$$

1.6 Reaction of Manganese and Calcium Hypochlorite

While $Ca(OH)_2$ is a precipitate Mn(OH)2 is react with HClO.

$$Mn(OH)_2 + 2HClO \rightarrow MnO_2 + CaCl_2 + 2H_2O$$

1.7 Objective of the Study

To investigate the behavior of pH and Eh of the solution while changing the different concentration of Calcium hypochlorite, in certain concentrations of the Manganese rich solution by studying the Manganese phase diagram.

To discover the particulate form of the Manganese species which can be removed by filtration.

2 Materials and Methodology

2.1 Analytical Techniques

2.1.1 Determination of pH Value

pH values of the water samples were directly measured by the pH meter. Before that, the pH electrode was removed from the storage solution, rinse with distilled water and dried with tissue paper. Then, the water sample was taken to the beaker and the pH electrode was placed in it and the value on the digital screen was recorded. This pH value is at fixed temperature.

2.1.2 Determination of Concentration of Manganese

UV spectrophotometer was used to measure the dissolved Manganese concentration of the sample. A program was selected to measure the concentration of Manganese and the wavelength was kept as 525 nm. Water sample was filled to the two 10 ml cells and one is kept as reference. A Manganese citrate buffer powder pillow was added to the other cell and shaken until it dissolved well. Then, Sodium periade reagent was added and kept for two minutes to complete the reaction. After two minutes, the reference sample was placed inside the hole of the UV spectrophotometer and calibrated to zero. After it became zero, the working cell was placed and the reading on the digital screen of UV spectrophotometer was recorded.

2.1.3 Determination of Oxidation Reduction Potential (ORP) Value

Same pH meter was used to measure the ORP value and it demonstrated the reference Eh value. To get the ORP value, the program was changed by pressing the 'Option' button. ORP value is record as millivolt.

2.1.4 Determination of Residual Chlorine (RCl) Value

Residual chlorine value gives the remaining free chlorine amount of the sample. It can be measured directly by using RCl meter. For that, the sample is filled into two cells and one was kept as reference. DPD free chlorine reagent was added to the other sample and shaken well until it fully dissolved. If Chlorine is present in the sample, light pink color appears. The reference sample was placed in the hole inside the RCl meter and calibrated to Zero and then the working sample was placed and the value given in the digital screen was recorded.

2.2 Preparation of Manganese Solution

5 L of 0.9 ppm Mn^{2+} solution was freshly prepared by dissolving 0.0124 g of $MnSO_4$ in to raw water sample collected from Maduru Oya. A raw water sample was used instead of tap water or distilled water, in order to obtain the natural environmental condition.

2.3 Preparation of Calcium Hypochlorite Solution Series

1L of 10 ppm Calcium hypochlorite solution was made by dissolving 0.01 g of Bleaching powder in to distilled water. Bleaching solution series was selected with respect to the Residual Chlorine value given from the stock sample.

2.4 Method

Initial concentration of the Mn^{2+} measured by using UV spectrophotometer and Bleaching solution series was added to the 4 beakers which were filled with 500 ml of Mn^{2+} sample. Each sample was mixed continuously using a shaker for almost 30 min. Finally pH, RCl, ORP and final concentration of Manganese was measured.

3 Result and Discussion

3.1 Calculating the Standard Potential

When considering the redox reactions of Mn^{2+} and ClO^-

Oxidation reaction of Mn^{2+}

$$Mn^{2+} + 2H_2O \rightarrow MnO_2 + 4H^+ + 2e^-$$

Reduction reaction of ClO^-

$$ClO^- + 2e^- + H_2O \rightarrow Cl^- + 2OH^-$$

Final reaction of Mn^{2+} and ClO^-

$$Mn^{2+} + ClO^- + H_2O \rightarrow MnO_2 + 2H^+ + Cl^-$$

According to Nernst equation

$$E = E^0 + \frac{RT}{2F}\ln\left(\frac{(H^+)^2(Cl^-)}{(Mn^{2+})(ClO^-)}\right)$$

The standard potential at 1 atm and 25 °C for oxidation reaction is +1.23 V, while for reduction reaction is +0.9 V.

To obtain E^0 for the final reaction, Standard potential of the reduction reaction should be subtracted from standard potential of the oxidation reaction. It is expressed as potential per electron.

Therefore standard potential of the final reaction is 1.23 V − 0.9 V = 0.33 V.

3.2 Removal Efficiency

Table 1 represents the removal efficiency of Manganese after adding the bleaching powder solution. Initial concentration of Manganese is 0.9 ppm

After 30 min, in each trial, Residual Chlorine value was zero. It indicated that at that moment, all of the bleaching powder was almost finished.

Table 1. Final concentration of Mn with respect to different dosages of ClO^- after 30 min of reaction

Trial	Concentration of ClO^-/ppm	Concentration of Final Mn^{2+}/ppm	Removal efficiency/%	Residual chlorine value/ppm
1	0.7	0.6	33.33	0
2	2	0.7	22.22	0
3	4	0.8	11.11	0
4	6	0.9	0	0

The variations of removal efficiency and the concentration of ClO^-] are shown in Fig. 2. It indicated that increasing concentration of ClO^- reduces the Manganese removal efficiency.

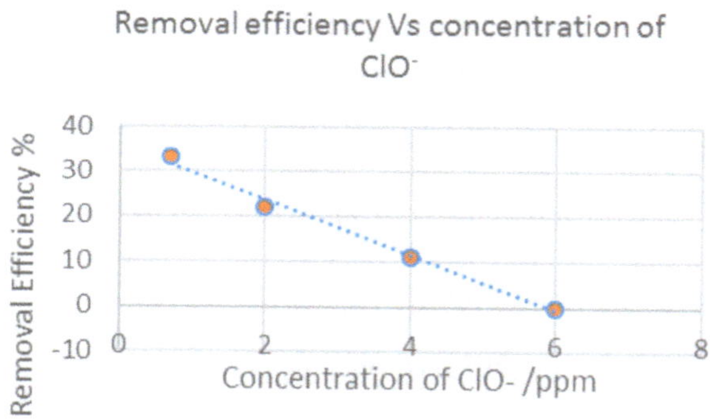

Fig. 2. Graph of removal efficiency vs concentration of ClO^-

Since this is the result of synthetic Manganese contained in raw water, it did not give the natural condition of raw water and it also influences this circumstance because in nature, Manganese mostly coexists with Ferrous (Kasim *et al.* 2017).

3.3 Relationship Between pH and ORP

Figure 3 presented the relation between pH and ORP showed in Table 2. Since the concentration of ClO^- is zero, only an oxidation reaction is occurring. According to that, the reaction produces H^+ and due to that, pH reduces. Therefore, a higher concentration of final Manganese has a lower pH.

Figure 3 indicates that the ORP value is decreasing with increasing pH. For the most part, ORP increased with increasing oxidant dosage (James *et al.* 2004). In this experiment, ClO^- is oxidant and from trial 1 to trial 4, concentration of the concentration of ClO^- is increased. Therefore, ORP value is increasing from trial 1 to trial 4.

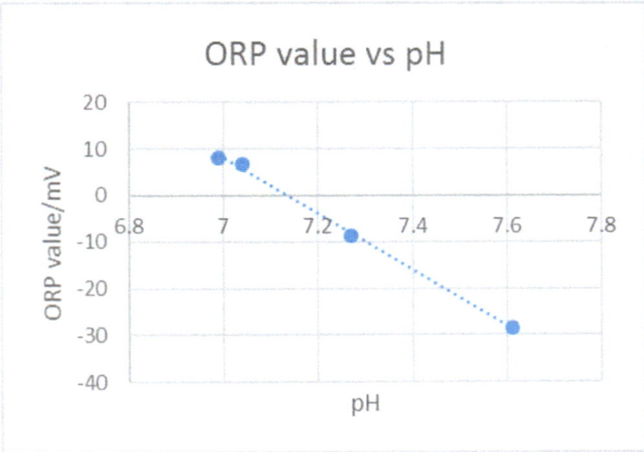

Fig. 3. Graph of ORP value and pH

Table 2. pH and ORP value of the each sample

Trial	Temperature/°c	pH	ORP [$E_{measured}$]/mV
1	25.3	7.61	−28.5
2	25.2	7.27	−8.6
3	25.2	7.04	6.7
4	25.3	6.99	8.2

3.4 Calculating $Eh_{reference}$

$Eh_{reference}$ value depends on the concentrations of Mn^{2+}, ClO^-, H^+, and Cl^-. Final concentration of Cl^- is zero, and under that condition $Eh_{reference}$ value of the each trial is equal to the standard potential of the reaction as shown in Table 3.

Table 3. $Eh_{reference}$ values calculated from Nernst equation for each trial

Trial	Concentration of Mn^{2+}/ppm	Concentration of ClO^-/ppm	Concentration of H^+/ppm	Concentration of Cl^-/ppm	Temperature/K	$Eh_{reference}$/V
1	0.6	0.7	10.22×10^{-8}	0	298.45	0.33
2	0.7	2.0	9.12×10^{-8}	0	298.35	0.33
3	0.8	4.0	5.37×10^{-8}	0	298.35	0.33
4	0.9	6.0	2.45×10^{-8}	0	298.45	0.33

Since the Eh value is the integration of both ORP value and the calculated Eh value from Nernst equation, Table 4 illustrates the final Eh value and pe value, which is acquired from the relationship of Eh and pe (Appelo and Pstma 2005).

Since the higher pe value has higher tendency to accept protons, according to the obtained pe, propensity to accepting protons is developed from trial 1 to trial 4 respectively as shown in Table 5. Therefore, the solution of trial 1 is more oxidized than the trial 2 solution.

Figure 4 shows that Eh is reduce with increasing pH of the solution.

Table 4. Indicate that $\mathbf{E_{reference}}$ values influence to the Eh more than $E_{measured}$ values.

Trial	$E_{measured}$/V	$E_{reference}$/V	Eh/V	pe/V
4	−0.0285	0.33	0.3015	5.0906
3	−0.0086	0.33	0.3214	5.4285
2	0.0067	0.33	0.3367	5.6869
1	0.0082	0.33	0.3382	5.7103

In this experiment, pH of the samples exist between 6.8 and 7.8. According to the Manganese phase diagram shown in Fig. 4, at that level, Manganese can exist as a dissolved species (Mn^{2+}) or as a precipitate such as Mn_3O_4, MnO_2, Mn_2O_3, and MnO_4 (Bruins *et al.* 2014).

Table 5. pH values and Eh values of each trial

Trial	pH	Eh/V
1	7.61	0.3015
2	7.27	0.3214
3	7.04	0.3367
4	6.99	0.3382

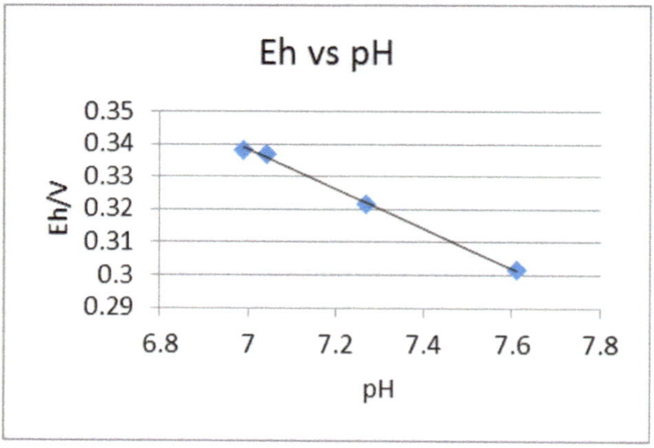

Fig. 4. Graph of Eh vs pH

Here, Eh value exists between 0.3 and 0.4. Type of the Manganese species to be obtained varies with the pH and the Eh as shown in Fig. 5. From the above result of Mn_3O_4, Hausmannite was obtained after adding bleaching powder solution between 0.7 ppm and 6 ppm to 0. 9 ppm Mn^{2+} solution.

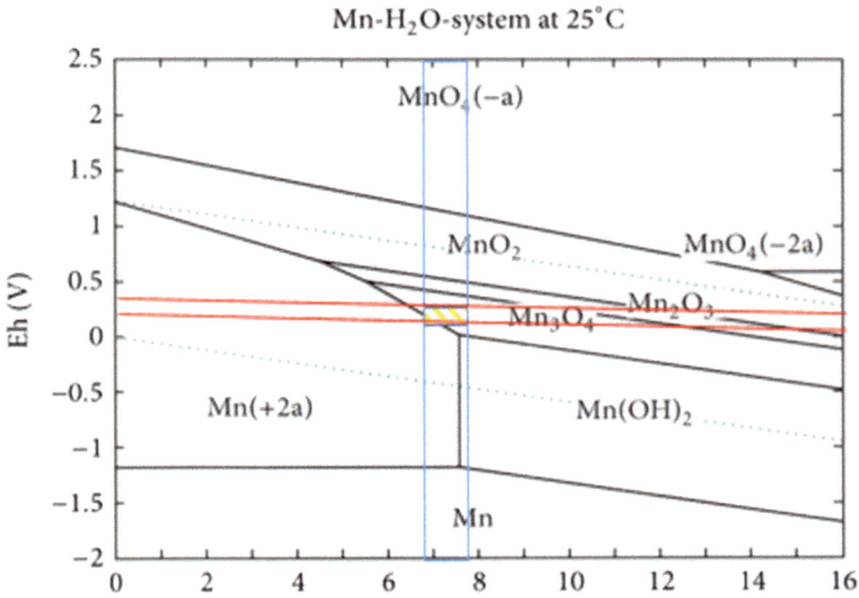

Fig. 5. Phase diagram with analyzed data

After obtaining the particles, they may be removed by using a filtration process (Isaeva and Castro 2011).

If water contains higher concentrations of Manganese as a dissolved form, before removal it should be converted to its particular form by changing the pH or Eh values. pH can be changed by adding diluted acid or base and Eh can be amended by changing the concentration of the initial concentration of the Manganese solution and the bleaching solution. Particulate form can be selected according to the Manganese phase diagram considering the adequate pH and Eh values.

4 Conclusion

Mn_3O_4, Hausmannite occurs as a final product when 0.9 ppm synthetically prepared Manganese solution reacts with different concentration of Calcium hypochlorite solutions. Final product can be observed after some time as a brown colour precipitate. This precipitate can be removed by using an adequate filtration method.

Acknowledgement. The authors would like to thank to staff of Water safety plan advisory unit of National Water Supply and Drainage board in Katugastota for financial support and laboratory support. Also give thanks to the Laboratory of the National Water Supply and Drainage board in Kegalle for help of the analysis.

References

Appelo CAJ, Postma D (2005) Redox processes. In: Geochemistry, groundwater and pollution, 2nd edn. A.A. Balkema Publishers, Leiden, pp 415–480. The Netherlands a member of Taylor & Francis Group plc

Barloková D, Ilavský J (2009) Removal of iron and manganese from water using filtration by natural materials. J Environ Stud 19:1118–1120

Bruins JH, Petrusevski B, Kruithof JC, Kennedy MD (2014) Manganese removal from groundwater: characterization of filter media coating. Desalin Water Treat 1–13

Christensen TH, Bjerg P, Banwart SA, Jakobsen R, Heron G, Albrechtsen HJ (2000) Characterization of redox conditions in groundwater contaminant plumes. J Contam Hydrol 45:165–241

Isaeva M, Castro NM (2011) Water treatment for the removal of iron and manganese. Bachelor Degree Project in Mechanical Engineering – Development Assistance, 30 ECTS

James CN, Copeland RC, Lytle DA (2004) Relationship between oxidation reduction potential, oxidant and pH in drinking water. In: American water works association WQTC conference

Kasim N, Mohammad AW (2017) Iron and manganese removal by nanofiltration and ultrafiltration membranes: influence of pH adjustment. Malays J Anal Sci 21:149–158

Rozainy ZMR, Jamil R, Adlan MH (2015) A review of removal iron and manganese by using cascade aeration systems. Sc Eng 74:69–76

Salem MG, El-Awady MH, Amin E (2012) Enhanced removal of dissolved iron and manganese from nonconventional water resources in Delta District, Egypt. J Energy Proc 18:983–985

Watershed Characterization of Demodara Reservoir with Special Attention on Soluble Manganese Concentration

W. G. Chathura Sanjeewa[1(✉)], R. A. C. B. Ranaweera[1], Ruwan Liyanage[1], W. B. M. M. W. Wijekoon[1], and G. P. R. Ireshika[2]

[1] National Water Supply and Drainage Board, Regional Support Center (Uva), Badulla, Sri Lanka
chatsanjwb@gmail.com
[2] Faculty of Science and Technology, Uva Wellassa University, Badulla, Sri Lanka

Abstract. To ensure smooth operation and to manage water quality of the Demodara Water Treatment Plant (DWTP), better knowledge of its watershed is required. In line with this evolving response to water quality management; the National Water Supply & Drainage Board is currently piloting the development of a Water Safety Plan for the Demodara water supply scheme. For the purpose, watershed assessment as a baseline study in a border context for the study area is addressed in this paper. The DWTP is fed by the Demodara reservoir. Water quality tests were conducted representing its catchment and none of its water quality parameters except manganese (Mn) shows critical levels in the catchment. The catchment was analysed to capture the Mn source. Results show that the geological formation of the catchment consists of MnO_2; especially in the Charnockitic gneisses layers. With rainfall, Mn^{+4} converts to dissolved Mn^{+2}. It then comes to the Demodara reservoir from surface runoff and, it remains as Mn^{+2} at the bottom levels of the reservoir due to anaerobic, reducing conditions. Raw water is pumped to the DWTP from low levels of the reservoir, and therefore Mn enters in to DWTP system. Manganese chemistry is very complex and consists of multiple oxidative states. At the DWTP, Mn is converted to insoluble manganese by oxidation using $KMnO_4$ followed by the removal of oxidized, insoluble manganese through filtration. However, manganese treatment does not completely remove all manganese from the treated water. It will allow some manganese to enter the distribution system, particularly in high rate water treatment plants. Literature reports emphasize that Mn has no negative health impacts at the levels commonly found in drinking water. However, aesthetic problems caused by Mn are a continuing source of concern for customers and DWTP operators. Therefore, treating Mn in raw water requires continuous monitoring by plant operators. For the smooth functioning of the DWTP, a number of recommendations are discussed in this paper.

Keywords: Manganese · Oxidization · Catchment · Distribution system

R. Dissanayake and P. Mendis (Eds.): ICSBE 2018, LNCE 44, pp. 147–161, 2020.
https://doi.org/10.1007/978-981-13-9749-3_14

1 Introduction

Badulla, Haliela and Ella integrated water supply scheme is mainly fed by the Demodara Water Treatment Plant (DWTP) which is owned and operated by the National Water Supply & Drainage Board (NWSDB), Sri Lanka. The DWTP was designed and built by Tetra Tech, USA and commissioned towards the end of 2017. It has a water treatment capacity of 15,000 m^3 per day and is able to supply safe drinking water to Badulla, Haliela and Ella demand centres and its suburbs. This integrated water treatment system includes a new dam, new and upgraded water treatment plant, storage tanks, pumping stations, and transmission and distribution pipelines. The water system originates at the new Demodara reservoir, located on the Badulu Oya about 12 km upstream of the city of Badulla. It is a reservoir which has a capacity of 130,000 m^3 to store raw water for the system.

To ensure smooth operation and to manage the quality of water of the DWTP, better knowledge on its watershed is required. In line with this evolving response to water quality management, the NWSDB is currently piloting the development of Water Safety Plan for the Demodara water supply scheme. For the purpose watershed assessment as a baseline study in a border context for the study area is addressed in this paper.

Special attention was given to soluble manganese concentration since the DWTP facility has experienced manganese (Mn^{+2}) levels up to 1,000 μg per litre in certain time periods. During the water treatment process and in the distribution system, the manganese oxidized and caused brown staining of the plumbing fixtures leading to public complaints. SLS 614 water quality standards has established a maximum contaminant level of total Mn as 100 μg per litre of treated water. However, the DWTP has experienced Mn problems from effluent concentrations which is lower than its stipulated standard limits.

Although Mn^{+2} oxidation has been practiced for years in water treatment, many studies reported in literature have not fully explored oxidation of low initial Mn^{+2} concentrations and issues that may be important to water treatment plants. This study also sought to address Mn related issues in the DWTP for its treatment.

2 Methodology

Location maps, catchment maps, elevation maps, land use patterns, population distribution, river morphology maps and geological maps of the catchment were developed using ArcGIS software. Rainfall data and river flow data were analysed using statistical methods.

Spatial and Temporal distribution of water quality at the Demodara catchment along the Badulu oya were analysed. Water samples were collected from 16 locations along the Badulu oya covering the whole catchment area. Physical, chemical and bacteriological parameters were checked for all samples and tests were carried out for both dry and wet seasons. Then water quality parameter distributions were mapped to get an average idea of the catchment. The Manganese variation of the catchment area is presented in this paper.

Water samples were checked at the Demodara reservoir and an X-ray Powder Diffraction (XRD) test was conducted for reservoir sediment samples. Phase identification was done by using reference XRD graphs.

Since there are no mining sites, industrial discharge or leachates from landfills in the Demodara catchment, the anticipated source for accumulated Mn in the Demodara reservoir is eroded natural deposits from unplanned upper catchment. Therefore, geological studies were done to identify Mn sources from the catchment. Fourteen rock samples were analysed chemically representing all soil types of the Demodara catchment area. Fertilizers used by farmers in the catchment were also studied.

To capture the correlation of raw water manganese level and upper catchment rainfall, correlation coefficient analysis was conducted. The Mn level at raw water and water at clear wells were analysed. The pH levels at raw water, tube settler surface, adsorption clarifier surface, filter outlet, contact tank and clear well were taken and analysed statistically.

The Demodara-Badulla main distribution pipe line (450 mm diameter HDPE) was flushed. After continuous supply of 15 days, manganese at four wash-out locations were checked in order to capture the ability of Mn accumulation in distribution lines.

2.1 Study Area

The Demodara dam catchment is entirely located in the Badulla district and is in the South-West of the Namunakula mountain range in the hill country of Sri Lanka (between latitude 6° 51′ 00″ to 6° 55′ 30″ and longitude 80° 58′ 30″ to 81° 7′ 00″) (Fig. 1). The Demodara reservoir is fed by the Badulu oya which is a major tributary of the Mahaweli river, the longest river in Sri Lanka.

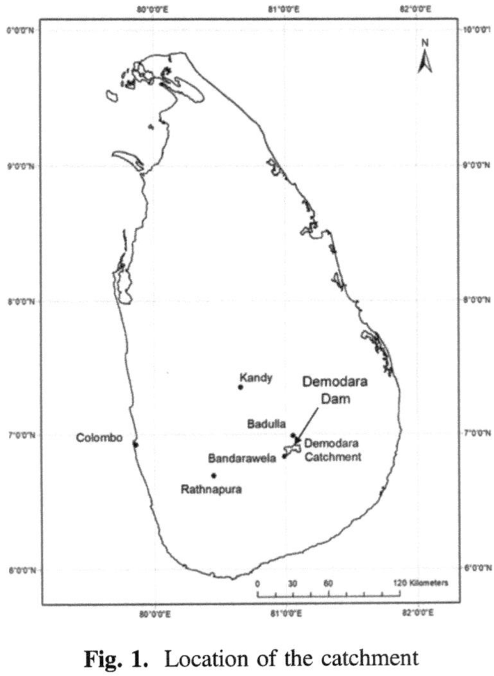

Fig. 1. Location of the catchment

The Demodara dam location is just downstream of the confluence of the Nawela oya and the Heel oya (Fig. 2). The Nawela Oya originates at an altitude of 1500 m from Andagala and it joins the Heel Oya at Demodara. Heel Oya originates around Bandarawela at an altitude of 1200 m.

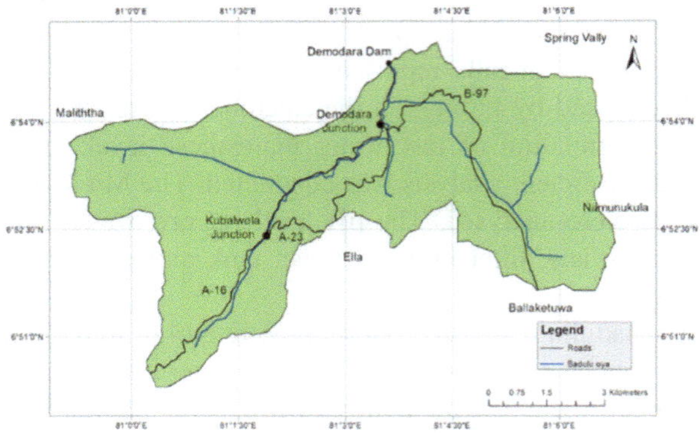

Fig. 2. Demodara catchment

The total catchment area of the Demodara dam is 59.1 km^2. Being located in an upland area with narrow, steep and confined valleys, the catchment reacts quickly. The terrain is generally mountainous with steep slopes where the elevation varies from 700 m to 1800 m. Soil cover in the basin is not deep and this will result in a quick runoff response. In general, mountainous catchments have higher specific runoff than lower regions due to high rainfall into the catchment. Furthermore, the storage capacity of the surface zone and root zone is less in the mountainous areas.

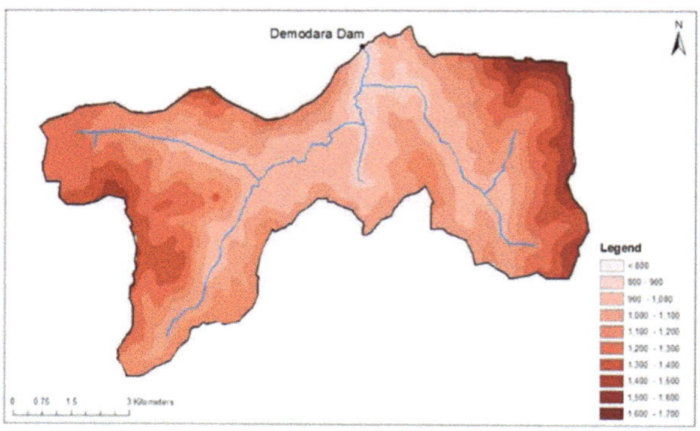

Fig. 3. Elevation variation of the catchment

2.2 Drainage Density

Drainage density (DD) relates to the number of streams in a particular drainage basin and can be measured by dividing total length of all streams in a basin (L) by its area (A). As a rule, the higher the drainage density (DD) the more quickly water drains to a river.

$$DD = \frac{\sum L}{A} \tag{1}$$

2.3 The Correlation Coefficient (CC) Test

Correlation coefficient denoted by ‘r’ is a statistical analysis method representing the degree of linear relationship between two variables (Taylor 1990). The correlation coefficient takes on values ranging between +1 and −1.

$$r = \frac{\sum_{i=1}^{n}((xi - \bar{x})(yi - \bar{y}))}{\sqrt{\sum_{i=1}^{n}(xi - \bar{x})^2 \sum_{i=1}^{n}(yi - \bar{y})^2}} \tag{2}$$

x_i denotes the rainfall of the Bandarawela meteorological station at i^{th} date. y_i denotes the manganese level of the incoming raw water to the DWTP at i^{th} date. $\bar{x}$ and $\bar{y}$ denotes average values of the corresponding group. The relationship between x and y is as follows according to r value (Table 1).

Table 1. Correlation coefficient value interpretation

‘r’ value	Interpretation
−1	Variables are completely anti-correlated
Between −0.7 to −1.0	Strong negative linear relationship
Between −0.3 to −0.7	Moderate negative linear relationship
Between 0 to −0.3	Weak negative linear relationship
0	Variables are completely uncorrelated
Between 0 to 0.3	Weak positive linear relationship
Between 0.3 to 0.7	Moderate positive linear relationship
Between 0.7 to 1.0	Strong positive linear relationship
1	Variables are absolutely correlated

3 Results

3.1 Land Use Pattern of the Catchment

Tea cultivation is the prominent land use pattern in the area, which is about 54% from the total land (Table 2).

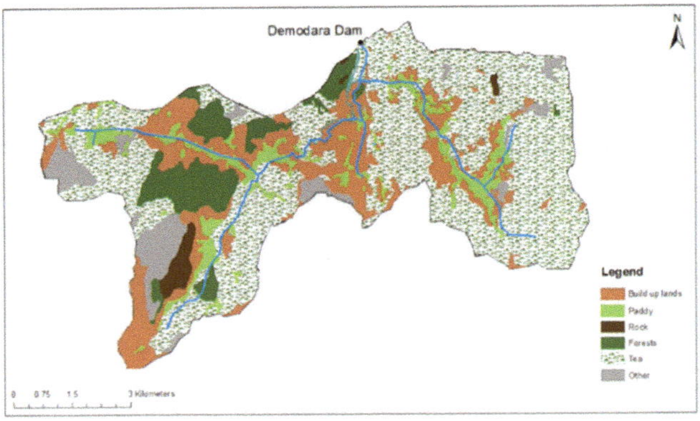

Fig. 4. Land use patterns in the catchment

Table 2. Land use of the catchment

Land use type	Percentage (%)
Tea	54
Build-up lands	21
Forests	9
Paddy	8
Rock	2
Other	8

3.2 Fertilizers Mainly Used for Cultivations

Most fertilizers used by farmers in the catchment consisted of calcium, nitrate, phosphate and sulphate. No fertilizers containing manganese were found during the study.

3.3 Drainage Density (DD) of the Catchment

In rainy season, the DD is nearly 1.5 and it means large channel frequency (tributaries). Therefore, in the rainy seasons, higher sediment yields at the Demodara reservoir is expected. However, all the eroded material does not enter the stream system as some of the material is deposited at natural or manmade barriers within the watershed. Soil erosion from the catchment also depends on rainfall intensity, soil type and geologic formation, ground cover, land use, topography, runoff, sediment characteristics-grain size, mineralogy, etc. and channel hydraulic characteristics which are interconnected.

3.4 General Geology of the Catchment

Rock types in the catchment area were Charnockitic gneisses, Garnet-sillimanite-biotite gneiss, Garnetiferous quartzofeldspathic gneiss, Marble, Quartzites and Walawe gneisses. Soluble Mn concentrations in each rock type are presented in Table 3.

Table 3. Rock types in the Demodara catchment and average soluble Mn in each rock type

Rock type	Dominant minerals & chemical compositions	Average soluble Mn level (mg/l)
Charnockitic gneisses	Hornblende [$Ca_2(Fe_2,Mg)_4(Al,Fe_3)(Si_7Al)O_{22}(OH,F)_2$], Graphite[C]	0.021
Garnet-sillimanite-bioite gneiss	Sillimanite [Al_2OSiO_5], Biotite [$KFe_3,Mg_3(AlSi_3O_{10})(OH,F)_2$] Graphite [C]	0.016
Garnetiferous quartzofeldspathic gneiss	Quartz [SiO_2]	0
Marble	Dolomite [$CaMg(CO_3)_2$], Calcite [$CaCO_3$]	0.003
Quartzites	Sillimanite [Al_2OSiO_5], Biotite [$KFe_3,Mg_3(AlSi_3O_{10})(OH, F)_2$]	0.007
Walawe gneisses	Hornblende [$Ca_2(Fe_2,Mg)_4(Al,Fe_3)(Si_7Al)O_{22}(OH,F)_2$], Graphite [C], Quartz [SiO_2], Biotite [$KFe_3,Mg_3(AlSi_3O_{10})(OH,F)_2$] Tourmaline [$FeWO_4$], Topaz [$Al_2SiO_4(F,OH)_2$]	0.011

Results show that the major contributor to Mn in the Demodara catchment is metamorphic Charnockitic gneisses rock type. Rock samples from Quartz and Marble layers shows very low amounts of soluble Mn (Fig. 5).

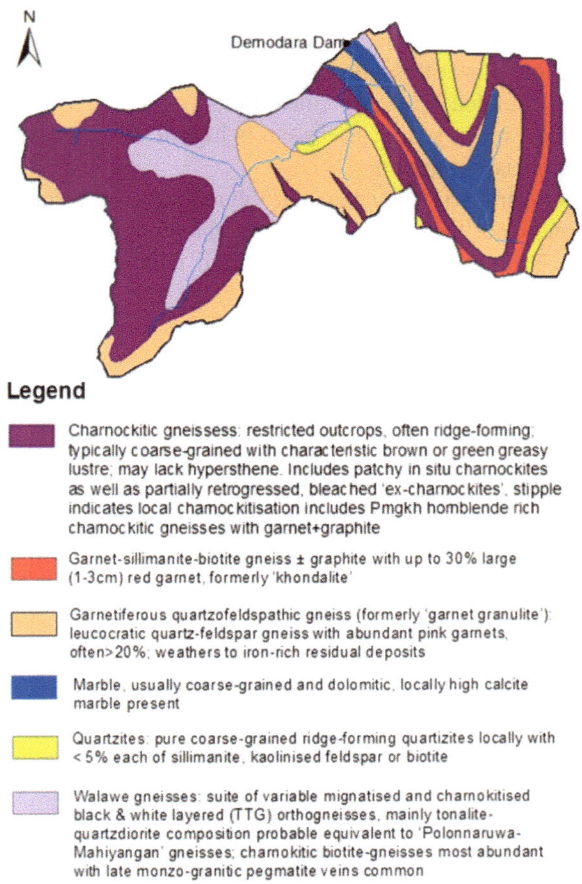

Fig. 5. Geological map of the catchment

3.5 Water Quality Parameters

Alkalinity, electrical conductivity, total hardness, phosphates, sulphates, nitrate, nitrite, manganese, iron and other chemical parameters where checked and none of the water quality parameters except that of manganese showed critical levels in the catchment. Also, Heavy metals were not found in the Demodara raw water.

3.6 Correlation Coefficient (CC) Test Results

Results show a correlation coefficient value of 0.66 which indicates that there is a moderate positive relationship with the DWTP raw water manganese levels and the rainfall to the upper catchment. Therefore, manganese in raw water is critical in the raw water of the DWTP.

3.7 Manganese Variation of the Catchment

According to the map, Mn concentration is comparatively high on the Bandarawela side. This is due to the presence of a Charnokitic gneisses rock layer. Intermediately, due to river flow, soluble Mn concentration is becoming low due to the self-aeration capability of the stream. Near the Demodara Sri Shailatharama Viharaya, again, Mn concentration is considerably high. Reviewing the characteristics of this area emphasized that soil erodibility near this area is high. The main reason for the determination is that this area is surrounded by paddy cultivations (refer Fig. 4 for land use pattern). During high rainfall events, sediment erosion is high in this area. Moreover, this area has a Charnokitic gneisses stratum.

From the Demodara Sri Shailatharama Viharaya to the Demodara reservoir the elevation variation is not high (refer Fig. 3 for the contour details). Therefore, self-aeration from the Badulu oya is not expected and therefore Mn will transfer to the Demodara reservoir (Fig. 6).

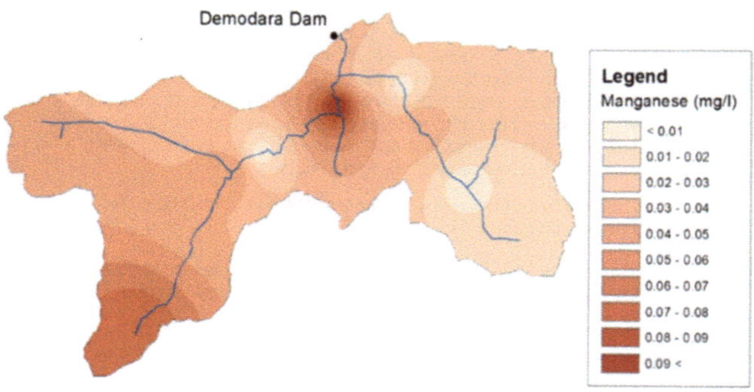

Fig. 6. Manganese variation in wet season

3.8 Presence of Manganese in Demodara Reservoir

Average pH, Mn and Fe levels of water samples collected from bottom of the reservoir were presented in Table 4.

Table 4. Soluble Mn and Fe in reservoir sediment samples

Parameter	Units	Average concentration
pH	At 25 °C ± 2 °C	7.10
Total Iron	mg/l as Fe	3.3
Manganese	mg/l as Mn	0.58

XRD result of the reservoir sediments is presented in Fig. 7.

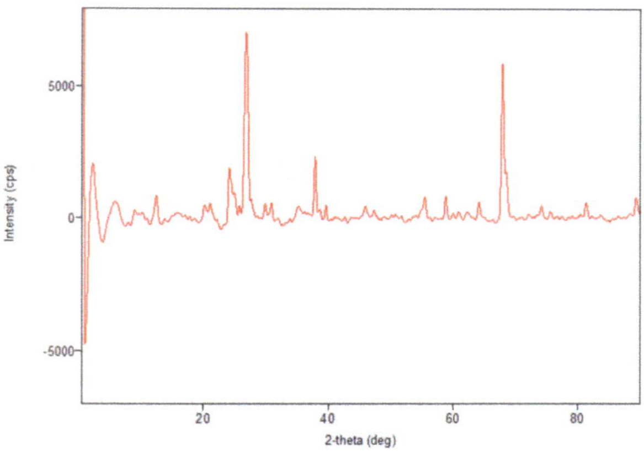

Fig. 7. XRD analysis results for reservoir sediments

According to the XRD results the dominant minerals contained in sediments are Gibbsite ($Al(OH)_3$), Kaolinite ($Al_2Si_2O_5(OH)_4$) and Quartz (SiO_2). Slight amount of MnO_2 was also present.

3.9 Scenario of Manganese Accumulation in Demodara Reservoir and Issues Faced by DWTP Due to Manganese

According to these results, during the rainy season the DD is nearly 1.5 in the Demodara catchment and it means large channel frequency (tributaries). Therefore, during rainy seasons, higher sediment yields at Demodara reservoir are expected. In addition to this, the pH in rainfall ranges from 4.46 to 6.89 in Sri Lanka (Silva and Manuweera 2004). The most stable form of manganese in natural soil in Mn^{+4} and rain provided an acidic media which converted Mn^{+4} to Mn^{+2} which is a soluble form in water. Therefore it flows with the surface runoff and accumulates, combined with the high sediment yields in the Demodara reservoir.

Manganese (Mn) often occurs naturally in the environment and is widely distributed in the soil, water and air (Brandhuber 2013). Literature reports emphasized that Mn has no negative health impacts at the levels commonly found in drinking water. However, aesthetic problems caused by Mn are a continuing source of concern for customers and DWTP operators. Problems associated with Mn includes water discoloration (usually black or dark red), clothes and fixture staining, turbid water, sediments and at very high levels, metallic taste. If Mn remains soluble it will be undetectable to the human eye. But if the customer washes clothes with bleach (NaClO), it will oxidize Mn and precipitate directly on to the surface of the clothing, leaving small black dots (Kohl and Medlar 2006). Manganese chemistry is very complex and consists of multiple oxidative states. In fact, the most common process to treat manganese in drinking water is to convert soluble manganese to insoluble manganese by oxidation and to remove the insoluble manganese through filtration or sedimentation. However, manganese treatment does not completely remove all manganese from the treated water. It will allow some manganese to enter the distribution system, typically solid MnO_2 which can form colloidal bunches, do not settle and are not easily captured onto a filter under moderate oxidation conditions (Brandhuber 2013). Increasing both the ORP (oxidation reduction potential) and the pH is the most viable approach for converting soluble manganese (Mn^{+2}) to insoluble manganese (Mn^{+4}) oxides for removal (Tobiason et al. 2016).

Fig. 8. Manganese staining inside pipe fixtures

3.10 Methods to Minimize the Existence of Manganese in Water Supply System

Methods to be adopted to minimize the existence of manganese in water supply system are;

1. Selecting an alternative source or blending with other water sources
2. Management of manganese source
3. River catchment management
4. Sequestration
5. Water treatment process

The only feasible option at the DWTP is remove Mn by treatment since the source of the Mn is in geological formation of the Demodara catchment and the Demodara reservoir is the only source which is capable in suppling 15,000 m^3/day to the area. Sequestration is a temporary control measure to limit the aesthetic water quality problems associated with oxidation of manganese (Kohl and Medlar 2006).

Removing manganese through water treatment process comprises two stages. The first stage is to oxidize soluble forms of Mn to insoluble precipitates and the precipitated material will be removed from the water through a solid separation process. The DWTP uses Potassium permanganate ($KMnO_4$) to oxidize manganese. The balanced reaction between soluble Mn and Potassium permanganate is;

$$2KMnO_{4(aq)} + 2H_2O_{(l)} + 3Mn^{+2}_{(aq)} \rightarrow 5MnO_{2(s)} + 4H^{+}_{(aq)} + 2K^{+}_{(aq)}$$

Potassium permanganate is more effective at oxidizing manganese than aeration or chlorination (MWWOM 2009). Maintaining the pH from 7.2 to 7.4 during the water treatment process has shown 85% removal of dissolved manganese from raw water. Graph 1 shows pH variation throughout the water treatment process at DWTP under similar ORP conditions and potassium permanganate dosage ratios (i.e.: amount of permanganate dosing per mg/l of soluble Mn and Fe).

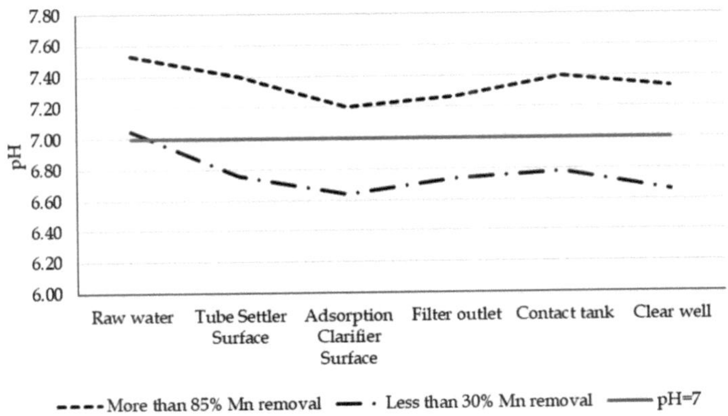

Graph 1. pH variation throughout the water treatment process at DWTP

3.11 Controlling Manganese After Entering the Distribution System

It was noticed that even at Mn levels lower than 0.1 mg/l at Demodara clear well, samples collected from the distribution pipe line contained a considerable amount of Mn. Table 5 shows that the Mn level of samples and distance from the DWTP to particularly wash out.

Table 5. Mn concentrations in distribution lines

Sample no.	Distance from DWTP (m)	Mn level (mg/l)
01	1550 (near Halpe Kovil)	0.08
02	1830 (near culvert 35/1)	0.21
03	2370 (near culvert 35/9)	0.16
04	3540 (near culvert 36/8)	0.02

To mitigate problems caused by Mn accumulation in the distribution systems, periodic flushing of the distribution system is recommended according to a recognized time period. However, minimizing the manganese concentrations in the lowest possible limit before entering the distribution system is the most effective strategy.

4 Conclusions and Recommendations

The total area of the catchment is 59.1 km^2 and terrain is generally mountainous with steep slopes. The dominant land use pattern is tea plantations. During the rainy seasons, due to high drainage density of the catchment, higher sediment yields at the Demodara reservoir is expected.

Water quality analysis has shown physical and bacteriological contamination of the reservoir to be considerably high due to poor catchment management practices as a result of human activities in the catchment. But no considerably stressing chemical contamination has occurred in the catchment except manganese which is a naturally occurring and widely distributed chemical in the soil, water and air. Literature reports emphasized that Mn has no negative health impacts at the levels commonly found in drinking water. However, aesthetic problems caused by Mn are a continuing source of concern for customers and for water utility.

The most common process to treat manganese in drinking water is to convert soluble manganese to insoluble manganese by oxidation and to remove insoluble manganese through filtration or sedimentation. Increasing both the ORP and pH is the most viable approach for this conversion.

The DWTP receives Mn from its raw water by its catchment's geological formation and land use pattern. Soluble Mn is rich in the Charnockitic gneisses, Garnet-sillimanite-biotite gneiss and Walawe gneisses geology layers in the catchment. Also, as a major Mn source, paddy fields near Sri Shailatharama Viharaya, Demodara have been highlighted both in the dry and wet seasons. Therefore, soil erosion at the particular location has to be minimized through appropriate tree plantations, soil retaining structures etc.

The correlation coefficient test result has shown that Mn level in the DWTP raw water, has a moderately positive relationship with rainfall amounts in the upper catchment. This happens due to insoluble Mn becoming soluble Mn with rainfall (pH of the rainfall <7). It then comes to the Demodara reservoir from the surface runoff and, it remains as Mn^{+2} at the bottom levels of the reservoir due to anaerobic conditions. Raw water is pumped to the DWTP from the lower level of the reservoir, and therefore Mn will enter in to the DWTP system.

However, manganese treatment does not completely remove all manganese from the treated water. It will allow some manganese to enter the distribution system, particularly in high rate water treatment plants. According to the Federal-Provincial-Territorial Committee on Drinking Water (2016); even manganese concentrations low as 0.02 mg/l can cause complaints about discoloured water, plumbing fixture and laundry staining, and general dissatisfaction with the water quality due to continued accumulation of Mn in the distribution system from low but constant manganese present in the source or treated water.

Therefore, treating Mn in raw water is very complex and required continues monitoring by plant operators. For smooth functioning of the DWTP, following recommendations could be given for different capacities from this study.

4.1 Recommendations for Catchment Management

- Soil stabilization has to be done in paddy fields near the Sri Shailatharama Viharaya, Demodara since both turbidity, colour and manganese levels of water flowing through the Badulu oya highlighted that contamination from this location is high in both the dry and wet seasons.
- From the Sri Shailatharama Viharaya, Demodara to the Demodara reservoir, the self-aeration capability of the Badulu oya is less due to less elevation variation in the river. Providing additional aeration facilities or pH adjustments in the river will be able to provide added advantages for the DWTP operation. But it requires more studies related to this and a holistic approach in conduct.

4.2 Recommendations for Plant Operators

- Continued monitoring of raw water quality is required to ensure proper dosing of $KMnO_4$. The dosage needs to be carefully controlled so that the $KMnO_4$ residues shall not remain in the water as Mn. $KMnO_4$ feeding rate should be equal to oxidization requirement of both Mn and Fe. Theoretical $KMnO_4$ dosing is 1.92 mg for 1.0 mg of Mn and 0.94 mg for 1.0 mg of Fe.
- Dosing $KMnO_4$ and NaOH to raw water is best at the initial stage to ensure sufficient contact time with $KMnO_4$ and will be benefited by an improved plant efficiency.

- Maintaining pH levels tube settler, adsorption clarifier, filter and contact tank at level 7.2 to 7.4 will ensure more than 85% of Mn removed from water with proper the permanganate dosage.
- Periodic flushing of the distribution lines has to be conducted according to the Mn levels at water even at the allowable levels.
- Mn intake to the WTP has been increased due to the fact the raw water line is fixed at a lower level of the reservoir. If water can be extracted from an upper level of the reservoir, especially during the wet season, the possibility of increased Mn concentration in raw water could be reduced.
- Air bubbling at a pre-sedimentation tank will be able to increase dissolved oxygen in raw water and therefore will increase Mn removal efficiency at the filter.
- To reduce Mn accumulation of the reservoir, frequent flushing out of the sediments trapped in the reservoir has to be done, especially during rainy seasons.

4.3 Recommendations for Packaged Water Treatment Plant Designers and Manufacturers

- Since continued monitoring of manganese level and iron levels should be carried out in order to ensure smooth operation of the plant including inline Mn and Fe measuring meters and accordingly continued adjustment of the $KMnO_4$ feeding system will be an additional value added design.
- It is recommended to increase the hydraulic retention time of packaged treatment units up to 15 min as per recommendations by researchers. It will provide additional contact time with $KMnO_4$.
- If the system can inject $KMnO_4$ and NaOH prior to alum or PAC it will then provide an additional benefit in pH support for Mn removal; since alum or PAC will reduce pH which is a disadvantage in $KMnO_4$ removal. However, PAC decreases pH less than alum.
- Addition of oxide coated media in filters will also effectively remove Mn and Fe.

4.4 Recommendations for Further Studies

- According to SLS 614: 2013, which is the standard for drinking water in Sri Lanka, stipulated Mn level of drinking water is 0.1 mg/l. This value is fairly high compared to aesthetic requirements of drinking water. In addition, Mn can be accumulated in distribution lines and therefore 0.1 mg/l level has to be modified according to the high rate advanced water treatment plant requirements.
- Buffer zone establishment, particularly for drinking water reservoirs, has to be done at a national level.

References

Taylor R (1990) Interpretation of the correlation coefficient: a basic review. J Diagn. Med. Sonogr 6(1):35–39

Silva EIL, Manuweera L (2004) Surface and rainwater chemistry in Sri Lanka-a risk of acidification. Asian J Water Environ Pollut 1(1 & 2):79–86

Kohl PM, Medlar SJ (2006) Occurrence of manganese in drinking water and manganese control. American Water Works Association, Washington, D.C., pp 1–9

Brandhuber P (2013) Legacy of manganese accumulation in water systems. Water Research Foundation

Tobiason JE, Bazilio A, Goodwill J, Mai X, Nguyen C (2016) Manganese removal from drinking water sources. Curr Pollut Rep 2(3):168–177

Minnesota water works operations manual (2009) Fourth Publication, Chapter 14, pp 1–6

Federal-Provincial-Territorial Committee on Drinking Water (2016) Manganese in drinking water. Health Canada

Application of Safety Plans for Rainwater Harvesting Systems in Sri Lanka

K. A. I. L. Rathnayake[1], S. A. C. Upeka[2(✉)], K. A. M. Kularathne[2], P. G. K. B. Ranathunga[2], T. D. Jayasinghe[3], W. B. M. L. I. Weerasekara[3], and S. K. Weragoda[3]

[1] Faculty of Science and Technology, Uva Wellassa University, Badulla, Sri Lanka
[2] China Sri Lanka Research Grant Project, Ministry of City Planning and Water Supply, Katugastota, Sri Lanka
chashinismw@gmail.com
[3] Water Safety Plan Advisory Unit, National Water Supply and Drainage Board, Katugastota, Sri Lanka

Abstract. Water is often considered the most basic need for sustaining life. As the world population increases, the demand for quality drinking water increases. Dug wells and tube wells are considered as the main water source among the rural community in the dry zone. Rainwater harvesting is receiving increased attention in the dry zone as an alternative source of drinking water. While rooftop rainwater harvesting is being promoted, little attention has been paid to the quality of collected rainwater until recently. It becomes nationally important to look for low cost options which could be managed and afforded by the communities themselves. Rain water systems are categorized as individual systems and there are no proper regulations for maintaining the rain water harvesting systems. Therefore, it is vital to develop a proper guideline for rain water harvesting systems. Kebethigollewa was selected as the sample location due to the high number of rain water harvesting systems available. A questionnaire survey and focus group discussions with the beneficiaries were conducted to collect data on different rain water harvesting systems, rain water usage pattern and health concerns due to consumption of rain water. Based on the outcome of the data from observations, questionnaire survey and sample analysis, cost effective and community manageable rain water harvesting safety plan were introduced. Rain water was mainly used for the drinking and cooking purposes and dug well water was used for other activities like bathing and washing. Over hanging vegetation was identified above the roof catchment which may attract animals. Fallen foliage which blocked the gutters were identified in most of the guttering channels. According to the analysis, samples shows slight acidity. Turbidity ranged from 0.22 to 2.80 and TDS varied from 5 to 120 mg/l. All the water samples have TDS, Nitrate, Sulphate and hardness concentration below the prescribed limit.

Keywords: Guideline · Focus group discussions · Rainwater harvesting · Rural community

R. Dissanayake and P. Mendis (Eds.): ICSBE 2018, LNCE 44, pp. 162–170, 2020.
https://doi.org/10.1007/978-981-13-9749-3_15

1 Introduction

Water is often considered the most basic need for sustaining life. As the world population increases, the demand increases for quality drinking water. A large population of the world do not have access to safe drinking water and is suffering from water scarcity problems. Water shortage issues are a growing concern in Sri Lanka as well, especially in the dry zone areas.

Dug wells and tube wells are the main water source among the rural community of the dry zone areas. With the existing questionable hypothesis (excessive hardness, fluoride and heavy metal contamination in groundwater) developed in the dry zone, which has susceptible factors leading towards the major health issue of Chronic Kidney Disease (CKD), the community is looking for alternative sources for drinking water. Rainwater harvesting is receiving increased attention in the dry zone as an alternative source of drinking water. Rainwater from individual household roof catchments is becoming a common practice by the rural community in the dry zone today. While rooftop rainwater harvesting is being promoted, little attention has been paid to the quality of collected rainwater until recently. It becomes nationally important to look for low cost options which could be managed and afforded by the communities themselves.

1.1 Problem Statement

Rooftop rain water harvesting systems can be contaminated at collection, conveyance, and storage system. As rain water is obtained as a source of drinking water, it is important to analyze the quality of the harvested rain water. Rain water systems are categorized as individual systems and there are no proper regulations for maintaining the rain water harvesting systems. Therefore, it is vital to develop a proper guideline for rain water harvesting systems.

1.2 Objectives

- To investigate physio-chemical and bacteriological quality of the harvested rain water.
- To study the water quality deterioration causes and pathways with the time period.
- To assess the health impact and social acceptability of rain water harvesting systems.
- To introduce water safety plan for rain water harvesting systems.

2 Methodology

Kebethigollewa was selected as the sample location due to high number of rain water harvesting systems available. A questionnaire survey and focus group discussions with the beneficiaries were conducted to collect data on different rain water harvesting systems, rain water usage patterns and health concerns due to consumption of rain water. Water samples were collected from each rain water harvesting unit and sample analysis was conducted for physicochemical parameters like turbidity, pH, electrical conductivity, hardness, total dissolved solids, nitrate, phosphate, sulfate, alkalinity and bacteriological parameters including total coliform and e-coli.

The water quality data was compared with the standards stipulated by SLS 614: 2013 for drinking water. Based on the outcomes of the data from observations, questionnaire survey and sample analysis cost effective and community manageable rain water harvesting safety plan was introduced.

3 Results and Discussion

The observed weather condition was dry during the inspection period and temperature ranged between 25 °C–30 °C. Rain water was mainly used for drinking and cooking purposes and dug well water was used for other activities like bathing and washing. Over hanging vegetation was identified above the roof catchment which may attract animals. Fallen foliage which blocked the gutters were identified in most of the guttering channels. Inadequately sealed storage tanks of the rain water harvesting units were also observed.

3.1 Water Quality Analysis

3.1.1 pH Value

The pH is a measure of the intensity of acidity or alkalinity and measures the concentration of hydrogen ions in water. It has no direct adverse effect on health. However, a low value, below 4.0 will produce sour taste and higher value above 8.5 shows alkaline taste. A pH range of 6.5–8.5 is normally acceptable as per guidelines suggested by Sri Lanka Standards for potable water (SLS 614: 2013). The pH values of the samples from Kebithigollewa area are in between 4.04–7.19. According to the study samples collected, the rain water of Kebithigollewa shows slight acidity (Fig. 1).

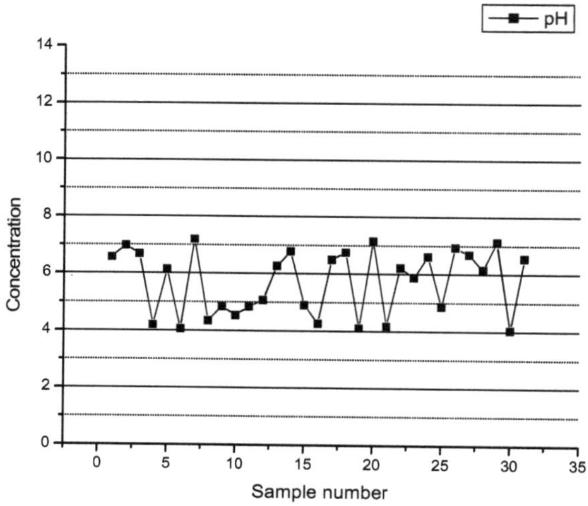

Fig. 1. pH variation

3.1.2 Turbidity

Measurement of Turbidity reflects the transparency in water. It is caused by the substances present in water in suspension. In natural water, it is caused by clay, silt, organic matter and other microscopic organisms. It ranged from 0.22 to 2.80 in the Kebithigollewa area. However, the prescribed limit of Turbidity for drinking water is 2 NTU (SLS 614: 2013). There are 4 water samples with higher turbidity values in the Kebithigollewa area (Fig. 2).

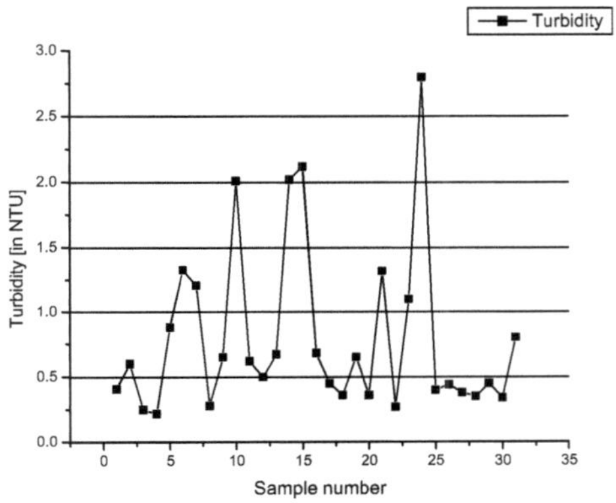

Fig. 2. Turbidity variation

3.1.3 EC

EC refers to the electrical conductivity of the water solution. When the water is pure without any dissolved solutes, it conducts electricity naturally. The amount of electricity that can pass through water is influenced by amount of dissolved ions in the water. According to Sri Lanka Standards value of electrical conductivity of drinking water should be lower than 750 mS/cm (Fig. 3).

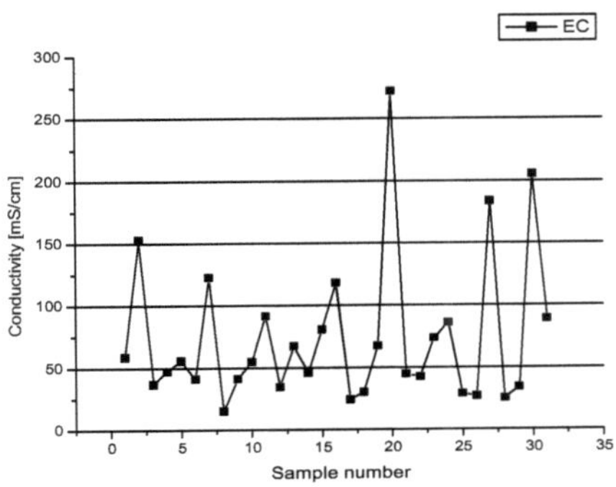

Fig. 3. EC variation

3.1.4 Total Dissolved Solids

Total Dissolved Solids may be considered as salinity indicator for water. The TDS in water is due to the presence of calcium, magnesium, sodium, potassium, bicarbonate, chloride and sulphate ions. In the study area TDS varied from 5 to 120 mg/l. As the prescribed limit of TDS for drinking water is 500 mg/l, all the water samples have TDS concentration below the prescribed limit (Fig. 4).

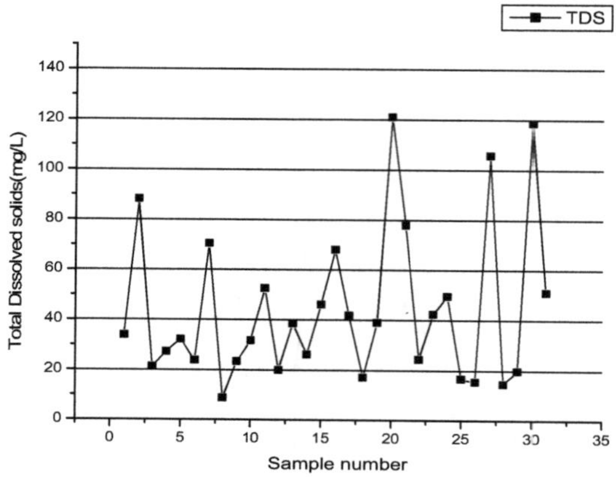

Fig. 4. TDS variation

3.1.5 Hardness

Hardness of water is objectionable from the view point of water use for laundry and domestic purposes since it consumes a large quantity of soap. Based on the present investigation, hardness is low. However, the permissible limit of Hardness for drinking water is 250 mg/l (SLS 614: 2013) (Fig. 5).

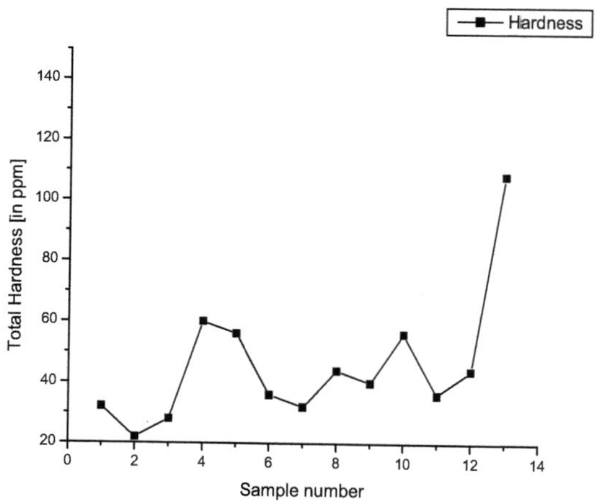

Fig. 5. Hardness variation

3.1.6 Nitrates

Nitrates are an essential source of nitrogen for plants. When nitrogen fertilizers are used to enrich soils, nitrates may be carried by rains. Nitrate levels in drinking water can be an indicator of overall water quality. According to Sri Lanka Standards for drinking water, Nitrates concentration should be less than 50 ppm (Fig. 6).

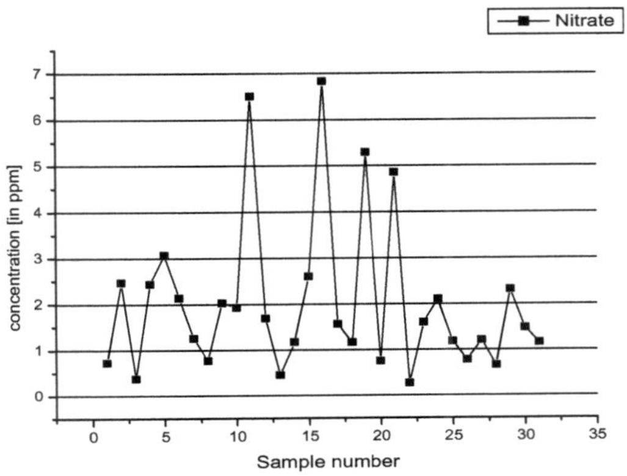

Fig. 6. Nitrate variation

3.1.7 Sulphate

Sulphate can be found in almost all natural water. The origin of most sulphate compounds is the oxidation of sulphate ores, the presence of shales, or industrial wastes. Sulphates is one of the major dissolved components of rain. According to Sri Lanka Standards for drinking water, Sulphate concentration should be less than 250 ppm (Fig. 7).

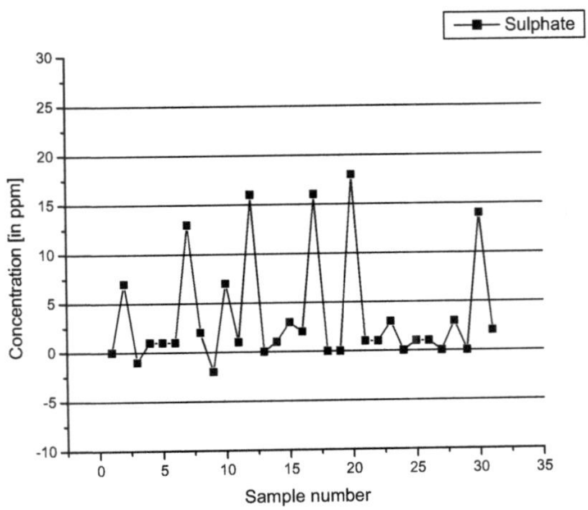

Fig. 7. Sulphate variation

3.2 Bacteriological Analysis

If coliform bacteria present in the water, it is not suitable for drinking purposes. (SLS 614: 2013). These samples are from Kebithigollewa from 5 through 11 are not suitable as potable water.

Location of the sample table	Coliform bacteria present in 100 ml sample at 37 °C	E coli in 100 ml of sample at 44 °C
Kebithigollewa	160	00
Rambewa	102	86
Rambewa	218	00
Rambewa	00	00
Kebithigollewa	00	00
Kebithigollewa	00	00
Kebithigollewa	00	00
Kebithigollewa	00	00
Rambewa	06	00
Rambewa	14	04
Rambewa	00	00

4 Recommendation

The results of the data indicates that all the water quality parameters are in acceptable level for drinking purposes but the considerations on poor maintenance practices of the rain water harvesting systems should be addressed.

The roof catchment and the guttering channels should be cleaned systematically because contamination of the roof and guttering channels may be washed into the storage tanks during the rainfall events and constitute a risk to water quality (When rain falls after a long dry period, or, any rainwater collected may carry with it significant amounts of contamination and debris which have accumulated on the roof and in the gutters. It is therefore recommended that the water running off the roof after the first storms of the season, and preferably for the first 5–10 min afterwards or until it runs clear, should be discarded or used for purposes other than drinking).

The overhanging vegetation near the rain water harvesting systems should be removed as the overhanging vegetation could attract animals that may contaminate the roof catchment with faecal material.

The gutters should be kept clean as the fallen foliage could also block the gutters and filters.

The filter box should be cleaned at least once in 6 months of time as clogged filters may result in contaminating the storage tank due to overflow.

The storage tank should be sealed/covered adequately as lizards and geckos may enter and produce elevated thermotolerant (faecal) coliform counts. (A fine mesh fitted to all openings to the tank will prevent the entry of organic debris).

Stagnant water of the water collection area should be cleaned as it may encourage mosquito breeding and increase the likelihood of contamination entering the storage tank.

An area of 10 m from the tank should be cleaned as the presence of waste material (household, agricultural, animal faeces) may constitute a risk to water quality.

- Activities that may result in airborne contaminants such as spray drifts from agricultural practices should be carried out at a safe distance from the water collection area because it may contaminate the roof catchment.
- The storage tank should be drained, cleaned and disinfected at least annually.

Stored rainwater should be safely stored and handled, and ideally disinfected prior to consumption (e.g. Chlorination, boiling or any other suitable household water treatment option).

The quality of the collected rainwater can also be improved by proper maintenance of the roof and gutters, and careful cleaning at the beginning of every wet season.

It is recommended to conduct an analysis determine the components of asbestos in water.

Use a food grade paint for the roofing.

Guideline to develop WSP for the Rain Water Harvesting System

- Modification of specified rain water collection removable guttering system to control contamination arising due to long dry seasons.
- Modification of specified rain water collection tanks with conical shape bottom for better flushing out performances.
- Educate people to construct at least one part of tile roof Need to facilitate with subsidy from the relevant authority, Establishment of new rules and regulations.
- The existing Asbestos roofs need to protected with food grade paint to minimise possible fibre contaminations.
- Introduce household treatment for drinking rain water: Boiling, Bacteriological filtration, Domestic Chlorination, Blending with existing ground water (10, 20%) for minimising possible ion deficiencies

5 Conclusion

The most significant impact of the rainwater harvesting system in Sri Lanka is the ensured supply of water in the homestead. Rain water was mainly used for drinking and cooking purposes and dug well water was used for other activities like bathing and washing. Over hanging vegetation was identified above the roof catchment which may attract animals. Fallen foliage which blocked the gutters was identified in most of the guttering channels. According to the analysis, samples shows slight acidity. Turbidity ranged from 0.22 to 2.80 and TDS varied from 5 to 120 mg/l. All the water samples have TDS, Nitrate, Sulphate and hardness concentration below the prescribed limit.

Acknowledgments. We express our sincere and honorable gratitude towards all the staff of water safety plan advisory unit - Katugasthota for providing the necessary support and guidance during the period of the research. Our sincere thanks go to the National Water Supply and Drainage Board (Anuradhapura and Kegalle) for being a supportive hand to conduct the study successfully. We would like to thank all the personalities including professionals and all other supportive hands who directed us to succeed in this study.

References

Abu-Zeid MA (1998) Water and sustainable development: the vision for world water, life and the environment1. Water Policy 1(1):9–19

Ariyananda T, Mawatha E (July 1999) Comparative review of drinking water quality from different rainwater harvesting systems in Sri Lanka. In: Proceedings of 9th IRCSA conference on Petrolina, Brazil

Coombes PJ, Kuczera G, Kalma JD, Dunstan HR (2000) Rainwater quality from roofs, tanks and hot water systems at Figtree Place. In: Proceedings of hydro 2000: interactive hydrology, p 1042

Despins C, Farahbakhsh K, Leidl C (2009) Assessment of rainwater quality from rainwater harvesting systems in Ontario, Canada. J Water Supply: Res Technol-Aqua 58(2):117–134

Hapugoda D (1995) Action research study on rainwater harvesting. Community water supply and sanitation project (CWSSP), Colombo, Sri Lanka

Morey A, Dhurve B, Haste V, Wasnik B (2016) Rainwater harvesting system. Int Res J Eng Technol 3(4):2158–2162

Yaziz MI, Gunting H, Sapari N, Ghazali AW (1989) Variations in rainwater quality from roof catchments. Water Res 23(6):761–765

Hardness and Fluoride Removal Efficiency of Groundwater: Use of Physiochemical Methods in Water Purification at Anuradhapura District Dry Zone in Sri Lanka

K. M. S. Kumara[1(✉)], K. A. M. Kularathne[2], P. G. K. B. Ranathunga[2], S. A. C. Upeka[2], D. C. Herath[2], and S. K. Weragoda[2]

[1] Department of Civil Engineering, General Sir John Kotelawala Defense University, Colombo, Sri Lanka
kmskumara@yahoo.com

[2] Water Safety Plan Advisory Unit, National Water Supply and Drainage Board, Katugastota, Sri Lanka

Abstract. The limitation of drinking water critically affects the day-to-day life of most of the community in Sri Lanka as well as in the world. The community in the dry zone of Sri Lanka has especially been facing this difficulty for a long period. It is identified that high concentration of Calcium, Magnesium and Fluoride in groundwater is the main reason for this situation. Even though there are various methods used for water purification, the problem remains the same for rural communities and some urban communities. This study is focused on installing the most suitable and effective water treatment pilot plant and the collection of data to design an industrial water treatment plant. Anuradhapura and Mihinthale were selected as the study area based on the data collected during field visits. A pilot plant with a Reverse Osmosis (RO) membrane was installed and operated for two months of a time period, at the army camp Padulagama while comparing it with an industrial treatment plant at Rajarata University. According to the physical and chemical parameter results it was revealed that 32.43% of the raw water was successfully purified. Hardness and Fluoride concentration could be reduced up to 42.07 mg/l and 0.06 mg/l respectively. The common parameters which were reduced due to the use of RO membrane were also analyzed and found acceptable. Operation lapses and difficulties in treatment plants were identified and rectifications were carried out successfully. Designing of a RO water treatment plant with a pre-treatment unit and maintenance, regeneration and replacements were conducted. Recommendations were made for appropriate discharging methods of the rejected water from the RO membrane.

Keywords: Hardness removal · Fluoride removal · Dry zone · Water purification · RO membrane

R. Dissanayake and P. Mendis (Eds.): ICSBE 2018, LNCE 44, pp. 171–182, 2020.
https://doi.org/10.1007/978-981-13-9749-3_16

1 Introduction

Generally, 70% of the earth's surface area is covered with water, but the availability of fresh water is less than 3% of the total global water. Most of it is in glaciers, ice caps, the atmosphere and deep groundwater reservoirs. Only 13% of the total fresh water is readily available to use for the human consumption. At the same time, water demand of the world becomes double every 21 years due to industrial, agricultural and domestic pollution of pure water. Water is becoming an increasingly precious resource which should be managed judiciously. Therefore, drinking water purification has become an essential requirement in the present world. Few decades ago in Sri Lanka, water at acceptable quality was readily available in many parts of the island. Most of the pure water sources have now been polluted and new solutions for accessibility of water in drinking condition are essential. However, in certain parts of the island such as the Northern and North central province (Dry Zone), drinking water under imposed standards are not available with respect to hardness and the presence of fluoride in the water.

The requirement of suitable drinking water for the community in the dry zone is becoming a critical issue and it has already affected the survival of the community. By now, it is a National problem of the country. The rural community of the dry zone are highly affected due to the use of contaminated drinking water. As a solution to this, there are various kind of drinking water treatment plants installed in several areas and purified water is distributed among the community in the rural areas in the north east. The final outcome of these plants is not guaranteed. Since it has become a fund-raising event, the situation has become worse. None of the owners of these drinking water treatment plants in the private sector and or government institutes has taken any precaution for the discharge of those plants. The discharges might be affecting the environment indirectly.

Therefore, correct identification of the drinking water treatment process very important. It is high time to find possible solution for groundwater water purification according to the ground water condition. This study is focused on the design a proper drinking water purification system considering the raw water quality and the required quantity while optimizing the available drinking water treatment systems.

2 Methodology

2.1 Overview

Primary data was obtained by testing the raw water samples collected in the initial field visit at Anuradhapura. The lab scale water treatment plant including the sand filter unit, granular activated carbon filter unit, ion exchange resin unit, RO membrane unit and UV chamber was installed at the army camp at Pandulagama - Anuradhapura (GR-8.331165, 80.362208) as shown in Fig. 1. For the operation of the lab scale water treatment plant, raw ground water dug from well source in Pandulagama Army Camp was used. Lab Scale reverse osmosis (RO) plant and industrial RO treatment plant were used for the study as shown in Figs. 2 and 3. To identify the performance, practical operating capability and maintenance of the lab scale RO treatment plant was compared with the Industrial RO treatment plant which was installed at the University of Rajarata in Mihinthale.

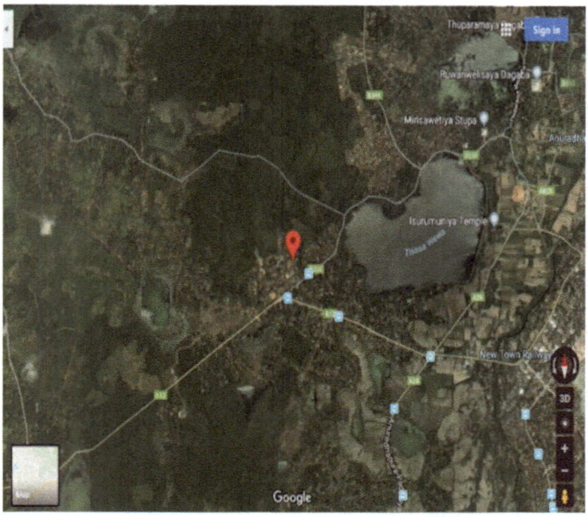

Fig. 1. Pilot plant site at Pandulagama Army camp (GR- 8.331165, 80.362208)

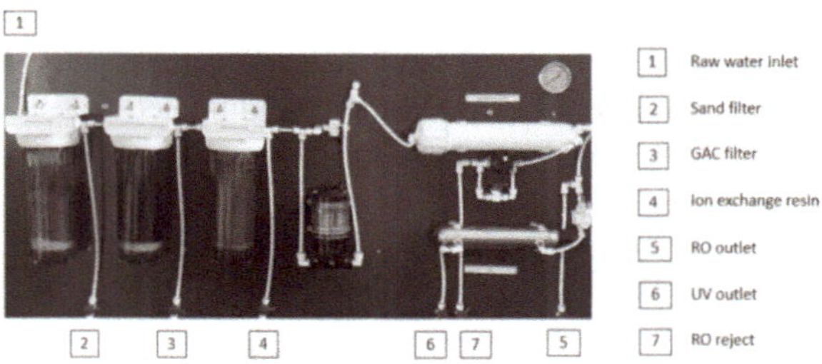

Fig. 2. Lab scale water treatment plant

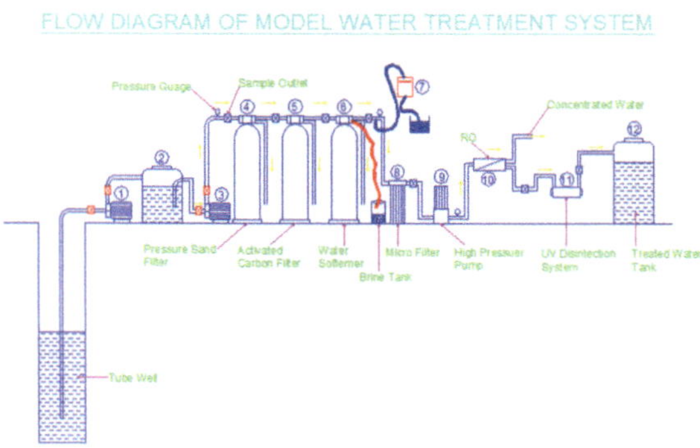

Fig. 3. Lab scale water treatment plant

Sampling of Analysis Water Quality Parameters

Water samples were collected into clean 250-ml bottles for a period of two months from raw water, sand filter, granular activated carbon filter unit, ion exchange resin unit, RO membrane unit, UV chamber and RO rejected of both lab and industrial scale RO water treatment plants. Physical water quality parameters such as pH, conductivity, and turbidity were measured onsite and laboratory experiments were done to measure the chemical water quality parameters such as Total Dissolved Solids (TDS), alkalinity, total hardness, fluoride and Total Organic Carbon (TOC).

2.2 Sand Filter for Turbidity Removal

Results and Analysis

The Lab Scale Water Treatment Plant (RO Plant)
Influent flow rate of the lab scale water treatment plant is 37 l/h
Permeable flow rate of the RO plant is 12 l/h
Purification efficiency = 12 × 100/37 = 32.43%

The Industrial Water Treatment Plant
Influent Flow rate of the Industrial water treatment plant is 566 l/h
Permeable flow rate of the plant is 148 l/h
Efficiency of the Industrial water treatment plant = 26%

Turbidity values of the first 3 days at the lab scale treatment plant are not of the expected manner. Turbidity of the treated water is higher than the raw water turbidity. Increasing of back washing period and having proper supervision, efficiency of sand filter can be increased. However, the raw turbidity and treated water turbidity are at the permissible range (<1 NTU) (Fig. 4).

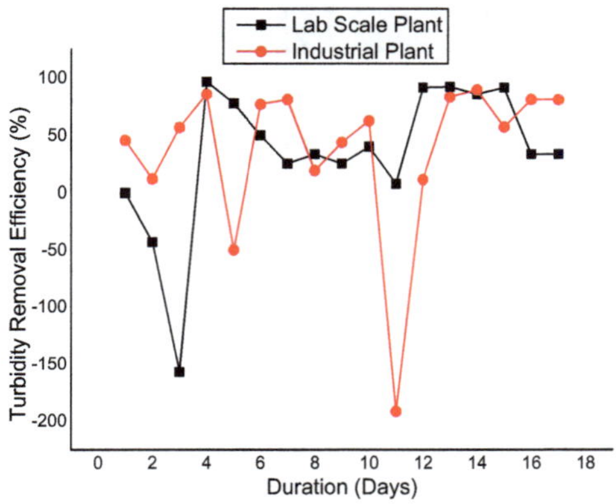

Fig. 4. Comparison of turbidity removal efficiency with the time using the sand filters in lab scale plant and industrial plant

2.3 Granular Activated Carbon (GAC) Filter for TOC Removal

Using the GAC filter, the main expectation is to reduce the TOC of the water. In this study total organic carbon of raw water and treated water was tested for both plants as follows (Fig. 5).

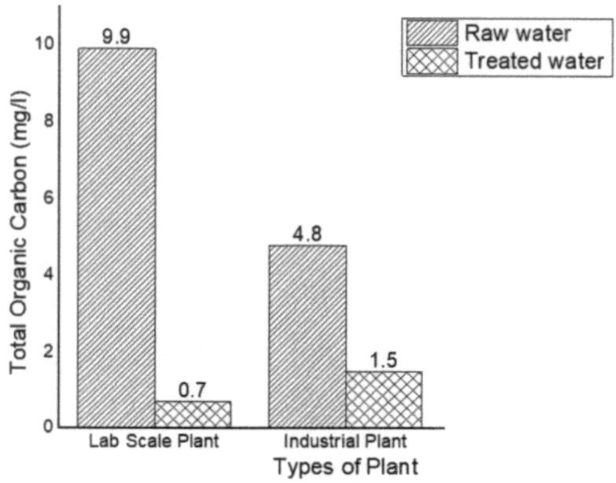

Fig. 5. Comparison of TOC removal in lab scale water treatment plant, industrial water treatment plant

Raw water TOC concentration is higher at the lab Scale plant. TOC reduction capacity is also high at the Lab Scale Plant. Where that is in the permissible range (<2 mg/l).

2.4 Ion Exchange Resin for Hardness Removal

According to the data, the reduction of hardness using ion exchange resin is considerably low. Improvement of that reduction is dependent on the maintenance of the ion exchange unit. Therefore, the regeneration process should be observed properly. There may be a practical problem of analyzing the data at the time the sample is taken, just after the regeneration process so that adjustment can be done for the quantity of resin, time of regeneration and frequency (Figs. 6 and 7).

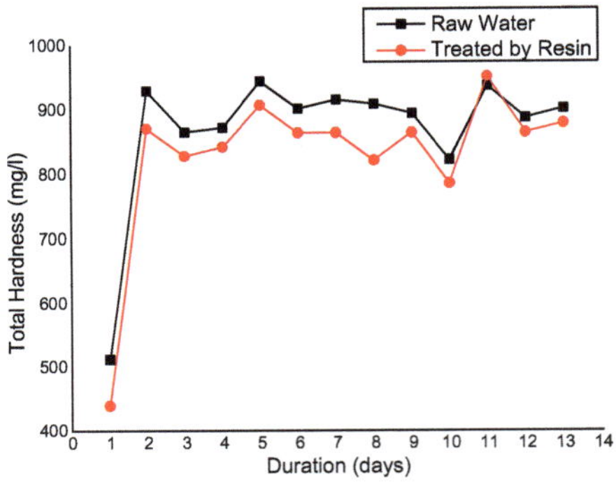

Fig. 6. Hardness of raw water and treated by ion exchange resin in lab scale water treatment plant

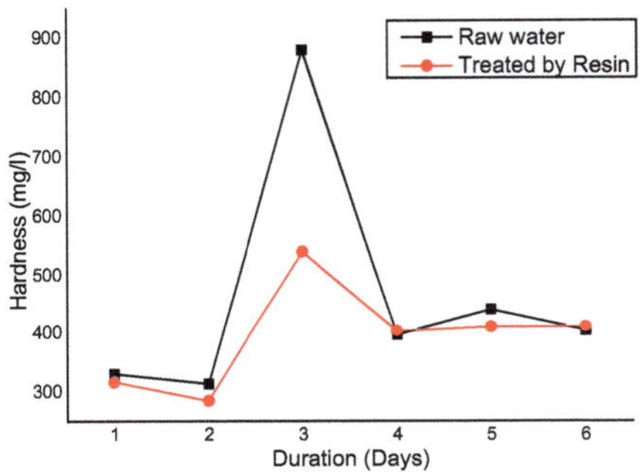

Fig. 7. Hardness of raw water and treated by ion exchange resin in industrial water treatment plant

2.5 RO Membrane

Even though this study mainly focuses on reduction of hardness and fluoride, with the use of an RO membrane, almost all other parameters are reduced considerably as well.

Generally, hardness removal efficiency of lab scale plants and industrial plants are same. But hardness concentration of a lab scale plant is significantly higher than the industrial plant. According to the data, it proves that hardness removal from both treatment plants depends on the raw water condition. Anyway, the removal of hardness using RO membrane is successful. Hardness of treated water is well below the permissible range (Fig. 8).

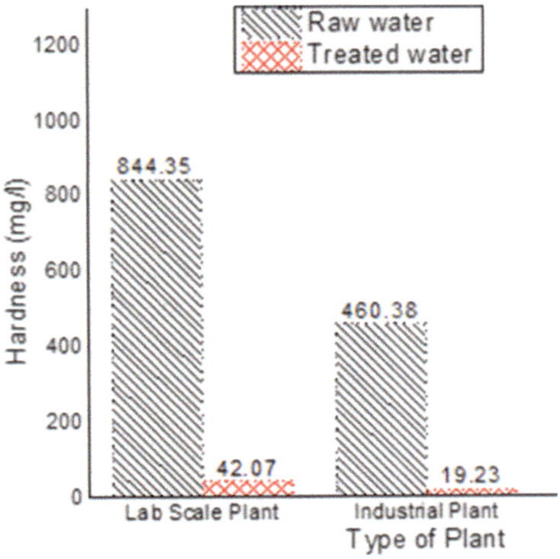

Fig. 8. Comparison of hardness removal capacity with lab scale water treatment plant industrial water treatment plant

2.6 Fluoride Removal

It is observed that the fluoride concentration of raw water at the industrial water treatment plant is already below the permissible range and the lab scale plant is marginal. Considering the fluoride removal efficiency of the treatment plants, the lab Scale plant shows higher efficiency than others. Fluoride concentration of both plants are successfully reduced and it is below of the permissible range (<1 mg/l) (Fig. 9).

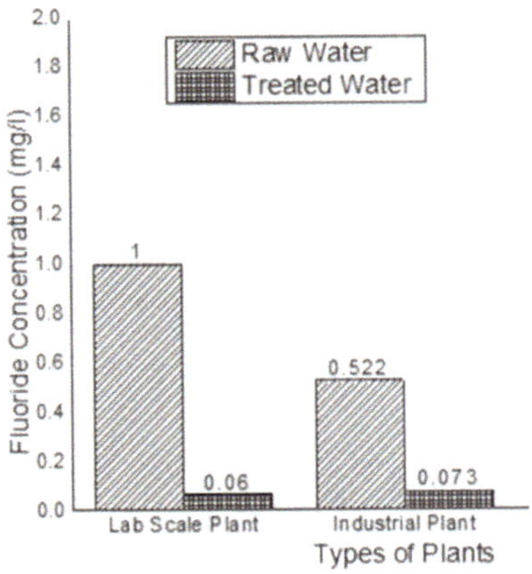

Fig. 9. Fluoride removal capacity in lab scale water treatment plant, industrial water treatment plant

2.7 Conductivity

According to the conductivity graph shown in Fig. 10, the maximum and minimum efficiencies of the lab scale plant for electrical conductivity are 96.6% and 84.5% respectively whereas the maximum and minimum electrical conductivity efficiency of industrial water treatment plants are 96.08% and 93.89% respectively. As per the figure, the efficiencies of conductivity reductions of both the plants varies between 96.2% to 83.9%. This shows the functions of the RO membranes of both plants which are gradually reducing. This capacity may be reduced due to the clogging of membranes. Furthermore, it shows that the efficiencies are varying within a short range. That may depend on the daily deposits of substance on the membrane.

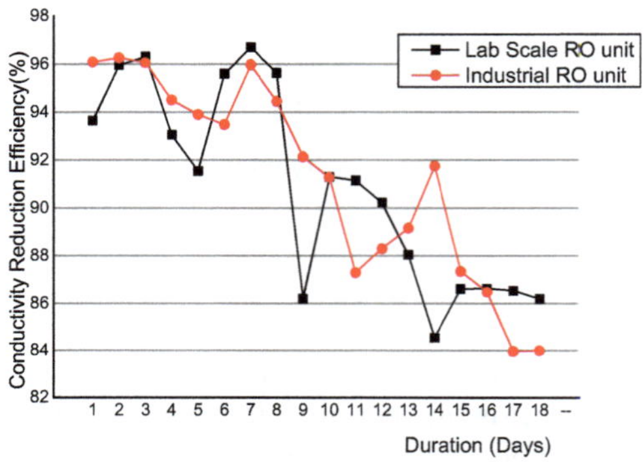

Fig. 10. Electric conductivity reduction efficiency of RO units in lab scale water treatment plants and industrial water treatment plants

2.8 pH Value

Generally, the average pH value of treated water in both the plants are within 6.5 to 8.5 (Figs. 11 and 12).

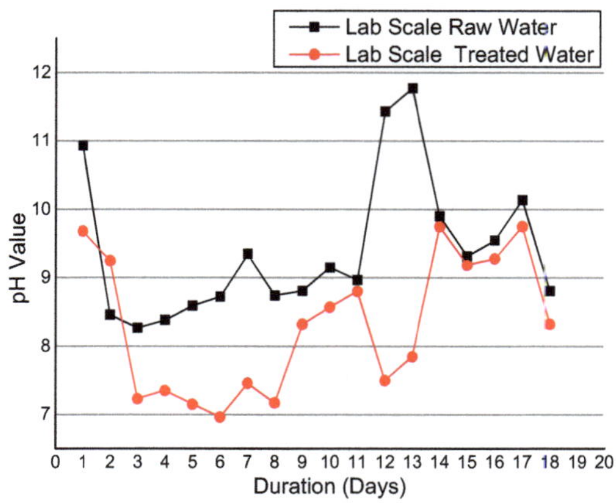

Fig. 11. pH variation of between raw water and purified water in lab scale water treatment plant

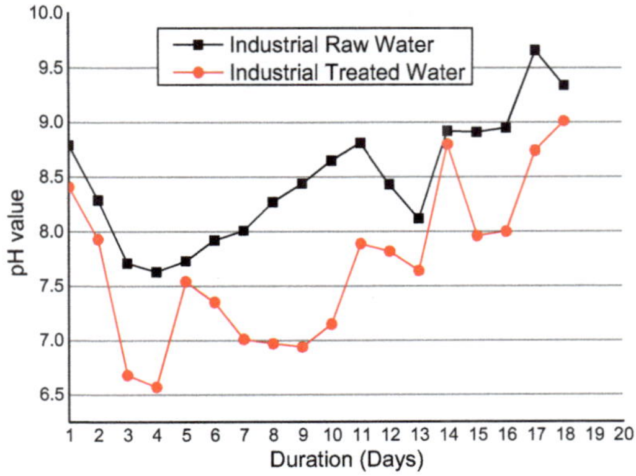

Fig. 12. pH variation of between raw water and purified water in industrial water treatment plant

2.9 Total Dissolve Solids

It is observed that the TDS concentration of raw water at the lab scale treatment plant is fairly higher than the TDS concentration of raw water at the industrial water treatment plant. The TDS concentration of raw water on the 10th day at the industrial treatment plant was unexpectedly higher than others. It may be a sample error or a testing error (Fig. 13).

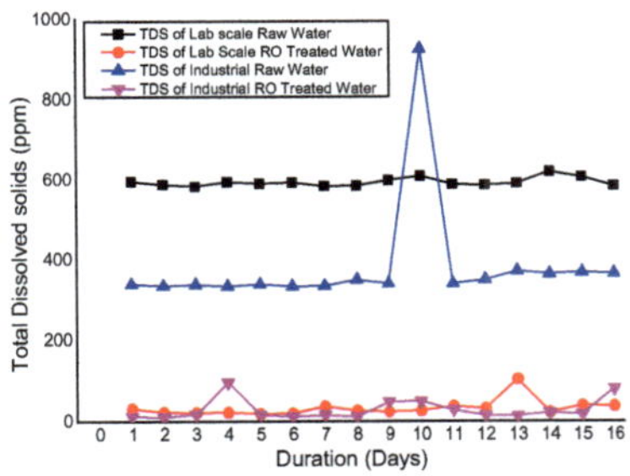

Fig. 13. Total dissolve solids of lab scale plant and industrial plant

2.10 Alkalinity

Alkalinity of the raw water at the Lab Scale plant is considerably higher than the raw water of the Industrial plant. But the alkalinity of treated water at the lab scale plant and the industrial plant has been reduced up to an average value of 6.38 mg/l and 7.43 mg/l respectively. Reduced values of alkalinity are below the permissible range, which is less than 200 mg/l. Alkalinity of raw water at the industrial plant is also less than the permissible value (Fig. 14).

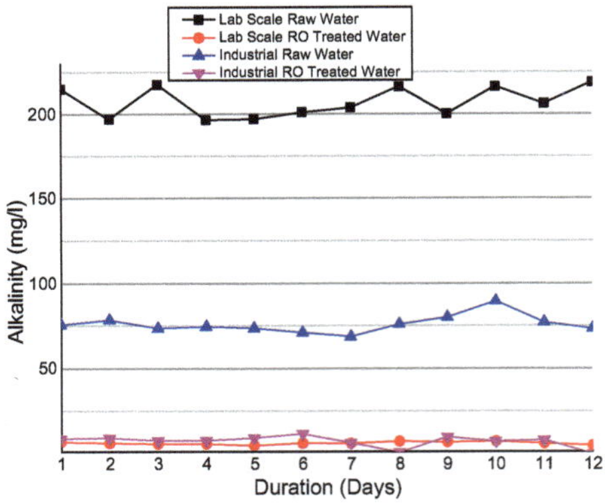

Fig. 14. Variation of alkalinity in raw water and treated water of lab scale water treatment plant and industrial water treatment plant

2.11 Ultra Violet Disinfection (UV)

An ultra violet lamp is used as a disinfection process at the lab scale plant and the industrial plant. Using the UV lamp, bacteria content is reduced up to zero as shown in Table 1.

Table 1. Removal of bacteria using uv chamber in lab scale treatment plant

Type of bacteria (per 100 ml)	Raw water	UV treated water
Coliformbacteria	<1000	0
E coli bacteria	320	0

3 Conclusion

In the dry zone of Sri Lanka, existing traditional systems of water purification are not capable enough to satisfy the requirement. Therefore, the latest technology should be used for water purification. This study was done in a selected area at Anuradhapura and Mihinthale to installing a pilot plant at Anuradhapura. The existing physicochemical parameters of the water in the selected area were examined and those are hardness 844.33 mg/l, fluoride 1 mg/l, alkalinity 207.08 mg/l, conductivity 1186.61 μS/cm, total dissolve solids 592.5 mg/l, total organic carbon 9.9 mg/l, pH 9, turbidity 0.11 mg/l.

Knowing the effects of the high concentration of fluoride and hardness in drinking water, ground water in the selected area of dry zone in Sri Lanka was treated using a RO membrane and pretreatment systems with combination of latest technology in order to select the suitable and effective purification process. The selected purification process of the RO membrane and the pre-treatment with sand filter, carbon filter, ion exchange resin and ultra violet chamber was able to achieve the requirement of reducing hardness and fluoride concentration of the ground water in the selected general area in the dry zone of Sri Lanka. Purification efficiency of hardness and fluoride are 95.17% and 94% respectively and physicochemical parameters of RO treated water are hardness 42.07 mg/l, fluoride 0.06 mg/l, alkalinity 6.38 mg/l, conductivity 108.61 μS/cm, total dissolve solids 32.31 mg/l, total organic carbon 0.7 mg/l, pH 8.5, turbidity 0.057 mg/l.

Considering the above data, collected and analyzed, the purification process of the RO membrane with pretreatment can be selected as the most effective and practical system of water purification for drinking water.

The literature survey and the laboratory experiments have indicated that each of the discussed techniques can remove fluoride and hardness under specified conditions. The fluoride and hardness removal capacity varies according to many site-specific chemical, geographical and economic conditions, so actual applications may vary from the generalizations made. Any particular process, which is suitable at a particular region may not meet the requirements at some other place. Therefore, any technology should be tested using the actual water to be treated before implementation in the field.

Acknowledgments. I would like to express my deepest gratitude to Sir John Kotelawala Defence University, Water safety Plan Advisory Unit, National Water Supply and Drainage Board, Katugastota, National Institute of Fundamental Studies and Faculty of Engineering, University of Peradeniya for giving the facilities to carry out experiments.

References

Kodikara1 DC, Kumara RDHA, Chaminda GT, Kawakami T (2015) Fluoride and hardness removal from groundwater by electrolysis. In: 3rd international symposium on advances in civil and environmental engineering

Degu G, Yigzaw T (2006) Research methodology. Ethiopia public health training initiative, Ethiopia

Dharmagunawardhane HA, Dissanayake CB (1991) Hydrogeochemistry of deeptube wells in the dry zone of Sri Lanka. In: The second international symposium on thegeology of Sri Lanka

Disanayaka C (2014) Water quality and dental health in the dry zone of Sri Lanka. Institute of Fundamental Studies, Hantana Road, Kandy, Sri Lanka: s.n

Dissanayake CB (2007) The fluoride problem in the ground water of Sri. Int J Environ Stud 20

Emamjomeh MM (2006) Electrocoegulation technology as a process for defluoridation in water treatment, s.l.: University of Wollongong thesis collection

Etiksson P (1988) Nanofiltration Extends the Range of membrane filtration. FilmTec Corporation, 7200 Ohms Lane: s.n

Freeman B, Sagle A (nd) Fundamentals of Membranes for Water, s.l.: s.n

Haftenberger M, Viergutz G, Neumeister V, Hetzer G (2001) Total fluoride intake and urinary excretion in German children. Fluoride

Herath KRPK, Ileperuma OA, Dharmagunawardhane HA (2005) Environmental health risk for the chronic renal failure in Sri Lanka. Faculty of Science, University of Peradeniya, Peradeniya, Sri Lanka: s.n

Hell F, Lahnsteiner J, Frischherzb H, Baumgartner G (1998) Experience with full-scale electrodialysis. Elsevier, Burgenland

Jayalath CPG, Miguntanna NS, Perera HAKC (2016) Burnt clay bricks as an alternative filter media for pebble matrix filters (PMF). Eng: J Inst Eng Sri Lanka 49:1–9

Jayasumana C (2013) Possible link of chronic arsenic toxicity with chronic kidney disease of unknown etiology in Sri Lanka. Rajarata University of Sri Lanka, Faculty of Medicine and Allied Sciences, s.n

Junco LI (1972) Renal failure and fluorosis. JAMA 222:783–785 Fluorine and dental health

Priyantha N, Keerthiratne S (1997) Removal of heavy metal ions from textile effluents. Ceylon J Sci: Phys Sci 4(1):77–85

Kothari C (2006) Reseach methodology; methods and techniques. New Age International Publishers, New Delhi

Kumar R (2014) Research Methodology. SAGA, New Delhi

Dapeng LI, Jiuhui QU (2009) The progress of catalytic technologies in water purification: a review. J Environ Sci 21:713–719

Dharma-Wardana MWC, Amarasiri SL (2014) Chronic kidney disease of unknown aetiology and. Environ Geochem Health

Land-Use Planning Model: A Case Study on Transit-Oriented Development (TOD)

G. A. M. A. S. B. Udapitiya[1], R. M. T. Rathnayake[1], and K. K. K. Sylva[2(✉)]

[1] Department of Civil Engineering/Faculty of Engineering, University of Peradeniya, Peradeniya, Sri Lanka
[2] Department of Engineering Management/Faculty of Engineering, University of Peradeniya, Peradeniya, Sri Lanka
kamanisylva@pdn.ac.lk

Abstract. Many communities are shifting their land-use approaches by moving towards cities for better access to transportation and other facilities resulting in densification, particularly around main streets and urban centres. These trends are driving the increased emphasis on better models of city planning for urban centres, urbanization of suburbs of cities and crowding of either side of minor and major highways of Sri Lanka. Transit Oriented Development (TOD), is found to be a concept that has succeeded in reaching better sustainability for both society and the environment closer to the main transit access. The concept has been a success in the reduction of individual vehicle usage and the practice in the usage of the main transit systems which is in lieu of the concept of sustainability. In this milieu, this study was aimed at creating a TOD model for a developing city as a case study to verify the pros and cons of the concept of TOD in a context in Sri Lanka. A qualitative study was done with in-depth interviews to verify the complementary factors of a selected village in Sri Lanka to be converted to a TOD model as a case study. This paper presents the conceptual model on TOD for Meewatura, Sri Lanka.

Keywords: Transit-Oriented development (TOD) · Land-Use planning · Model · Sustainability

1 Introduction

In recent decades, cities have expanded dramatically due to rapid urbanization processes (Motieyan and Mesgar 2017). Consequently, several issues associated with the management of urban built environments such as unplanned urban sprawl, unfair distribution of land uses, and inappropriate utilization of infrastructures, have emerged. As Motieyan and Mesgar (2017) points out a significant rise in the use of private cars over public transit is one of the most conspicuous issues in many cities. According to them, this issue can lead to both environmentally and socially harmful consequences, such as traffic congestion, global warming, climate change, environmental pollution, and socio-economic problems. On the other hand, in most developing countries, the existing public transportation services are unsuccessful in attracting people because

R. Dissanayake and P. Mendis (Eds.): ICSBE 2018, LNCE 44, pp. 183–192, 2020.
https://doi.org/10.1007/978-981-13-9749-3_17

land use characteristics are not considered when planning and designing public transit for easy access of the public. According to Motieyan and Mesgar (2017) to facilitate decision-making processes in urban areas, it is, therefore, necessary to integrate land use and public transportation planning into a comprehensive index.

In order to achieve a favourable environment and a good image, there should be a balanced built environment in a city. As Pathinayaka and Wijesundara (2015) state, "cities are specially planned to enhance the functionality while balancing civic, commercial and recreational activities". With the failure of city planning models such as sprawl development, models such as TOD and Business Improvement Districts (BID) have been practiced in the world to achieve a better sustainability (Knowles 2012; Hesse 2009 and Cervero 2004; Hoyt 2004).

Sri Lanka is in the process of rapid development and population towards cities is increasing making the cities unbearably crowded. This uncoordinated growth results in an unplanned development of infrastructure and simultaneously increasing pollution. Many recent incidents such as flooding in cities and problems with solid waste management are evidence for this uncoordinated and unplanned urbanization. When more small villages add up to the same process of urbanization along major and minor highways in Sri Lanka the problems are aggravated. It is, therefore, necessary to introduce better land-use planning models to Sri Lankan villages on either side of the major and minor highways. Introducing such models would enable 'feed-in' (Sylva 2018) solutions rather than concentrating on end-of-pipe solutions to the problems created by improper planning such as solid waste management and dealing with traffic related problems and pollution.

With the aim of reviewing better practices for community development and related land use planning for small villages on either side of the minor and major highways of Sri Lanka, this research study focuses on developing a model on Land-Use planning based on the concept of TOD. It is expected to uplift the quality of life of the people while maintaining sustainable practices by such a model. Meewathura village, with similar characteristics depicted to the TOD concept, which is located in the Central Province of Sri Lanka, is selected as the case study, for applying TOD concept to research into the strengths and weaknesses of the application of the concept in Sri Lanka.

A conceptual model for Meewathura, Sri Lanka, was developed through a comprehensive literature survey on the concept of TOD taking into consideration the features of the concept. This paper presents the background to the study, the features of the concept of TOD and the development of the conceptual model to Meewathura. The paper aims to disseminate present practices in the sustainable built environment to the research community specifically focused on city planning for residential communities. Since there are no residential city models developed with the concept of TOD in Sri Lanka this model will be an eye-opener for the planning sector or the public authorities to create better models for villages developing into cities on either side of the major and minor highways. Ultimately the society as a whole and the environment too will be benefitted with the application of such sustainable city models in Sri Lanka.

2 City Development Models

Main city development models found in the literature are sprawl, TOD, and BID for applications in different contexts (Knowles 2012; Hesse 2009 and Cervero 2004; Hoyt 2004). According to Knowles (2012), Hesse (2009) and Cervero (2004), sprawl development, which encouraged human population away from urban centers into low-density areas, was highly accepted in the past. Sprawl created monofunctional and usually car-dependent communities, in a process called 'suburbanization' (Cervero and Day 2008). This was the same reason for the failure of the sprawl concept; due to the high dependency on private vehicles causing high expenses and high levels of environmental pollution related to emissions.

With the identification of negative consequences of sprawl development models such as TOD has been chosen in spite of other alternatives because of its broader benefits. TOD is a type of community development that includes a mixture of housing, office and other facilities integrated into a walkable neighbourhood and located within a half-mile of quality public transportation (Cervero 1997). According to Cervero (1997), TOD allows households to live in neighbourhoods where they can walk, bike, or take transit to get around. With transportation, the second highest household expense after housing, living near transit makes it easier for people to forego the high cost of auto ownership. This not only benefits new and existing residents but also businesses, transit agencies, local governments, merchants and developers are made to get positive effects by mixing of these communities (Lu et al. 2018). BID, on the other hand, is a sustainable concept introduced to improve business districts ranging from few blocks to many miles of streetscape (Hoyt 2004). BID is mainly targeted on developing business cities to fulfill the requirements of such a community with tax collection from the community to run the city and a Public-Private Partnership to maintain the cities.

2.1 The Concept of TOD

TOD has emerged as one possible solution for sustainable urban transport, with supporters claiming that it can help to reshape the quality and form of urban growth towards enhanced accessibility and mobility, pedestrian friendliness, increased sustainability and potentially a higher degree of human interaction (Curtis et al. 2009). By definition (Knowles 2012; Hesse 2009 and Cervero 2004), a TOD community (or neighbourhood) means creating a pedestrian-friendly built environment, in which the centrally located rail or bus stations are surrounded by relatively high-density and mix-used commercial and residential developments. According to Dittmar and Ohland (2004) and Cervero and Day (2008), TOD communities are generally located within a radius of about 600 m from a transit stop, as this is considered to be an appropriate scale for pedestrians. A TOD city refers to the built urban forms that are composed of numerous TOD communities built on the transit lines (Dittmar and Ohland 2004; Cervero and Day 2008).

According to Cervero (1997), TOD is a type of community development that includes a mixture of housing, office and other facilities integrated into a walkable neighbourhood and located within a half-mile of quality public transportation. Research by Hasibuana et al. (2013) found that the majority TOD is a concept of managing urban growth in transit corridor which has characteristics of mixed land use, compact, walking distance, and development focused around public transit area. Further research (Shohani et al. 2016) introduce TOD as a process of reconnecting transportation and land use and can be considered as a potential solution for revamping suburban neighbourhoods by developing a new alternative transit system conforming to the principles of TOD. They also suggest that TOD is a viable tool for achieving a sustainable future. They too highlighted that transit can enable a city to use its market forces to increase densities near stations where most services are situated, thereby creating more efficient sub-centers and reducing urban sprawl, rediscovering ways in which planning and design can better incorporate less automobile-dependent land use. Thus, at its core, TOD is a process of reconnecting transportation and land use.

According to Shohani et al. (2016), on the basis of the extent of the area served, TOD is identified as either Neighbourhood TODs or Regional TODs. Focusing on Neighbourhood TODs the emphasis is on residential use with locally-oriented shopping in facilities that are sized to serve the population living in close proximity. Calthorpe (n.d.) has identified TOD as a mixed-use community within an average one-fourth mile walking distance of a transit stops and core commercial area. Calthorpe (n. d) further added more specifics to the definition as the design, configuration, and mix of uses emphasize a pedestrian-oriented environment and reinforce the use of public transportation; TOD is a mix residential, retail, office, open space, and public uses within comfortable walking distance, making it convenient for residents and employees to travel by transit, bicycle or foot, as well as by car.

Zeigler (2009) showed the practical implications of TOD as a metropolitan sustainable development governing framework for growth management in the twenty-first century is essential for a sustainable future. This includes Higher density urban centers and a change in public attitude away from "not in my backyard" thinking.

2.2 Application of TOD

TOD allows efficient land use, energy and resources efficiency and helps to conserve open space. Shohani et al. (2016) identified other benefits for the community of TOD application as encouraging walking, a cleaner environment, and minimising traffic. Referring to rail accessibility, Papa and Bertolini (n.d.) carried out a study to investigate how TOD structure affects accessibility in cities and found that the accessibility increases in urban development clustering along rail corridors and stations. Lund et al. (2004) studied the travel behaviour in California TODs. According to them, TOD residents are more likely to use rail transit "if there is less of a time benefit for traveling via highways (compared to transit), if there is good pedestrian connectivity at the destination if they are allowed flexible work hours, and if they have limited vehicle availability." But they found that TOD residents are less likely to use transit if the trip involved multiple stops or "trip chaining", if there is good job accessibility via highways, if they can park for free at their workplace, and if their employer helps to pay vehicle expenses such as tolls, fuel, etc. Yap and Goh (2017) listed the consideration factors to determine the potential requirements of TOD. Amenities related to jobs and family for baby boomers, time-saving, location and house should be located in relation to working place, school, etc.

Sung and Oh (2011) stated TOD development will depend upon the courage of the political leaders and the effectiveness of the targeted policies. These will include reshaping monotonous land use, comprehensive planning with a greater focus upon the transport network, restricting the total freedom for citizens to own and drive their cars, and focusing the power of real estate companies to meet wider sustainable transport investments. According to them, these areas of policy will attract opposition and in some cases go against the interest of filling the government coffers. According to Sung and Oh (2011), it is expected that Chinese mayors with vision, courage and endurance are likely to go this extra mile; in applying TOD in big Chinese cities it can only be hoped that many of them indeed have these qualities.

According to Massachusetts Bay transportation authority & Massachusetts department of transportation (2017), TOD means more than locating individual buildings or projects near transit. As the department explains, it is a pattern of land use that ties development to transit in a synergistic way at the station level, the corridor level, and even the regional level. Every place is unique in a particular degree and the department has identified four foundational principles for successful TOD's.

- *Density and a mix of uses:* the first foundational principle of TOD is the concentration of dense, mixed-use development within walking distance of stations. Density places more homes, jobs, stores, offices, and recreational activities within walking distance of transit which could relatively change with place and context.
- *Equitable development:* affordable and workforce housing is identified to be offsetting gentrifying pressure on community residents and businesses since TOD can connect housing with jobs, reducing a household's combined cost of housing and commuting. As employment becomes more decentralized, jobs located near transit remain accessible to households that cannot afford to own one automobile per worker
- *A great public realm*: Every transit rider is a pedestrian, and successful TOD is pedestrian-friendly. This means that station areas have safe, interconnected walking environments with sidewalks, plazas, lighting, way-finding, and amenities.
- *A TOD approach to parking*: "right-size" overall parking supply in TOD settings encourages shared parking, and ensure that parking is designed so as not to dominate the streetscape or permanently block development on sites most convenient to the station.

Hasibuana et al. (2013) have proposed ecological footprint, carbon emission, and green open space carrying capacity as sustainable urban development indicators to measure the success of TODs.

3 Development of TOD Model City for Meewathura

In the development of the conceptual model for Meewathura in relation to the concept of TOD, first, a criteria comparison was done. The criteria comparison for Meewathura through the research study is presented in Table 1. Then relevant areas to be developed as residential and business, employment opportunities for residents and wildlife habitats and parks were identified from the satellite map of Meewathura obtained from Grama Niladari Boundary Map of Sri Lanka Survey Department. The Satellite Map of Meewathura is presented in Fig. 1. The model was verified with a quantitative study.

Table 1. Criteria comparison for Meewathura TOD model

Features of the concept of TOD to be taken as indicators for development	Features of Meewathura comparable to TOD
Walkable neighbourhood and located within a half-mile of quality public transportation (Cervero 1997)	Meewathura is a small village of an area of around 80.5 ha and a walkable distance of a quarter to half a kilometer
Pedestrian-friendly built environment, in which the centrally located rail or bus stations are surrounded by relatively high-density and mix-used commercial and residential developments (Knowles 2012; Hesse 2009 and Cervero 2004) Communities are generally located within a radius of about 600 m from a transit stop (Dittmar and Ohland 2004; Cervero and Day 2008)	The A5 road running through Meewathura and is currently being improved as a major connection between Colombo and Nuwaraeliya. Main bus stop of Penideniya and railway station of Peradeniya are located in the visinity of the village at a walking distance of around a quarter to half a kilometer
Mixed-use community within an average one-fourth mile walking distance of a transit stops and core commercial area Calthorpe (n.d.)	The community in Meewatura has both residential and business mixed environment which could be improved further with the development of the A5 road
Sustainable with more efficient use of land, energy and resources, helps conserve open space, encourages walking, cleaner environment, Shohani et al. (2016)	There are areas which are yet not subjected to urbanisation which could be developed as park-lands and wild-life habitats allowing connectivity along the river flowing adjacent to the village and along the railway track where foot-paths could be located

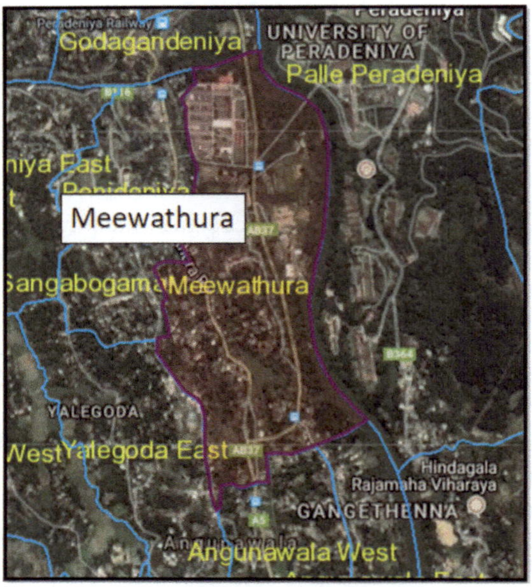

Fig. 1. The satellite map of Meewathura

Connectivity for the wildlife habitats was identified for the sustainability of the city either from ground level or other as appropriate owing to the indicators suggested by Hasibuana et al. (2013) in order to reduce the ecological footprint and have more green open space carrying capacity for sustainability. Some connectivity paths had to be proposed above ground level or below ground level. Footpaths were identified along the rail-track and major highways which could also be used as jogging tracks or cycling lanes for residents. These footpaths were connected to railway stations and major bus stops.

The principles identified by Massachusetts Bay transportation authority & Massachusetts department of transportation (2017) were used as the main criteria in selecting spaces for proposed development which are presented in Table 2.

Table 2. Principles used in the development of TOD for Meewathura

Principle	Description
Density and mis of uses	• Mixed-use development • Homes, jobs, stores, offices, and recreational activities within walking distance of transit
Equitable development	• Affordable and workforce housing • Connect housing with jobs • Decentralized employment opportunities • Jobs near transit accessible to households that cannot afford to own one automobile per worker
A greater public realm	• Pedestrian-friendly • Safe, interconnected walking environments with sidewalks, plazas, lighting, way-finding, and amenities
A TOD approach to parking	• "Right-size" overall parking supply • Shared parking not to dominate the streetscape or permanently block development on sites most convenient to the station

A proposal on a two-dimensional model was developed by identifying essential features of TOD and trying to achieve these with the minimum disturbance to existing features in Meewathura. A mixed housing, commercial and parking were proposed with employment opportunities identified in the commercial and nearby university premises for people who are unskilled and households that cannot afford to own one automobile per worker. Transit access for other workers to move away from the city for employment was identified within walking distance to main mass transit systems. Safe walking paths preventing the illegal crossing of rail tracks were identified on either side of the rail track with safe connectivity. The TOD model for Meewathura is presented in Fig. 2.

Legend	
Notation	Description
	Foot path/ Jogging track
	Railway line
	Residential and business areas with parking
	Recreational areas and wild- life habitats
	Ground level, above or below ground level connectivity for natural habitats
	Employment opportunities with parking
	Overhead footpath
	Bus Stops and Railway Stations

Fig. 2. The TOD model for Meewathura

4 Conclusions

The city model developed for Meewathura using the concept of TOD is expected to be a sustainable model in terms of economy, society and environment. The economic factors of the society are expected to be developed by providing employment opportunities to the city dwellers in the vicinity and allowing transit access to employment opportunities to others who are travelling out of the village. The society will benefit from a cleaner and nature-friendly environment which allows public recreational areas and walkways. The environment will benefit from the regeneration of natural habitats and environmentally friendly living styles of people. The green connectivity proposed will enhance the sustainability of biodiversity.

A mix-use development of residential and commercial activities will enable the community to reduce transit requirements and this would reduce environmental impacts related to settlements. In general, the sustainability of the city will be improved in the application of a model built on the concept of TOD. The expected results are to be verified after the application of the model.

References

Calthorpe P The next American metropolis: ecology, community, and the American dream. Princeton Architectural Press, New York

Cervero R (1997) Transit-oriented development in the United States: experiences, challenges, and prospects, vol 102. Transportation research board, Washington

Cervero R, Day J (2008) Suburbanization and transit-oriented development in China. Transp Policy 15(5):315–323

Curtis C, Scheurer J (2017) Performance measures for public transport accessibility: learning from international practice. J Trans Land Use 10(1):1–26

Dittmar H, Ohland G (2004) The new transit town. Best practices in transit-oriented development. Island Press, London

Grama niladari boundary map of Sri Lanka. Survey department, Sri Lanka. https://www.survey.gov.lk/gn_updating/. Accessed 1 May 2018

Hasibuana HS, Soemardia TP, Koestoerb R, Moersidik S (2013) The role of transit oriented development in constructing urban environment sustainability, the case of Jabodetabek, no. 4. Post Graduate of Environmental Science, University of Indonesia

Hesse M (2009) Transit oriented development: making it happen. J Trans Geogr. 17(6):510. https://doi.org/10.1016/j.jtrangeo.2009.07.007

Hoyt LM (2005) Do business improvement district organizations make a difference? Crime in and around commercial areas in Philadelphia

Knowles RD (2012) Transit oriented development in Copenhagen, Denmark: from the Finger Plan to Ørestad. J Trans Geogr 22:251–261. https://doi.org/10.1016/j.jtrangeo.2012.01.009

Lu Y, Gou Z, Xiao Y, Sakar C, Zacharias J (2018) Do transit-oriented developments (TODs) and established urban neighborhoods have similar walking levels in Hong Kon'. Int J Environ Res Public Health 15:555

Lund HM, Cervero PR, Willson RW (2004) Travel characteristics of transit-oriented development in California. https://www.bart.gov/sites/default/files/docs/Travel_of_TOD.pdf

Massachusetts Bay Transportation Authority & Massachusetts Department of Transportation (2017) TOD Policies and Guidelines

Motieyan H, Mesgar MS (2017) Towards sustainable urban planning, through transit-oriented development (a case study: Tehran). Int J Geo-Inf 6:402

Papa E, Bertolini L (n.d.) Accessibility impacts of TOD experiences in European metropolitan areas. https://biblio.ugent.be/publication/5789325/file/5789360. Accessed 05 May 2018

Pathinayaka AD, Wijesundara J (2015) Imaging cities; social perception and physical composition; case study of Pettah, Colombo. J Cities People Places 1(1):47–60

Sung H, Oh J (2011) Transit-oriented development in a high-density city: identifying its association with transit ridership in Seoul. J. Science Direct Cities. 28:70–82

Shohani AV, Thomas M, Rao KVK (2016) World conference on transport research, masters student. Civil Engineering Department, Bombay, Mumbai 400 076, India

Sylva KKK (2018) A circular economic model for a sustainable city in South Asia. In: Leal Filho W, Rogers J, Iyer-Raniga U (eds) Sustainable development research in the Asia-Pacific region. World Sustainability Series. Springer, Cham, pp 345–359

Yap JBH, Goh SV (2017) Determining the potential and requirements or transit oriented development (TOD): the case of Malaysia. Property Manag 35(4):394–413. https://doi.org/10.1108/PM-06-2016-0030

Zeigler EH (2009) The case for megapolitan growth management in the 21st century: regional urban planning and sustainable development in the united states. Int J Law Built Environ 1(2)

Effective Sizing of Natural Systems for Grey Water Treatment Using Locally Derived Parameters

S. Karunaratne(✉)

Department of Civil Engineering, Faculty of Engineering,
Sri Lanka Institute of Information Technology, Malabe, Sri Lanka
shiromi.k@sliit.lk

Abstract. When properly managed, grey water can be a valuable resource, which agricultural and horticultural practitioners as well as home gardeners can benefit from using to provide an excellent nutrient sources for vegetation. Natural treatment systems such as constructed wetlands emphasizing on sustainability and wastewater reuse typically for agricultural irrigation/home gardening can be considered as an ideal alternative to close the loop in terms of nutrient capture and water conservation. This paper presents a part of an analysis done to estimate the locally derived treatment wetland parameters based on BOD_5 removal mechanisms of a pilot scale subsurface horizontal flow constructed wetland (SSHF CW) treating greywater. A pilot scale SSHF CW with a surface area of 8.64 m^2 and a substrate depth (with 20 mm aggregates) of 0.7 m was monitored. This is a part of a long term study carried out to test the viability of SSHFCW for treating grey water for agricultural reuse. The wastewater flow measurements were taken from the inlet and the outlet of the constructed wetland while monitoring the hydraulic (HRT) and organic loading rates. Water samples from four sampling points were collected on a long term basis. Equations based on a kinetic model of a first-order piston flow reactor was used to calculate the relationship among the design parameters appropriate for local conditions. The HRT varied from 3.3 to 4.7 d for maximum and minimum inflow loading rates observed. Using the data from the current study, a resizing exercise for a subsurface horizontal flow constructed wetland was carried out. Accordingly, a household with four people, discharging 720 L/d greywater with an average influent BOD_5 of 300 mg/L could be treated up a BOD_5 concentration of 25 mg/L using an area of 3.60 m^2 of SSHF CW, implying only 0.975 m^2 of per capita land usage.

Keywords: Hydraulic retention time · Inflow rate · Per capita land usage · Subsurface horizontal flow · Treatment efficiency

1 Introduction

According to the Sri Lanka Standard code of practise for the design and construction of septic tanks and associated effluent disposal systems SLS 745-Part 2, generally black water comprises of only 25% of combined (all) waste water. However, this small

R. Dissanayake and P. Mendis (Eds.): ICSBE 2018, LNCE 44, pp. 193–201, 2020.
https://doi.org/10.1007/978-981-13-9749-3_18

portion of black water contains 90% of Nitrogen, 60% of COD and 50% of P. On the other hand, 75% of all wastewater contains relatively a small amount of pollutants. Furthermore, the nitrogen found in grey water can be filtered out and used by plants readily. However, grey water, if left untreated for a few days, will behave like sewage, becoming malodorous becoming anaerobic. Also, grey water decomposes much faster than black water.

When properly managed, grey water can be a valuable resource, which agricultural and horticultural practitioners as well as home gardeners can benefit from providing an excellent nutrient source for vegetation.

Grey water irrigation has long been practiced in areas where water is in short supply. The engineering of these systems is still a relatively young technology; but it is one making rapid progress. As these systems utilize the nutrient (potential pollutant) content in the effluent, they constitute a real solution to the treatment/management of grey water.

A largely undocumented practice which occurs in the country (Sri Lanka) in most semi-urban and rural areas is the separation of black water from grey water. Usually the toilet waste goes to a septic system while all grey water is discharged mostly untreated directly into the environment, especially in rural areas or soakage pits in urban and semi urban areas. This pre-existing culture of source separation has the potential for the development of sustainable wastewater treatment systems based on the principles of natural treatment or ecological sanitation. Natural treatment systems such as constructed wetlands emphasizing on sustainability and wastewater reuse typically for agricultural irrigation/home gardening can be considered as an ideal alternative to close the loop in terms of nutrient capture and water conservation.

The demand for treated water is reaching a level which might not be satisfied with the available resources even in places where there is no physical lack of it. In such situations, the necessity of effective and efficient wastewater treatment and promotion of water reuse would be highly productive. Hence, the use of highly favoured treatment systems such as constructed wetlands which are associated with natural removal mechanisms of pollutant removal could be successful applications, especially in developing countries since these systems combine low-cost, low-maintenance, simple and reliable operations, and high removal efficiency (Economopoulou and Tsihrintzis 2003).

In general, Subsurface Flow (SF) constructed wetlands can treat a variety of wastewaters including domestic (black water and greywater) and municipal wastewater, industrial wastewater, agricultural and urban storm water runoff, combined sewer overflow and landfill leachate. Amongst most popular types of constructed wetlands, the Subsurface Horizontal Flow Constructed Wetlands (SHFCW) are considerably efficient in the removal of pollutants and eliminate some of the common problems such as bad odour, growth of mosquitoes, etc. Relatively, less research has been done so far on the use of constructed wetlands for greywater treatment in tropical areas and the design parameters still remain as black boxes due to the complexity of treatment mechanisms (Economopoulou and Tsihrintzis 2003). In Sri Lanka too, constructed wetlands have not yet been identified as a very efficient and cost effective wastewater treatment method especially due to the notion of the high land demand, which is prohibitive in urban and peri-urban areas. The main reason for this fact could be the insufficient research work on constructed wetlands, specifically under the local environmental conditions of the country. The design information available in the temperate region of the world cannot be translated to Sri Lanka for optimization of pollutant removal mechanisms and design criterion of constructed wetlands due to the varying climatic conditions. Therefore, it is an essential requirement to estimate both design and operational parameters of the constructed wetland under the specific climatic and environmental conditions in the country.

Few researches have been carried out focussing on parameter estimation and optimization of SF constructed wetland systems in Europe and United States. Economopoulou and Tsihrintzis (2003) have presented a simple criterion including the guidelines and mathematical models for the preliminary sizing of subsurface horizontal flow constructed wetland systems by considering the conditions of definite seasonal variations. But, this research work still has some limitations to apply the methodology in a tropical climate with the values given for certain parameters. However, the methodology could provide relevant guidelines for the estimation of parameters in a tropical climate and determining removal efficiencies of existing systems.

Wynn and Liehr (2001a, b) developed a mechanistic compartmental simulation to model and predict seasonal trends in the removal efficiencies of HFCWs. The model consists of six sub-models representing the carbon cycle, the nitrogen cycle, an oxygen balance, autotrophic and heterotrophic bacteria growth, and a water budget. Therefore, this could make a close relationship with the project under concern.

Vymazal (2005) has provided a very important discussion on subsurface horizontal flow and hybrid constructed wetlands including the pollutant removal mechanisms and their efficiencies with experimental results. Since the experimental data and analyses are based on the temperate continental climate conditions, it would not be possible to make a direct relationship with the local conditions.

Werner and Kadlec (2000) and Chazarenc et al. (2003) modelled wetland residence time distributions in temperate climes. Garcia et al. (2004) and Maloszewski et al. (2006) used tracer experiments to determine hydraulic parameters with different filter materials.

In Sri Lanka, the acceptance of constructed wetlands as an efficient and cost effective wastewater treatment method is not yet established. This may be especially due to the scarcity of the land due to high population density as well as the insufficient research work on constructed wetlands, specifically under the local conditions. Therefore, it is an essential requirement to estimate both design and operational parameters of the constructed wetland under the specific climatic and environmental conditions in the country (Vymazal 2011).

Treatment systems based on natural degradation processes, such as stabilization ponds and constructed wetlands, are particularly suited for domestic wastewater treatment and they require little or no energy, are relatively simple to operate, and show reliable treatment performance.

This paper presents a part of an analysis done to estimate the locally derived treatment wetland parameters based on BOD_5 removal mechanisms of a pilot scale subsurface horizontal flow constructed wetland (SSHF CW) treating greywater. This is a part of a long term study carried out to test the viability of SSHFCW for treating grey water.

2 Methodology

2.1 Pilot Scale Subsurface Flow Constructed Wetland

The subsurface flow constructed wetland system consisted of two settling tanks with a grease trap for the pre-treatment (Fig. 1).

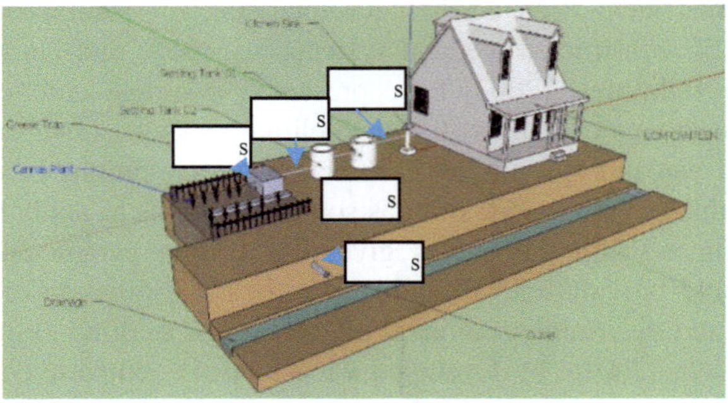

Fig. 1. Configuration of pilot scale subsurface flow constructed wetland (SSFCW) indicating the sampling points (SP 1-4).

The pre-treated water was introduced into the wetland cell through a 50 mm perforated pipe and both inlet and outlet partitions were filled with 40 mm aggregates. The water passes through a 20 mm aggregate bed with baffles inside the wetland cell. Wetland cell was partitioned width-wise to extend the flow path creating a long and narrow treatment cell. Wetland cell maintained a bed slope of 1% with the cannas as the vegetation.

Table 1 shows the characteristics of the pilot scale subsurface flow constructed wetland (fed with grey water discharged from a canteen) used for the study.

Table 1. Characteristics of the pilot scale subsurface flow constructed wetland

Parameter	Value
Φ - porosity	0.435
d - substrate depth	0.70 m
L - length	4.00 m
W - width	2.16 m
V_e - effective volume	2.643 m^3
A - surface area	8.64 m^2

2.2 Flow Measurement, Sample Collection and Analysis

The wastewater flow measurements were taken from the inlet and the outlet of the constructed wetland while monitoring the loading rate. Water samples from four sampling points (at the sink outlet, after primary settling, at inlet to the CW and outlet of the CW) were collected on 1–2 week time intervals and analyzed for pH, Temperature, Turbidity, Conductivity, Biological Oxygen Demand (BOD_5), Chemical Oxygen Demand (COD), Nitrate Nitrogen, Nitrite Nitrogen, Total Phosphorous, Total Coliforms and Suspended Solids. However the present analysis concentrates mostly on the BOD_5 measurement data.

The efficiency of the concentration abatement of the pollutants were calculated using Eq. 1.

$$\text{Percentage abatement } (\%) = (Co - Ce)/Co \quad (1)$$

Similar to Andreo-Martínez et al. (2016), equations derived by Reed et al. (1995) based on a kinetic model of a first-order piston flow reactor proposed was used to calculate the relationship among the design parameters appropriate for local conditions (Eq. 2). By using Eq. 2, the theoretical wetland surface areas were calculated for the measured data.

$$A_s = [Q(\ln C_0 - \ln C_e)]/K_t.d.n \quad (2)$$

Where, As-Surface wetland area (m^2), Co-inlet BOD_5 concentration, Ce- Outlet BOD_5 concentration, Q-inflow rate (m^3/d), K_t temperature dependent reaction coefficient (as given in Eq. 3), d- treatment wetland depth in m, n- porosity of the substrate media.

$$K_t = (1.104).(1.06)^{(T_o - 20)} \quad (3)$$

Where, K_t rate is constant corresponding water temperature in wetland (d^{-1}), T_o is the water temperature in the treatment wetland, which was taken as 25 °C.

Hydraulic retention time (HRT) for corresponding observed data was calculated by using Eq. 4.

$$HRT = (As .d.n)/Q \quad (4)$$

Previous studies (Andreo-Martínez et al. 2016; Karunaratne et al. 2011; Udukumburage et al. 2013) have shown that, BOD_5 can be considered as the governing parameter for the surface area optimisation of SSHF CW design when treated for reuse in irrigation as P and N removal is not an objective. P and N could be retained as plant nutrient.

Accordingly, using the observed BOD_5 data, relationships between the optimum wetland surface area, inflow rates and hydraulic retention time were calculated using Eq. 2 and trend lines were established using power functions to make predictions.

3 Results and Discussion

The ratio of BOD_5/COD is related to the biodegradability of the organic matter present in the water. In the present study, the ratio BOD_5/COD for influent wastewaters had an average of 0.74 ± 0.08 in raw wastewater, a characteristic value for wastewater of high biodegradability. The average BOD_5 of the influent was 892 ± 18 mg/L, a relatively higher value than the usual domestic grey water in Sri Lanka. This higher value is mainly due to the food varieties, mostly Sri Lankan style menus served in the staff canteen. Average PO43-, NO3-and BOD of the influent greywater over the study period was 1.79, 42.62 and 892.19 mg/L respectively.

Figure 2 shows the variation of treatment efficiencies with varying loading rates. Inflow rate ranged from 0.56 m^3/day to 1.55 m^3/day.

At all loading rates, treatment efficiencies of BOD_5 did not fall below 65%. At the minimum loading rate, all parameters measured displayed efficiencies ranging from 62–88%, removal of P showing the minimum while BOD_5 showing the highest. Same trend was observed generally at all loading rates. Even at the highest loading rate of 1.55 m^3/d, N removal efficiency of more than 50% was recorded.

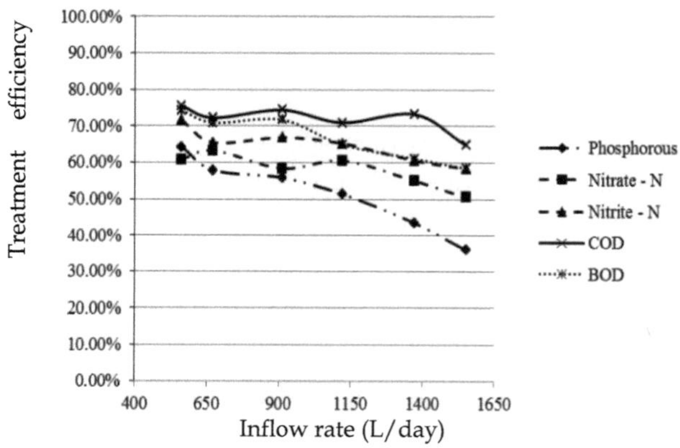

Fig. 2. Variation of treatment efficiency of pollutants with varying inflow rate of the subsurface horizontal flow constructed wetland

The optimum retention time that would yield the minimum required treatment level was calculated using the measured data and the temperature dependent reaction rate constant for BOD_5 was calculated based on the Eqs. 2 and 3. The HRT varied from 3.3 to 4.7 d for maximum and minimum inflow loading rates observed. This is comparatively lesser than the HRT reported from temperate region but reasonable when considering the higher temperature observed in Sri Lanka (Andreo-Martínez et al. 2016).

Generally, sizing of wetlands treating greywater is performed based on the COD or BOD_5 removal efficiencies. Targeting higher removal efficiencies of N and P through constructed wetlands demands higher land requirements, hence, making such targets impractical. It was observed that very few models have been developed to simulate pollutant removal processes in SFCWs and only a fewer number of models of these are applicable for SSHFCWs (Langergraber 2008). In fact, many of those models are applicable for a particular climate or a clime where the model has been developed. Such models should be modified before applying for different conditions. Modification of an existing model can be rather difficult and complex than developing a new model in relation with the local conditions. Modelling of pollutant removal mechanisms of a SSHFCW in a particular climate and sizing of the wetland systems would be a challenge because, the procedure should consist of the formulation of removal processes, the estimation of parameters in the model and the sizing of the system.

It is known that efficiency of wastewater treatment depends on the influent water quality, climate conditions as well as substrate type and plants used. The reaction rate constants yielded in this study were calculated under high loading rates (i.e. three times the original designed). Therefore, influent water quality had a higher variability over the prolonged study period. Hence, it was important to calculate the reaction rate constants under varied loading rates and compare the same for better accuracy. K_t calculated at 25 °C was 1.5 d^{-1}.

Figure 3 shows the relationship between the inflow BOD loading (in g) of greywater and the load reduction per unit surface area (kg m^2).

Figure 4 shows the relationship between the wetland surface area and the hydraulic retention time calculated for the SSHF CW in consideration. Equations 2 and 3 were used to calculate the HRT and wetland surface areas based on the observed data. As mentioned earlier, the estimations of hydraulic retention time for specific treatment levels was based on the BOD_5 treatment efficiency only. Figures 3 and 4 can be used with reasonable accuracy to size treatment wetland cells to treat greywater discharged from households. The relevant estimation of flowrates can be based on SLS 745 (Part 2) standards.

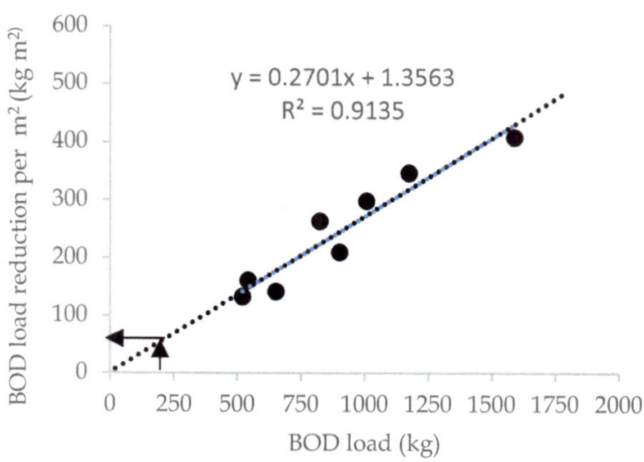

Fig. 3. The relationship between the inflow BOD loading (in kgs) of wastewater and the load reduction per unit surface area (kg m^2).

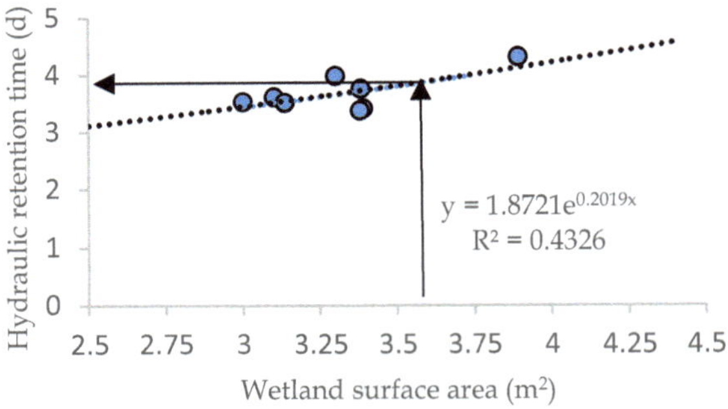

Fig. 4. Relation between the wetland surface area (A_s) and the HRT.

Using the outcomes from the current study, a resizing exercise of SSHFCW was carried out as shown below;

A household with 4 people was considered. As per the SLS standards 745 (part 2), greywater discharge of 180 L/p/d accounts for 720 L/d of grey water from the household. An average influent BOD_5 of 300 mg/L (according to literature it can vary from 200–500) and effluent concentration of 25 mg/L was considered. Hence, BOD load will be 216 g/d (720 * 300/1000). A more stringent discharge standard was considered due to the undiluted discharge conditions prevailing at domestic environments for irrigation purposes.

Using Fig. 3, BOD_5 load reduction per m^2 corresponding to 216 g of BOD_5 load can be estimated to be 59.68 BOD/m^2. As such, the surface area required is 3.6 m^2 (216/59.68). Using Fig. 4, HRT corresponding to 3.6 m^2 of wetland surface area is estimated as 3.9 days. According to the present calculation, average strength grey water, usually discharged from a typical Sri Lankan household can be treated using an area of 0.975 m^2/person surface area of a SSHF CW.

If a ratio L/W of 3:1 is assumed (Andreo-Martínez et al. 2016), L and W can be calculated as, W = 1.10 m while L = 3.30 m, respectively. The plan dimensions can even be readjusted to maintain the total surface area depending on the space availability of the location. The depth of the substrate is 0.7 m. This can be considered as a size which can be conveniently accommodated even in an urban setup and mimic a flower trough if plants such as Cannas are used.

4 Conclusion

An analysis was done to estimate the locally derived treatment wetland parameters based on BOD_5 removal mechanisms of a pilot scale subsurface horizontal flow constructed wetland (SSHF CW) treating greywater. The treatment wetland resizing exercise for a subsurface horizontal flow constructed wetland revealed that for a household with 4 people discharging 720 L/d greywater with an average influent BOD_5 of 300 mg/L could be treated up a BOD_5 concentration of 25 mg/L using an area of 3.6 m^2, implying only 0.975 m^2 of per capita land usage.

Acknowledgements. The author would like to acknowledge the contributions by Udukumburage RS, Ariyasena MADN, Rupasinghe GGMBMB, Wijesiri, B. S., Jayasooriya, V. M. for the field work as well as laboratory analysis done in relation to this work.

References

Andreo-Martínez P, García-Martínez N, Almela L (2016) Domestic wastewater depuration using a horizontal subsurface flow constructed wetland and theoretical surface optimization: a case study under dry Mediterranean climate. Water 8:434. https://doi.org/10.3390/w8100434

Chazarenc F, Merlin G, Gonthier Y (2003) Hydrodynamics of horizontal subsurface flow constructed wetlands. Ecol Eng 21:165–173

Economopoulo MA, Tsihrintzis VA (2003) Design methodology and area sensitivity analysis of horizontal subsurface flow constructed wetlands. Water Resour Manage 17:147–162

García J, Chiva J, Aguirre P, Alvarez E, Sierra JP, Mujeriego R (2004) Hydraulic behaviour of horizontal subsurface flow constructed wetlands with different aspect ratio and granular medium size. Ecol Eng 23:177

Langergraber G (2008) Modeling of processes in subsurface flow constructed wetlands—a review. Vadoze Zone J 7(2):830–842

Karunaratne S, Wijesiri BS, Jayasooriya VM (2011) Estimation of reaction rate constants of pollution removal for subsurface flow constructed wetlands treating grey water. In: Proceedings of IESL-SSMS joint international symposium, SSMS 2011, 14th - 16th September 2011, Colombo, Sri Lanka, pp 289–293

Małoszewski P, Wachniew P, Czupryński P (2006) Study of hydraulic parameters in heterogeneous gravel beds: constructed wetland in Nowa Słupia (Poland). J Hydrol 331:630–642

Reed SC, Crites RW, Middlebrooks EJ (1995) Natural systems for waste management and treatment, 2nd edn. McGraw-Hill, New York, p 433

Sri Lanka standard code of practise for the design and construction of septic tanks and associated effluent disposal systems SLS 745-Part 2

Udukumburage RS, Ariyasena MADN, Rupasinghe GGMBMB, Karunarathne S (2013) Effect of hydraulic loading rate on treatment efficiency of a subsurface horizontal flow constructed wetland. In: Proceedings of Sri Lanka water convention 2013, 20–21 March 2013, Colombo, Sri Lanka

Vymazal J (2005) Horizontal sub-surface flow and hybrid constructed wetlands systems for wastewater treatment. Ecol Eng 25(5):478–490. https://doi.org/10.1016/j.ecoleng.2005.07.010

Vymazal J (2011) Constructed wetlands for wastewater treatment. Environ Sci Technol 45:61–69

Werner TM, Kadlec RH (2000) Wetland residence time modeling. Ecol Eng 15:77–90

Wynn TM, Liehr SK (2001a) Development of a constructed subsurface flow wetland simulation model. Ecol Eng 16:519–536

Wynn TM, Liehr SK (2001b) Development of a constructed sub-surface flow wetland simulation model. Elsevier Science, Amsterdam, pp 1–18

Sustainability via ERP and BIM Integration

F. S. T. Hewavitharana(✉) and A. A. D. A. J. Perera

Department of Civil Engineering, University of Moratuwa, Katubedda, Sri Lanka
sachindrahewavitharana@gmail.com

Abstract. Sustainability is a paramount factor in the present construction industry which enhances the quality of life, thus allowing people to live in a healthy environment and improves social, economic and environmental conditions for present and future generations. Therefore, lots of construction companies have attempted to use different kinds of software for the development of sustainable structures. Two of those major software are Building Information Modelling system (BIM) and Enterprise Resource Planning system (ERP). This review seeks to identify how ERP and BIM addresses sustainability at the current level, a way of integrating BIM and ERP and how the integration of these two leads to sustainability of the projects according to the current literature. It is identified that there is a significant impact on sustainability by integrating ERP and BIM together than implementing them separately.

Keywords: Sustainability · ERP · BIM · Integration

1 Introduction

The term sustainable development can be described as enhancing the quality of life, allowing people to live in a healthy environment and improving social, economic and environmental conditions for present and future generations [11]. More than ever, the improvement of these indicators is drawing the attention of the construction industry, which is a globally emerging sector and a highly active industry in both developed and developing countries [6]. Because of this rapid development in the construction industry, lots of sustainable issues are created. Some of them are problems related to designs, over usage of energy and resources, problems related to the integration of building systems, non-adherence to ethical standards, negative environmental impacts, and excess utilization of money [1]. Thus, to overcome these negative impacts, sustainability practices should be carried out during project design, construction, usage, maintenance, and demolition periods respectively [20].

In the current construction practice, although BIM is developing a sustainable design in the design phase of the project there is a lack of maintaining the sustainability aspects throughout the operation. To continue that sustainable practice throughout the project lifecycle, ERP will act as a bridge which connects sustainable design data with the operation of activities.

In the design stage, BIM can assist in performing complex building performance analyses to ensure an optimized sustainable building design due to its representation of the building as an integrated database of coordinated information [12, 17]. Further,

R. Dissanayake and P. Mendis (Eds.): ICSBE 2018, LNCE 44, pp. 202–210, 2020.
https://doi.org/10.1007/978-981-13-9749-3_19

BIM can manipulate energy performance data such as power consumption, temperature, CO2 emission and environmental performance data such as indoor air quality [20]. But the issue is, BIM is not capable of conducting operations sustainably in the construction phase.

However, in the construction stage, ERP can integrate and streamline the business processes, reduce costs and improve the quality which leads to economic improvements, social equity & justice and environmental protection [5, 13]. Further, as each ERP module provides the standard business processes with an integrated set of functions, it facilitates the organization to improve and optimize current business processes. But the issue is, ERP is not capable of designing sustainable drawings in the design stage.

Hence, it is clear that there is a gap in operationalization of business functions at the design stage and the construction stage of the construction industry. The bridging of this two software would be the best solution to fill this gap.

However, the integration of these two software is challengeable as they are not compatible with each other with the functional structure. Although the integration of different architectural features in BIM and ERP are difficult, there are significant benefits which can be achieved through this integration [3, 4]. Therefore, this paper aims to contribute to the body of knowledge by reviewing existing literature on this specific topic. The objective of this review is multifold, first, to identify and summarize the contribution of ERP and BIM for the sustainability aspects in the current construction industry. Second, to identify the most effective way of integrating ERP and BIM. Third, to identify how BIM and ERP integration can support the sustainable development of the construction industry.

2 Methodology

Literature review can be named as a well-established method for accumulating existing knowledge within a domain of interest [21]. As the methodology for this article, a systematic review approach is expected to be applied. This approach is characterized by adopting explicit procedures and conditions which minimize bias. To find relevant articles for this review, various keywords searches have been conducted in Science Direct. In Table 1 an overview of the included articles in this research is presented. The period of publishing was set to the years from 2000–2017 in order to narrow down the search. A combination of different keywords such as "ERP and Sustainability", "BIM and Sustainability", "ERP and BIM integration", "Sustainability with ERP and BIM integration" were used to find relevant articles to choose the topic. However, few articles were identified relating to the targeted topic in specific. Therefore, the topics were broadened, and article's abstracts were carefully read to check their relevance.

Table 1. Overview of the articles

Author(s)	Title	Year	Journal
Hande Aladaga, Gökhan Demirdögenb, Zeynep Isıkc	Building Information Modelling (BIM) use in Turkish construction industry	2016	Procedia Engineering
Ibrahim Motawa, Kate Carterb	Sustainable BIM-based Evaluation of Buildings	2012	Procedia social and behavioural sciences
Donya Mehran	Exploring the Adoption of BIM in the UAE construction industry for AEC firms	2016	Procedia Engineering
Eng.Parisa Esmaeili, Moakher and Prof. Dr. S. S. Pimplikar	Building Information Modelling (BIM) and Sustainability – Using Design Technology in Energy Efficient Modelling	2012	*IOSR Journal of Mechanical and Civil Engineering*
Saumyendu Ghosh, Sam Negahban, Young Hoon Kwak, and Miroslaw J. Skibniewski	Impact of Sustainability on Integration and Interoperability between BIM and ERP–A Governance Framework	2008	Ph.D. Thesis
Yen-Ray CHENab and H. Ping TSERNGb	An Integrated Methodology for Construction BIM & ERP by Using UML Tool	2017	(ISARC 2017)
Mohamed Abdalla, Mohamed Badreldin Lotfy	Sustainability of Enterprise Resource Planning (ERP) Benefits Post-implementation: An Individual User Perspective	2015	Ph.D. Thesis

3 Finding and Discussion

BIM and Sustainability

Previously, traditional computer-aided design (CAD) tools are used in modelling the construction designs [2]. Further, energy simulation packages such as Energy plus, Ecotect and IEs virtual environment were used to consider design features such as Thermal insulation, Climatic response glazing, Shading, Solar gain, Solar penetration, Airtightness, Natural ventilation, and Mechanical ventilation [17].

Today these multiple operations are replaced by one system called BIM which improves the performance through the whole life cycle of building [1]. Based on these understandings, BIM can be used to minimize energy consumption, reduce water consumption, reduce carbon footprint, select sustainable materials, reduce waste, conserve the natural environment and safeguard human health and wellbeing [8, 12]. For this type of sustainability assessment, it is essential to incorporate the green building assessment criteria into BIM [11]. At the present condition, BIM is mostly utilized in the design phase of the construction project, and it is a well-recognized tool in sustainable building design [1]. However, at the current level, BIM is unable to help with functionalities of the project at the operational stage. Therefore, there should be separate software to carry out those operations. In that case, ERP would be the best solution to integrate with the BIM.

When considering the evolution of BIM, it is useful to see the prediction which has been made at the initial stage. Now, most of those predictions have become real. According to the UK records, the government construction strategy published in May 2011 states that Government will require fully collaborative 3D BIM (with all project and asset information, documentation and data being electronic) as a minimum by 2016". Also, in the reports published by McGraw-Hill in 2010 more than one-third of western European construction industry participants (36%) reported that they started to use BIM technologies by the year 2010 [1]. Thus, it is clear the BIM has gained a significant place in the construction industry.

Despite all the advantages, BIM implementation faces some difficulties. Some of them are ignorant about the added value of the BIM in the industry, lack of Standards/Legislations, non-coherence with organization culture, high initial investment, technological deficiencies of the stakes holders [1]. Mainly, these issues are due to a lack of BIM standards, lack of BIM awareness and resistance to change [4]. Thus, if these problems are sorted out by an organization, BIM will be an ideal solution to address sustainable issues at the designing stage.

ERP and Sustainability

ERP integrates the functions of an organization and provides a set of flexible best practices to improve and optimize current business processes through re-engineering the current business processes [13]. It helps with cost reduction, integration of resources, sharing of efficient information, flexibility, agility and improved performance [5]. Thus, it impacts the individual, workgroup, management and the whole organization in a positive manner. Further, when considering the reduction of using hard copies and the documentation level due to the ERP implementation and the efficiency and productivity obtained through ERP implementation, it is very clear that ERP can handle all three pillars of sustainability. (People, Environment, Profit).

On the other hand, in the current construction industry, ERP is upgraded as S-ERP (Sustainable ERP) with sustainable features [9]. The concept of a sustainable ERP system can be named as a holistic and integrative information system that is driven by sustainability concerns that cover all aspects [10]. However, introduction of S-ERP system is in the primary stage as per the current situation [5]. The indicators embedded in S-ERP system under the economic category are financial health, performance, quality, cost, and flexibility. Under the social category, internal human resources and under environmental category air resources, land resources, mineral and material resources, energy resources, waste impact and environmental commitments are considered [5].

Therefore, it is clear that the ERP system is capable of addressing sustainability parameters when conducting the project. But the problem is, it is not capable of manipulating sustainable design for the project. Thus, BIM should also be incorporated in the design phase with ERP.

However, despite the benefits, ERP implementation shows some issues and challenges. The implementation can be complex and expensive due to the re-engineering. Further, it needs organizational change which needs considerable time and top management commitment towards the implementation. Also, infrastructure may be a barrier for the ERP implementation [5]. If these challenges are successfully achieved then ERP would be the best tool to address sustainability concepts in the operation phase of the construction projects.

Integration of BIM and ERP

BIM is a concept that uses digitalized and parametric form design to build an integrated module [7]. On the other hand, ERP is a concept which integrates all operations in to one place [7]. Thus, at the design stage, BIM can create sustainable design while in the operational phase ERP is capable of carrying out the operation properly [7]. Therefore, the combination of this BIM and ERP will provide the best practices for the construction organizations.

The integration of BIM and ERP should happen in three ways. Namely, Knowledge integration, Data integration and Component integration. Knowledge integration is the task of identifying how new and prior knowledge interacts while incorporating new information into a knowledge base. Data integration involves the combining of data from these two separate systems and provides a unified view of the data. Component integration means the identification of the way of integrating hardware and software according to the requirement.

Subsequently, customization should be involved in the integration of BIM and ERP. According to the literature described by Chen and Tserng, combining BIM into ERP is much easier than combining ERP into BIM [7]. This is because customization of ERP can be easily incorporated into the BIM design data rather than incorporating ERP concepts into BIM. This ERP customization can include changing application parameters, (Sustainability parameters), adjusting workflow models (Data flow from BIM to ERP) and choosing which modules will integrate [7].

The integration between BIM and various components of the ERP systems such as estimates, controls and schedules can result in flexibility and availability of real-time information required to execute complex and huge projects while maintaining integrity between operations and back office automation systems. Thus, in order to establish a cohesive time-sensitive information flow throughout their operation, many construction organizations that utilize both BIM and ERP would help the two systems to control project management tasks [9].

Sustainability via BIM-ERP Integration

It can be assumed that ERP and BIM integration will be a novel experience for construction companies. While BIM allows constructing a building virtually inside a computer, ERP will help in understanding how much it will cost and how to account for it [14]. A BIM model includes every element along with geometric and parametric information, provides an accurate quantity of materials and takes the guesswork out of cost estimation. It will generate a bill of quantities (BOQ) automatically and produce accurate material quantities [20]. On the other hand, ERP takes it from there and enables the management of inventory, invoices and financial reporting efficiently with lower cost and lower resources [15, 16]. Benefits of BIM-ERP integration will be further explained in the sections below of the study.

The environment issues related to construction practices can be evaluated through leadership in Energy and Environmental Design (LEED) which was introduced by

United States Green Building Council (USGBC). To fulfill those requirements, documentation will be required in energy modeling water use, daylighting, and sun shading. For that, proper documentation should prevail. Correct data flow will ensure that through ERP with related to the information generated by BIM.

Sustainability of the design can be incorporated into BIM and ERP integration. The sustainability issues related to actual construction like carbon emissions, heating/ cooling loads, ventilation/air flow, the data could be compared related to the project easily by using the project BIM model [17].

The integration of building systems like rainwater collection, usage of renewable energy, utilization of sustainable materials, and energy modeling can be related easily with the projects based on the schedules. That comparison of schedules can be easily handled by ERP. If there is a variation in the schedule, BIM allows flexible modifications [17]. Findings of the literature review can be summarized as below (Figs. 1 and 2).

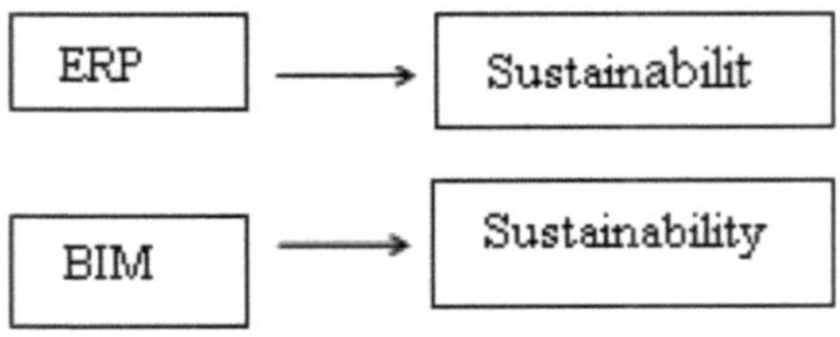

Fig. 1. Current situation of BIM and ERP on sustainability

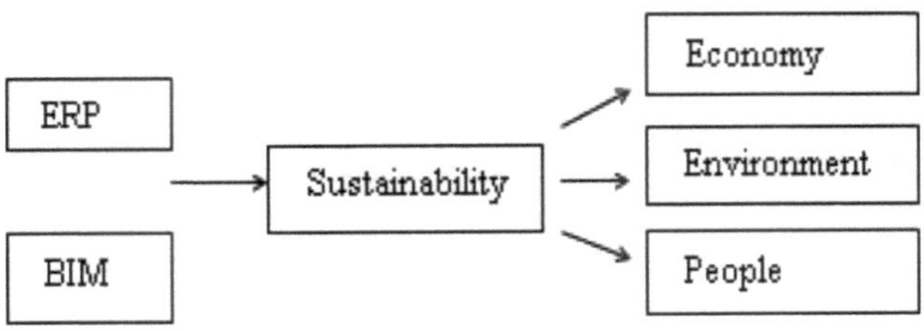

Fig. 2. Future situation of combined BIM and ERP on sustainability

The rapid development of the construction industry creates a lot of sustainable issues, now more than ever. Some of the main issues are; overutilization of energy and resources, negative environmental impacts, excess use of money, non-adherence to standards [1]. Thus, new technologies are concentrated on solving these issues. BIM and ERP are two such systems discussed throughout this paper.

The fundamental low of true sustainability is the integration of all building systems within themselves as well as with the external economic and environmental realities of the project. To fulfill that goal, understanding the global sustainability issues and addressing them in BIM is not the final solution if it is not possible to integrate with another system like ERP [22].

In the present construction industry, BIM is developing sustainable designs at the design phase but cannot carry out operations in a sustainable manner. Thus, to continue the sustainable aspects, ERP can act as a bridge which connects sustainable design data throughout the operations. Hence, the integration of these two will create great benefits for the construction industry.

In the design stage, BIM helps with performing complex building performance analyses while in the construction stage, ERP helps with integrating and streamlining the business processes, cost reduction, and quality improvements. Hence, this integration allows addressing sustainability aspects in both phases.

BIM and ERP integration will ideally suit the delivery of information needed for the improved design and building performance. The two most significant benefits of BIM–ERP integration would be integrated project delivery and design, operation optimization [14].

The three bottom lines of sustainability will be assured by the BIM–ERP integration as follows.

Environment Aspect:
The integration will eliminate the extra cost of design change during the period, enhancing the delivery culture in the future. Design optimization will reduce the use of material and energy significantly so that energy, water and material, wastage due to inefficiencies or mistakes in the project delivery processes can also be reduced [18, 19].

Social Aspect:
Integrated project delivery will reduce the risk in project delivery with collaboration and proper communication. It will also improve the safety by solving the anticipated problems earlier at the planning stage. The quality of sustainable design will improve the living environment of the people, enhancing living standards. Also, this integration will create a number of new job opportunities with the ERP and BIM integration [14].

Economic Aspect:
The undesirable wastage throughout the planning stage and operation stage can be eliminated through BIM–ERP integration. Further, most of the operations through BIM and ERP reduce much of paperwork. Moreover, optimum design which comes out of this will reduce the capital and lifetime cost through the improved material and resources [2].

However, it is not easy to integrate ERP and BIM as their design architecture is different from one another [17]. Therefore, the integration of these two software should be carefully monitored. On the other hand, if ERP and BIM have not been integrated, their performance will limit into one stage of construction. So, the integration of BIM and ERP is a very much beneficial to facilitate sustainability practices throughout the project.

Potential Benefits of the Research

In the current construction industry, ERP and BIM are being used for construction activities separately [17]. Though ERP has a proper way of integrating all the functionalities across the organization, there is no proper way of addressing the building models for the analysis. Hence, BIM needs to be used to analyse the building model [19]. On the other hand, BIM has no power of integrating generated building structure with other functionalities of the system. Therefore, this BIM-ERP integration will combine these two processes. Further, addressing of sustainability issues in construction by an integrated system will be more efficient rather than having two separate systems. Moreover, these findings will motivate construction organizations to implement more efficient and productive systems to their organizations. It will lead to a more sustainable construction industry in the near future.

Future Work

According to the literature review, it is established that there are huge benefits from the integration of the ERP and BIM. Hence, it will be interesting to carry out future research work on this literature review. So further research analysis should be carried out under following objectives.

1. Identifying a method to incorporate a sustainable BIM design into an ERP system.
2. Listing out the sustainable variables that should be included in ERP to combine the BIM design and ERP.
3. Establishing the measurements to justify the impact on sustainability through BIM and ERP integration.

References

1. Aladag H, Demirdögen G, Isık Z (2016) Building information modeling (BIM) use in Turkish construction industry. Proc Eng 161:174–179
2. Motawa I, Carter K (2013) Sustainable BIM-based evaluation of buildings. Proc - Soc Behav Sci 74:419–428
3. Migilinskas D, Popov V, Juocevicius V, Ustinovichius L (2013) The benefits, obstacles and problems of practical BIM implementation. Proc Eng 57:767–774
4. Mehran D (2016) Exploring the adoption of BIM in the UAE construction industry for AEC firms. Proc Eng 145:1110–1118
5. Lotfy MAMB (2015) Sustainability of enterprise resource planning (ERP) benefits post implementation: an individual user perspective (Ph.D). College of Management and Technology
6. Ortiz O, Castells F, Sonnemann G (2009) Sustainability in the construction industry: a review of recent developments based on LCA. Constr Build Mater 23(1):28–39
7. Chen Y, Tserng H (2017) An integrated methodology for construction BIM & ERP by using UML tool. In: International symposium on automation and robotics in construction, Taiwan, p 6
8. Chong H, Lee C, Wang X (2017) A mixed review of the adoption of building information modelling (BIM) for sustainability. J Clean Prod 142:4114–4126
9. Chofreh A, Goni F, Shaharoun A, Ismail S, Klemeš J (2014) Sustainable enterprise resource planning: imperatives and research directions. J Clean Prod 71:139–147
10. Hasan M, Ebrahim Z, Mahmood W, Rahman M (2017) Sustainable-ERP system: a preliminary study on sustainability indicators. J Adv Manuf Technol 11:61–74
11. Wong K, Fan Q (2013) Building information modeling (BIM) for sustainable building design. Facilities 31(3/4):138–157
12. Love P, Matthews J, Simpson I, Hill A, Olatunji O (2014) A benefits realization management building information modeling framework for asset owners. Autom Constr 37:1–10
13. Gholamzadeh Chofreh A, Goni F, Ismail S, Mohamed Shaharoun A, Klemeš J, Zeinalnezhad M (2016) A master plan for the implementation of sustainable enterprise resource planning systems (part I): concept and methodology. J Clean Prod 136:176–182

14. Lindblad H, Vass S (2015) BIM implementation and organizational change: a case study of a large swedish public client. Proc Econ Finan 21:178–184
15. Babič N, Podbreznik P, Rebolj D (2010) Integrating resource production and construction using BIM. Autom Constr 19(5):539–543
16. Hessari P, Kirkegaard P, Zandieh M, Kani I (2016) Adoption of BIM systems in the AEC industry. PONTE Int Sci Res J 72(9):123–133
17. Ghosh S, Negahban S, Hoon Kwak Y, Skibniewski M (2011) Impact of Sustainability on Integration and Interoperability between BIM and ERP – a governance framework
18. Schwegler B (2010) Green BIM: successful sustainable design with building information modeling by Eddy Krygiel and Bradley Nies. J Ind Ecol 14(5):859–860
19. Chen C, Chen S, Li S, Chiu H (2017) Green BIM-based building energy performance analysis. Comput-Aided Des Appl 14(5):650–660
20. Moakher E (2012) Building information modeling (BIM) and sustainability – using design technology in energy efficient modeling. IOSR J Mech Civ Eng 1(2):10–21
21. Saunders M, Lewis, P, Thornhill A (n.d.) Research methods for business students
22. Bonenberg W, Wei X (2015) Green BIM in sustainable infrastructure. Proc Manuf 3:1654–1659

A Comparative Study of Speed and Flow Measurements Methods as Applied to Four Lane Dual Carriageway Roads

W. U. A. Lowe(✉), H. S. A. Mendis, and I. M. S. Sathyaprasad

Department of Civil Engineering, Faculty of Engineering,
University of Peradeniya, Peradeniya, Sri Lanka
umeshlowe@gmail.com

Abstract. Flow-density-speed relationships play an important role in traffic flow modelling and provides basis to illustrate traffic flow behaviour under different traffic conditions. Analyses of this traffic condition critically depends on the collected traffic data. Traffic data collection under mixed traffic conditions is one of the major problems faced by researchers as well as the traffic regulatory authorities. Therefore a comparative study on traffic data measurement techniques has been carried out in this study. The traffic data can be collected using different kinds of techniques. In this study, traffic flow and speed measurements have been collected using a pneumatic tube detector method, video capturing method, moving observer method and manual classified vehicle counting method. By comparing these methods, suitable methods are suggested for application depending on the existing conditions. This study concentrated on traffic data collection along four lane dual carriageway roads and they are William Gopallawa Mawatha (AB42) and Gannoruwa road (B365). Along these roads, study areas have been selected such that traffic flow is uninterrupted. Tube detectors have been installed and data has been analysed using Metrocount software. For other methods data have been analysed manually. The flow measurements and speed measurements graphs have been obtained for each direction in both four lane roads. These graphs have been drawn such that they represent both forced flow conditions and free flow conditions. Comparison has been done suing graphs obtained.

Keywords: Speed · Flow · Four lane dual carriageway roads

1 Introduction

The fundamentals of traffic flow and their characteristics have become more important and form the foundation for all theories, techniques and procedures that are being applied in the design, operation and development of road transportation systems. The three most important characteristics of traffic are flow, speed and density. These characteristics can be observed and studied at microscopic level and macroscopic level. Flow is the number of vehicles passing a point on a road during stated period of time. The scientific study of traffic flow had begun in the early 80's with the study of models relating volume and speed. Measurement of speed requires observation over both time and space. There are two methods of calculating average traffic stream speed: a time

R. Dissanayake and P. Mendis (Eds.): ICSBE 2018, LNCE 44, pp. 211–223, 2020.
https://doi.org/10.1007/978-981-13-9749-3_20

mean speed and a space mean speed. The space-mean speed reflects the average speed over a spatial section of roadway, while the time-mean speed reflects the average speed of the traffic passing a specified point over a specified time. In this study both speeds will be calculated. There are a lot of methods available to collect traffic data. Each of these methods have their pros and cons. Identifying these pros and cons is very important for transportation engineers because it reveals which method is more suitable depending on the nature of the study. The considered methods in this study are pneumatic tube detector method, video capturing method, moving observer method and manual classified vehicle count (CVC). Each of these techniques is briefly described in following section. This study it is limited to four-lane dual carriageway roads in Kandy. This type of roads consists of four lanes (two lanes per each direction) and they are divided by a central reservation. Even though relatively high speeds can be observed in four-lane dual carriageway roads, they are much safer when comparing with two-lane dual carriageway roads. Head-on collisions, slide swipe collisions (in same direction) are completely eliminated less in four-lane dual carriageway roads because of the central reservation. Even though study areas are the same, observation points are quite different in each method. In the study areas road stretches have been selected such that traffic flow is uninterrupted. In Gannoruwa road (B365) stretch of 200.0 m is selected and in William Gopallawa Mawatha (AB42), road stretch is 268.0 m. At the end of the study traffic flow and speed measured using several methods have been compared.

1.1 Pneumatic Tube Detector Method

With the aid of the Road Development Authority (RDA), tube detectors have been installed in the middle of each road stretch. In the Gannoruwa road it has been installed 100.0 m from the Kandy side edge and in the William Gopallawa Mawatha it is 134.0 m from the Kandy side edge. To observe the speed measurements two tube detectors are required. Tubes are fixed at 1.0 m apart. Since the roads selected are four lane roads, to get the traffic flow by lane, four tubes have been installed in one direction. So, there were four data loggers and 8 tubes collecting data along each road stretch. Data has been collected for a Period of one week (Fig. 1).

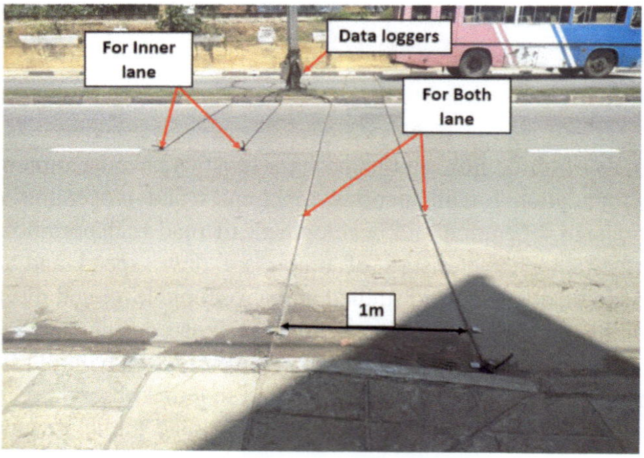

Fig. 1. Pneumatic tube detectors

1.2 Moving Observer Method

In this method an observer witnessing vehicle travel with traffic flow collects data. Two observers were in the vehicle since data collecting has to be done covering both through traffic and against traffic. When the vehicle entered into the road stretch the time is recorded by one observer and same done as the vehicle leaving the road stretch. While the vehicle moved with traffic one observer recorded the number of vehicles in the against traffic (vehicle in the opposite direction) while the other one recorded the number of vehicles over taken by the observer vehicle and number of vehicles over take the observer vehicle. Once this process is done in one direction, to complete one cycle the same procedure has to be repeated in the opposite direction too. For a complete data set 8–10 cycles have been done for one road stretch at a time.

1.3 Video Capturing Method

In this method videos have been captured covering the selected road stretches. All the cameras have been mounted at the edges of each road stretch. Four cameras have been used per one direction which had been synchronized earlier. This is because, the speed measurement travel time of each vehicle is required and to do so each vehicle should be identified when it is entering into the road stretch and when it is leaving. Two cameras captured the inner lane flow while the other two captured the outer lane flow. For each camera, a reference line has been marked 5 m away from the point the cameras are mounted. When a vehicle crosses the reference line, the time, vehicle type and the registration plate number are recorded. The same vehicle is identified from the other camera which is mounted at the other end of the road stretch and time difference is calculated.

1.4 Classified Vehicle Count

In Sri Lanka, the RDA uses 15 classes of vehicle for Classified Volume Count surveys, it is shown in Table 1. CVC of the traffic flow is done manually with the aid data loggers, which involves a group of people recording the number of vehicles passing on a pre-determined location. These locations are installed 400 m away from the tube installed location. It automatically gives CVC in 15 min intervals.

Table 1. Vehicle classification system used by RDA

Type of vehicle	Class	Description
Motor cycle	MCL	All type of motor cycles and scooters
Three wheelers	TW	All type of three wheelers
Car	CAR	All type of cars, Jeeps, cabs, dual cabs, micro van and van with three rows of seats
Van	VAN	Van with seats more than three rows
Buses	MBUS	Buses with lower 26 seats
	LBU	Buses with more than 26 seats

(*continued*)

Table 1. (*continued*)

Type of vehicle	Class	Description
Light goods vehicle	LGV	Small lorries with 2 axels and 4 tyres
Medium goods vehicle	MG1	Small lorries with 2 axel and 6 tyres
	MG2	Large lorries with 2 axel and 6 tyres
High goods vehicles	HG3	Large lorries with 1 chacy with 3 axel
Articulated vehicle	AG3	3 axels with tailor
	AG4	4 axels with tailor
	AG5	5 axels with tailor
	AG6	6 axels with tailor
Farm vehicle	FVH	Tractor with 4 tyres with tailor, Tractor with 2 tyres with tailor

2 Literature Review

2.1 Introduction

There are many researches available on traffic flow analyses. But only some of them have been carried out for mixed traffic condition. Mixed traffic condition consists of different kinds of vehicle categories. For an example car, vans, buses, trucks, three wheelers, bikes and etc. Since in Sri Lanka also has a mixed traffic condition, the focus of this section goes to the literature connected to mixed traffic conditions. Traffic flow analysis can be used to simulate traffic. For example, to evaluate a new part of the infrastructure, traffic flow models can be used. Also, traffic flow analysis can be used to improve road safety.

2.2 Traffic Flow and Speed Measurement

There are a lot of methods available to collect traffic data. But when it comes to heterogeneous traffic automatic collection methods seem to be more suitable. Most of the time, video capturing method has been utilized. Sharma et al. (2011) used video capturing method. In their study, they had considered the heterogeneous mix vehicles along an undivided two-lane road facility. This study of traffic had not been influenced by any nearby intersections or road geometry (Gradient). Therefore, this traffic is considered as an uninterrupted flow of traffic. Also, verbal recordings of vehicle categories and license plate number were done. Certain key features were noted down so that the identification of vehicles could be done. But some researchers have used software to do that. Mallikarjuna and Rao (2011) have used this method to collect traffic flow data under heterogeneous conditions. They have used video data to extract speed, flow, density, lateral gap and several other parameters. There can be many other lateral movements but, in this study, it is restricted to five movements. They have also used a software named TRAZER to track vehicles and get their trajectory data over a road length of 50 m. This software is capable of classifying vehicles capturing the lateral movement of vehicles. Traffic speeds had been obtained using the trajectory data. Also, Mallikarjuna et al. (2009) have used the same method to collect traffic data under mixed traffic conditions. In their study the same software has been used. The velocities and acceleration values were obtained by performing first and second order differentiation on trajectory equations. In their study they have proved that TRAZER is capable of detecting vehicles with an average accuracy of 95% and classifying with an average accuracy of 85%. Using the software for traffic flow detection seems to be easy and more useful in gathering other information too. But Sharma et al. (2011) have used FMV6 (full motion video) player for video data extraction. This software is capable of fragmenting the video into frames and uses that to obtain frame readings when traffic vehicle just crosses the reference point. Ranthatiyawa et al. (2017) have also utilized the video capturing method. But in that study, roads selected were residential roads. They have used the video data for traffic flow counting and have manually counted the traffic flow using the videos. For their study not using software was possible since they only need the traffic flow.

Other than the video capturing method, the loop detector method has also been used to collect traffic data. Stathopoulos and Karlaftis (2003) have used the loop detector method. Speeds and travel times along selected routes have been obtained using flow and occupancy data collected from the loop detectors. They had used these data to develop ITS (intelligent transportation system). For this, six origin points had been defined at major city entrances, from which a total of 17 possible routes had been considered. Traffic data along these 17 routes had been collected directly from the controllers installed at 144 locations of the down-town urban network. Also, Hueper et al. (n. d) have used traffic flow and occupancy data from loop detectors. They have used the collected data to calibrate their model. In their study, they had considered a Californian freeway. They have also used PeMS database to achieve traffic measurements like occupancy, flow, speed, delay, vehicle hours travelled and productivity loss etc. The only direct measurements taken from the loop detections are the number of cars which cross a detector station and the fraction of time a vehicle is present over the loop. These researches used loop detector method because in their context it was more convenient. And also, for a mixed traffic condition it is a very effective method. Therefore, choosing video capturing method seemed to be a wise choice.

3 Methodology

3.1 Site Selection

Firstly, suitable sites have been selected. Few major aspects were paid attention to when selecting the sites. First, was to avoid any junctions along the road section. Suitable lengths for the road sections have also been selected to get accurate data. The road stretches have been selected such that they do not consist of central openings, pedestrian crossings and bus stops. The locations for tube detector installation were selected such that it is in the middle of the road stretch. The places to video camera mounting were selected such that they had the capability to capture the vehicle registration number clearly. On the Gannoruwa road, a stretch of 200 m is selected and on the William Gopallawa Mawatha, the road stretch is 268 m (Figs. 2 and 3).

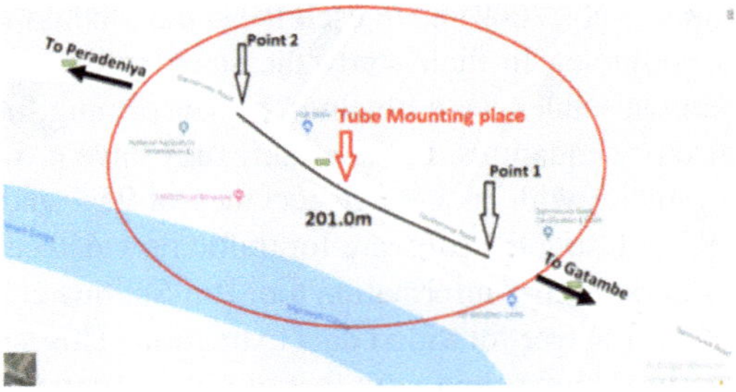

Fig. 2. Survey location at Gannoruwa Road

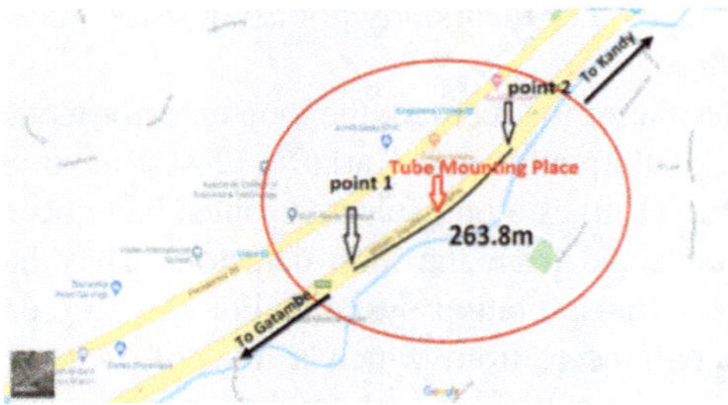

Fig. 3. Survey location at William Gopallawa Mawatha

3.2 Field Data Collection

In the video survey method, the distance between two camera locations and widths of the road sections were measured using a road tracer.

In the tube detector method, tubes have been installed for a week with aid of the RDA. In the same week, Moving observer method also was carried out. Manual CVC was done on the same day. But the video capturing was done in a different day. In manual CVC it was done from 6.00 am to 22.00 pm. In the video capturing method the videos have been captured for two hours (7.00 am–9.00 am) so that it covered the peak hours. Peak hours were firstly identified using tube detector data.

3.3 Data Extraction and Analysis

The data extraction was done only for the video capturing method because in all the other methods recorded data can be directly retrieved in an Excel file format. In the video capturing method, form each video data had to be fed into a computer database with parameters such as vehicle registration plate number, time stamp and vehicle type when the vehicle passed the reference point. In the tube detector method, recorded data was converted into an Excel file format using Metrocount software. This software is used for tube data extraction and analysis. In manual CVC also, recorded data was downloaded from data loggers. It was directly collected as an Excel file format. Analysis of the data collected is different in each method and it is briefly discussed in following sections.

3.3.1 Pneumatic Tube Detector Method

In the tube detector method, data analysis was done using Metrocount software. From each data file speed and flow is analysed along each road for each direction. To obtain lane wise traffic flow, traffic flow obtained for inner lane was deducted from the traffic flow obtained for both lanes. From the Metrocount software data files are converted in to excel files as it directly gives the flow and speed data separately. Time interval which data are required can be changed accordingly. Classification systems are also can be changed. Classification systems in the Metrocount software are quite different and none of them comply with RDA classification system. They are ARX +, Scheme F +, Scheme F2 +, Scheme F99 +, AustRoads94 +, NAASRA +, Swedish +, TNZ1999 +, Scheme axle +. Under each of these classification systems flow data has been plotted and compared with the RDA classification system. Also, flow variation in the week has been plotted along with lane distribution, flow variation in city-bound and out-bound directions. Traffic flow and speed data have been calculated and analysed for both lanes for each direction.

3.3.2 Video Capturing Method

In the video capturing method, the speed of each vehicle has been calculated using time measurement extracted from the videos and length between the specified points.

$$V = \frac{L}{t_1 - t_2} \tag{1}$$

Where,

V: Speed of the vehicle
L: Distance between two survey points
t_1: Time of entering of the vehicle
t_2: Time of leaving of the vehicle

Since flow measurements are recorded according to the RDA classification, they are directly taken from the extracted data. After obtaining the flow data, they were converted into passenger car units by using the relevant passenger car equivalency factors. Flow, speed calculation and analysis have been carried out only in William Gopallawa Mawatha towards Kandy direction. Plots have been developed to represent the speed-flow variation.

3.3.3 Moving Observer Method

In the moving observer method, the following equations have been used. When observer is moving with the stream, let $\boldsymbol{m_0}$ be the vehicles overtake the observer and $\boldsymbol{m_p}$ be the vehicles overtaken by the observer. Let the difference be $\boldsymbol{m}$. Then,

$$\boldsymbol{m} = \boldsymbol{m_0} - \boldsymbol{m_p} \tag{2}$$

$$\boldsymbol{m} = (\boldsymbol{q} * \boldsymbol{t}) - (\boldsymbol{k} * \boldsymbol{v_0} * \boldsymbol{t}) \tag{3}$$

Here $\boldsymbol{m}, \boldsymbol{t}$ and $\boldsymbol{v}_0$ can be obtained from the survey. $\boldsymbol{q}$ can be obtained using below equation.

$$\boldsymbol{q} = \frac{\boldsymbol{m}_w + \boldsymbol{m}_a}{\boldsymbol{t}_w + \boldsymbol{t}_a} \tag{4}$$

Where,

m_w: Vehicle difference when observer vehicle is going with the traffic flow
m_a: Vehicle difference when observer vehicle is going against the traffic flow
t_w: Time taken to travel road stretch with the traffic flow
t_a: Time taken to travel road stretch against the traffic flow

Mean stream speed $\boldsymbol{v}_s$ can be obtained using below equation.

$$\boldsymbol{v}_s = \frac{\boldsymbol{l}}{\boldsymbol{t}_w - \frac{\boldsymbol{m}_w}{\boldsymbol{q}}} \tag{5}$$

Where,
$\boldsymbol{v}_s$: mean stream speed
Flow and speed calculations have been carried out for both lanes for both directions. Plots have been developed to represent flow-speed variation.

4 Results and Discussion

Figure 4 shows the traffic flow variation along William Gopallawa Mawatha for the city bound direction with different data collection methods. The methods considered are pneumatic tube detector method, moving observer method and manual CVC.

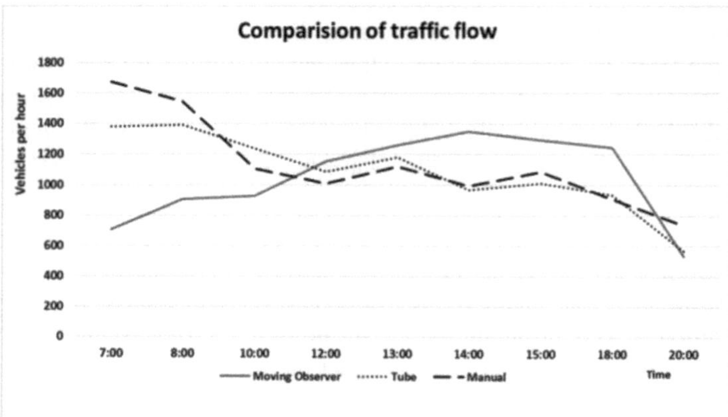

Fig. 4. Traffic flow comparison using different methods

According to Fig. 4, during peak hours manual count shows an over estimation compared to tube detector data while moving observer method shows an under estimation. However, during low flow conditions it is quite different. Moving observer method shows over estimation compared both other methods. Even though tube data show an underestimation during low flow condition, it almost complies with the manual count data. Here the reason for the over estimation could be traffic congestion. Since the tube detector method is an automated method, when it detects vehicles during congestion, errors could occur due to vehicles over lapping. Because vehicles move closely to each other during congestion, tube detectors could detect them as one vehicle. In the moving observer method, these flows are calculated using 8-10 cycles. But the cycle time is less than one minute since the road stretch is small. The recommended time period to take a flow measurement mentioned in highway manual is 15 min. So, the results obtained using moving observer method could differ from the manual count data.

Figure 5 shows the traffic speed variation along William Gopallawa Mawatha in the city bound direction. The data represented here are collected using pneumatic tube detector method and moving observer method.

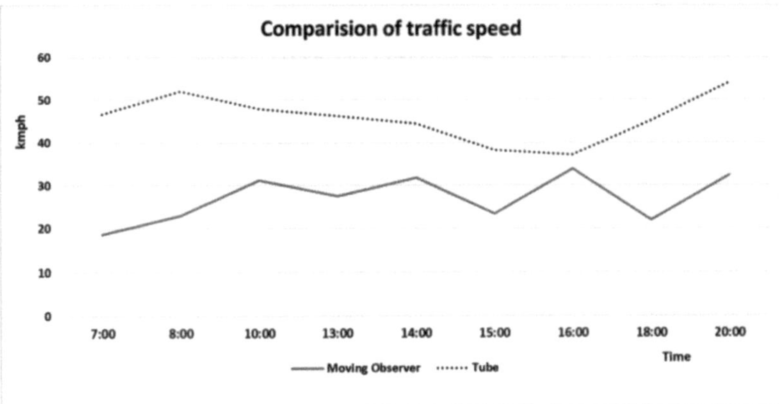

Fig. 5. Traffic flow comparison using different method

According to Fig. 5, in the moving observer method speed variation is between 15–35 kmph while it is in tube detector method 35–55 kmph. Even though speed variation during the day doesn't show any significant difference, the moving observer method always shows an underestimation of speed. The reason could be the same as mentioned above.

Figure 6 shows the shows the comparison of speed measurements using video capturing method and tube detector method in the William Gopallawa Mawatha in the city bound direction. As clearly seen in the figure tube detector method underestimates the speed measurements. This is mainly because of the survey location. Since tube detector was located in the middle of the road stretch, it gives the spot speed at the middle of the road stretch. But in the video method speed measurements have been calculated considering the whole road stretch and therefore it gives an average value for the speed along the road stretch.

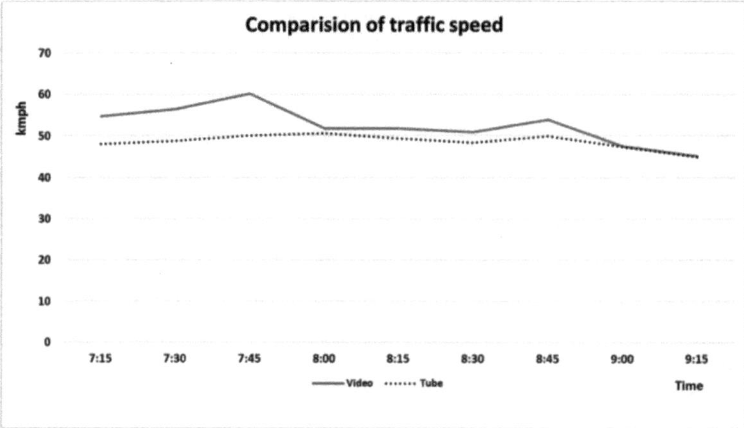

Fig. 6. Speed comparison between video and tube method

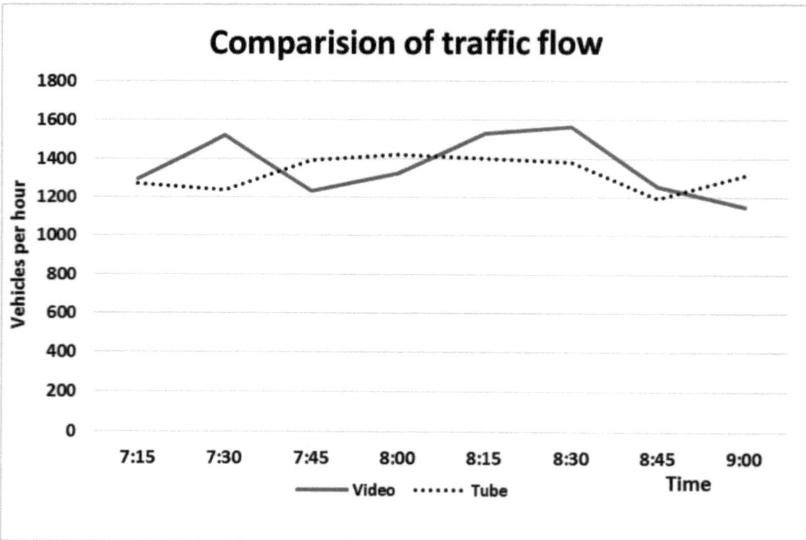

Fig. 7. Flow comparison between video and tube method

The Fig. 7 shows the flow measurement comparison using video method and tube detector method in the William Gopallawa Mawatha in the outbound direction. So, the figure shows zigzag variation pattern between the two methods. Since the video has been taken from 7.00 a.m. to 9.00 a.m., it includes the peak hour. So, during the peak condition, tube data gives an underestimation most of the time. The reason could be the same as mentioned above.

Figure 8 plot shows the lane wise traffic flow variation along William Gopallawa Mawatha. This data has been obtained using pneumatic tube detector method. Since other methods cannot be used to collect lane distribution, here, explanation is done only considering tube detector method.

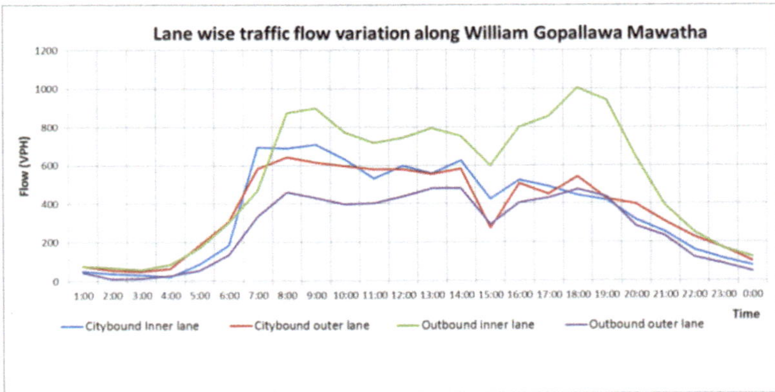

Fig. 8. Lane distribution of traffic flow

A clear difference of lane usage can be obtained in the outbound. Whether it is low traffic condition or high, it is always greater in the inner lane. The main reason can be the commercial buildings located along outer lane side towards outer bound. Since the vehicles that come to these buildings use the outer lane, through traffic allure to use the inner lane. To the city bound it varies. During peak hours inner lane usage is higher while during off peak hours it is the other way around.

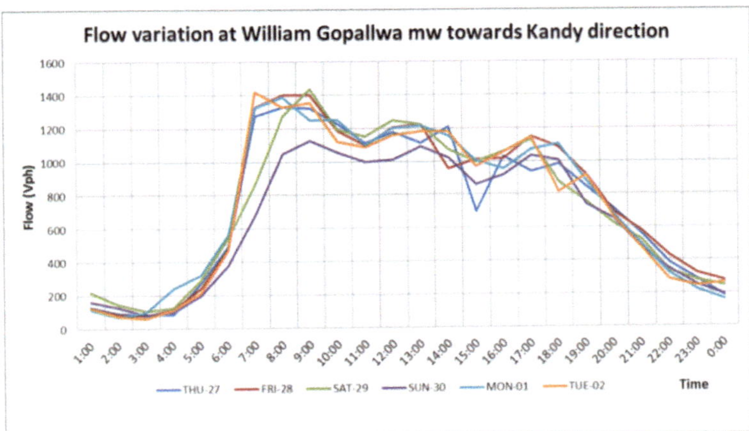

Fig. 9. Flow variation at William Gopallawa Mawatha in city Bound direction

Since other methods hadn't been carried out for a week, weekly flow variation is described here only using pneumatic tube detector data. Graph in Fig. 9 has been plotted showing the 24-h flow variation representing week days as well as weekends.

Following graph shows flow variation along the William Gopallawa Mawatha towards the city (Table 2).

Table 2. Peak hour details using tube detector method

	City-Bound			Out-Bound		
Day	Peak hour	Peak hour volume (vph)	Total daily volume	Peak hour	Peak hour volume (vph)	Total daily volume
Thu	7–8 a.m.	1,330	17,526	17–18 p.m.	1,476	19,066
Fri	8–9 a.m.	1,401	18,368	17–18 p.m.	1,413	19,253
Sat	8–9 a.m.	1,434	17,687	15–16 p.m.	1,290	18,610
Sun	8–9 a.m.	1,124	15,539	17–18 p.m.	1,253	16,600
Mon	7–8 a.m.	1,390	18,187	17–18 p.m.	1,504	19,611
Tue	8–9 a.m.	1,352	17,679	-	-	-

Flow variation during weekdays does not show any significant difference. But flow on Sunday is quite lower than other days. It doesn't mean weekend flow is lower than weekdays. Saturday flow's variation is more similar to the weekdays flow. Peak hour, during weekdays is from 8.00 a.m. to 9.00 a.m. except on Mondays, where it is from 7.00 a.m. to 8.00 a.m. During the weekend it is from 7.00 a.m. to 8.00 a.m. During the low flow condition, flow variations of all days are very similar.

According to Fig. 10, in the outbound direction, flow variation is quite different from it is for the city bound. In the city bound weekends flow is lower than weekdays flow. The major change is in the peak hour. Even though the morning peak hour remains the same for the weekdays, it varies on the weekend. The out bound peak hour is in the evening and is from 5.00 p.m. to 6.00 p.m. except for Saturday. On Saturday it is from 3.00 p.m. 4.00 p.m. This is mainly because the city consists of a lot of commercial buildings and schools. Therefore, during the morning flow it is higher in the city bound direction while it is higher in the afternoon for the outbound.

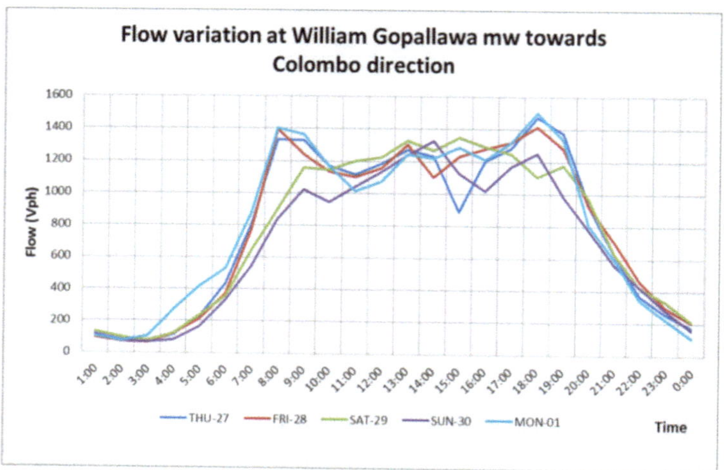

Fig. 10. Flow variation at William Gopallawa Mawatha - Outbound direction

So, in conclusion, the video capturing method and Manual counting method seem more accurate than the tube detectors because its shows different results when it comes congestion condition. Also, for a short road stretch, the moving observer method does not seem suitable.

Acknowledgements. The authors acknowledge the continuing support from the Road Development Authority and the assistance by the staff of the surveying laboratory.

References

Arasan T, Gnanavelu D (2009) Concentration of heterogeneous road traffic. In: Jayanthakumaran K (ed) Advance technologies. InTech, London, pp 503–530

Gerlough DL, Huber MJ (1975) Traffic flow theory. Transportation Research Board, Washingtion

Mallikarjuna C, Rao KR (2011) Heterogeneous traffic flow modelling: complete methodology. Transportmetrica 7:321–345

Bains MS, Ponnu B, Arkatkar SS (2012) Modeling of traffic flow on Indian expressway using simulation technique, Changsha, s.n

Ranthatiyawa RDMDR, Perera BVLS, Sathayaprasad IMS, Nandalal HK (2017) Use of remote-sensed spatial data in estimating traffic flows from small residential road catchments, Kandy, s.n

Sharma N, Arkatkar SS, Sarkar AK (2011) Study on heterogeneous traffic flow characteristics of a two-lane road. Transport 26(2):185–196

Stathopoulos A, Karlaftis MG (2003) A multivariate state space approach for urban traffic flow modeling and prediction. Transp Res Part C 11:121–135

Transportation Reasearch Board (1995) Highway capacity manual. National Research Council, Washington, D.C

Green Technology and Energy

A Stochastic Differential Equation Based Wind Speed Forecasting Model

H. M. D. P. Bandarathilake[1(✉)] and G. W. R. M. R. Palamakumbura[2]

[1] Sri Lanka Technological Campus, Ingiriya Road, Padukka, Sri Lanka
darshanab@sltc.edu.lk
[2] Department of Engineering Mathematics, Faculty of Engineering, University of Peradeniya, Peradeniya, Sri Lanka

Abstract. Stochastic differential equation (SDE) based approaches have demonstrated an improved capability in predicting temporal wind speed patterns that have statistical properties which are very similar to those observed in reality. However, no standard approach for deriving such models exist. One difficulty in arriving at a unified framework is the presence of a seemingly wide variation in the temporal statistical properties that one observes from location to location. In this work, we propose an SDE model for short term wind speed pattern predictions for Kokkilai, an area that is located in the North-Eastern coastal region of Sri Lanka. Here, we propose a systematic approach for modelling short term temporal wind speed variations, using a 1-dimensional stationary stochastic differential equation along with an approximated marginal distribution for the long term empirical wind speed distribution and the corresponding autocorrelation function. The model parameters were estimated to fit the empirical wind speed distribution of wind speed data that has been recorded at the wind measurement center of Kokkilai, from February 2015 to February 2016. The normal Gaussian distribution turned out to be the best fitted marginal distribution, in terms of maximum likelihood, for long term wind speed data. Then, an Ornstein-Uhlenbeck based SDE was obtained using the parameters of the fitted normal Gaussian distribution and the empirical autocorrelation function. Subsequently, considering special features such as cyclic behavior and certain trends of the empirical autocorrelation function, a set of new parameters were introduced to the drift and diffusion terms of the SDE. The obtained model was capable of generating a short-term wind speed forecasting trajectories that fall within an acceptable confidence level.

Keywords: Stochastic differential equation · Drift · Diffusion · Autocorrelation function

1 Introduction

During the last few years, the contribution of wind based electricity generation to the global green energy industry has shown a remarkable increase. Wind is an energy source that is inherently volatile and uncertain in nature. Therefore, an efficient wind speed forecasting mechanism developed by means of mathematical/statistical modelling, can play a pivotal role in the analysis of many aspects related to power systems

R. Dissanayake and P. Mendis (Eds.): ICSBE 2018, LNCE 44, pp. 227–238, 2020.
https://doi.org/10.1007/978-981-13-9749-3_21

such as power generation economics, evaluation of the generation capacity and ensuring reliability. Time series based methods, probability distribution based models, four component composite models and Kalman filter approaches (Carta et al. 2009) are some of the mathematical/statistical approaches that have been used to forecast wind speeds in both metrology and the energy industry (Zhu and Genton 2012). However, the development of an accurate and efficient forecasting modelling still remains a real challenge.

In comparison with traditional modelling approaches of wind speed, stochastic differential equations based methods have gained popularity in the field of wind power forecasting during the last few years. The novel stochastic differential equations (SDE) based approaches have indicated an increased capability of sufficiently and accurately capturing the inherent stochastic nature of wind.

Due to the topographical dependence of wind flow patterns, the wind speed always indicates high variations from one location to another location. Therefore, most studies have only concentrated on studying the wind speed forecasting for a given location. Two major types of wind speed forecasting exist, namely, probabilistic forecasting and point-forecasting.

In the case of probabilistic forecasting, wind speed distributions are generated while in point-forecasting wind speed predictions are explicitly provided for a certain time period ahead. Point-forecasting models can be also split into models that generate long term forecasting trajectories and short-term forecasting trajectories.

In (Zarate-Minano et al. 2016) (Zárate-Miñano and Milano 2016) (Calif 2012), a one dimensional stationary stochastic differential equation has been employed to obtain wind speed distributions. In these cases, the empirical wind speed distribution and the corresponding autocorrelation function, of a moderately large wind speed data set, are used to obtain the forecasting trajectories. According to the method employed in (Zarate-Minano et al. 2016) (Zárate-Miñano and Milano 2016), the empirical wind speed distribution has been approximated by a known standard probability distribution and the exponentially decaying region of the empirical autocorrelation function, is used to model the forecasting trajectories using the principles of Ito calculus as pointed out in (Zárate-Miñano and Milano 2016). In (Calif 2012), the forecasted wind speed trajectories have been presented as wind speed density functions. In (Zarate-Minano et al. 2016) where the marginal distribution was selected as a Weibull distribution, the model provides probabilistic forecasting of long term wind speed variation together with short-term autocorrelation forecasting. The work presented in (Calif 2012 and Zarate-Minano et al. 2016), has indicated a situation where the empirical wind speed distribution was approximated by a Gaussian mixture distribution. In addition to the above cases, (Iversen *et al.* 2015) describes a model that refines a given numerical wind speed prediction using a one dimensional SDE. In (Iversen *et al.* 2015), the developed SDE, has generated accurate wind speed forecasting by modifying an SDE which generates an Ornstein-Uhlenbeck process model by considering the cyclic variation of wind speed patterns. Consequently, it provides a more accurate short-term wind speed forecasting model in comparison with the numerical wind speed predictions.

In this paper, we present a short-term wind speed forecasting model which is obtained by modifying an SDE model based on an approximated marginal distribution and the corresponding autocorrelation function.

In this work, we have considered the wind speed data that has been recorded at Kokkilai, in the North-Eastern coast of Sri Lanka, from February, 2015 to February, 2016. The data set was recorded under a feasibility assessment of the "Quantum Leap Wind Power Development" project funded by the Asian Development Bank (http://www.energy.gov.lk/ta/node/380).

1.1 Methodology

A differential equation which consists of one or more stochastic terms is identified as a stochastic differential equation. A one dimensional non-stationary Ito stochastic differential equation has the following form,

$$d(x(t)) = a(x(t),t)dt + b(x(t),t)dW(t) \tag{1}$$

Where $a(x(t),t)$ and $b(x(t),t)$ terms are called drift and diffusion coefficients respectively. Here, $W(t)$ denotes the standard Wiener process (Øksendal 2003). The integral form of Eq. (1) takes the form,

$$x(t) - x_0 = \int_0^t a(x(u),u)du + \int_0^t b(x(u),u)dW(u) \tag{2}$$

In the above Eq. (2), the first integral which is associated with the drift term can be solved by means of ordinary Riemann-Stieltjes integration. The second integral, which is termed the diffusion term, should be integrated using stochastic calculus (Øksendal 2003) (Higham and Higham 2001).

The wind speed data $\{v_t : t = 1,2,3,\ldots\}$ can be regarded as one realization of an underlying stochastic process $\{V_t : t = 1,2,3,\ldots\}$. This says, for each time point t, v_t is treated as an observation of the random variable V_t. Such process can consist of a deterministic part and a random component. Thus stochastic differential equations are a natural choice for modeling wind speed data.

Provided that the stationarity conditions are met, statistics such as the autocorrelation function are not only useful as a descriptive tool of the process but can also can be used to make inferences about the underlying process of which the observed series is a sample (Brett and Tuller 1991). In this context, in addition to the selection of model for forecasting, analytical methods based on the empirical autocorrelation function are also employed to identify important periodicities in the underlying process.

In this work, we are interested in reproducing the wind speed values with statistical properties similar to that of the empirical data, by means of a SDE of the form of Eq. (1). Here, we have assumed that the considered wind speed variation preserves the stationarity property. The stationary form of the Eq. (1) is given by

$$d(x(t)) = a(x(t))dt + b(x(t))dW(t) \tag{3}$$

and will be employed to obtain the forecasting trajectories.

In order to solve the stationary SDE in Eq. (3), the corresponding drift and diffusion terms should be estimated. In the following, we present two approaches for doing this. The first method is based on the estimation of the marginal probability distribution function for the process while in the second method we explicitly account for diurnal and other periodic effects on wind speed.

Method 1

Following the approach presented in (Zárate-Miñano and Milano 2016), the solutions of the SDE are assumed to be distributed according to a known probability distribution function. Then, by applying the Ito's lemma to the stationary SDE, the corresponding stationary Fokker-Planck equation associated with the considered SDE can be obtained as follows,

$$0 = -a(x(t))p(x(t)) + \frac{1}{2}\frac{\partial}{\partial x(t)}\left[b^2(x(t))p(x(t))\right] \tag{4}$$

where $p(x(t))$ is the probability distribution function of the stochastic process that is generated by the SDE in Eq. (3). Since the wind speed distribution of the model generated values should be the same as that of the empirical wind speed distribution, the best fitted marginal distribution to the empirical wind speed distribution will be selected for $p(x(t))$.

Then, the Fokker-Planck equation in Eq. (4), can be solved for $a(x(t))$ and $b(x(t))$ in terms of this estimate as follows,

$$a(x(t)) = b(x(t))\frac{\partial b(x(t))}{\partial x(t)} + \frac{1}{2}b^2(x(t))\frac{\partial}{\partial x(t)}(\log p(x(t))) \tag{5}$$

$$b^2(x(t)) = \frac{2}{p(x(t))}\int_{-\infty}^{x(t)} a(z(t))p(z(t))dz(t) \tag{6}$$

where $p(x(t)) \neq 0$ and $b(x(t)) = 0$ if $p(x(t)) = 0$.

In order to require the desired drift term, we consider the regression theorem as in (Zárate-Miñano and Milano 2016) Consequently, the drift term is obtained in the form:

$$a(x(t)) = -\alpha(x(t) - \mu) \tag{7}$$

where μ is the mean of the considered probability distribution $p(x(t))$ and in order to obtain the value of α, the exponentially decaying part of the autocorrelation function of the wind speed data is fitted with the function of the form, (here α and c are constants)

$$f(x) = ce^{-\alpha x} \tag{8}$$

It should be noted that the type SDE in Eq. (3), always generates a stochastic process with exponentially decaying autocorrelation function. Therefore, the parameter α plays a crucial role in the evaluation of the drift and the diffusion terms as shown in (Zárate-Miñano and Milano 2016). Then using the obtained drift term, diffusion term is calculated using Eq. (6). In our case, the empirical distribution is best approximated by a normal Gaussian distribution. So, taking

$$p(x(t)) = \frac{1}{\sqrt{2\pi\sigma^2}} exp(-\frac{(x-\mu)^2}{2\sigma^2}) \tag{9}$$

we have that the drift and diffusion terms must take the form,

$$a(x) = -\alpha(x-\mu) \tag{10}$$

$$b(x) = \sqrt{2\alpha}\sigma \tag{11}$$

Therefore, the above drift and diffusion terms generate an Ornstein-Uhlenbeck process of the form,

$$dx = -\alpha(x-\mu)dt + \sqrt{2\alpha}\sigma dW \tag{12}$$

The parameters μ, σ and α are estimated so that Eq. (9) best fits the marginal probability density of the empirical data and Eq. (8) fits the initial exponentially decaying part of the empirical autocorrelation function. Using these parameters, the solutions of Eq. (12) can be used to provide a point wise estimate of wind speeds. Notice that the autocorrelation of the so modeled process is necessarily exponentially decaying and cannot handle periodic variation as observed in reality.

Method 2

In (Iversen *et al.* 2015), they have incorporated cyclic variation of wind pattern by modifying the drift term of an SDE model based on numerical weather prediction which generates an Ornstein-Uhlenbeck process. The purpose of short-term model in (Iversen *et al.* 2015), was to refine the numerical wind speed prediction which also provides forecasts for short periods ahead. In this paper, we introduce a short-term forecasting model which can be validated for several selected portions of data based on the different features exhibiting in the autocorrelation function. After incorporating these in the drift and the diffusion terms of Eq. (12), we propose the following model for the selected portions of the series.

$$\boldsymbol{dX}(t) = \left(\left(\boldsymbol{\beta}\sin\left(\frac{\boldsymbol{\pi}}{\boldsymbol{T}}\boldsymbol{t}+\boldsymbol{\varphi}\right)\right)\left(1-\boldsymbol{e}^{-\boldsymbol{X}(t)}\right)+\boldsymbol{\alpha}(\boldsymbol{\mu}-\boldsymbol{X}(\boldsymbol{t}))\right)\boldsymbol{dt}+\boldsymbol{\sigma X}^{\gamma}(\boldsymbol{t})\boldsymbol{dW}(\boldsymbol{t}) \tag{13}$$

Here, $\left(\beta Sin\left(\frac{\pi}{T}t+\varphi\right)\right)$ term has been used to capture the sinusoidal behavior of wind speed variation and $\left(1-e^{-X(t)}\right)$ can avoid generating negative values. In addition to that, value of T should be chosen to capture the cyclic variation of the considered region. The term σX^{γ} replaced the constant value of the diffusion term of Eq. (12) in order to capture the trends of the resultant process.

2 Results and Discussion

2.1 Estimation of the Parameters for the SDE Model Proposed in Method 1

Wind speed data which had been recorded as ten-minute average values, were converted to hourly mean wind speed values and the obtained dataset of hourly average wind speed consisted of 8419 observations. Resulting empirical wind speed distribution is shown in Fig. 1. In order to find the best fitted marginal distribution for the empirical wind speed distribution, the maximum log likelihood values of each marginal distribution was obtained as in Table 1. Figure 2, indicates, graphically, some approximations of the marginal distributions for the empirical wind speed distributions.

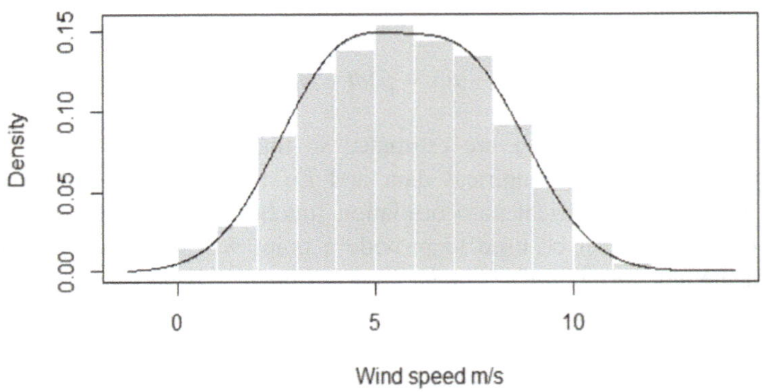

Fig. 1. Empirical wind speed distribution of Kokkilai

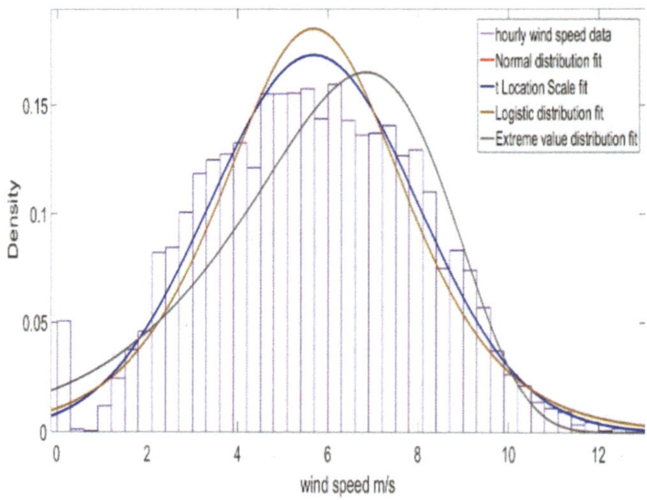

Fig. 2. Approximation of empirical density function of hourly average wind speed of Kokkilai

Table 1. Log likelihood values

Log-likelihood values	Probability distribution			
	Logistic	Extreme value	t-Location scale	Normal
Log-likelihood	−19173.7	−19522.9	−18977.8	−18977.8

According to the Table 1 and Fig. 3, the normal Gaussian distribution was selected as the most suitable marginal distribution with the parameter values, $\mu = 5.68398$ and $\sigma = 2.30547$.

The value of the parameter α was obtained by the empirical autocorrelation function. For this case, the exponentially decaying region of the empirical autocorrelation function should be identified. The empirical autocorrelation function of Kokkilai which was constructed by considering the entire period, is given in the Fig. 3. It indicates that the autocorrelation coefficient decays exponentially only for a 12-hour time interval from the beginning and then it indicates a complex cyclic behavior.

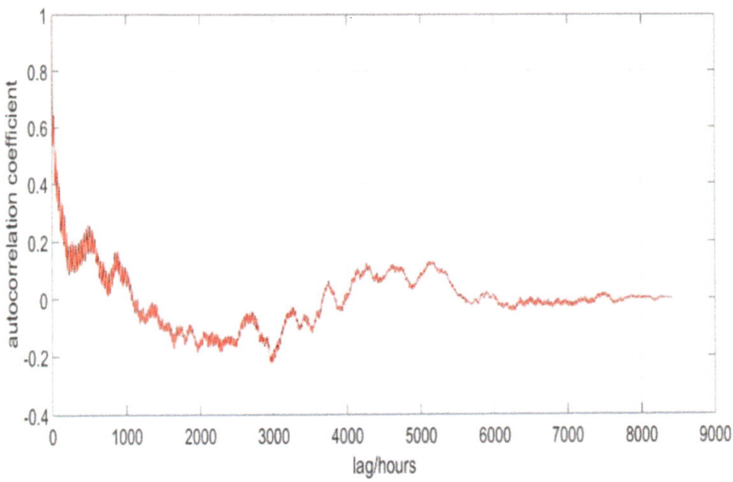

Fig. 3. Autocorrelation function of hourly wind speed of Kokkilai for entire time domain

Using the method of least squares, the parameter c and α in Eq. (8) were estimated within a 95% confidence interval and estimated values were found to be $c = 0.6678$ and $\alpha = 0.2204$ (Fig. 4).

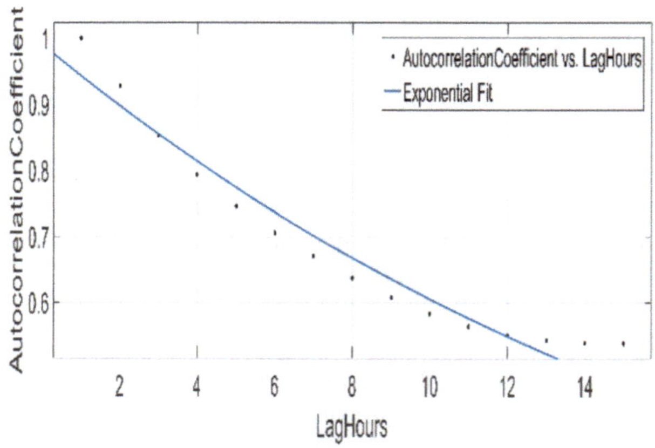

Fig. 4. Curve fitting of autocorrelation function with 12 lags

Finally, the SDE model of Eq. (12) was simulated with obtained parameter values by using the Euler-Maruyama numerical scheme with the initial value $x(0) = v_0 = 5.428$ and the wind speed forecast for the entire period was obtained. The wind speed distribution of the model generated values for several simulations are compared with the empirical wind speed distribution in Fig. 5.

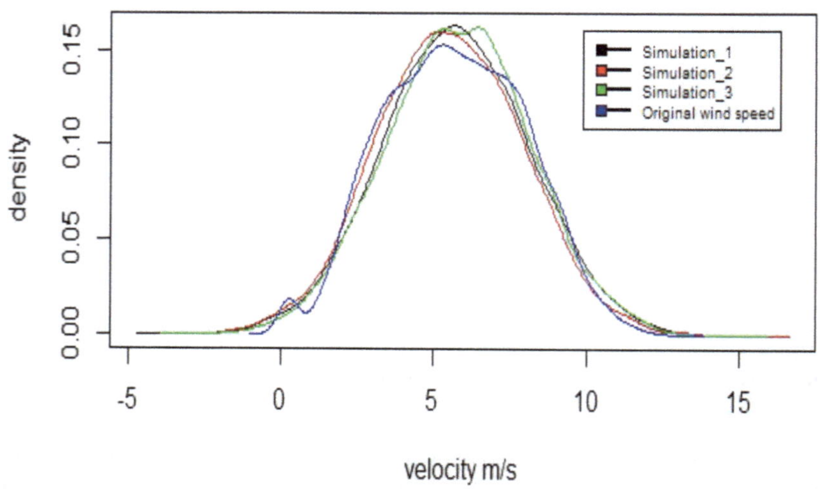

Fig. 5. Comparison of original & initial SDE model generated wind speed distributions obtained by kernel density estimation of Kokkilai

Figure 5 indicates that the model generated wind speed values and original wind speed data have high deviation in the median velocity values. So, it is obvious that even though the model has been capable of generating a wind speed distribution that has a similar shape to that of the original data, it is incapable of generating a long term prediction of wind speed with considerable accuracy.

2.2 Estimation of the Parameters for the SDE Model Proposed in Method 2

The autocorrelation function in Fig. 3, indicates a complex cyclic behavior accompanied with a short period of exponential decay. The periods of the cyclic patterns indicated in the autocorrelation vary with the time. Therefore, by observing these patterns in the autocorrelation function, the wind speed data was grouped and labeled as portion-1_1, portion-1_2, portion-2_1 and portion-2_2 accordingly. In this work, portions 1_1, 2_1 and 2_2 were chosen for forecasting since they exhibit sinusoidal behaviors with almost constant periods. For the three cases the model (13) was to estimate the parameters of for each respective portion (Fig. 6).

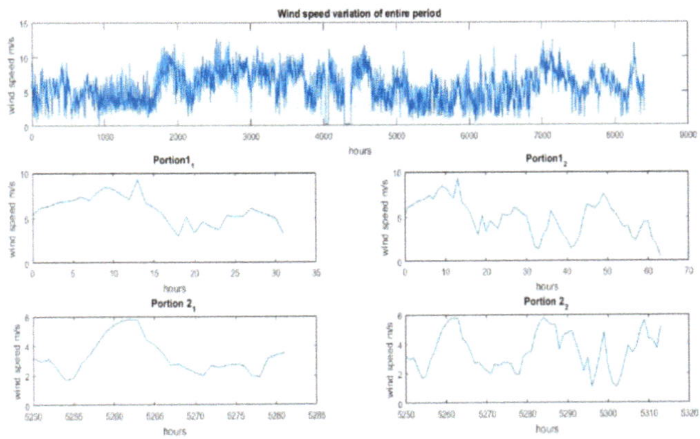

Fig. 6. Selected portions of the main dataset for incorporation of short-term models

In constructing the models for the three cases, the proposed model (13) was implemented. The parameters were estimated by implementing the pseudo-maximum likelihood estimation method. With estimated parameters, the SDE model was simulated using the Euler-Maruyama numerical scheme with 100 simulations and mean forecasting trajectory was obtained by taking the mean of all simulated trajectories for all three cases. These are shown in Figs. 7, 8 and 9 respectively. The estimated parameter values for the respective portions are in Tables 2, 3 and 4.

The corresponding parameter values of the SDE model, estimated for the portion-2_2 and portion-2_1 which have been indicated in Tables 3 and 4, display much similarity. It shows us that the wind speed data of a total of 64 h including the data of portion-2_1, consisted of almost constant periodic variation. Therefore, it has ensured a comparatively longer period of time in generating a mean forecasting trajectory within the accepted confidence levels. In addition, it should be observed that the corresponding parameter values which are involved with the diffusion term have taken similar values. According to the figures, the mean forecasting trajectory of portion 2_1, can be identified as the best.

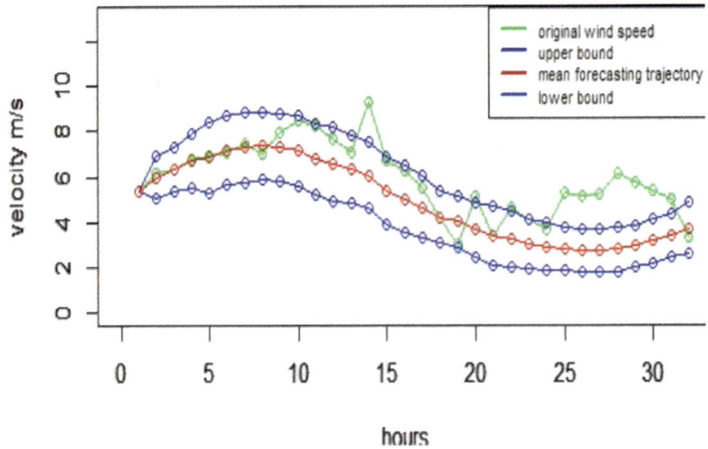

Fig. 7. Mean forecasting trajectory of portion-1_1 (within a considered confidence interval)

Table 2. Parameter values of short-term model of portion-1_1

Parameter value	Parameter						
	T	α	β	γ	σ	φ	μ
Value	19	0.2204	0.6500	0.4230	0.4210	1.0472	4.9500

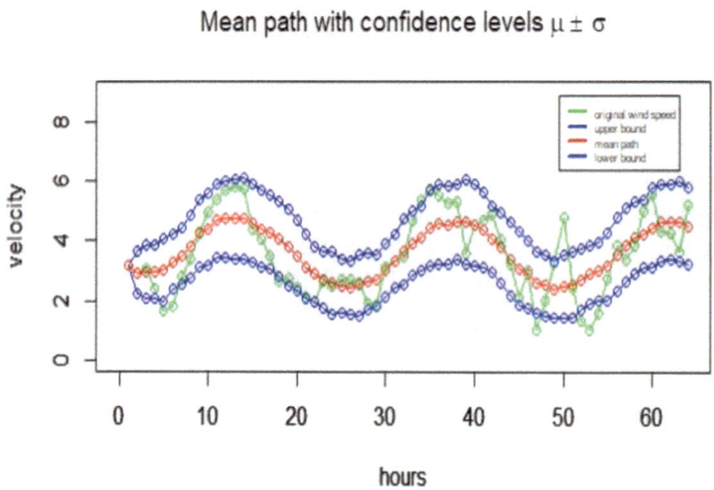

Fig. 8. Mean forecasting trajectory of portion 2_2 (within a considered confidence interval)

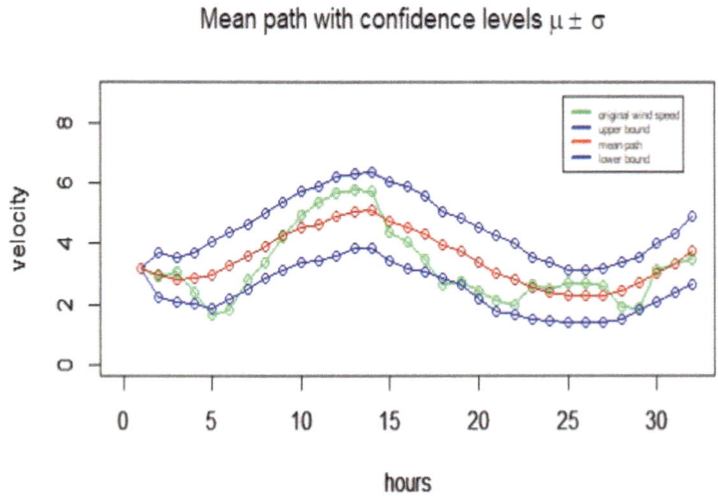

Fig. 9. Mean forecasting trajectory of portion 2_1 (within a considered confidence interval)

Table 3. Parameter values of short-term model of portion-2_2

Parameter value	Parameter						
	T	α	β	γ	σ	φ	μ
Value	12	0.2204	0.4900	0.4230	0.4210	1.0472	3.5200

Table 4. Parameter values of short-term model of portion-2_1

Parameter value	Parameter						
	T	α	β	γ	σ	φ	μ
Value	12	0.2204	0.3900	0.4230	0.4210	1.0472	3.5200

2.3 Conclusion

In this work, we present a short-term wind speed forecasting model which is capable of generating a mean forecasting trajectory that agrees well with the empirical data for a considerable period of time (32 h and 64 h) into the future. Considering the cyclic variation and the diffusion process of the selected portions of data, a new set of parameters were initially introduced to the SDE which generates a well-known Ornstein-Uhlenbeck process. In addition to the new set of parameters, an exponential expression was also added to the drift term of the SDE to avoid the generation of negative values that can be observed as a drawback in previous SDE models on wind speed forecasting. Then, using a pseudo-maximum likelihood estimation method the parameter values were estimated for the considered portion of data. In all three portion of data, the corresponding region of autocorrelation function has exhibited a sinusoidal variation of almost constant period. Constant cyclic variation in autocorrelation function confirms the cyclic variation in wind speed data in respective regions. Results of the all three cases, demonstrate that the mean forecasting trajectory coincides very well with the original wind speed values upto 10 h ahead. Of these three cases, the mean forecasting trajectory obtained for portion 2_1, has been totally confined to the confidence levels because it is the region where constant periodic behavior has been greatly indicated.

References

Brett AC, Tuller SE (1991) The autocorrelation of hourly wind speed observations. J Appl Meteorol 30:823–833

Calif R (2012) PDF models and synthetic model for the wind speed fluctuations based on the resolution of Langevin equation. Appl Energy 99:173–182

Carta JA, Ramirez P, Velazquez S (2009) A review of wind speed probability distributions used in wind energy analysis. Renew Sustain Energy Rev 13(5):933–955

Higham DJ, Higham DJ (2001) An algorithmic introduction to numerical simulation of stochastic differential equations. SIAM Rev 43(3):525–546

Iversen EB et al (2015) Short-term probabilistic forecasting of wind speed using stochastic differential equations. Int J Forecast 32(3):981–990. https://doi.org/10.1016/j.ijforecast.2015.03.001

Øksendal B (2003) Stochastic differential equations: an introduction with applications, New York. https://doi.org/10.1109/TAC.2006.882767

Zarate-Minano R, Mele FM, Milano, F (2016) SDE-based wind speed models with Weibull distribution and exponential autocorrelation. In: IEEE power and energy society general meeting. https://doi.org/10.1109/PESGM.2016.7741754

Zárate-Miñano R, Milano F (2016) Construction of SDE-based wind speed models with exponentially decaying autocorrelation. Renew Energy 94:186–196. https://doi.org/10.1016/j.renene.2016.03.026

Zhu X, Genton MG (2012) Short-term wind speed forecasting for power system operations. Int Stat Rev 80:2–23

Experimental Investigation to Achieve Minimum Surface Roughness in Wire EDM Process

K. P. M. Priyankara(✉) and G. I. P. Perera

Department of Mechanical and Manufacturing Engineering,
Faculty of Engineering, University of Ruhuna, Matara, Sri Lanka
kpmpeter@gmail.com

Abstract. This paper presents the behaviors and evaluation of Wire Electrical Discharge Machining (Wire EDM) manufacturing process. Wire EDM is also known as Wire-cut EDM, which uses a dielectric fluid submerged thin metallic wire to cut the conductive work piece material. Taguchi orthogonal array was used for planning the experiments to optimize the control variables for better surface roughness while machining AISI 420 modified metallic material. Pulse-on/off Time, Wire Feed, Spark Voltage and Peak Current were selected as the control variables. AISI 420 metal, also called as Stavax®ESR has a higher metal strength and is used for all types of moulds and moulding tools in the industry. The experiments were carried out by changing the aforesaid control variable parameters and TR200 handheld roughness tester was used to measure the surface roughness of metal cuts. Three surface roughness measurements were taken for each machined section and averaged. Plotting method and Analysis of Variance (ANOVA) were used for the analysis and interpretation of the results. Error variances were estimated by using Pooling-up strategy. The influences of each control parameter on surface roughness were estimated. Then, the results were used to develop a mathematical model. The developed mathematical model can be used to minimize surface roughness. Control variable parameter settings for the optimized surface roughness can be obtained by using the mathematical model.

Keywords: Wire EDM · ANOVA · Surface roughness · Optimization

1 Introduction

Wire electric discharge machine (WEDM) is well known for its ability to machine complex features with tough metals which normally cannot be easily machined in transitional machines. Thus, EDM is an ideal process for producing cutting tools, dies, and moulds for injection moulding, etc. (Chen and Cox 2017) since using the WEDM for the above applications may give better surface finish for the work material. But to achieve a better surface finish for the work a better selection of parameter control should be included. Then, the selection of optimum machining parameters in WEDM is an important step. Improperly selected parameters may result in serious problem on the work surface damage which imposes certain limits on the product quality (Muthu Kumar and Suresh Babu 2010). It involves the removal of metal by discharging an

R. Dissanayake and P. Mendis (Eds.): ICSBE 2018, LNCE 44, pp. 239–251, 2020.
https://doi.org/10.1007/978-981-13-9749-3_22

electrical current from a pulsating Direct Current (DC) power supply across a thin inter-electrode gap between the tool and the work piece. It is most commonly used for machining hard and difficult to machine materials with very close tolerances. Generally, WEDM is perceived to be an extremely actuate process and there are various reasons for this perception. Firstly, in WEDM, no direct contact takes place between the cutting tool (electrode) and the work piece; as a result, the adverse effects such as mechanical stresses, chatter, and vibrations normally present in traditional machining are eliminated. Secondly, the wire used as a cutting tool has high mechanical properties and small diameter 0.076 to 0.30 mm that produces very fine, precise, clean cuts. Finally, in WEDM, the movements of the work piece during cutting are controlled by a highly accurate Computer Numerical Control (CNC) system (Kumar et al. 2013). A schematic diagram of WEDM is shown in Fig. 1. Here, the work material is vaporized to perform a thin cut though the workpiece.

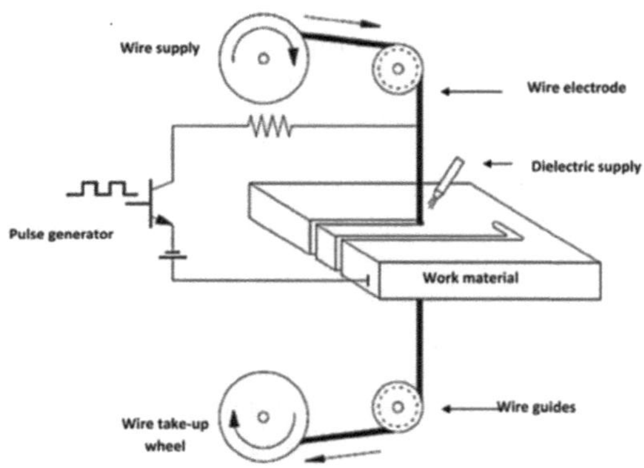

Fig. 1. Schematic diagram of WEDM (Goyal 2017)

2 Literature Review

To predict the Surface roughness and Kerf width while machining Aluminum and mild steel, (Tilekar and Das 2014) developed an empirical model. The analysis of variance resulted that spark on time has major influence on the surface roughness in the case of aluminum. On the other hand, input current has major influence in case of mild steel. In the case of Kerf width, wire feed rate and spark on time have significant effect on aluminum and mild steel respectively. (Lodhi and Agarwal 2014) made an investigation to study the Pulse-on time, Pulse-off time, wire feed and Peak current on Surface roughness. Lower amounts of surface roughness shows the high productivity of Wire EDM. (Babu et al. 2017) studied the optimization of the trim cutting operation of WEDM of AISI A3 tool steel for given machining conditions by the Taguchi technique coupled with Grey relational analysis to optimize the input variables of WEDM on Material Removal Rate (MRR), Surface Roughness (SR) and Gap Current (GC) of ballistic grade aluminum alloy. Results confirm that Ton, peak current and spark voltage are significant variables to the Grey relational grade.

(Prajapati et al. 2013) made an attempt to determine the important machining parameters for performance measures like SR, MRR, GC and kerf separately in the WEDM process. Response surface methodology experimental design method is used to obtain optimum parameter combination for maximization of MRR, SR as well as minimization of kerf while conditioning GC. The conclusion is made that cutting rate and surface roughness, the pulse on and pulse off time is most significant. The spark gape set voltage is significant for kerf. (Pasam et al. 2010) proposed, and investigated the Optimizing Surface Finish in WEDM using the Taguchi Parameter Design Method. In this work, an attempt is made to determine the optimal parameters for surface finish under varying conditions using the Taguchi parameter design process in the WEDM of Ti6Al4V alloy. This process was applied using a specific set of control parameters, ignition pulse current, short pulse duration, time between two pulses, servo speed, servo reference voltage, injection pressure, wire speed and wire tension with a response variable of surface roughness. In this research it was observed that increase in spark time meant discharge will remain for a longer time and leads to high cutting speed and high discharge energy. This improves the MRR and worsens the surface finish. Hence, small spark time gives better surface finish. To achieve better surface finish lower values of delay time should be selected.

(Azhiri et al. 2014) proposed a new approach, for the application of Taguchi, ANFIS and grey relational analysis for studying, modeling and optimization of wire EDM process while using gaseous media. The machining parameters, namely; pulse on time, pulse off time, wire feed rate, and wire tension and voltage are optimized with consideration of performance characteristics SR and Cutting Velocity (CV) improving the machining efficiency. Here, to correlate the relationship between process inputs and responses, adaptive neuro-fuzzy inference system has been utilized. In the end, a grey relational analysis has been used to maximize CV and minimize SR simultaneously. Results indicated that oxygen gas and brass wire guarantee superior cutting velocity. Also, according to Analysis of Variance (ANOVA), pulse on time and current were found to have significant effect on CV and SR.

(Bobbili et al. 2013) used the Experimental results showing that both approaches can be optimized in machining parameters effectively and input variables of WEDM on MRR and SR of high strength armour steel have been studied. Design of Experiments (DOEs) and regression techniques have been adopted to optimize the response characteristics. The conclusions are as Pulse-on Time (Ton), Pulse-off Time (Toff), and Servo Voltage (SV) are significant variables to the MRR and SR of the material. Also, optimum process parameter combinations for better MRR and SR are achieved by employing ANOVA. This technique is cost effective and time saving in evaluating the machining parameters without conducting a large number of experiments. Kumar et al. (2013) developed optimization of process parameters based on response surface methodology for pure titanium using WEDM process. The experimental plan is based on the Box–Behnken design. In this experiment, six set parameters, i.e., pulse on time, pulse off time, peak current, spark gap voltage, wire feed and wire tension have been varied to investigate their effect on output responses. Then concluded that SR and MRR are highly depend on pulse on time. Also, the result says that the dimensional deviation depends on moderate input current.

(Kumar and Ravikumar 2013) conducted an experiment on modeling and optimization of the wire EDM process by optimizing the parameters of the (WEDM) process. In the article, an attempt was made to determine the significant machining parameters for performance measures surface roughness in the WEDM process. Factors like speed, feed, Time on and Time off have been found to play a significant role for MRR and surface roughness. Muthu Kumar and Suresh Babu (2010) proposed a new approach, for optimization of the WEDM parameters on machining incoloy800 super alloy with multiple quality characteristics. In this paper, an application of combined Taguchi Method and Grey Relational Analysis made to improve the multi-response characteristics of MRR, Surface roughness in the Wire-Cut EDM of Inconel 800 has been reported. While applying the Grey-Taguchi method, The Material Removal Rate shows an increased value of 0.05351 g/min to 0.05765 g/min, the Surface Roughness shows a reduced value of 3.31 μm to 3.10 μm and the Kerf width shows a reduced value of 0.324 to 0.296 mm respectively, which are positive indicators of efficiency in the machining process.

(Maher et al. 2015) stated that the effect of process parameters on SR, Heat Affected Zone (HAZ) and CS in WEDM varying pulse on time, pulse off time and wire tension to reveal their impact on optimization parameters using Adaptive Neuro-Fuzzy Inference System (ANFIS). This study concludes that the peak current and pulse on time are the most significant parameters affecting the cutting speed, surface roughness and heat affected zone. The wire tension has minor effect on the cutting speed and heat affected zone, but it has great effect on the surface roughness. (Goyal 2017) identified significance of input variables of WEDM on MRR and SR of Inconel 625 super alloy. DOEs and regression techniques have been adopted to optimize the responses characteristics. The main conclusions are that current and pulse on time are the most influential parameters that affect the Material removal rate and surface roughness. Also, cryogenic tool electrode provides better machining performance (maximum MRR and better surface roughness) when compared to the normal tool electrode.

(Bobbili et al. 2015) performed modelling and analysis of material removal rate and surface roughness in wire-cut EDM of armour materials which used Buckingham pi theorem to model the input variables of WEDM on MRR and Ra of Al 7017 and Rolled Homogeneous Armour (RHA) steel alloy. The article says that the dependence of thermo-physical parameters on the mechanism of MRR and Ra has been interpreted, and owing to the thermal properties of WEDM, MRR of machining is higher for materials with a low melting temperature and specific heat, the rise in pulse-on time from 0.85 ms to 1.25 ms causes improvement in MRR and deterioration of surface finish and the wear rate of brass wire increases with increase in input energy, leading to wire breakage.

(Kubade et al. 2015) Parametric Study and Optimization of WEDM Parameters for Titanium diboride TiB2. The effect of pulse-on time, pulse-off time and wire feed are investigated in machining of TiB2 by keeping other process parameters constant. The experiment gave results that the pulse-on time and pulse-off time are most significant whereas wire feed rate is insignificant for material removal rate. Also the research proposes a suitable set of variable combinations or better output. (Babu et al. 2017) conducted an experiment on multi response optimization of wire-EDM process parameters of ballastic grade aluminum alloy based on Taguchi method for Inconel 625 super alloy. Results confirm that TON, IP and SV are significant variables. Mathematical models were developed using response surface method for MRR, SR and IG to determine the relation between machine variables and performance measures. Optimum response characteristics such as MRR, SR and GC are improved with 6% error. To predict the optimization of process parameters for surface roughness of Inconel 625 in Wire EDM, using Taguchi and ANOVA method is noted.

(Babu and Soni 2017) developed the empirical model by taking different process parameters like pulse on time, pulse off time and peak current for the surface roughness. The results of the experiments say that surface roughness increases with increase on pulse on time and decrease with increase in pulse off time. For surface roughness, the most significant factor is pulse on time followed by pulse off time and peak current is an insignificant factor for surface roughness. Taguchi-based optimization of surface roughness and dimensional accuracy in wire EDM process with S7 heat treated steel is new research which has been done by (Chen and Cox 2017). This research focuses on the use of the Taguchi method to reduce the surface roughness and improve dimensional accuracy by optimizing the pulse on time, pulse off time, servo voltage, voltage, servo feed, wire tension, and wire speed. The researcher has given sets of parameters for better optimization of output while proposing that this systematic Taguchi approach can be further implemented to more complex manufacturing processes such as 3D printing or injection moulding processes (Table 1).

Table 1. Conclusion of literature review

Reference	Metal	Used methods	Expt. variables	Control variables
(Tilekar and Das 2014)	Aluminum and mild steel	Taguchi method	Ton, Toff, Peak Current (I), Wire Feed (Fw)	SR, Kerf Width (KW)
(Lodhi and Agarwal 2014)	AISI D3 steel	Taguchi method	Ton, Toff, I, Spark Voltage (Fw)	SR
(Babu et al. 2017)	Ballistic grade aluminum	Taguchi method	Ton, Toff, I, V	SR, Gap Current (GC), MRR
(Prajapati et al. 2013)	AISI A3 tool steel	Response surface methodology	Ton, Toff, V, Fw, Sw	SR, MRR, KW, GC
(Pasam et al. 2010)	Titanium alloy	Taguchi method	Ton, Toff, Tg, I, V, Fw, Pd	SR
(Azhiri et al. 2014)	Al/SI metal matrix	Taguchi method	Ton, Toff, V, Sw, Fw	SR, CV
(Bobbili et al. 2013)	Nickel based super alloy	Taguchi method	Ton, Toff, Fw, Pf, V, Sw	SR, MRR
(Kumar et al. 2013)	Pure titanium	Response surface methodology	Ton, Toff, I, V, Fw, Sw	SR, MRR, DD

(*continued*)

Table 1. *(continued)*

Reference	Metal	Used methods	Expt. variables	Control variables
(Kumar and Ravikumar 2013)	Al-SiC	Taguchi method	Ton, Toff, Fw	MRR, SR
(Muthu Kumar and Suresh Babu 2010)	Inconel 800 super alloy	Taguchi method	Ton, Toff, Fw, V	MRR, SR, KW
(Maher et al. 2015)	AISI 1050 Carbone steel	ANFIS	Ton, Toff, Sw	SR, CS, HAZ
(Goyal 2017)	Inconel 625 super alloy	Taguchi method, ANOVA	Ton, Toff, I, Fw, Sw	SR, MRR
(Bobbili et al. 2015)	Al, RHA	Buckingham pi theorem	Ton, I*V, Pf, S, H	MRR, SR
(Kubade et al. 2015)	Titanium Dibromide	Taguchi method	Ton, Toff, Fw	MRR, SR, OC
(Babu et al. 2017)	Inconel 625 super alloy	Taguchi method	Ton, Toff, I, V	SR, GC, MRR
(Babu and Soni 2017)	S7 heat treated steel	Taguchi method	Ton, Toff, I	SR
(Chen and Cox 2017)	Ballistic grade aluminum alloy	Taguchi method	Ton, Toff, Fw, Sw, V	SR, DD

3 Materials and Method

The machines, equipment and materials which are used for the experiment may be discussed here in this chapter with the method.

3.1 Material List of the Experiment

- Wire EDM machine - BMXP wire cut EDM
- Surface roughness tester – Time TR200
- Workpiece material - Stavax ESR (AISI 420 modified stainless tool steel)
- Wire material – Copper

3.2 Methods, Levels and Experiment Arrangement

Method: Here, the experiment is arranged by considering Taguchi's orthogonal array method and the experiment outcomes are analysed by ANOVA (Analysis of variance) method. Error variances were estimated by using Pooling-up strategy.

The five variables of the experiments are put into three levels to arrange the experiments. Each parameter has three levels except voltage. Voltage has two levels in which the levels are limited by machine capability (Table 2).

Table 2. Levels of variable parameters

No.	Process parameter	Unit	Level 1	Level 2	Level 3
1	V	V	Low	-	High
2	Ton	μs	48	55	62
3	Toff	μs	2	5	8
4	Fw	mm/min	1	2	3
5	I	A	2	4	6

The assigned levels of parameters are used to plan the experiment setups by using L18 Taguchi's orthogonal array. Then, there are 18 experiments conducted according to the Table 3 arrangement of variable parameters.

Table 3. Experimental arrangement

Expt. no	Parameter				
	V (H/L)	Ton (μs)	Toff (μs)	Fw (mm/min)	I (A)
1	L	48	2	1	2
2	L	48	5	2	4
3	L	48	8	3	6
4	L	55	2	2	6
5	L	55	5	3	2
6	L	55	8	1	4
7	L	62	2	3	4
8	L	62	5	1	6
9	L	62	8	2	2
10	H	48	2	1	2
11	H	48	5	2	4
12	H	48	8	3	6
13	H	55	2	2	6
14	H	55	5	3	2
15	H	55	8	1	4
16	H	62	2	3	4
17	H	62	5	1	6
18	H	62	8	2	2

4 Results and Discussion

The results of the experiments and results of the analysis are presented in this section. The analysed results are presented by using plotted graphs to get a better understanding on variations of controlled parameters. A mathematical model is also developed for surface roughness by using results. Then, the discussion of the experiment result is further discussed.

4.1 Experiment Outcomes

Three roughness values are taken for each experiment set to obtain an accurate result. Then, the mean value of measurements is taken as the roughness value which considers an analysis of the experiment outcomes. We can clearly identify the variations of outputs as follows in the Table 4.

Table 4. Experimental results

Sample no.	Measurements (μm)			Average (μm)
	No. 01	No. 02	No. 03	
1	5.953	5.927	6.086	5.989
2	5.803	5.403	5.865	5.690
3	6.474	6.249	6.192	6.305
4	5.824	5.894	6.018	6.088
5	5.173	5.453	5.301	5.309
6	6.952	7.070	6.850	6.957
7	5.811	5.744	5.800	5.785
8	6.287	6.293	7.118	6.566
9	6.328	6.289	6.240	6.286
10	5.082	5.357	5.762	5.400
11	6.937	6.999	6.839	6.880
12	7.225	6.467	6.838	6.843
13	6.525	6.181	6.428	6.378
14	5.321	5.647	5.261	5.410
15	7.217	7.465	6.975	7.219
16	5.348	5.566	5.440	5.451
17	7.068	6.381	7.419	6.956
18	6.211	5.779	6.205	6.065

The results are analysed, and the contribution factors of each variable is generated (Table 5).

Table 5. Contribution factors of parameters to SR

Parameter	Contribution factor (%)
V	2.8
Ton	-
Toff	34.0
Fw	25.4
I	37.7

The roughness values of experiment outcomes are plotted with respect to levels of each process parameter in identifying the variation of surface finish considering the changes of process parameter values. Figure 2 shows that the surface roughness increases when spark voltage is shift to the higher values. Figure 3 indicates that the variation of surface roughness for three levels of pulse-on time is quite similar and have a smooth peak at the middle level. There is dramatic increment of surface roughness with respect to the increment of pulse-off time according to Fig. 4. Figure 5 shows a strong negative relationship between surface roughness and the wire feed. The relationship between peak current and surface roughness is strong and positive according to Fig. 6.

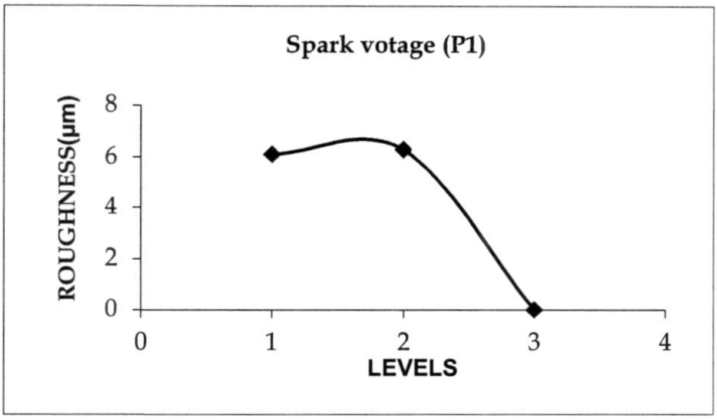

Fig. 2. Levels of voltage Vs SR

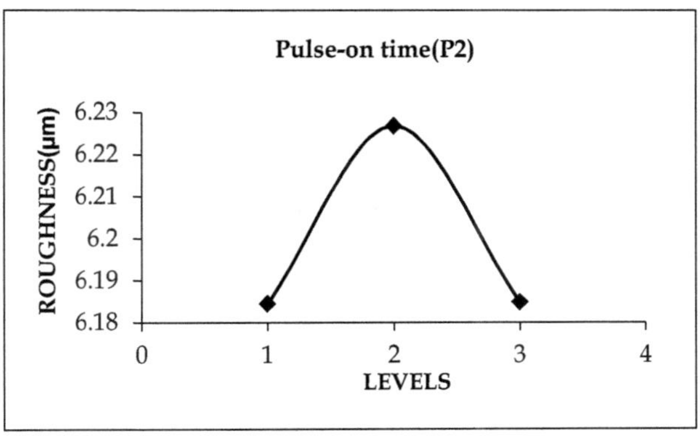

Fig. 3. Levels of pulse-on time Vs SR

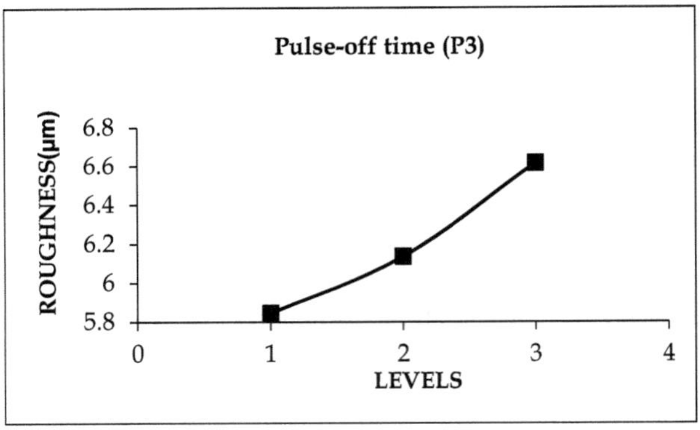

Fig. 4. Levels of pulse-off time Vs SR

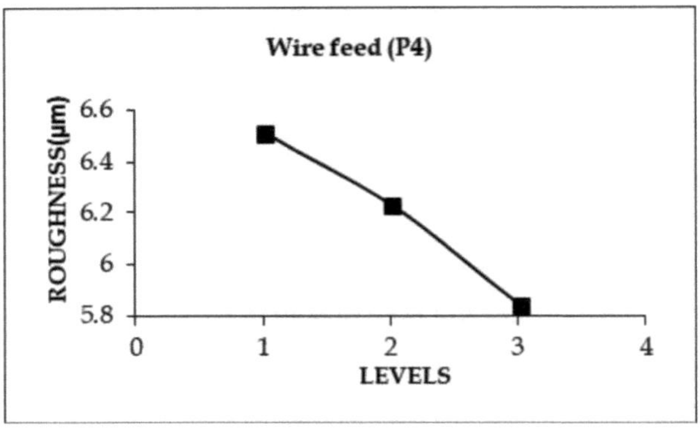

Fig. 5. Levels of wire feed Vs SR

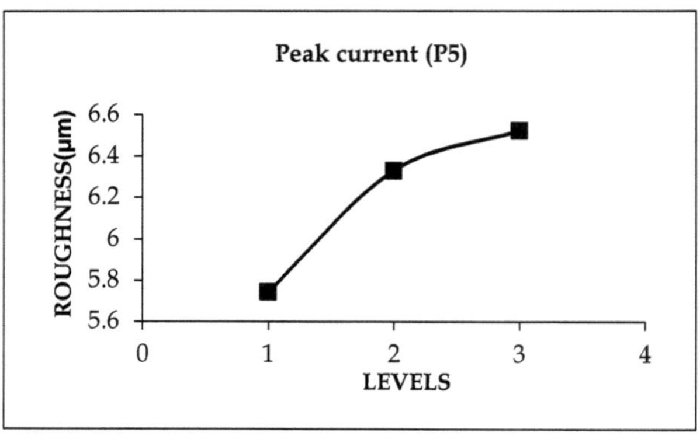

Fig. 6. Levels of peak current Vs SR

The following figure (Fig. 7) shows the change of roughness value with different parameters while the levels of the parameters are in consideration.

The graph shows that the roughness values of the five process parameters are almost close to 6 μm while the individual roughness value for different levels of the same input parameter interpret some variations. The variation of surface roughness is negligible when considering the variable parameter as Pulse-on time.

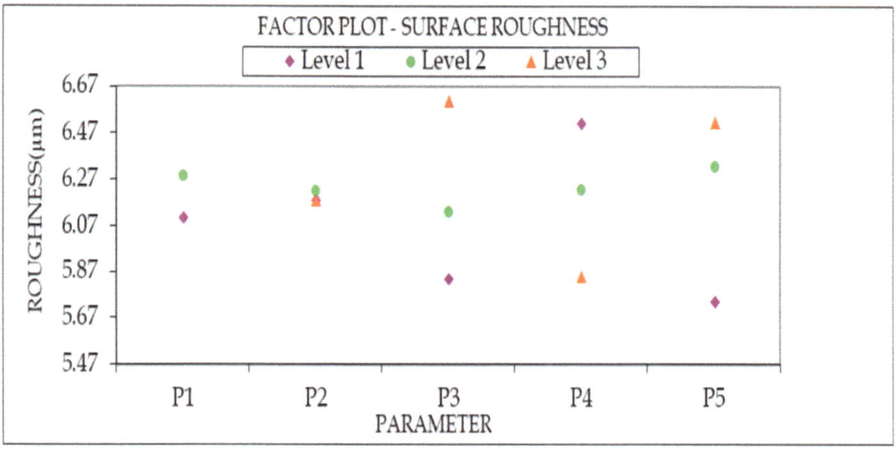

Fig. 7. Factor plot – surface roughness

4.2 Mathematical Model

$$SR = -61.1852 + 117.4868V - 34.0843V^2 + 0.0214T_{off} + 0.0106\,[T_{off}]^2 - 0.1373F_w - 0.0487\,[F_w]^2 + 0.5897I - 0.0494I^2 \quad (1)$$

The above equation shows the mathematical model of surface roughness. However, it needs to be verified. Thus, if any errors occur at the verification, the mathematical model will have to be modified to minimize the error.

5 Conclusion

The Taguchi orthogonal array was used for planning the experiments to optimize the control variables for better surface roughness while machining AISI 420 modified metallic material. Pulse-on/off Time, Wire Feed, Spark Voltage and Peak Current were selected as the control variables. AISI 420 metal, also called as Stavax®ESR has a higher metal strength and is used for all types of moulds and moulding tools in the industry. The experiments were carried out by changing the aforesaid control variable parameters and TR200 handheld roughness tester was used to measure the surface roughness of the metal cuts. Three surface roughness measurements were taken for each machined section and averaged.

Plotting method and Analysis of Variance (ANOVA) were used for the analysis and interpretation of the results. Error variances were estimated by using Pooling-up strategy. The influences of each control parameter on surface roughness were estimated. Then, the results were used to develop a mathematical model. The developed mathematical model can be used to minimize surface roughness. Control variable parameter settings for the optimized surface roughness can be obtained.

Acknowledgments. We would like to thank Department of Mechanical and Manufacturing Engineering, Faculty of Engineering, University of Ruhuna to providing necessary facilities for this research work. Finally, we would give our greater gratitude to the free education system of Sri Lanka.

References

Azhiri RB, Teimouri R, Baboly MG, Leseman Z (2014) Application of Taguchi, ANFIS and grey relational analysis for studying, modeling and optimization of wire EDM process while using gaseous media. Int J Adv Manuf Technol 71:279–295

Babu TV, Kumar SA, Soni JS (2017) Multi response optimization of wire-EDM process parameters of aluminum alloy 6063 using Taguchi method. Int J Curr Eng Technol 7 (3):1040–1045

Babu TV, Soni JS (2017) Optimization of process parameters for surface roughness of Inconel 625 in wire EDM by using Taguchi and ANOVA method. Int J Curr Eng Technol 7(3):1127–1131

Bobbili R et al (2013) Effect of wire-EDM machining parameters on surface roughness and material removal rate of high strength armor steel. Mater Manuf Process 28:364–368 6914 (August 2017)

Bobbili R, Madhu V, Gogia AK (2015) Modelling and analysis of material removal rate and surface roughness in wire-cut EDM of armour materials. Eng Sci Technol Int J 18(4):664–668. https://doi.org/10.1016/j.jestch.2015.03.014

Chen JC, Cox J (2017) Taguchi-based optimization of surface roughness and dimensional accuracy in wire EDM process with S7 heat treated steel. World Acad Sci Eng Technol Int J Chem Mol Nucl Mater Metall Eng 11(8):551–556

Goyal A (2017) Investigation of material removal rate and surface roughness during wire electrical discharge machining (WEDM) of Inconel 625 super alloy by cryogenic treated tool electrode. J King Saud Univ - Sci 0–7. http://dx.doi.org/10.1016/j.jksus.2017.06.005

Kubade PR et al (2015) Parametric study and optimization of WEDM parameters for titanium. Int Res J Eng Technol 2:1657–1661

Kumar A, Kumar V, Kumar J (2013) Multi-response optimization of process parameters based on response surface methodology for pure titanium using WEDM process. Int J Adv Manuf Technol 68:2645–2646

Kumar K, Ravikumar R (2013) Modeling and optimization of wire EDM process. Int J Mod Eng Res 3:1645–1648

Lodhi BK, Agarwal S (2014) Optimization of machining parameters in WEDM of AISI D3 steel using Taguchi technique. Proc CIRP 14:194–199. https://doi.org/10.1016/j.procir.2014.03.080

Maher I et al (2015) ScienceDirect improve wire EDM performance at different machining parameters – ANFIS modeling. IFAC-PapersOnLine 48(1):105–110. https://doi.org/10.1016/j.ifacol.2015.05.109

Muthu Kumar V, Suresh Babu A, Venkatasamy R, Raajenthiren M (2010) Optimization of the WEDM parameters on machining Incoloy800 super alloy with multiple quality characteristics. Optimization 2(6):1538–1547

Pasam VK, Madar Valli P, Swapna M (2010) Optimizing surface finish in WEDM using the Taguchi parameter design method. J Braz Soc Mech Sci Eng XXXII(2):107–113

Prajapati SB et al (2013) Effect of process parameters on performance measures of wire EDM for AISI A2 tool steel. Int J Comput Eng Res 3:274–278

Tilekar S, Das SS, Patowari PK (2014) Process parameter optimization of Wire EDM on aluminum and mild steel by using Taguchi method. Proc Mater Sci 5:2577–2584

Eco-Friendly Porous Carbon Materials for Wastewater Treatment

C. J. Thambiliyagodage[1(✉)], V. Y. Cooray[2], I. N. Perera[2], and R. D. Wijesekera[2]

[1] Faculty of Humanities and Sciences, Sri Lanka Institute of Information Technology, New Kandy Road, Malabe, Sri Lanka
charitha.t@sliit.lk

[2] Department of Chemistry, Faculty of Science, University of Colombo, Colombo 03, Sri Lanka

Abstract. Different porous carbon materials were applied to remove organic compounds and heavy metals from wastewater. Hierarchically, porous nitrogen rich monolithic carbon (HPC) was synthesized through the sol-gel synthesis method and carbonization process. Rice husk (RH), an agricultural waste material followed by carbonization at two different temperatures (RH400 and RH600) was used as the naturally generated porous carbon composite. Activated carbon (AC) was used as the control. According to the FT-IR spectra, different functional groups are present in all adsorbents. Scanning electron microscopic (SEM) images show an irregular shaped, random disordered macropore structure in RH and a robust sponge-like disordered macropore structure with ligaments in HPC. Transmission electron microscopic (TEM) images of these materials show a disordered mesopore network. Adsorption capacities of these porous carbon materials were determined for organic dyes and metal ions. Methylene blue and Methyl orange were used as the model organic compounds and Pb^{2+}, Cu^{2+}, Ni^{2+}, and Cd^{2+} ions were selected as the metal ions. The experimental data demonstrate that the adsorption capacity of AC for Methyl orange (5.88 mg/g) is much higher than that of the alternatives (RH400–1.97 mg/g, RH600–0.69 mg/g, and HPC – 1.14 mg/g). Adsorption capacity of all the adsorbents for Methylene blue is quite similar. Adsorption capacity of RH400 for Pb^{2+} is much higher than that of RH600, AC and HPC. RH400 has a greater adsorption for Cu^{2+} and Ni^{2+} than the other adsorbents. All adsorbents show a low adsorption capacity for Cd^{2+}. RH400 is a promising adsorbent for wastewater purification.

Keywords: Rice husk · Hierarchically porous carbon · Organic compounds · Metal ions · Adsorbance

1 Introduction

Organic pollutants (Li *et al.* 2006) and heavy metals (Jaishankar *et al.* 2014) in wastewater are toxic and non–biodegradable environmental hazards that can affect human health. The volume of wastewater generated by the industrial parks of Sri Lanka has been estimated as 30 million cubic meters per year but the extent of wastewater

R. Dissanayake and P. Mendis (Eds.): ICSBE 2018, LNCE 44, pp. 252–260, 2020.
https://doi.org/10.1007/978-981-13-9749-3_23

used for agricultural activities has not been determined as yet (Senarath and Visvanathan 2001). The chronic kidney disease of unknown etiology (CKDU) which is prevalent in the North-central area of the country has become one of the major health issues related to water pollution of agricultural wastewaters (Janakan et al. 2009). Organic pollutants harm ecosystems by causing acute and chronic health effects, and by destroying aquatic life etc. (Jayasumana et al. 2014). Hence, it is necessary to remove these toxic materials from wastewater. Various methods have been used to remove pollutants, such as precipitation (Kim et al. 2002), membrane filtration (Nasirabadi et al. 2016), oxidation (Ebrahiem et al. 2017) and adsorption (Sophia and Lima 2018). Adsorption technology is rapidly gaining prominence, and activated carbon (AC) is the most popular adsorbent due to its porous structure, high specific area and high adsorption capacity (Dias *et al.* 2007). However, the high cost of commercially available AC (~US $90/kg) has limited its usage. The main objective of the present study is to investigate potential alternatives to AC. Rice husk was selected as the agricultural waste due to its abundance at low cost and the presence of silica in addition to carbon after carbonization. In addition to making use of an agricultural waste, synthesis of hierarchically porous monoliths was also of interest to investigate the effect of organized interconnected porosity on adsorption.

2 Materials and Reagents

Rice husk was collected from a rice mill at Thalawathugoda, Sri Lanka. Methylene Blue (Basic Blue 9) and Methyl Orange were purchased from Eurostar Ltd, UK, Copper (II) nitrate trihydrate, A.R was purchased from Himedia Laboratories, India. Lead acetate was purchased from British Drug Houses, England. Cadmium acetate was purchased from Hopkin and Williams, UK. Ethyl alcohol (95%), formaldehyde (34.5 wt%), *o*-phenylenediamine (99.5%) and polyethylene glycol (PEG, Mol. Wt: 6000) were purchased from Sigma-Aldrich. Resorcinol recryst. extra pure was purchased from Merk Limited Mumbai.

3 Material Characterization

FT-IR spectra of the samples were collected by a Varian 600 FT-IR spectrometer. For FT-IR studies, samples and KBr were mixed in the ratio of 1:10 and ground to make a pellet. The morphology of the samples was analyzed by Hitachi SU6600 Analytical Variable Pressure FE-SEM and Oxford instruments EDX with AZtec software **scanning electron microscope operated at 10 kV. Transmission electron microscope images of the samples were collected with a** EOL - JEM – 2100 High resolution transmission electron microscope**. The acceleration voltage was 200 kV. Samples were prepared on a lacey carbon film supported on a Cu grid.** The adsorption capacity of the adsorbents on metal ions was investigated using an Atomic Absorption Spectrometer (GBS 932 plus AAS). The adsorption of organic dyes was measured on a GENESYS 10S UV-VIS spectrophotometer.

4 Methodology

Rice husk (RH) was washed with distilled water and ethanol followed by drying at 75 °C. The sample was carbonized separately at 400 °C and 600 °C for 2 h in air (RH400 and RH600 respectively). Hierarchically porous carbon (HPC) was prepared by dissolving resorcinol (3.0 g) and polyethylene glycol 6000 (1.25 g) in a solvent mixture of ethanol (11.4 mL) and distilled water (9 mL) with magnetic stirring at 50 °C. Afterward, *o*-phenylenediamine (0.1 g) was added to the above solution and stirred for 30 min at 50 °C. Formalin containing formaldehyde (4.42 g, 35 wt%) was quickly added to the solution. The reaction mixture was stirred at 50 °C for another 10 min. The homogenous emulsion was sealed and transferred to an oven and heated at 90 °C for 4 h. It was dried at 50 °C for 24 h followed by carbonization at 600 °C for 2 h in an oxygen restricted environment. Both RH and HPC were separately crushed to a fine powder and sieved through a 500 μm mesh.

Batch adsorption studies were carried out by stirring 20 mL of dye solution (9.46 mg/L Methylene blue and 3.10 mg/L Methyl orange, separately) and 40 mL of metal ion solution (6.05 mg/L Cu^{2+}, 6.88 mg/L Pb^{2+}, 1.20 mg/L Cd^{2+} and 8.23 mg/L Ni^{2+}, separately) with varying weights of adsorbent (0.025 g and 0.05 g) at 300 rpm for 1 h. The adsorption was determined by UV-visible spectrophotometry (organic dyes) and AAS (metal ions).

5 Results and Discussion

Rice husk is a naturally available agricultural waste material which is highly abundant in rice cultivating countries. Rice husk contains mainly carbon based compounds such as cellulose, hemicellulose and lignin, and silicon (Chuah *et al.* 2005). In addition, compounds of Mg, K and Ca are also present (Chuah *et al.* 2005). Upon carbonization, the above mentioned carbon based compounds convert to carbon which is rich with functional groups. Silica is also produced while metal ions form their oxides. The carbon and silica components function as adsorbents in this study. Hierarchically porous monolithic carbon was synthesized by the sol-gel method where the solution containing resorcinol and formaldehyde turned to a gel upon completion of the polymerization. Hydrophilic polymer PEG, acts as the porogen. *O*-phenylenediamine catalyzes the polymerization of resorcinol and formaldehyde by providing the required alkalinity for the reaction. The authors assume that it also acts as a crosslinker in the polymerization through the Mannich reaction, as with diaminohexane as reported by Hao *et al.* (2011). The reaction mixture was heated at 50 °C to provide the required activation energy.

Table 1. FT-IR frequencies RH400, RH600, HPC and AC

Sample	RH400	RH600	HPC	AC
IR frequencies (cm^{-1})	3400–3800 ~1620 1095 798.4 465	3400–3800 ~1620 1095 798.4 465	3400–3800 1500–1600	3400–3800 ~1620 1500–1600

The chemical functionalities of the prepared porous carbon materials RH400, RH600, HPC and AC were characterized by FT-IR spectroscopy as given in Table 1. The bands in the region of 3400 cm^{-1}–3800 cm^{-1} are ascribable to vibrations in non-bonded O-H groups for alcohols, phenols and carboxylic acids (Allouche et al. 2015). The band appearing at around 1620 cm^{-1} corresponds to the C = O vibration of lactonic, carboxyl, or anhydride groups in AC and RH (Allouche et al. 2015). These functional groups are absent in HPC. The bands in the region of 1500 cm^{-1}−1600 cm^{-1} which are more apparent in AC and HPC are ascribable to aromatic ring C = C stretching vibrations (Mopoung *et al.* 2015), and these are hidden in RH400 and RH600 due to the presence of a strong band around 1620 cm^{-1}. In RH400 and RH600, the band at 1095 cm^{-1} is assigned to Si-O-Si stretching, the band at 798.4 cm^{-1} to Si-O bending and the band at 465 cm^{-1} to Si-O out of plane deformation (Shokri et al. 2009).

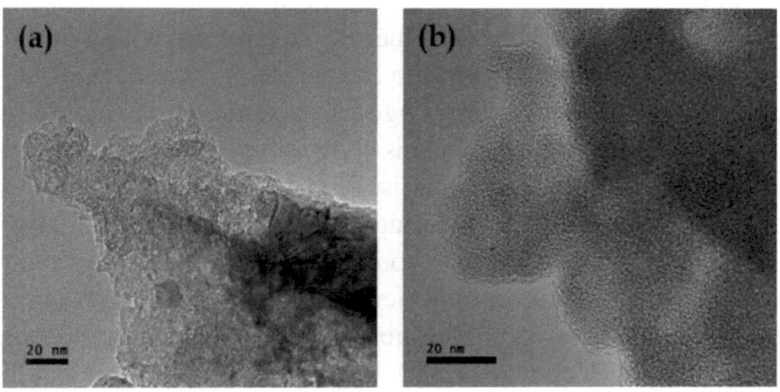

Fig. 1. TEM images of (a) RH (b) HPC.

TEM images of RH and HPC (Fig. 1 (a) and (b)) Polymers such as cellulose, hemicellulose and lignin are subject to chain scissions or depolymerization, breaking of C-C and C-O bonds within ring units and aromatization, and ultimately forming graphite like structures. This process creates more mesopores and micropores (Galarneau *et al.* 2016). The disordered mesopore structure of HPC is created mainly by the thermal decomposition of PEG during the carbonization (Hou *et al.* 2004). PEG molecules leave the carbon matrix leaving voids which remain in the structure as mesopores.

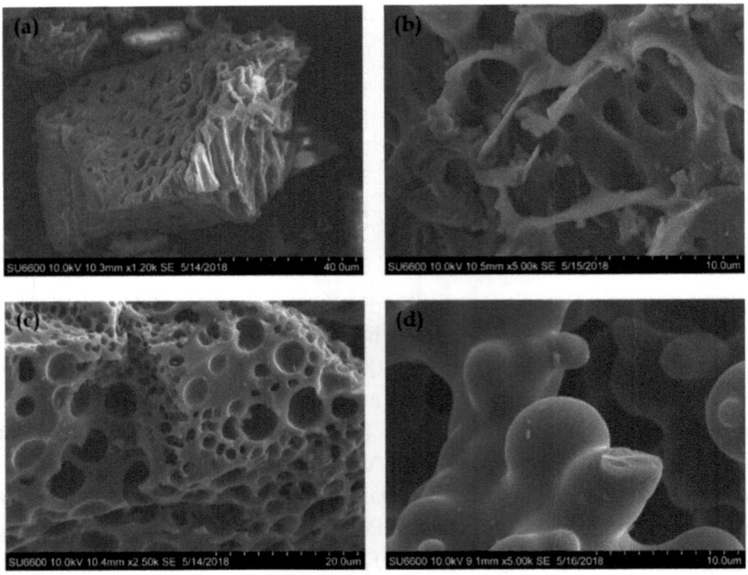

Fig. 2. SEM images of (a) RH × 1200 (b) RH × 5000 (c) HPC × 2500 (d) HPC × 5000.

The macropore structure of the prepared materials was studied by SEM. The irregular shaped, disordered, naturally originated macropore structure could be seen in RH 400 (Fig. 2(a)). Figure 2(b) shows the connectivity of the macropores providing macro channels which facilitate the movement of electrolytes within the composite and hence enhance the interaction between the electrolyte and the composite. SEM image of HPC (Fig. 2(c)) shows the robust sponge-like disordered macropore structure with ligaments. The structure of the ligaments is shown in Fig. 2(d). Ligaments are formed during the phase separation of the synthesis. PEG acts as the porogen of this synthetic procedure. The macropore structure is formed due to the polymerization-induced phase separation (Mopoung *et al.* 2015). Resorcinol and formaldehyde monomers were mixed with thermally unstable polyethyleneglycol polymer to produce a well distributed xerogel. Mesoporous structure are caused by the thermal decomposition of PEG at high temperature (Hou *et al.* 2004). Resorcinol, PEG and water/ethanol solvent is a homogeneous mixture where resorcinol molecules are surrounded by PEG molecules. When formaldehyde is added to the mixture, base catalyzed resorcinol formaldehyde condensation polymerization occurs. As the polymerization progresses, hydrophilic groups of the resorcinol formaldehyde polymer interact with PEG through H-bonding and form aggregates. This is the starting point of spinodal decomposition, leading to a phase separation of the polymer rich phase and water rich phase. The phase separation ends at the sol-gel transition. The macropore size will be determined by the sol-gel transition kinetics, which in turn is controlled by the size of the polymer, the ratio of PEG/polymer, the amount and the type of the polymer and the temperature (Aghababazadeh *et al.* 2007). The SEM image (Fig. 2(c)), shows a heterogeneous distribution of the macropore system. The amount of PEG and base, the amount and type of polymer, and temperature are constant during the synthesis and hence, sol gel transition kinetics is constant. During the synthesis, PEG 6000 was used. Lower molecular weight PEG interacts more strongly with the oligomers of the resorcinol-formaldehyde polymer *via* multiple interactions, adsorb on the surface of the oligomers and cover the particle leading to faster condensation of the oligomers. This favors the sol-gel transition resulting in smaller macropores. However, during the synthesis when more and more oligomers are produced, the number of PEG molecules in the reaction mixture is insufficient to interact with the oligomers and hence, incomplete condensation results. Eventually, the sol gel transition decreases producing larger pores (Galarneau *et al.* 2016).

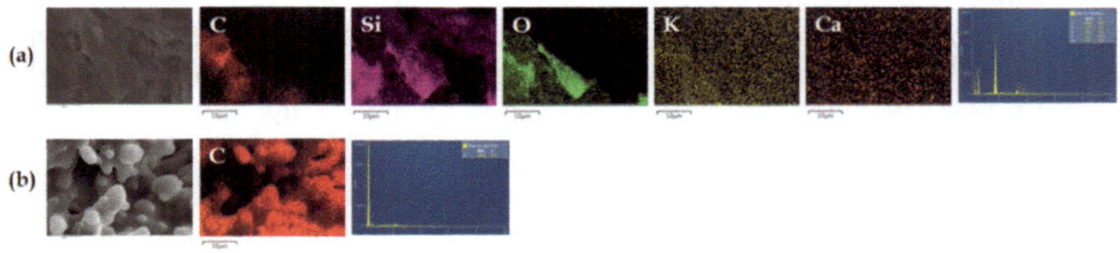

Fig. 3. SEM image, element maps and EDX spectrum of (a) RH and (b) HPC.

Samples were analyzed by SEM-EDS in order to qualitatively determine the elemental composition. Figure 3 shows the element maps and the energy dispersive X-ray spectra of RH (Fig. 3(a)) and HPC (Fig. 3(b)). The element maps and EDX spectra show that RH is rich in carbon and silica and that K, and Ca are also present. These elements are probably present as their oxides (K_2O and CaO respectively) as supported by the presence of oxygen in the element maps and the spectrum. HPC on the other hand consists mainly of carbon.

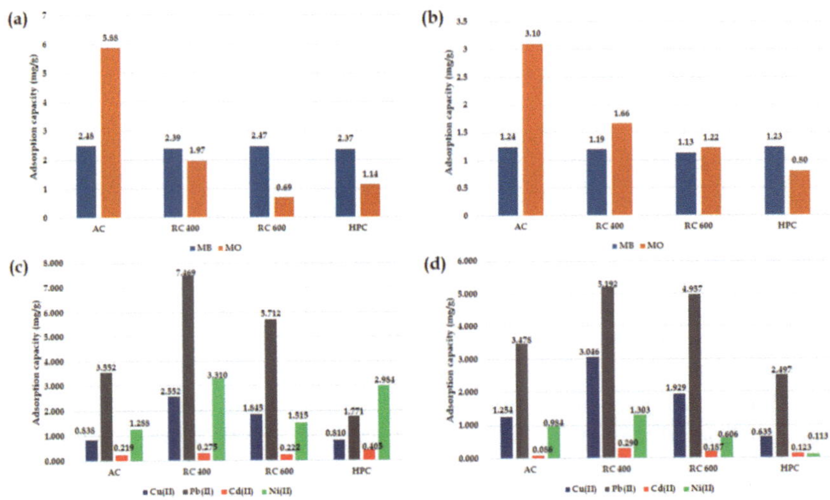

Fig. 4. Adsorption capacity of organic dyes, adsorbent dosage (a) 0.025 g (b) 0.050 g, and the adsorption capacity of metal ions, adsorbent dosage (c) 0.025 g and (d) 0.050 g.

The adsorption capacities of the adsorbents (adsorbent dosage as 0.025 g and 0.05 g) for organic dyes and metal ions are given in Fig. 4(a)–(d). The adsorption capacity of AC for Methyl orange is much higher than that of the examined alternatives, RH400, RH600 and HPC, irrespective of the adsorbent dosage used. However, the adsorption capacity of all the adsorbents is almost the same for Methylene blue. The mechanism of dye adsorption depends on many factors such as the physical and chemical nature of the adsorbent and the mass-transfer process (Gong *et al.* 2013). Methyl Orange and Methylene Blue are planar molecules and can be readily adsorbed through π-π interactions between the aromatic rings of the dye molecules and the hexagons of AC, RH and HPC (Gong *et al.* 2013). The higher adsorption capacity of AC for Methyl Orange than the other adsorbents suggest that AC has more rings and other functional groups than the other adsorbents. However, the adsorption of Methylene Blue by these adsorbents is almost the same. Methyl Orange is an anionic dye and Methylene Blue is a positively charged dye at the working pH (6.5). Methylene Blue interacts electrostatically with negatively charged and electron rich functional groups of the adsorbents. Negatively charged silica is found in RH but not in AC and HPC. However, all of these adsorbents have electron rich functional groups. The net result is that Methylene Blue would interact with all the adsorbents in a similar manner. Interaction of Methylene Blue and Methyl Orange with the carbon structure is shown in Fig. 5.

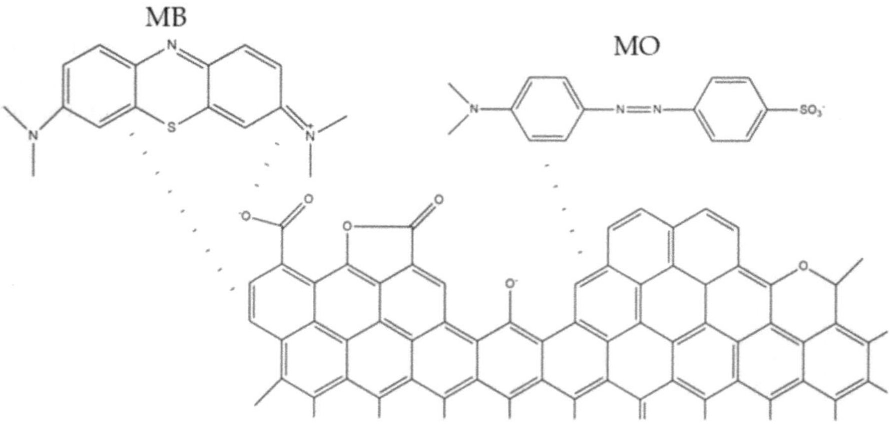

Fig. 5. Interactions of Methylene Blue and Methyl Orange with the carbon structure

The adsorption capacity of RH400 for all the metal ions selected for the study is higher than for the other adsorbents, the highest being for Pb^{2+} (7.47 mg/g). In general, RH600 showed a higher adsorption capacity than AC and HPC. All the examined adsorbents adsorbed the other tested metals, Cu^{2+}, Cd^{2+}, and Ni^{2+} the lowest being for Cd^{2+} (0.275 mg/g by RH400). Metal ions are positively charged and they can interact only *via* electrostatic interactions. The negatively charged silica and electron rich functional groups in RH400 can readily interact with metal ions. RH600 has silica as well but the carbon content and concentration of functional groups is low due to the higher temperature of pyrolysis. Functional groups leave as gaseous molecules with increasing temperature of pyrolysis (Moldoveanu 1998). AC and HPC have only the electron rich functional groups which can interact with metal ions resulting in low adsorption capacities. Though HPC has a well-defined porous structure its adsorption capacity is lower than that of the other adsorbents. The results obtained indicate that the surface functionality of the adsorbent is more critical for adsorption of the studied organic dyes and metal ions than the surface area and porosity. The adsorption capacity of both dyes and metal ions studied decreased with increasing adsorbent mass. Adsorption sites which adsorb the adsorbates remain saturated and at higher adsorbent dosage aggregation of particles take place resulting in a lower surface area and hence leads to lower adsorbent capacities.

6 Conclusions

Hierarchically porous nitrogen rich monolithic carbon could be successfully synthesized by sol-gel synthesis. Rice husk based porous composites were also prepared as the naturally available low cost adsorbent. HPC possesses a defined macropore and mesopore structure and RH has a naturally originated porous structure.

Methyl Orange was removed more successfully by AC (77.5%) than the other adsorbents. The adsorption of Methylene Blue was similar with all the adsorbents (>90%). All the tested metal ions were removed most efficiently by RH400. The adsorption capacities of the metal ions ranged from ~ 0.28 mg/g for Cd^{2+} to 7.5 mg/g for Pb^{2+} by RH400. RH400 is a promising adsorbent for removing Methylene Blue and Pb^{2+}.

Acknowledgements. Authors acknowledge Sri Lanka Institute of Nanotechnology (SLINTEC) for providing facilities for SEM and TEM analysis. This work was supported by a grant from Sri Lanka Institute of Information Technology (SLIIT).

References

Sophia AC, Lima EC (2018) Removal of emerging contaminants from the environment by adsorption. Ecotoxicol Environ Saf 150:1–17. https://doi.org/10.1016/j.ecoenv.2017.12.026

Aghababazadeh R et al (2007) Synthesis and characterization of silica nanoparticles. Solid State Phenom 121–123:49–52. https://doi.org/10.4028/www.scientific.net/SSP.121-123.49

Allouche F-N, Yassaa N, Lounici H (2015) Sorption of methyl orange from aqueous solution on Chitosan biomass. Proc Earth Planet Sci 15:596–601. https://doi.org/10.1016/j.proeps.2015.08.109

Chuah TG et al (2005) Rice husk as a potentially low-cost biosorbent for heavy metal and dye removal: an overview. Desalination 175(3):305–316. https://doi.org/10.1016/j.desal.2004.10.014

Dias JM et al (2007) Waste materials for activated carbon preparation and its use in aqueous-phase treatment: a review. J Environ Manag 85(4):833–846. https://doi.org/10.1016/j.jenvman.2007.07.031

Ebrahiem EE, Al-Maghrabi MN, Mobarki AR (2017) Removal of organic pollutants from industrial wastewater by applying photo-Fenton oxidation technology. Arab J Chem 10: S1674–S1679. https://doi.org/10.1016/j.arabjc.2013.06.012

Galarneau A et al (2016) Synthesis and textural characterization of mesoporous and meso-/macroporous silica monoliths obtained by spinodal decomposition. Inorganics 4(2):9. https://doi.org/10.3390/inorganics4020009

Gong R et al (2013) Adsorptive removal of Methyl orange and Methylene blue from aqueous solution with finger-citron-residue-based activated carbon. Ind Eng Chem Res 52(39):14297–14303. https://doi.org/10.1021/ie402138w

Hao G-P et al (2011) Structurally designed synthesis of mechanically stable poly (benzoxazine-co-resol)-based porous carbon monoliths and their application as high-performance CO_2 capture sorbents. J Am Chem Soc 133(29):11378–11388. https://doi.org/10.1021/ja203857g

Hou ZH et al (2004) Influence of polyethylene glycol on pore structure and electric double-layer capacitance of carbon xerogel. J Cent South Univ Technol (Engl Ed) 11(3):255–260. https://doi.org/10.1007/s11771-004-0052-z

Jaishankar M et al (2014) Toxicity, mechanism and health effects of some heavy metals. Interdisc Toxicol 7(2):60–72. https://doi.org/10.2478/intox-2014-0009

Janakan N, Dissanayake LJ, Hewawasam S (2009) Chronic kidney disease of unknown aetiology in Sri Lanka. Weekly Epidemiol Rep 36(November/December):1–4. http://www.epid.gov.lk

Jayasumana C, Gunatilake S, Senanayake P (2014) Glyphosate, hard water and nephrotoxic metals: are they the culprits behind the epidemic of chronic kidney disease of unknown etiology in Sri Lanka? Int J Environ Res Public Health 11(2):2125–2147. https://doi.org/10.3390/ijerph110202125

Kim Y, Osako M, Lee D (2002) Removal of hydrophobic organic pollutants by coagulation-precipitation process with dissolved humic matter. Waste Manag Res 20(4):341–349. https://doi.org/10.1177/0734247X0202000405

Li QQ et al (2006) Persistent Organic Pollutants and Adverse Health Effects in Humans. J Toxicol Environ Health Part A 69(21):1987–2005. https://doi.org/10.1080/15287390600751447

Moldoveanu SC (ed) (1998) The chemistry of the pyrolytic process. In: Analytical pyrolysis of natural organic polymers. Elsevier (Techniques and Instrumentation in Analytical Chemistry), pp 9–32. https://doi.org/10.1016/S0167-9244(98)80023-1

Mopoung S et al (2015) Characterization and properties of activated carbon prepared from tamarind seeds by KOH activation for Fe (III) adsorption from aqueous solution. Sci World J 2015:1–9. https://doi.org/10.1155/2015/415961

Nasirabadi PS, Saljoughi E, Mousavi SM (2016) Membrane processes used for removal of pharmaceuticals, hormones, endocrine disruptors and their metabolites from wastewaters: a review. Desalin Water Treat 57(51):24146–24175. https://doi.org/10.1080/19443994.2016.1140081

Senarath U, Visvanathan C (2001) Environmental issues in brackish water shrimp aquaculture in Sri Lanka. Environ Manag 27(3):335–348. https://doi.org/10.1007/s002670010153

Shokri B, Firouzjah MA, Hosseini SI (2009) FTIR analysis of silicon dioxide thin film deposited by metal organic-based PECVD. In: Proceedings of 19th international plasma chemistry society, pp 1–4. www.ispc-conference.org

Identification of Joint Efficiencies in 13 mm Finger Jointed Timber Species Used in Sri Lanka

C. K. Muthumala[1(✉)], Sudhira De Silva[2], K. K. I. U. Arunakumara[3], and P. L. A. G. Alwis[4]

[1] Research, Development and Training Division, State Timber Corporation, Battaramulla, Sri Lanka
ck_muthumala@yahoo.com

[2] Department of Civil and Environmental Engineering, Faculty of Engineering, University of Ruhuna, Galle, Sri Lanka

[3] Department of Crop Science, Faculty of Agriculture, University of Ruhuna, Kamburupitiya, Sri Lanka

[4] Department of Agricultural Engineering, Faculty of Agriculture, University of Ruhuna, Kamburupitiya, Sri Lanka

Abstract. When using timber in construction and the furniture manufacturing industry, waste timber materials and short length sections of timbers which are dumped by sawmills are considered to be matter of concern. 'Finger joint' is a recognized technique connecting two small pieces of waste timber together to ensure their sustainable utilization. Currently, the technique is used in Sri Lanka for non-structural purposes such as making timber boards and furniture. However, issues related with the strength of the joints have not fully investigated in Sri Lanka. This study was undertaken to determine the tensile strength performance of seven timber species in both unjointed (clear) and finger-jointed methods with 13 mm finger lengths and 4 mm finger pitch. An SWR adhesive (PVA) type was used as bonding material at normal exposure conditions. Well-seasoned Grandis, Jack, Kumbuk, Mahogany, Pine, Satin and Teak timber materials were used for the study.

BS 373: 1957 and BS EN 15497:2014 were used as standards for tests. The tests for tensile properties were performed using the Universal Testing Machine (UTM 100 PC) with a loading plate moving speed of 01 mm/min. Maximum load was read on graph to calculate the ultimate tensile strength.

The highest ultimate average tensile strength values were recorded in control specimens and different joint efficiencies were recorded in different finger jointed species. The highest mean finger joint efficiency percentage was obtained from Mahogany timber species and the least mean finger joint efficiency percentage was recorded in Kumbuk timber species.

Keywords: Finger joint · Tensile strength · SWR glue type · Joint efficiencies

R. Dissanayake and P. Mendis (Eds.): ICSBE 2018, LNCE 44, pp. 261–267, 2020.
https://doi.org/10.1007/978-981-13-9749-3_24

1 Introduction

The history of wood architecture of Sri Lanka has revealed that architects and craftsman with the natural creative skills have existed since the beginning of the country's long history and civilization. As a result of low quality and low strength of the local wood used by some of these ancestral architects and craftsmen, most wood creations have not lasted for long and have perished with time (Ruwanpathirana and Muthumala 2010).

The forests of Sri Lanka have nearly 450 types of hardwood timber species which are used for structural and non-structural purposes. The wood is excellent material for roofing and other construction works, furniture, interior decorations, doors and window frames, panelling, partition borders, floorings, wood carvings, musical instruments etc.

Physical and mechanical properties are very important in structural and non-structural purposes of timber. The properties not only vary with the species but also with factors such as the moisture content and defects of wood.

Finger joint is a sustainable, eco-friendly and economically viable concept for the furniture industry. It ensures the sustainable utilization of small wood cut pieces which are removed as waste (Sandika 2017). Finger joint timber manufacturing is considered to be a viable solution for minimizing the waste production in furniture manufacturing activities. Types of glue applied for the joining process of finger joint is one of the key factors which determines the strength of the product. The highest tensile strength of finger jointed timber solids was recorded in the SWR glue type in Sri Lanka (Muthumala et al. 2018a, b).

However, issues related with the strength of the joints have still not been fully investigated in Sri Lanka. This study was undertaken to determine the tensile strength performance of both unjointed (solid) and finger-jointed in seven species of timber with 13 mm finger lengths, which is the commonly used finger type in Sri Lankan furniture manufacturing.

2 Materials and Methods

Unjointed and jointed finger joint samples were cut from seasoned, defect-free wood timber to calculate tensile strength properties. Finger joint specimens were made at the finger joint factory at Boossa in the State Timber Corporation (STC) using 13 mm finger joint length cutters and an assembling pressure of 6 MPa was used in this study. Clear timber specimens were taken as control specimens and finger jointed timber specimens with the same dimensions were made with constant finger geometry such as 13 mm Finger Length, 1 mm Tip width and 4 mm finger pitch. Finger joint geometry is shown in Fig. 1. Polyvinyl acetate (PVA) SWR adhesive was used as a bonding agent (Glue type) for finger jointed wood.

Table 1 shows that timber species mainly used in furniture industries in Sri Lanka. In this study those timber species were used to calculate tensile strength properties.

The specimens were prepared from defect – free, sawn woods and dimensions of the sample prepared for above test is shown in Fig. 2.

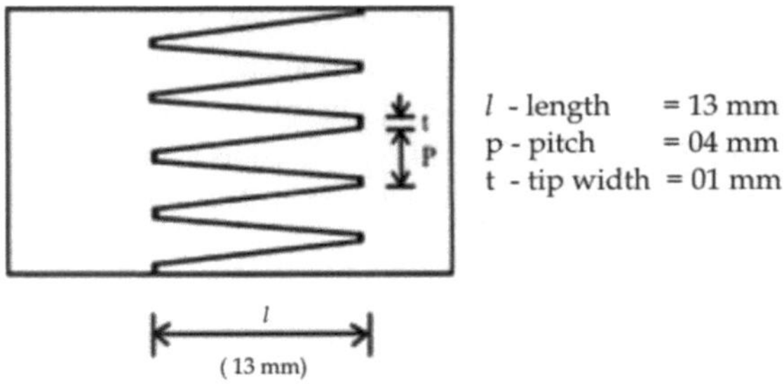

Fig. 1. Finger joint geometry

Table 1. Timber species mainly used in the furniture manufacturing industry in Sri Lanka.

Timber species	Botanical name	Timber class (STC)
Grandis	*Eucalyptus grandis*	Class-II
Jack	*Artocarpas heteroplyllus*	Luxury
Kumbuk	*Terminalia arjuna*	Special
Mahogany	*Swietenia macrophylla*	*Luxury*
Pine	*Pinus caribaea*	*Class-III*
Satin	*Chloroxylon swietenia*	*Luxury*
Teak	*Tectona grandis*	*Supper Luxury*

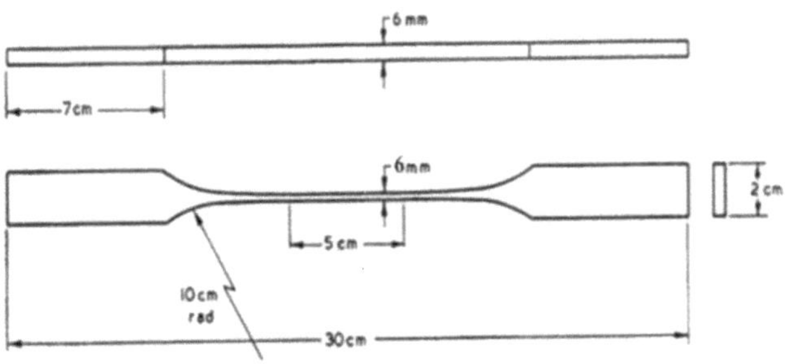

Fig. 2. The dimensions of the samples prepared for tension test.

The density values were calculated for seven timber species using following equation (Eq. 1). Dry weight of the timber samples was taken by placing them in the electric oven for a period of 48 h at 105 °C (BS EN 373:1957).

$$\text{Density} = \frac{\text{Weight of oven dry wood (kg)}}{\text{Volume of wood } (\text{m}^3)} \tag{1}$$

Determination of basic density was done based on the green volume and oven-dry weight using a water displacement method. Moisture contents were measured using a moisture meter.

Ten replicates were made for each timber species. Samples which were placed in normal room temperature conditions showed good structural performance compared to hot and wet conditioned ones (Vievek et al. 2016).

BS 373: 1957 and BS EN 15497:2014 were used as standards for tests. The test for tensile strength were performed by the Universal Testing Machine (UTM 100 PC) (Figs. 3 and 4).

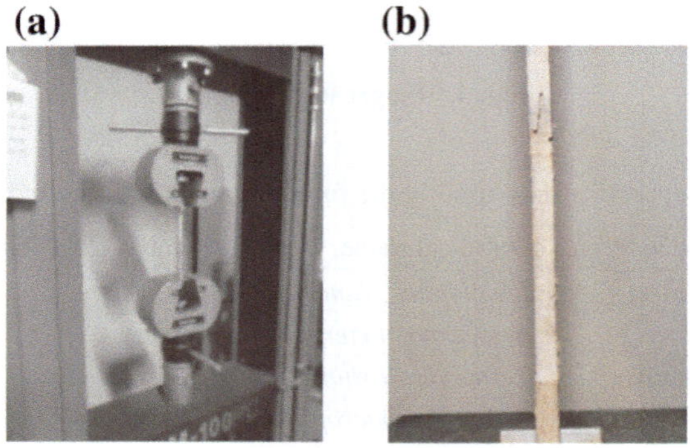

Fig. 3. a. The loading setup b. Finger joint specimen

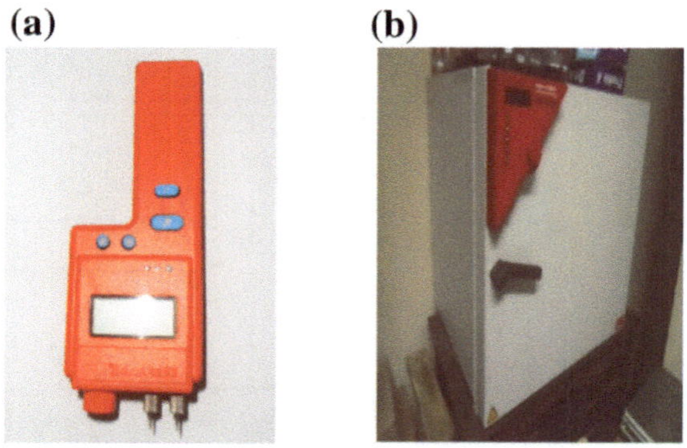

Fig. 4. a. Moisture meter b. Elec. Oven

Determination of the Tensile Strength

Maximum load act in timber section was taken into calculation. The following equation was used to calculate the Tensile strength (Eq. 2).

$$\text{Tensile strength of timber} = \frac{\text{Maximum Load}}{\text{Average cross section area of specimen}} \text{N/mm}^2 \quad (2)$$

3 Results and Discussion

Dry density values and moisture content of the timber species are shown in Table 2 (Fig. 5).

Table 2. Dry density values and moisture content of timber species.

Timber species	Density (kg/m^3)	Moisture content %
Grandis	570 ± 5	11 ± 2
Jack	645 ± 3	12 ± 2
Kumbuk	750 ± 4	11 ± 2
Mahogany	570 ± 3	12 ± 1
Pine	460 ± 2	12 ± 2
Satin	975 ± 5	12 ± 1
Teak	720 ± 5	12 ± 2

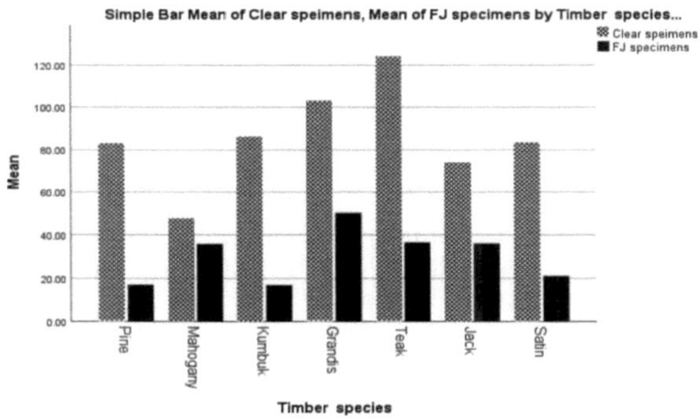

Fig. 5. Tensile strengths of clear and finger jointed specimens

According to Table 3, clear specimens and finger jointed specimens have statistically significant effect. $P < 0.0005$

Table 3. Statistical data from SPSS for tensile strength in both c lear and finger-jointed specimens

ANOVA						
		Sum of squares	df	Mean square	F	Sig.
Clear specimens	Between groups	33890.571	6	5648.429	7.372	0.000
	Within groups	48272.864	63	766.236		
	Total	82163.435	69			
FJ specimens	Between groups	9385.919	6	1564.320	14.253	0.000
	Within groups	6914.491	63	109.754		
	Total	16300.410	69			

According to the overall results of the tests, control (unjointed) samples gave significantly ($p < 0.005$) higher tensile strength than 13 mm finger jointed specimens. The highest tensile strength was recorded in clear specimens of Teak and the least tensile strength was showed by finger jointed Kumbuk.

Clear specimen of Teak showed the highest tensile strength followed by Grandis, Kumbuk, Satin, Pine, Jack, while the lowest recorded was Mahogany. Finger joint specimen of Grandis showed the highest tensile strength followed by Teak, Jack, Mahogany, Satin, Pine, while the lowest was Kumbuk.

According to Table 4, the highest mean efficiency percentage (75.08%) for finger joint was obtained from Mahogany and the least mean efficiency percentage for finger joint (19.60%) was recorded in Kumbuk. Mean finger joint efficiency percentage varied as, Mahogany > Jack > Grandis > Teak > Satin > Pine > Kumbuk.

Table 4. Finger joint efficiencies of seven timber species

Timber species	Mean tensile strength of clear specimens (N/mm^2)	Mean tensile strength of finger-joint specimens (N/mm^2)	Strength difference (Clear - joint) (N/mm^2)	Joint efficiency %
Grandis	103.16	50.24	52.92	48.70
Jack	73.88	36.06	37.82	48.80
Kumbuk	86.13	16.88	69.25	19.60
Mahogany	47.64	35.77	11.87	75.08
Pine	83.23	17.04	66.19	20.47
Satin	83.57	21.13	62.44	25.28
Teak	124.25	36.47	87.78	29.35

Yeh et al. (2011) reported that no significant difference in the tensile strength of finger-jointed lumber was found between horizontal and vertical finger formation.

4 Summary and Conclusion

The following conclusions can be drawn from this study.

Control (unjointed) samples gave higher tensile strength than 13 mm finger jointed specimens.

(1) The highest tensile strength was recorded in clear specimens of Teak and the least tensile strength was showed by finger jointed Kumbuk.
(2) The highest mean efficiency percentage (75.08%) in finger joint was obtained from Mahogany and the least mean efficiency percentage (19.60%) in finger joint was recorded in Kumbuk.
(3) The study illustrates the joint efficiency of seven finger jointed timber species. Finger joint manufacturers can use these results in increasing their production efficiency.

References

British Standards Institution (1957) BS 373: 1957, Methods of testing small clear specimens of timber. British Standards Institution, London

British Standards Institution (2014) BS EN 15497:2014, Structural finger jointed solid timber-performance requirements and minimum production requirements. European Committee for Standardization, B 1000 Brussels

Yeh M-C, Lin Y-L, Huang Y-C (2011) Evaluation of the tensile strength of structural finger-jointed lumber. Taiwan J Sci 26(1):59–70. www.airtilibrary.com. Accessed 3 Aug 2018

Muthumala CK, De Siva S, Alwis PLAG, Arunakumara KKIU (2018a) Factors affecting the glue strength of finger-joints in commonly used timber species in Sri Lanka. In: International symposium on agriculture and environment. Faculty of Agriculture, University of Ruhuna, Sri Lanka, pp 126–128

Muthumala CK, De Siva S, Alwis PLAG, Arunakumara KKIU (2018b) Investigate the most suitable glue type for finger-joints production in Sri Lanka. Res J Agric For Sci 6(11):6–9

Ruwanpathirana ND, Muthumala CK (2010) Wooden wonders of Sri Lanka. State Timber Corporation, Sri Lanka

Sandika AL, Pathirana GDPS, Muthumala CK (2017) Finger joint timber products for effective utilization of natural resources: an analysis of physical properties, Economic factors and Consumers' perception. In: International symposium on agriculture and environment, University of Ruhuna, Sri Lanka, pp 109–111

Vievek S, De Silva S, De Silva S, Muthumala CK (2016) Finger joint and their structural performance in different exposure conditions. In: 7th international conference on sustainable build environment, Kandy, Sri Lanka, pp 207–210

Experimental Investigation of Top Hinged Flap Type Wave Energy Device with a Bottom Gap – Preliminary Experimentations

L. U. Bakmeedeniya(✉) and S. D. G. S. P. Gunawardane

Department of Mechanical Engineering, University of Peradeniya, Peradeniya, Sri Lanka
lekhab@pdn.ac.lk

Abstract. The "Pendulor" type wave energy converting system essentially consists of a caisson, a water chamber facing the offshore open, and a flap that is hung down in that. The flap is driven by the oscillatory motion of the standing waves generated inside the caisson. The flap then drives a generator through a hydraulic power transmission system. The main objective of the proposed study is to study the hydrodynamics of the device with a gap between the bottom of the flap and the caisson. As an initiative, this paper describes the test facility arrangements with wave tank modification, calibrations and preliminary model test results. Accordingly, the experimental facility, the wave tank at the Faculty of Engineering, University of Peradeniya was modified to comply with a scale models of (1/40) of the proposed device configuration for typical wave climatic conditions in which the wave length is 100 m, maximum wave height is 6 m and the power density is 20 kW/m.

Keywords: Wave energy · Power capture · Flap-type converter · Hydrodynamic characteristics · Caisson · Bottom gap

1 Introduction

1.1 Significance of the Research Work

The energy from ocean waves is the most conspicuous form of ocean energy, possibly because of often spectacular but destructive effects. The waves are produced by wind action and are therefore an indirect form of solar energy. The potential for extracting and using the energy in ocean waves has been recognized for more than 200 years with the first patent for a wave energy converter being submitted by a father and a son named Girad in 1799 in Germany [1]. More than thousand patents had been registered by 1980 and the number has increased markedly since then but no viable technology has gained merit as a fully functional commercial product to harness wave energy. After the first oil crisis in 1974, significant investments were done in wave energy research. As a result, during 1973–1984 a number of large scale (2 GW) wave energy concepts were developed. By the time of the oil crisis many investors were inspired by the possibility of converting wave energy into usable energy, but in over three decades there are not many commercial uses compared to other renewables. Up until the year

R. Dissanayake and P. Mendis (Eds.): ICSBE 2018, LNCE 44, pp. 268–281, 2020.
https://doi.org/10.1007/978-981-13-9749-3_25

2000, there was a development of small scale wave energy projects. By the year 2000 and moving to the 21st century world awakened with the key challenges of the growing concerns regarding global warming, the need for a sustainable energy form and the utilisation of renewable form. Wave energy is a non-polluting and renewable source of energy, created by natural conversion of part of the wind energy above the oceans. The wave power density is higher than the wind and solar power since it is a concentrated form of solar energy through wind energy. Just below the ocean water surface the wave energy flow, is typically five times denser than the wind energy flow 20 m above the sea surface, and 10 to 30 times denser than the solar energy flow. As a non-polluting renewable source of energy it can give consistent power for low operating costs. Also it will provide the least visual impact in the point of view of structure since most of the systems will be fully submerged. More importantly these systems can be used efficiently and as shoreline protection systems as well.

Having considered the fact that there is significant potential for extracting and using the energy in ocean waves the world's attention now has focused on the possibilities of extracting power from wave energy. The "pendulor" device discussed in this paper invented in Japan is one of the pioneers among wave energy converters (WEC) which is the attention of this research. This was first invented in Japan at Muroran Institute of Technology as a result of work done in the early 1980s and has proven 40–50% of wave energy conversion efficiency at the sea of Muroran.

1.2 Background

The situation on wave energy in Europe was changed from 1990s and in northern America by 2000. Since wave energy absorption is a hydrodynamic process of considerable theoretical difficulty, much research needed to be done to get maximum benefit. Due to the problems that lie in the variability of the energy flux absorbed from the waves, in several time scales; sea states and seasonable variations and the survivability in extreme conditions are major issues. In general development, from concept to commercial stage in wave energy utilization has found to be difficult, slow and expensive process. This research is primarily focused on a top hinged flap type wave energy device which is known as Pendulor (Fig. 1(a)) invented in Japan with a bottom gap. Experimental sea trails at the Muroran sea, has shown that overall conversion efficiency is about 40–50%, while the theoretical efficiency of this type device is 100% (regular wave assumption without gaps) under the normal running conditions [2, 4]. However related experimental studies are still limited on testing the energy capture of Pendulor wave energy devices through hydrodynamic investigations. There is not much research done to explore the avenues in this area in Sri Lanka although a substantial progress has been achieved in the theoretical and numerical modelling of WECs [2]. Model testing in wave basins and evaluation of optimum conditions is still essential and finally, testing in real sea conditions [3].

1.2.1 Operation of Pendulor Wave Energy Conversion Device

In Fig. 1(a) basic components of a "Pendulor" type energy device are shown. A flap hung inside the caisson which faces to the ocean waves is driven by the horizontal water particle motion of the waves (Fig. 1(b)). The device operates on standing waves

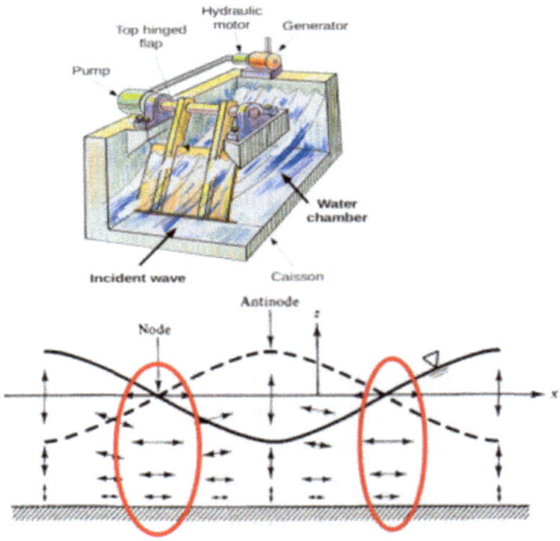

Fig. 1. (a) and (b) Basic "Pendulor type energy conversion device configuration and movement of water particle in the wave

which are created by superposition of the incident waves and reflected waves from the back wall. The excitation moment of the flap depends on the caisson configuration as well as the wave frequency and the wave height. The pendulum is placed at the nodal point of the standing waves Fig. 1(b) where the water particle motion is almost horizontal and to extract the maximum power output (where the kinetic energy is maximum). The nodal point is at a distance 'd' where d = L/4 for a wave inside a caisson, (L = wave length) for regular waves.

Inside the caisson, the pendulum is placed. The natural frequency (ω_n) of the pendulum and the water chamber (water quantity inside the caisson) is set to the incident wave frequency, to obtain the maximum efficiency of the device. The pendulum drives a rotary vane pump and the wave energy is converted into pressure energy of a fluid by vane pump. This pressurized fluid is then used to fulfil any energy requirement.

2 Objectives

In the ocean, the wave's frequency is varied according to local wave climate as well as global wave climate. To absorb the maximum wave energy, most of the device should resonate with the wave frequency. In the case of Pendulor device, various researches have been done to develop mathematical models and experiments to improve the performances [3, 5, 6]. Based on those models, analysis of motion response and corresponding power conversion capability of the device were conducted and discussed for regular and irregular wave climate in Sri Lanka [2]. The related theory and most of the research reveals that the performance of the device highly depends on the caisson configurations. These configurations involve the placement of the flap inside the caisson and it is highly necessary to understand the hydrodynamics to optimize the configuration [14]. No research has been done to look into the hydrodynamic characteristics of caisson with gaps around the flap. Therefore in this research a caisson with a flap hung on from the top with a bottom gap was tested for maximum power capture under different wave frequencies.

3 Research Design and Methodology

3.1 Development of the Scale Model of 1/40

A scale model of 1/40 was designed and constructed to test with some modifications to the existing towing tank. The experimental setup is shown in the Fig. 2 below.

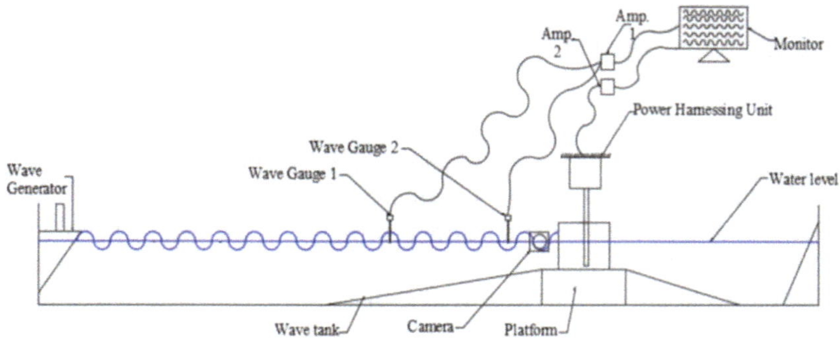

Fig. 2. The profile of the final test model in the wave tank

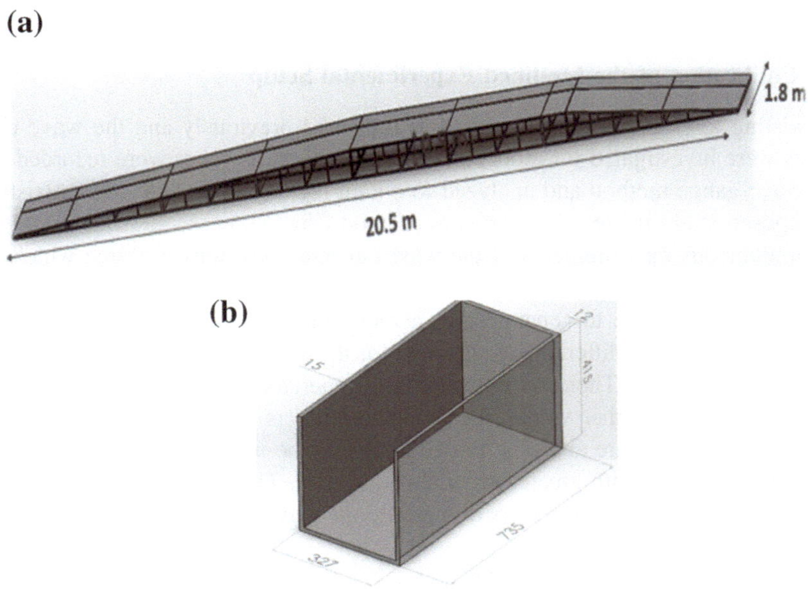

Fig. 3. (a) The platform model, (b) The caisson model

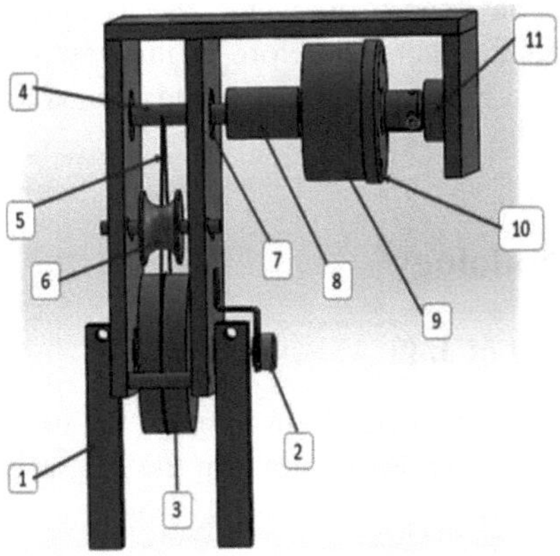

1. Flap Arms
2. Potentiometer
3. Main Pulley
4. Main shaft
5. Cable
6. Tension pulley
7. Main bearings
8. Flexible coupling
9. Magnetic particle brake
10. Rigid coupling
11. Torque sensor

Fig. 4. Components of power harnessing unit

The test model scale (1/40) has been selected mainly to match the width of the wave channel to achieve a 3D effect around the caisson (the model is nearly 1/3 rd of the width of the channel). To match the required water depth, for the model testing the tank needed to be modified using a platform as shown in the Fig. 2. The slope of the bottom was kept at 1/20 at seaward and 1/10 in the leeward direction. The dimensions of the platform and the basic caisson configuration with dimensions are shown separately in Fig. 3(a) and (b).

The caisson is placed at the top of the platform and the energy absorbed by the flap is transferred to the power harnessing unit. The main features of the power harnessing unit are shown in Fig. 4.

3.2 Calibration of the Modified Experimental Setup

The existing towing tank was modified as indicated previously and the wave characteristics were investigated for model study. The generated waves were recorded by the three wave gauge method and analyzed to obtain the required wave characteristics for the proposed study using a least squares method (Fig. 5).

Simultaneous measurements of the waves at three positions in flume which are in reasonable proximity to each other and are in a line parallel to the direction of wave propagation were done to separate the incident and reflected a spectra of generated irregular sea state (for different frequencies) for the experiment. Such measured waves are shown in Fig. 6. Then, to compare the generated waves with the anticipated sinusoidal waves, in other words, the fluid motion was done using a 3-point method which uses a least square analysis for decomposing the measured spectra into incident and reflected spectra with greater accuracy and range. Thereby, the wave channel was calibrated with the analyzed results to measure the wave heights required for proposed WEC model study (Table 1).

Fig. 5. Wave gauge assembly

Table 1. Incident wave height results for wave generation

	Stroke/cm , Wave Height/mm								
PT	19	16	15	14	13	11	9	8	6
0.9									
1									55
1.1							70	60	50
1.2						65	55	50	40
1.3					70	60	50		
1.4				65	58	50			
1.5		70	62	57	52				
1.6	75	60	57	55	50				
1.7	70	60	55	50					
1.8	65	55							
1.86	55								

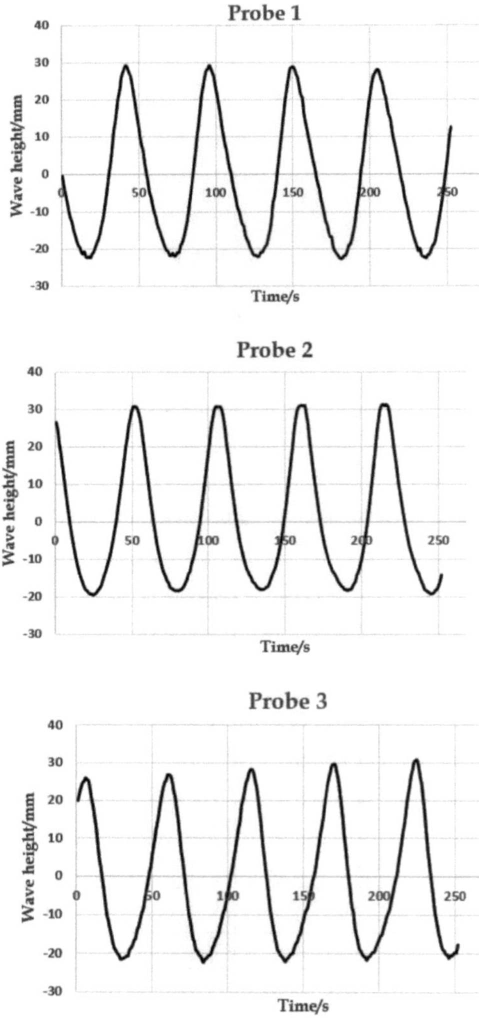

Fig. 6. Wave profiles for the three wave probes at periodic time 1.1 s

3.3 Power Takeoff System (PTO) to Measure the Power Capture

The arrangement shown in Fig. 4 is the PTO measuring system developed for the model. A torque sensor model **TQM 202-0.7** which has a sensitivity of 2.4435 mV/V has the range from 0–0.7 Nm is used to measure the excitation torque by the waves. A reaction torque, magnetic particle **BRAKE B35** model which has a torque range of 0.07–4 Nm is used to apply the load to the system and a high precision potentiometer was used to measure the angle of rotation of the flap (**Model 157** from Vishay Spectral can rotate 360º). Installation and measurements in action are shown in Fig. 7(a) and (b).

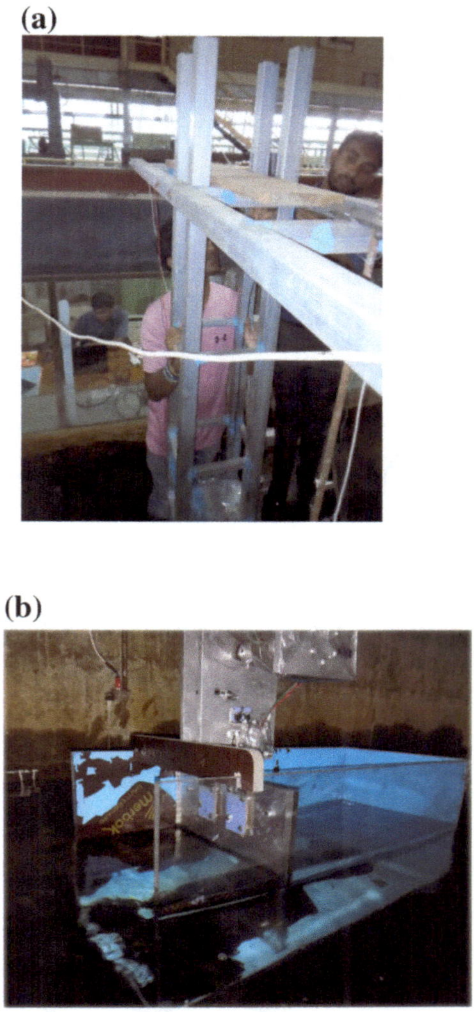

Fig. 7. (a). Assembling and assembled unit for preliminary model testing (b). Assembling and assembled unit for preliminary model testing

3.4 Measurements and Calculations

Measurements were obtained through a LABVIEW program as real time data for Torque on the shaft due to wave power, flap rotation angle and the incident wave height. A video camera was used as shown in the Fig. 2 to capture the motion of the flap and the water inside the caisson. The two signals from the power harnessing unit (torque and angle) will lead to interpret the wave energy absorbed by the flap which is generated through the torque sensor and the potentiometer. A magnetic particle brake was used to adjust the damping torque of the power harnessing unit (Fig. 8).

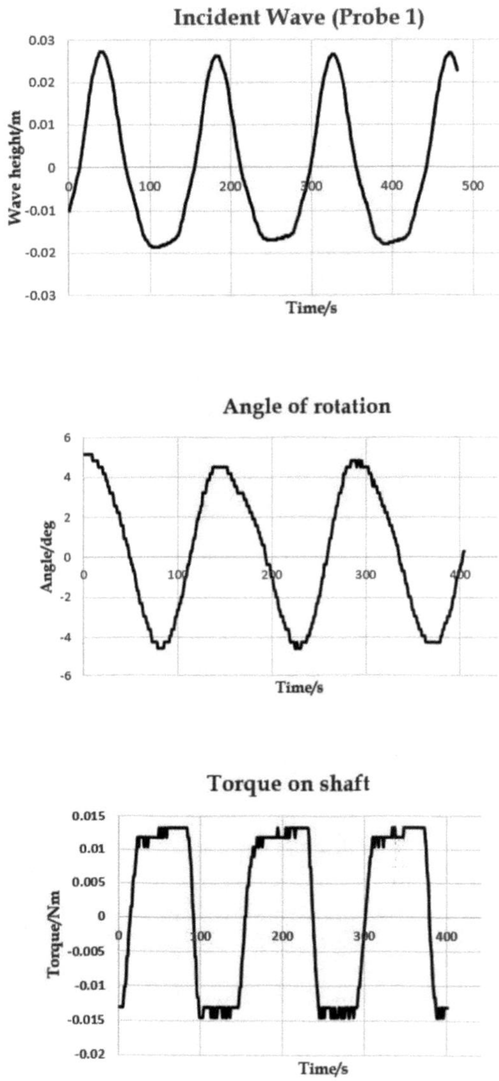

Fig. 8. Graphical representation for incident wave, angle of rotation and the torque

4 Results

4.1 Mathematical Model for the Flap Dynamics

The equation of motion of the flap for the constant damping (Coulomb damping N_m) brake torque device is given as the following single degree of freedom Coulomb damped forced oscillation system.

$$\sum I\ddot{\theta} + [N_h(\omega)].\dot{\theta} + N_m + \sum K.\theta = M_0 \sin \omega t \quad (1)$$

Where $\sum I = I_h(\omega) + I_m$ and $\sum K = K_h(\omega) + K_m$

I_m : moment of inertia of the flap, $I_h(\omega)$: moment of inertia of the added water mass, N_m: Coulomb damping torque, N_h: Hydrodynamic damping torque, K_m: Coefficient of restoring moment by the flap gravitation, $K_h(\omega)$: Coefficient of restoring moment by the water mass elevation, M_0: Wave excitation moment. For a straight caisson, the relationship between the above parameters and the wave parameters were discussed in detail in literature [2].

And the Power capture is given by

$$P_{max}(\omega) = N_m\left(\dot{\theta}\right) \quad (2)$$

Where $\Theta(\omega)$: Complex amplitude of flap rotation. The normalized power capture is the capture factor C_f defined by the ratio of the power capture divided by the wave power incident on the width of the flap.

$$\tilde{P}_{\max}(\omega) = C_f(\omega) = P_{max}(\omega)/[P_i(\omega)] \quad (3)$$

$PTO = \Lambda(\omega) = \left(\sum_{t_1}^{t_2} \frac{Torque \times Angle\, difference}{Time}\right)$ is the power take off (power capture) from the system. And the wave power is given by

$$P_w = \frac{1}{4} b \left(\frac{H}{2}\right)^2 \rho g \frac{\omega}{k_0}\left(1 + \frac{2k_0 h}{sinh(2k_0 h)}\right) \quad (4)$$

where b: flap width, H: incident wave height, h: water height, ω: wave frequency, k_0 is the wave number which can be calculated from the dispersion relationship given by $\omega^2/g = k_0 \tan(k_0 h)$.

4.2 Experimental Analysis of the Model

Initially A 1/40th scale model caisson configuration for a water depth of 25 cm and a wave length of 2 m, was considered as a preliminary test flap arrangement without a bottom gap for wave periods 1.1(ω = 5.712), 1.4(ω = 4.488), 1.6(ω = 3.927) and 1.86 (ω = 3.378). Figure 9 shows the variation of average power capture and conversion efficiency with PTO damping for different wave periods. It indicates that the peak for efficiency will be beyond the experimented range and some modifications (large damping) to the PTO damping is required.

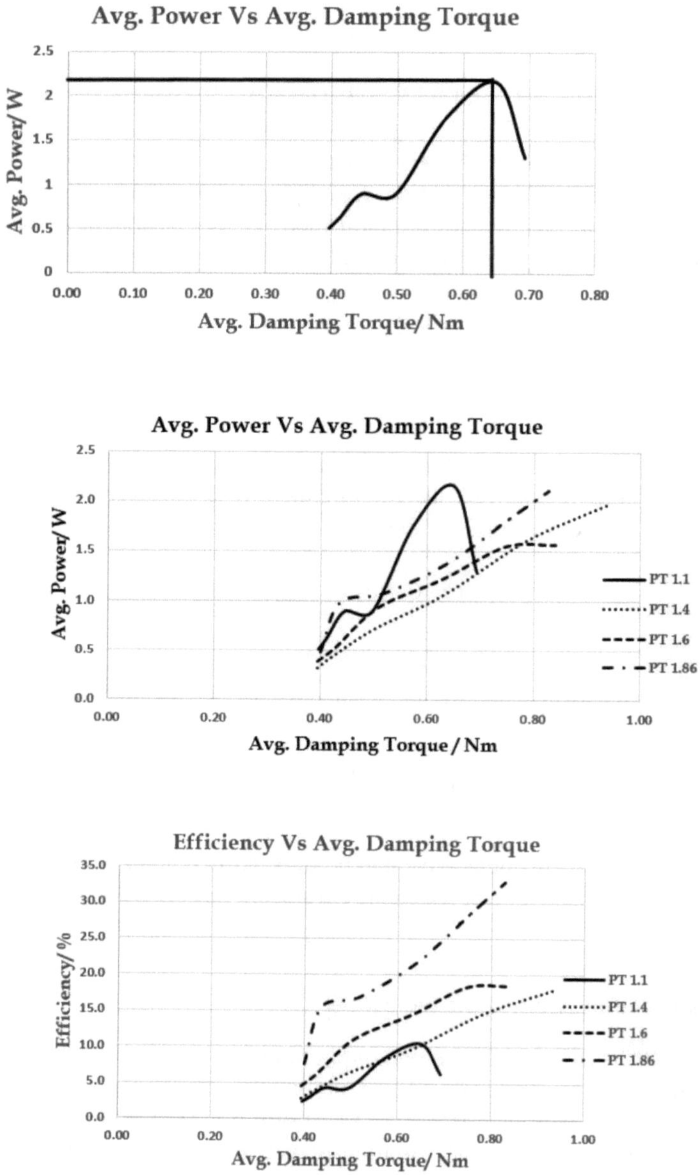

Fig. 9. Variation of average power and efficiency with average torque for different frequencies

Then, a 1/40th scale model caisson configuration for the same water depth of 25 cm and a wave length of 2 m, was considered as a preliminary test flap arrangement with a bottom gap of 25 mm. Two positions were considered in the experiments as the flap placed at distances of one quarter and one-eighth of a wave length from the back wall of the caisson.

Experiments were conducted for the periodic times 1.1($\omega = 5.712$), 1.2($\omega = 5.236$), 1.3($\omega = 4.833$).

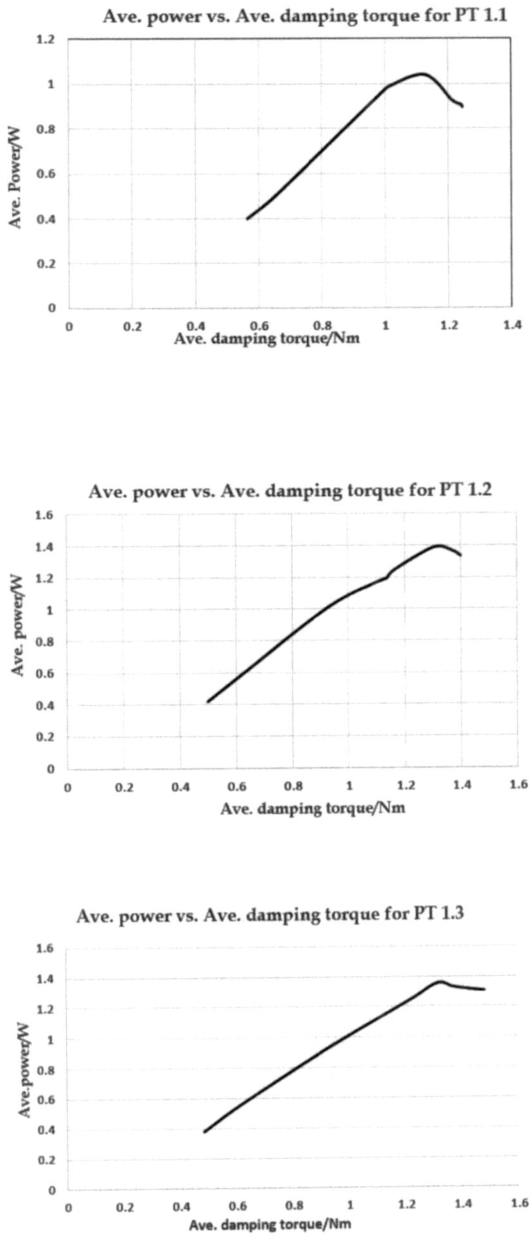

Fig. 10. Variation of average power with average torque for periodic times 1.1, 1.2 and 1.3 (Flap placed at quarter of a wave length with a bottom gap)

5 Results and Discussion

The conventional model without a gap that is the sealed caisson chamber in the first set of tests indicates a peak as can be seen in Fig. 9 in the power capture with a value of 0.4 W for the wave frequency of 5.7 rad/s (PT 1.1). That is, it produces an average power of 2.25 W when there is a damping torque of 0.65 Nm applied, resulting with an efficiency of 10.4%. A peak can be anticipated from the trend lines for the periodic times of 1.4, 1.6 and 1.86 as can be seen in the Fig. 9. It is observed that there is a trend to increase the capture efficiency with the increase of torque. Nevertheless, in this stage of the experimentation, the applied torque had to be limited due to the limitations of the PTO system. In this case, the peak in wave torque occurs as a standing wave is generated in front of the flap and then the wave torque reduces as diffraction begins to reduce the wave torque.

Then, the variation of the average power and the damping torque for the settings of the flap with a bottom gap of 25 mm in the scale model which was placed at a distance of a quarter of a wave length from the back wall is shown in Fig. 10. In both cases of the caisson without a gap and caisson with a bottom gap the chamber length was set at 50 cm (that is, the flap was placed at a distance of 50 cm from the back wall-a quarter of a wave length) and the chamber width was 30 cm in 1/40 scale. As can be seen in the figures there are peaks in the capture factor for the vented caisson too. Also, repeated results for the flap at the distance of approximately one-fourth and one-eighth of a wave length indicates a maximum power capture as expected. These maxima occur as the sloshing mode is excited by the incident wave resulting in the formation of a standing wave within the chamber. Although the tank tests carried out so far have not been extensive, the analysis shows promising results. However, extensive tank experiments are needed to validate and come to a better conclusion with the numerical model. Further work is required of wave tank experiments with utilizing PIV and video analyzing to validate the numerical models.

6 Conclusion

The objective of the experimental work is to investigate the hydrodynamics of a top hinged wave energy converter with a bottom gap and to compare with the numerical model. The experimental setup was established in the existing towing tank by modifying the tank to match with the testing requirements. Preliminary test results showed that the experimental setup needs further modifications to capture the essential characteristics of the device (peak efficiency etc.) for broader spectrum of cases.

References

1. de Antonio FO (2010) Wave energy utilization: a review of the technologies. Renew Sustain Energy Rev 14:899–918
2. Watabe T, Yokouchi H, Gunawardana SDGSP, Obeysekara BRK, Dissanayake UI (2001) Preliminary study on wave energy utilization in Sri Lanka. In: Proceedings of ISOPE 2001, Kithakyushu, Japan
3. Gunawardane SP, Kankanamge CJ (2016) Study on the performance of the "Pendulor" wave energy converter in an array configuration. Energies 9(4):282. https://doi.org/10.3390/en9040282
4. Watabe T, Kondo H, Matsuda T, Yano K, Dote Y, Takagi M (1984) Method and apparatus for generating electric power by waves, U.S. Patent 4,490,621, 25 December 1984
5. Watabe T, Kondo H, Yano K (1985) Characteristic of a pendulor wave energy converter. Ocean Eng 12:475–598 1st symposium of wave energy utilization
6. López I, Andreu J, Ceballos S, de Alegría IM, Kortabarria I (2013) Review of wave energy technologies and the necessary power equipment. Renew Sustain Energy Rev 27:413–434
7. Wan Z, Zheng H, Sun K, Zhou K (2017) A model and experimental study of an improved pendulor wave energy converter. Energy Procedia 105:283–288 8th international conference on applied energy 2016
8. Whittaker T, Folley M, Osterried M The oscillating wave surge converter. https://www.researchgate.net/publications/266404049
9. Whittaker T, Folley M Nearshore oscillating wave surge converters and the development of Oyster, Philos Trans Roy Soc. http://rsta.royalsocietypublishing.org. Accessed 31 July 2017
10. Wilkinson L, Whittaker TJT, Thies PR, Day S, Ingram D (2017) The power-capture of a nearshore, modular, flap-type wave energy converter in regular waves. Ocean Eng 137:394–403
11. Porter R, Biggs RT (2012) Wave energy absorption by a flap-type oscillating wave surge converter. https://people.maths.bris.ac.lk/~marp/abstracts/single-flap-110912.pdf. Accessed 27 July 2017
12. Renzi E, Dias F (2013) Hydrodynamics of the oscillating wave surge converter in the open ocean. Eur J Mech B/Fluids 41:1–10
13. Sarkar D, Renzi E, Dias F (2015) Effect of straight coast on the hydrodynamics and performance of the oscillating wave surge converter. Ocean Eng 105:25–32
14. Gunawardane SDGSP, Folley M, Kankanamge CJ (2019) Analysis of the hydrodynamics of four different oscillating wave surge converter concepts. Renew Energy. https://doi.org/10.1016/j.renene.2018.06.115
15. Dean RG, Dalrymple RA (1991) Water wave mechanics for engineers and scientists. World Scientific Publishing Co. Pte. Ltd., Singapore ISBN 9810204205

Increasing the Efficiency of a Dye-Sensitized Solid-State Solar Cell by Iodine Elimination Process in Hole Conductor Material

A. W. M. V. Ekanayake[1(✉)], G. R. A. Kumara[2], R. M. G. Rajapaksa[3], and A. Pallegedara[4]

[1] Postgraduate Institute of Science, University of Peradeniya, Peradeniya, Sri Lanka
venikaekanayake@gmail.com
[2] Institute of Fundamental Studies, Kandy, Sri Lanka
[3] Department of Chemistry, University of Peradeniya, Peradeniya, Sri Lanka
[4] Department of Chemical and Process Engineering, University of Peradeniya, Peradeniya, Sri Lanka

Abstract. The use of silicon in photovoltaics are popular these days. Besides the use of silicon, the use of titanium dioxide as an n-type semiconductor material is trending in the emerging dye-sensitized solar cells due to its low cost, nontoxicity and ease in preparation when compared to silicon. The objective of this research is to find avenues to enhance the efficiency of a dye-sensitized solid state solar cell, fabricated using D-149 and N-719 dyes as sensitizers and excess iodine-eliminated cuprous iodide hole conductor. P-25 Degussa titanium dioxide in a titanium dioxide colloidal suspension was used to deposit titanium dioxide thin films with a thickness less than 20 μm on fluorine-doped tin oxide glasses. Titanium dioxide films were then immersed in the dye solution for 12 h to enable dye adsorption. Acetonitrile solution containing refined copper powder treated cuprous iodide was used to deposit excess iodine-eliminated cuprous iodide on to the working electrode. Optimum efficiency of 2.11% was obtained for the dye-sensitized solid-state solar cell fabricated using D-149 dye and 1.6% was obtained for N-719 dye. The highest solar parameters obtained were shown for the dye-sensitized solid-state solar cell fabricated for D-149 dye which were an open circuit photo-voltage of 475 mV and short circuit current density of 12.7 A cm^{-2}.

Keywords: Iodine · Titanium dioxide · Dye · Solar cells

1 Introduction

Without any controversy, it is an acceptable fact that world relies on energy [1]. As the population of the world increases, the supply of energy should be equal or more, corresponding to the demand. Most abundant energy sources come either directly or indirectly from the sun's energy [2]. As an alternative to the dwindling energy sources, solar energy can be used to heat buildings, generate electricity, heat water and a variety of other domestic and industrial uses. Solar energy provides a tremendous resource for

R. Dissanayake and P. Mendis (Eds.): ICSBE 2018, LNCE 44, pp. 282–287, 2020.
https://doi.org/10.1007/978-981-13-9749-3_26

generating clean and sustainable electricity with no air and water pollution [3], no global warming, no risk of electricity price hikes and no threats to public health. The principle behind every solar cell today is the photovoltaic effect [4] which produces an electric current and a voltage in a material upon exposure to solar radiation. Belonging to the third generation [5] of solar cells, dye-sensitized solid-state solar cells (DSSCs) mimics the action of photosynthesis [6]. DSSC configuration sandwiches a monolayer of dye in between two inorganic wide band gap semiconductor materials [7], one with a p-type hole conducting mechanism and the other one exhibiting n-type conduction mechanism. Here, the purpose is to enhance the efficiency of the DSSC by changing the properties of the p-type CuI semiconductor material.

2 Materials and Methods

The following procedure was used to prepare TiO_2 colloidal suspension solution. First, 5.0 ml of Titanium Tetraisopropoxide (Sigma Aldrich) was mixed with 5.5 ml of glacial acetic acid (Sigma Aldrich) and 20.0 ml of iso-propyl alcohol (Sigma Aldrich) and ground in a ceramic motor. Then, 5.0 ml of distilled water was added to the solution and a gelatinous mixture was obtained. Afterwards 0.650 g of P25 Degussa TiO_2 powder was added to the gelatinous solution and ground well. Afterwards, fluorine doped tin oxide (FTO) glasses were immersed in a detergent solution and ultrasonicated for 15 min. TiO_2 films were prepared on FTO glass substrates by using the drop coating method. Cleaned FTO glass substrates were kept on a hot plate at 150 °C with the conducting side facing upwards. TiO_2 colloidal suspension paste was deposited by using a dropper. Due to the heat, the liquid component of the thick paste may evaporate leaving a brownish residue on the glass surface. This brownish residue was blown off using a jetstream of air. After repeating this process for several times, the plates were sintered at 500 °C for 2 min and then allowed to cool. After following this process for a few rounds, a uniform layer of white TiO_2 is formed. Plates were sintered in a furnace at 500 °C for 10 min.

Indoline (D-149) & N-719 dyes were used as the sensitizer dyes in this research. 100 ml of solvent was prepared by mixing acetonitrile and tert-butyl alcohol in the ratio of 1:1. Then 0.3 mM of D-149 & N-719 dyes were dissolved in separate solvents to prepare separate dye solutions.

TiO_2 coated electrodes were preheated to 80 °C and dipped in a newly prepared dye solution for 12 h and allowed to deposit a monolayer of dye on the TiO_2 surface. The electrodes were taken out of the dye solution and washed with acetonitrile to remove the excess dye molecules on the TiO_2 surface. Then, the dye-sensitized TiO_2 plates were dried using a hot air-gun.

Initially, copper oxide powder was washed with dilute HNO_3 (0.1M) in a Buchner funnel. Next, it was washed with iso-propyl alcohol. Finally, copper powder was dried at 150 °C for 30 min. Purified copper powder was stored in a desiccator.

0.60 g of CuI powder was dissolved in a $\sim$10 ml of moisture free acetonitrile solution and stirred for 10 min to obtain a saturated solution of CuI. Then, it was transferred into a flat bottom flask and 3 g of oxide-free refined copper powder was added to the saturated CuI solution and stirred overnight in the dark. Finally, clear colorless supernatant was separated using filter paper.

Dye-coated TiO_2 electrodes were placed on a hot plate at 80 °C. Excess iodine free CuI solution was spread over the exposed area of 0.25 cm^2 while rest of the area was covered with a mask. This procedure was repeated until the CuI layer was filled just above the TiO_2 surface while measuring the I-V characteristics for the best result. After the deposition of CuI on the dye-coated TiO_2 film, CuI layer on the working electrode was sandwiched between the dyed TiO_2 layer and the platinum counter electrode. Then current-voltage (I-V) characteristics of DSSC at 1.5 AM (100 mW cm^{-2}) simulated sunlight irradiation were recorded using a calibrated solar-cell evaluation system.

3 Results

It indicates that, the prepared solid-state solar cells sensitized with D-149 show an overall efficiency (η) of 2.11% for Iodine free CuI over CuI with Iodine. Short circuit current density (Jsc) was 12.76 mA cm^{-2} for Iodide free CuI which was higher than the short circuit current density of (Jsc) for CuI (Table 1 and Fig. 1).

Table 1. Comparison of parameters of DSSC constructed using Iodine free CuI and CuI for D-149 dye

Hole conducting material	J_{sc}/mA cm^{-2}	V_{oc}/V	FF	η/%
Iodine free CuI	12.7	0.475	0.35	2.11
CuI with Iodine	11.3	0.348	0.31	1.22

J_{sc} - Short circuit current density
V_{oc} - Open circuit voltage
FF - Fill factor
η - Efficiency

Fig. 1. I-V characteristics of DSSC sensitized with D-149 at highest conversion efficiency

According Table 2 and Fig. 2, prepared solid-state solar cells sensitized with N-719 show an open circuit voltage (Voc) of 425 mV, Short circuit current density (Jsc) of 11.2 mA cm^{-2} and fill factor (FF) of 0.34 with an overall efficiency of 1.6% for Iodide free CuI over pure CuI.

Table 2. Comparison of the variation of the solar cell parameters with the DSSC produced using N-719 dye.

Hole conducting material	J_{sc}/mA cm^{-2}	V_{oc}/V	FF	η/%
Iodine free CuI	11.20	0.425	0.34	1.60
CuI	6.23	0.416	0.43	1.11

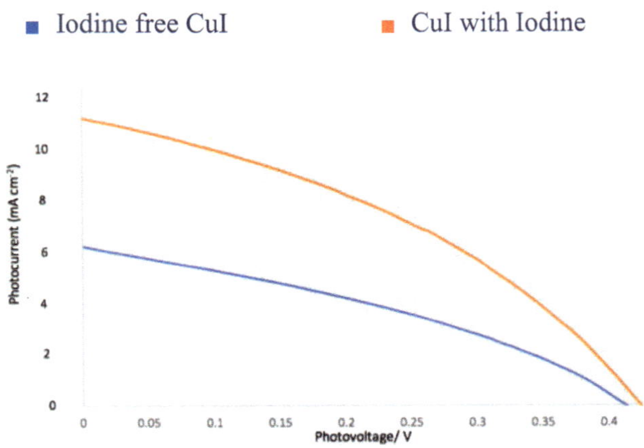

Fig. 2. I-V characteristics of DSSC sensitized with N-719 at highest conversion efficiency

This shows that the removal of excess Iodine in CuI leads to an increase in current. In other terms, the charge mobility of the Iodine free CuI is higher than CuI with Iodine.

The SEM image in Fig. 3(b) shows Iodine free CuI crystal structure with a highly porous structure since the size of the crystals have decreased as shown in the Fig. 3(b) compared to Fig. 3(a). Since excess iodine blended with CuI crystals are now detached, pure CuI crystal surfaces are free to attract holes and transport the charges efficiently via the crystal chains as there is no insulating impurity to block the pathways of the charges (Fig. 4).

Both the XRD patterns corresponding to CuI and Iodine free CuI give comparable peak positions. Thus, it confirms that the removal of excess iodine in the system preserved the crystalline structure of pure CuI.

(a) CuI crystals with Iodine

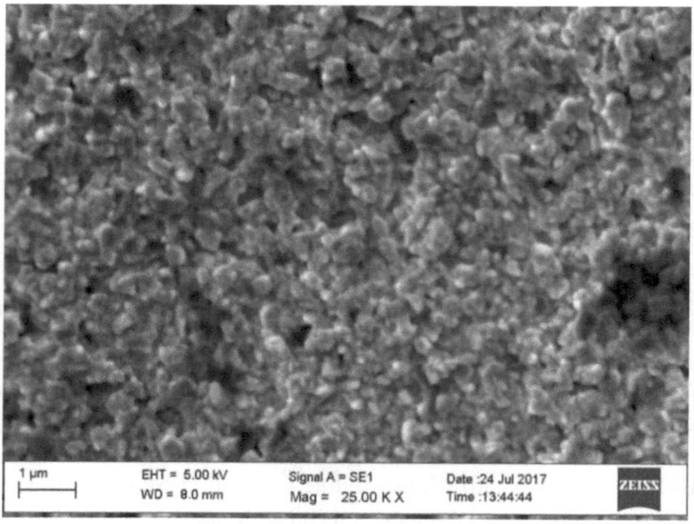

(b) Iodine free CuI crystals

Fig. 3. Scanning electron microscope images of CuI

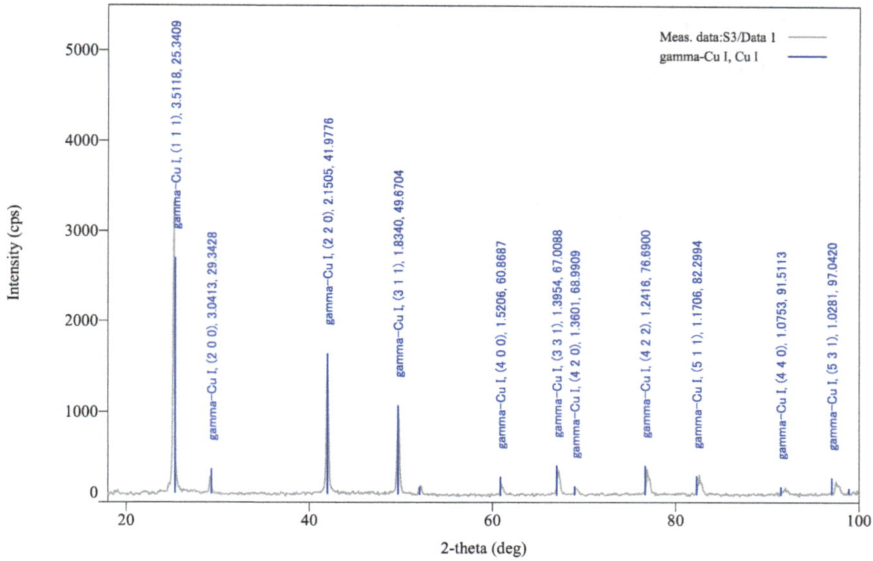

Fig. 4. XRD spectrum of Iodine free CuI

4 Conclusion

Optimum efficiency (η) of 2.11% and the short circuit current density (J_{sc}) of 12.7 mA cm^{-2} were obtained for the DSSC built for Indoline dye (D-149) with excess iodine eliminated CuI saturated acetonitrile solution.

References

1. Poizot P, Dolhem F (2011) Energy Environ Sci 4:2003–2019
2. Manahan SE (2008) Fundamentals of environmental chemistry, vol 3, pp 632–951
3. Suttie JM, McDowell RW (2008) Environmental impact of pasture-based farming, vol 45, pp 1–3
4. Dincer I (2000) Renew Sustain Energy Rev 4:157–175
5. Carlson DE, Wronski CR (1976) Appl Phys Lett 28:671–673
6. Green MA (2004) Sol Energy 76:3–8
7. Nozik AJ (1978) Annu Rev Phys Chem 29:189–222

Monitoring Volatile Fatty Acids in an Anaerobic Process with Microbial Fuel Cell

H. M. K. S. B. Thennakoon(✉), S. M. W. T. P. K. T. Ariyarathne, and M. Danthurebandara

Department of Chemical and Process Engineering, University of Peradeniya, Peradeniya, Sri Lanka
Kasun.thennakoon@pdn.ac.lk

Abstract. Volatile fatty acids (VFAs) concentration is a good indicator to control and optimize the anaerobic digestion process. Though VFA concentration is a high sensitive parameter, most of the methods of measuring it are complex and require time. Therefore this study is to find out a reliable, quick method to monitor VFA concentration in the process. In this study, a microbial fuel cell (MFC) was developed using synthetic samples of VFA. As the first part of the study, the correlation between voltage and the acetic acid concentration in the anaerobic digestion process was developed using developed MFC. Microbes to digest the acetic acid were grown in the coir brush. That coir brush was used as a bio film in the MFC. The salt bridge, which is made using 1 M KCl and agar was used as a proton exchange membrane in the MFC. The relationship was observed between voltage and acetic acid concentration with acetic acid level from 0.09 M to 0.01 M.

Keywords: Microbial fuel cell · Volatile fatty acid · Salt bridge

1 Introduction

The growth of population with the growing level of modernization presents a significant challenge on energy consumption. Eighty six percent of the global energy requirement is covered using fossil fuels (Parkash 2015). But, it is a declining natural resource and the burning of fossil fuels accounts for a lot of negative environmental issues. Then, there is a high demand for the search for new materials and technology to enhance power generation. Out of these, the energy resources of Bio fuels, solar, nuclear, hydraulic, wind, geo thermal, tidal power, wave and ocean currents play the major roles.

Currently, by going through various research directions, biogas has been identified as a renewable and sustainable energy source since there are a lot of bio degradable waste in Nature. But bio gas production is challenged by the instability of the Anaerobic Digestion (AD) process. Therefore, it is not easy to maintain an optimum biogas production and the overall anaerobic digestion process should be monitored to reduce the instability of anaerobic digestion process. The following parameters: pH

R. Dissanayake and P. Mendis (Eds.): ICSBE 2018, LNCE 44, pp. 288–294, 2020.
https://doi.org/10.1007/978-981-13-9749-3_27

level, Alkalinity, volume of biogas generation, concentration of Methane and CO_2 will determine the state of the process. However, several investigators have pointed out disadvantages of these parameters as process indicators due to low sensitivity and reliability. But researchers have found Volatile fatty acids (VFA) as intermediate products of the fermentation stage. These are substrate of the methanogenesis stage and more sensitive reliable parameters to monitor the AD process. Most researchers have tried to use VFA as an indicator in anaerobic digestion systems. They have tried to correlate current to VFA concentration. Also, traditional off line methods of VFA quantitative measurement have been tried in last few years. Most of the methods of measuring VFA concentration are complex and require more time. For example, Gas chromatographic (GC), high pressure liquid chromatographic (HPLC), UV absorption are advanced methods but depend on expensive instruments, skilled operators and so on. Therefore, researchers have been moved to find easy and reliable methods to measure the VFA concentration in the AD process. The literature so found explains a way of using Microbial Fuel Cell (MFC) to monitor the VFA. Accordingly, it has been found out that correlation can be developed between electrical signal that generates from MFC and VFA concentration in AD process (Jin et al. 2016).

Literature reveals that there are researches on this area and they have experimented to build a correlation between voltages generated from MFC with total volatile fatty acid concentrations (Lee et al. 2015). But it has been identified that it is difficult to find the exact cause for the instability of anaerobic digestion systems only with those correlations. The instability depends on individual VFAs and monitoring of overall VFAs generation would not help much. Correlation of individual fatty acid with voltage may be more advantageous in finding the instability in the process. Therefore, this study is to find out voltage variation with selected single acid concentration.

2 Materials and Method

Fabricating Microbial Fuel Cell

MFC was fabricated according to the Method given by Parkash in 2015 (Parkash 2015).

Two commonly available (500 ml) plastic containers were selected as the cathode and anode chambers of the MFC. Two holes of diameter 6 mm and 1.5 mm were made on each of the lids to insert the salt bridge and electrodes. The anode chamber was filled with 300 ml of anaerobic digestion solution. The cathode chamber was filled with 300 ml 0.05 mM NaCl solution. Only the anode chamber was closed tightly and sealed with tape. Anode and cathode electrodes were connected using copper wire through a millimeter.

Electrodes

Graphite rods of 10 cm length and 0.5 mm diameter was used as electrodes. Initially, a carbon rod which is inside the HB pencil was used as the graphite rod in the cell. The selected graphite rods were heated in 100% Ethanol for 30 min to remove the impurities (Parkash 2015). Then, the electrodes were washed using 1 M hydrochloric acid followed by 1 M Sodium hydroxide, each for 1 h to neutralize. Electrodes were stored in distilled water until use.

Anode Solution

In order to obtain a relationship among VFA concentration and generated voltage a solution of acetic acid with 0.1 M was used.

Bio Film Preparation

Firstly, discarded vegetables and fruits were collected from a stall. It was blended together to form a slurry with a domestic blender. Following the method carried out by Prakash in 2016, a sugar solution was added to the slurry to make it nutrient rich (Parkash 2015). Then, anaerobic bacteria sample collected from wastewater treatment plant at Kandy Citadel hotel was added to the slurry. Attached growth anaerobic bacteria were cultivated by dipping coir brush to the mixture. The coir brush was kept in the slurry under anaerobic condition for enrichment of bacteria for 2 months. The enriched coir brush (bio film) was set aside to starve for about 1 day (Jin et al. 2016).

Salt Bridge Preparation

Salt bridge was made with 5 mm diameter level tube as shown in Fig. 1. The salt bridge was prepared using 1 M Potassium chloride with 5% Agar. Initially 7.43 g of KCl was added to the 100 ml of distilled water to prepare a 1 M KCl solution. Then 0.5 g of agar powder was added to the KCl solution and mixture was boiled for 10 min. Then, the mixture was sucked into the level tube and allowed to solidify. After solidifying this salt, the bridge was connected to the corresponding MFC and sealed with tape as shown in Fig. 1.

Fig. 1. Salt bridge connection experiments

Experiment 1

The bio film (Coir brush) and 0.1 M acetic acid 300 ml was added to the anode chamber and it was sealed properly using tape (Parkash 2015). Then, the anode and cathode electrodes were connected using copper wire through a millimeter. After that, the generated voltage was recorded with 1 h time intervals and the corresponding acetic acid concentrations in the anode chamber were measured using a titration method (Titrations.info 2018).

Then, measured acetic acid concentration was plotted against the generated voltage and the relation between them was obtained.

Experiment 2
Experiment 1 was repeated with duplicates. In this experiment, generated voltage and the acetic acid concentration was recorded in 30 min intervals as in the first trial.

3 Results and Discussion

Experiment 1
Table 1 shows the gathered result of 1st experiment. In experiment 1, the MFC was run only for 4 h. During this time period only 3 readings were able to be taken. With these observation, the correlations of generated voltage and acetic acid concentration were developed and the relevant graph is shown in Fig. 2. There were not enough observations that could be collected to come to a conclusion and therefore experiment 2 was designed by minimizing the errors of experiment 1.

Table 1. Results of Experiment 1

Hours	pH	Voltage/mV	Volume of sample/ml	0.01 M NaOH volume/ml	Concentration/M
1	4.15	0.3	9.8	28	0.0286
3	4.05	0.1	9.4	18	0.0191
4	3.7	0	9.4	13	0.0138

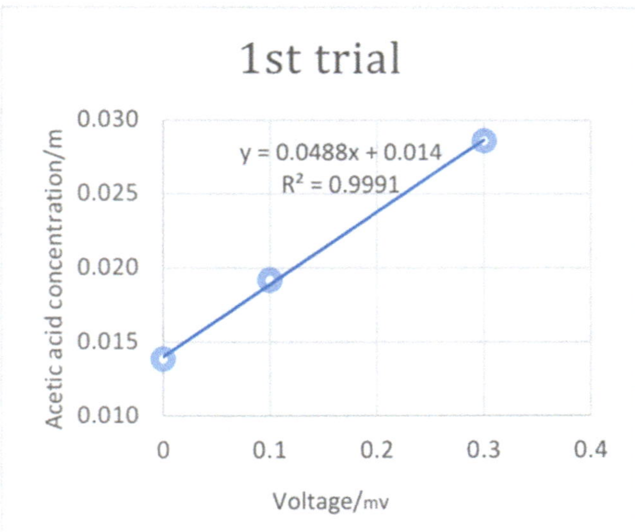

Fig. 2. Acetic acid concentration vs voltage in Experiment 1

Table 2 shows the errors encounter in experiment 1 and the remedies that were followed to develop the experiment 2.

Table 2. Errors encountered in Experiment 1 and remedies taken to develop

Errors encountered	Remedies
No voltage after 4 h. Therefore couldn't take more readings by taking reading in 1 h time intervals	Planned to take readings in 30 min time interval
No replicates to compare the results	Planned to duplicate the experiment

Experiment 2
Table 3 shows the result of experiment 2.

Table 3. Results of Experiment 2

Hours	pH	Voltage/mV	Volume of sample/ml	0.01 M NaOH volume/ml	Concentration/M
Sample 1					
0.5	3.65	0.1	10	52.5	0.0525
1	3.75	0	10	50.5	0.0505
1.5	3.91	0	10	50.3	0.0503
Sample 2					
0.5	3.44	0.1	10	52	0.0520
1	3.65	0	10	50.3	0.0503
1.5	3.7	0	10	50.2	0.0502

In this trial, the generated voltage has been reduced within 1 h. Concentration also has not been changed after 1 h. Relevant graphs obtained from experiment 2 are shown Fig. 3.

Causes to Obtain Different Operating Times of MFC's

Sudden pH Drop

In the first trial pH of bio film was 4.88 and the pH of the three samples were 4.18, 4.05 and 3.70. There was no significant pH difference that could be observed with the samples and the bio film generated. But, in the second trial, the pH of bio film was 4.72 and the pH of the three samples was the same around 3.7. There were significant pH differences between these two and it may suppress the microbial activity in the unit.

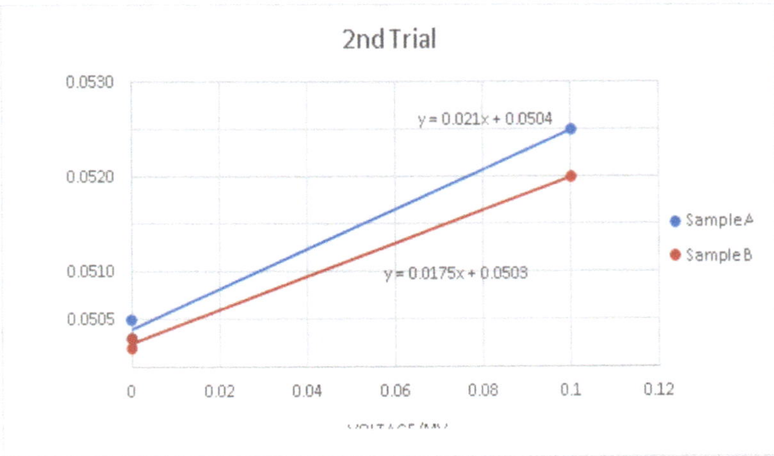

Fig. 3. Acetic acid concentration vs voltage in Experiment 2

Reduction of Microbes in Coir Brush and Microbes Decay with Time

Figure 4 explains the microbe growth pattern with time. According to that, it follows four phases as lag, log, stationary and death. Time between 1st and 2nd experiment is around 2 weeks. It can be assumed that coir brush which used for 1st experiment has taken out from the slurry (Mentioned in bio film preparation) when it was in log or stationary phase. At that time microbes were very strong and active. It could have potential to resist the sudden pH drop. After two-week time microbes in slurry could be in death phase. Therefore, are not rich enough to resist sudden pH drop. Therefore, microbes could be under stress when it exposes to the environment which have low pH value.

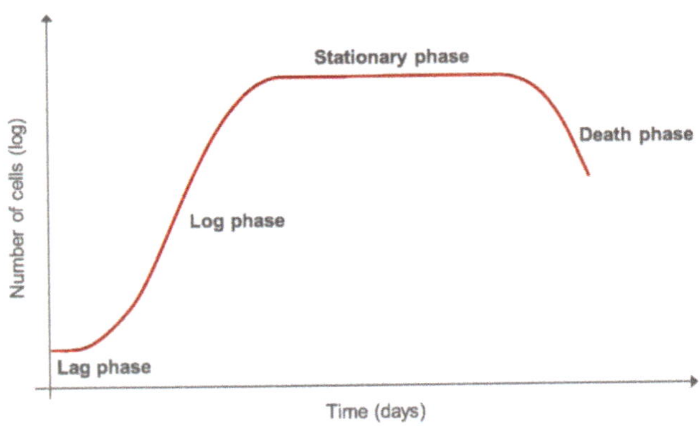

Fig. 4. Cell growth vs time

4 Further Developments

This research is a preliminary study to verify the method that has been developed using previous researches. Because of this, only acetic acid was used to develop the correlation. But there were many experimental difficulties identified with the study. In this present study, two trials were completed and the problems encountered identified. Future studies were planned to repeat the experiments as follows with remedies taken to overcome the problems.

Observe the voltage variation vs different initial acetic acid concentrations at 10 min intervals.

At that next step, the three main VFA with the following concentration will be fed into the anode chamber.

Acetic acid - 5431 mg/L
Butyric acid - 6014 mg/L
Propionic acid - 2248 mg/L (Lee et al. 2015).

Take ammeter readings at 30 min intervals and corresponding individual VFA concentrations will be taken using gas chromatography.

References

Titrations.info (2018) Determination of acetic acid in vinegar by titration. http://www.titrations.info/acid-base-titration-acetic-acid-in-vinegar. Accessed 9 Feb 2018

Jin X, Angelidaki I, Zhang Y (2016) Microbial electrochemical monitoring of volatile fatty acids during anaerobic digestion. Environ Sci Technol 50(8):4422–4429. https://doi.org/10.1021/acs.est.5b05267

Lee DJ et al (2015) Effect of volatile fatty acid concentration on anaerobic degradation rate from field anaerobic digestion facilities treating food waste leachate in South Korea. J Chem. https://doi.org/10.1155/2015/640717

Parkash A (2015) Design and fabrication of a double chamber microbial fuel cell for voltage generation from biowaste. J Bioprocess Biotech. https://doi.org/10.4172/2155-9821.1000246

Embodied Carbon Reduction Strategies for Buildings

L. M. T. Kumari, U. Kulatunga(✉), N. Madusanka, and N. Jayasena

Department of Building Economics, University of Moratuwa, Moratuwa, Sri Lanka
ukulatunga@uom.lk

Abstract. Global warming has become a major environmental issue faced by the international community. Greenhouse gases (GHG) have been suspected to be the source of global warming and carbon emission has identified as the main contributor for global warming. CO_2 could be released due to activities carried out by human such as burning fossil fuels, electricity consumption and transportation. Hence, the carbon footprint is the total amount of greenhouse gases produced directly or indirectly as a result of an activity. Up to 40% of all energy have been consumed by the building sector and it has contributed up to 30% of global annual GHG emissions. Therefore operational and embodied carbon were identified as two main general groups of carbon emissions related to buildings. Numerous researchers have paid their attention on operational carbon reduction and there are limited study on embodied carbon reduction. Therefore the researcher aims to identify embodied carbon reduction strategies for buildings. After a comprehensive literature review, selection of low carbon materials at the design stage of the building, reuse and recycling of carbon intensive materials, material minimization, optimum building design, local sourcing of materials, transportation minimization, efficient construction processes, policies and regulations by government, adaptive reuse of buildings and carbon labelling schemes were identified as the global embodied carbon reduction strategies for buildings.

Keywords: Embodied carbon · High-rise buildings · Strategies

1 Introduction

Global warming has become a significant environmental issue and, it is considered as one of the biggest challenges faced by the community (Lai et al. 2012). Methane, carbon dioxide, ozone, water vapour, nitrous oxide, and halogenated hydroxide have been identified as greenhouse gases (GHG) that are suspected to be the sources of global warming (Gajewski et al. 2013). According to Ozkaymak et al. (2017), CO_2 has been recognized as the main contributor for global warming and Gajewski et al. (2013) have revealed that CO_2 accounted for 9–26% of greenhouse effect.

Material extraction, material processing and component fabrication, construction and assembly, operation and service phase and end-of-life phase have been identified as

R. Dissanayake and P. Mendis (Eds.): ICSBE 2018, LNCE 44, pp. 295–308, 2020.
https://doi.org/10.1007/978-981-13-9749-3_28

the different life cycle stages of a building and energy use and carbon emissions can occur in each stage (Akbarnezhad and Xiao 2017).

According to RICS information by Lockie and Berebecki (2012), operational carbon (OC) and embodied carbon (EC) have been identified as the two main general groups of carbon emissions. Kang et al. (2015) recognized that the CO_2 emitted from the buildings from the use of heating, cooling and lighting as the operational carbon of a particular building. Kang et al. (2015) have further revealed that, CO_2 emitted during manufacturing, transportation and construction of building materials as the embodied carbon.

According to Crawford (2011), the lack of consideration and effort have been directed towards embodied carbon emissions occurring from construction materials while significant effort has been placed on measuring and reducing the impact of emissions occurring in the building operational stage. Further, Crawford (2011) has recognized that major impact is created by embodied emissions of the building and that it accounts for more than 50% of whole the life impacts of buildings towards global GHG emissions.

According to Wheating (2017), the building sector has been recognized as the major contributor for GHG emissions and the building sector has been identified as the sector with major opportunities for emission reduction strategies and mitigation of GHG emissions that contribution to climate change.

Most building owners tend to reduce operational carbon of the building in the operational stage with the aim of reducing the operational cost by using different strategies to save energy. However, there are limited previous studies for the reduction embodied carbon in buildings (Pomponi and Moncaster 2016). This situation has given rise to the need to conduct an analysis to identify suitable embodied carbon reduction strategies for buildings.

Therefore, the aim of this research paper is to review embodied carbon reduction strategies for buildings.

2 Research Methodology

A comprehensive literature survey was carried out to identify the concept of carbon footprint, carbon emission in building life cycles, embodied carbon emission and global embodied carbon reduction strategies.

3 Carbon Emissions in Building Life Cycle

Typically, a building life cycle could be divided into stages such as material manufacturing, transportation of building materials, construction of the building, operation and maintenance and demolition (Gustavsson et al. 2010). Further, according to the views of Monahan and Powell (2011), energy and carbon emissions related to these stages could be considered.

According to the review of the building life cycle assessment by Ibn-Mohammed et al. (2013), it has been proven that embodied carbon can account for anywhere between 2% and 80% of whole-life carbon emissions.

4 Embodied Carbon Emission

Embodied carbon emission is identified as the energy consumption associated with the mining, manufacturing, transportation, assembly, replacement and deconstruction of construction materials or products and generally is expressed in kilograms (Kg) of CO_2e per kilogram of product or material (Lockie and Berebecki 2012). Further authors stated that, all the activities relevant to the construction and demolition of a building will result in embodied emission. According to the Embodied Carbon Assessment, carbon footprint of a material is defined as embodied carbon (Anon n.d.). Embodied carbon can be measured from cradle-to-gate, cradle-to-site, and cradle-to-end of construction, cradle-to-grave, or cradle-to-cradle (Lockie and Berebecki 2012). Further, Lockie and Berbecki (2012) stated that, cradle-to-gate is the typical embodied carbon data sets.

Initial embodied emission and recurring embodied emission are identified as the two main categories of embodied emissions (Ramesh et al. 2010; Chen et al. 2001). Ding (2004) defined the initial embodied emission as the emissions associated with the acquisition of raw materials, their processing stage, transportation to the site and construction. Further, Ding (2004) stated that the two major categories of initial embodied emissions could be identified as direct emissions and indirect emissions. Accordingly, the emissions associated with the transportation of building materials or products to the construction site and the emissions related to the construction of the building were revealed as direct initial embodied emissions. According to Chen et al. (2001), the emissions incurred with the acquiring, processing and manufacturing of building materials including transportation linked to these activities were defined as the indirect initial embodied emission of a building.

Different types of materials are being used in building construction and various materials have a low life span compared to the building life span. Therefore, new materials are required to be replaced at the operation stage of the building (Ramesh et al. 2010). Accordingly, regular annual maintenance should be carried out in the building for its proper operation. Therefore, the emissions incurred with building maintenance, repair, restore, refurbishments or replacement of a system, components or a system during the effective life of the building were defined as the recurring embodied emissions (Ramesh et al. 2010; Chen et al. 2001; Ding 2004). Ibn-Mohammed et al. (2013), stated of recurring embodied emissions as the total emissions associated with the materials used in the rehabilitation and maintenance.

5 Importance of Embodied Carbon Reduction

According to Dutil et al. (2011) the importance of embodied carbon emission has increased and further, (Anon n.d.) mentioned that the construction sector has contributed for a large portion of embodied emissions. Due to this, embodied carbon has gained higher attention from both the industry and the government. Reducing embodied energy, energy consumption of buildings and switching to low carbon fuels were identified as the measures to reduce GHG emissions in buildings (IPCC 2007). According to Monahan and Powell (2011), most buildings have used less energy in their operational phase and have a lower GHG emission over the life span of the building. Therefore, a considerable amount of embodied carbon emissions have been included in the whole life carbon emission of the building and the importance of the GHG emissions associated with the building materials has increased.

In addition to operational emission reductions, the green construction board low carbon route map for the built environment defined that, embodied carbon reduction by 39% in 2050 is essential compared to the 2010 baseline (Green Construction Board 2013). Therefore, nine strategies were identified through a comprehensive literature review.

6 Global Strategies Used for Embodied Carbon Reduction

6.1 Use of Low Energy and Low Carbon Materials

According to the study carried out by Monahan and Powell (2011), cement, sand, gravel and concrete products were identified as the types of materials, which includes minerals. Further, minerals were identified as the most important material types, which was accounted for 45% of material related embodied carbon. High or low embodied carbon footprint of a building is decided according to the amount of carbon that exists in the selected building material (Wheating 2017). According to Monahan and Powell (2011) up to 82% of the total amount of embodied carbon is found in construction materials. Further, Strain, Simonen, Yang, and Webster (as cited in Wheating 2017) stated that building materials with highest embodied carbon have high energy consumption during their production. Depending on the type and amount of energy used for the building materials' manufacturing processes, the embodied carbon of a particular building material would increase (Gonzalez and Navarro 2006). Kwok et al. (2016) revealed that embodied carbon of building materials is controllable at its production sources.

Globally 60% of raw materials are being used for civil works and building construction. Out of this volume, the construction sector has consumed 40% and it contributes 40%–50% GHG emissions (Huang et al. 2017; Bribian et al. 2011). Gonzalez and Navarro (2006) found that a varied selection of materials with low environmental impact could reduced up to a maximum of 30% of carbon emissions. After evaluating embodied carbon of several office buildings constructed with different building materials, the study revealed that the highest portion of embodied energy in these buildings relates to structural materials such as concrete and reinforced steel which have accounted for 50%–66% of total embodied energy of a building (Akbarnezhad and Xiao 2017).

Therefore, according to the literature, Concrete could be identified as the major and widely used construction material for buildings and other infrastructures (Chen and Ng 2015). Further, Imbabi et al. (2012) revealed that 9 kg of CO_2 will be released as a byproduct of 1 kg of ordinary Portland cement. According to Imbabi et al. (2012), the construction industry has investigated alternative materials, which do not produce CO_2 in the process, to replace the raw materials.

Supplementary cementation materials (SCM) which are the byproducts of other industries could be used for restructuring the concrete mixes with the aim of reducing the carbon footprint of concrete (Lee et al. 2011). According to Imbabi et al. (2012), fly ash (a byproduct of coal combustion), blast furnace slag (a byproduct from iron smelting) and silica fumes (a byproduct of silicon production) are the best common identified SCM substitutes for traditional mixes.

Application of alternative fuels for the processing stage of cement could be identified as another strategy used to reduce the carbon footprint (Imbabi et al. 2012; Higuchi et al. 2014). Biomass is identified as an alternative fuel for cement clinker kilns where kilns have been traditionally fired using coal, petroleum coke, gas or fuel oil. Moreover, the author stated that emission from a biomass could reduce 20%–25% of the emissions from coal combustion (Imbabi et al. 2012).

There is a new strategy to convert production emissions into an aggregate and form artificial limestone aggregates which could be used as SCMs to reduce carbon emissions in a carbon emitting facility. This has been identified by (Blue Planet Company n. d.). CO_2 could be captured through the extension attached to the exhaust of the power plant and the cement manufacturing facility and finally, it is converted into calcium carbonate mineral. Moreover, they have stated that this strategy could tackle the two largest causes of embodied emissions in concrete, which could reduce the emissions from cement production and emissions from the burning of fuel.

According to Zhong and Wu (2015), concrete and steel have been identified as the most common building materials which frequently account for the highest proportion of materials in the building. Further, Chen et al. (2001) mentioned that steel and aluminium have accounted for more than three quarters of total embodied emission of the building. Compared to reinforced concrete and composite materials, structural steel is identified as a more carbon intensive material (Gan et al. 2017).

When compared to the masonry cavity structure, a timber frame has provided lighter weight (Monahan and Powell 2011). Further to the authors, materials for the sub structural supports could be reduced due to the lighter structure and accordingly, the foundation materials and high embodied building materials such as concrete and steel reinforcing could be reduced. There is low carbon emission in bamboo structures compared to brick concrete buildings (Yu et al. 2011).

According to studies carried out by Reddy (2009), stabilized mud blocks (SMB) could be an alternative for burnt clay bricks and it is a product of compacting a mixture of sand, soil, water and a stabilizer such as cement or limestone. Moreover, Reddy (2009) has mentioned that, these blocks could be used for wall construction and these are eco-friendly and energy efficient. Due to the fact that construction waste such as stone quarry dust and fly ash could be utilized in manufacturing process of SMB, there is no need to burn them in a clinker kilns, and it is possible to save 60%–70% of energy compared to burnt clay brick and it incurs less transportation emission because, these SMD could be easily produced at the construction site itself.

Ortiz et al. (2009) investigated that timber, glass, aluminum, concrete, slate, plasterboard, ceramic tiles, damp course and mortar are in a single family house and defined that concrete, timber and ceramic tiles have embodied emission of 61%, 14%, and 15% respectively. Further, Rai et al. (2011) mentioned that, using timber instead of steel cladding in distribution warehouse could reduce CO_2 emissions by 18%.

6.2 Reduce, Reuse and Recycling of Carbon Intensive Construction Materials

Metal, masonry, carpet, concrete, acoustic tiles, ceramic tiles and insulations are identified as the commonly used materials which have recycled content and these materials have provided an opportunity to reduce carbon and energy emissions in the initial phase of a building (Yu et al. 2011; Saghafi and Teshnizi 2011). Recycling of construction materials could be identified as the highest embodied emission saving potential, which accounts for 53% and reusing of construction materials has 6.2% potentials of embodied emission reduction (Ng and Chau 2015).

Therefore, recycling and closing the materials loop are being identified as the efficient and effective strategies for reduce GHG emissions of a building (Saghafi and Teshnizi 2011). Further, authors have mentioned that converting already made products consumes less energy than obtaining raw materials from scratch and producing a totally new product. According to Ng and Chau (2015), 2%–3% of GHG reduction could be gained by increasing the recycling rate of concrete from 27%–50% from deconstructed buildings. According to the studies carried out by Thormark (2002), it was found that 35%–40% of embodied energy could be save through recycling building waste. Concrete, steel, wood and sand are the main construction and demolition waste that could be found in building constructions (Thongkamsuk et al. 2017).

There is less volume of waste generated in precast and prefabricated structures due to designs that are detailed and manufacturing process has generated a less volume of waste (Thongkamsuk et al. 2017; Martos et al. 2018). Moreover, Martos et al. (2018) have explained that the amount of construction and demolition waste generated in reinforced or concrete structures vary compared to the prefabricated structures.

Chau et al. (2012) mentioned that reduction in quantity of material used and construction waste could be achieved through maintaining the existing building elements. This could be reduced CO_2 emission of the building. Further Chau et al. (2012) have revealed that, maintaining existing walls, roofs and floors could maximize the benefits of material reduction and construction waste reduction by reusing the whole building. Moreover, the authors have revealed that, properly maintained and stored construction materials and components from previously demolished from sites could be recovered for the current project site. Therefore, the embodied emissions could be reduced due to less construction materials.

Basbagill et al. (2013) defined that optimizing key parameters anywhere in the building such as thickness of piles and footings, external and internal walls accounts for 63% to 75% reduction in total embodied impact of the building.

Therefore, reusing and recycling building materials was considered during the primary evaluation for installation on a building (Ng and Chau 2015). Further to authors, at the end of life span of an initial recycled value and recyclability product could result in considerable reductions of embodied carbon value throughout the total lifecycle of a building.

6.3 Material Minimization

The usage of more materials results in the highest carbon footprint of a building because, embodied carbon footprint of a building is directly proportionate to the quantity of materials used in that building (Akbarnezhad and Xiao 2017). Therefore, according to Akbarnezhad and Xiao (2017), the reduction in the quantity of materials used would be quick method for embodied carbon reduction in a building. Further, authors have mentioned that it is necessary to measure the total amount of materials at the design stage without comparing materials with their embodied carbon value separately. According to Nadoushani and Akbarnezhad (2015), the type of construction materials used, height of the structure and selected structural system for the building were identified as the main factors which could affect the total amount of materials used in a building structure.

Yeo and Gabbai (2011) revealed that avoiding overdesign and use of optimum design results in a significant amount of reduction in the total amount of materials used and therefore, the total embodied carbon of the building could be reduced. Further, authors have mentioned that, minimization of material usage should be accomplished while retaining the capability of the structure to encounter all other performance and technical requirements.

According to studies carried out by Rashid and Yusoff (2015), they have identified that precast concrete floors are capable of having an extended life between beams. Therefore, the total amount of concrete used in precast concrete buildings could be reduced by using less amount of columns and footing needed for the building. According to studies carried out by Lopez-Mesa et al. (2009), it was identified that a considerable amount of embodied emission could be reduced by using hollow core concrete for load bearing constructions than using solid reinforced.

6.4 Optimizing Building Design

According to Iwaro and Mwasha (2013), sustainable development considerations have to be considered in the early design stage because the highest impact to the final impact of a building takes place at the early design stage. Therefore, construction of sustainable buildings could be appropriately integrated with this concept (Hakkinen et al. 2015). Further, Hakkinen et al. (2015) have mentioned that clear goals could be recognized during the design stage for environmental impact restrictions and building performance, which could be guided in the material selection process. Dixit et al. (2010) revealed that the selection of low energy and carbon intensive materials and products in the early stage of a building design has resulted in the reduction of embodied emission of a building. According to Shi and Yang (2013), the decisions taken in the conceptual design phase has the greatest impact on the total building performance.

6.5 Local Sourcing of Materials and Transport Minimization

Transportation activities have accounted for 16.5% of the total embodied carbon emissions in the construction of buildings (Serrano and Alvarez 2016). The transport requirements of a construction site is determined according to the selection of the location (Akbarnezhad and Xiao 2017). Moreover to the authors, according to the location, the transportation distance and transportation mode would be changed.

Numerous researchers have stated that increasing local material usage would reduce the embodied carbon emission through the reduction of transportation impacts (Gustavsson et al. 2010; Asdrubali et al. 2015; Chou and Yeh 2015). According to Gonzalez and Navarro (2006), the mode of transport, number of trips, country of origin and distance to be traveled have been identified as the main factors affecting transport emission of materials. It is not enough to reduce transport distance to achieve low carbon emission, the mode of transportation has a greater impact on the amount of embodied carbon in construction material (Weight et al. n.d.).

According to Hammad et al. (2016), site layout planning problem (SLPP) and container loading problem (CLP) have been identified as the two main problems related to carbon reduction in transportation. Under SLPP, optimization of tower crane locations on construction sites would account for low carbon emissions due to minimization of their operations. Further to Hammad et al. (2016), all the off-site transportation activities related to moving construction materials from manufacturers to suppliers and from suppliers to construction sites have considered under CLP and an optimized framework has been proposed in order to reduce embodied carbon emissions in constructions.

6.6 Policies and Regulations by Government

Policies and regulations implemented or reviewed by the government could be identified as another frequently cited strategy for the reduction of embodied carbon emissions (Dakwale et al. 2011; Foraboschi et al. 2014). Giesekam et al. (2015) mentioned that, this strategy has supported the other reduction strategies. For example, Giesekam et al. (2015) revealed policies related to the broader use of low embodied carbon materials. According to the studies carried out for China and Japan by Dhakal (2010), due to the influence of policies and regulations on design and construction practices 50% of CO_2 could be reduced.

6.7 Efficient Construction Processes and Techniques

Increase of efficiency in the construction sector was identified as a vital opportunity for embodied carbon reduction (Monahan and Powell 2011; Upton et al. 2008). Further, the efficiency in building materials manufacturing, use of new construction processes, reducing wastages during the construction stage or the combination of all these things have been identified as strategies that would enhance the efficiency of the construction process. According to Sandanayake et al. (2016), carbon emissions from materials, equipment usage and transportation were recognized as sources of emissions and finally concluded that emissions from equipment usage and transportation would contribute to the highest portion of embodied emission in construction of foundations. Therefore, concentrating on equipment usage and transportation could reduce the emissions in the foundation of constructions.

According to Junnila and Horvath (2003), the best practices in operation of construction equipment and use of temporary construction materials such as formwork, temporary structures which are not attached to the building permanently were identified as the opportunities to reduce embodied carbon emission of constructions. Selection of optimum equipment for the construction process, reduction in idle time of construction equipment to enhance the construction process, enhancement of the operation of equipment and reduction of both horizontal and vertical transportation within the site were identified as approaches to be used in order to reduce embodied carbon emission in the construction phase (Lewis et al. 2009; Avetisyan et al. 2012). According to the studies carried out by Carmichael et al. (2012), optimum earthmoving operations has resulted in the reduction of embodied carbon emission related to the construction process.

Nadoushani et al. (2017) revealed that the idle time of vehicles (trucks) and pumps could be reduced through optimizing the concreting operation on construction sites in order to reduce carbon emission occurred in the construction. Further, Nadoushani et al. (2017) have mentioned that a significant amount of emission could be reduced with the optimum use of cranes in the construction site. Operation time of the crane could be minimized by enhancing the places of supply and demand points in the construction site to reduce carbon emission.

6.8 Adaptive Reuse of Buildings

Moving to reusing and adapting of buildings has become a growing style in last decade (Wilkinson et al. 2009; Bueren and Jong 2007; Bradley and Kohler 2007). According to the Department of the Environmental and Heritage (2004), adaptive reuse was defined as a process that converts unproductive items into a new product which could be used for an altered purpose. Buildings could contribute to climate change and sustainability via the reduction of carbon emission by adapting this adaptive reuse process (Bullen 2007). Moreover to previous author, reduction of material usage, transportation and energy consumption could be achieved through extending the life span of a current building. It has contributed to the enhancement of the sustainability and carbon reduction of a building.

6.9 Carbon Labelling Schemes

Carbon labelling schemes are designed for construction materials in order to reduce the impact to the global climate (Wu et al. 2014). Further, according to Wu et al. (2014), appraising the inputs of raw materials, energy demand and emission to air as an output of production of a product or a service, consists in the total labelling process and environmental impact of a product or a material is represented in a single score.

Wu et al. (2015) revealed that this carbon labelling scheme is only focused on global climate change. According to Boardman (2008), to provide information on low carbon products to consumers to take correct decisions on their selections, to inspire sellers to sell non carbon intensive products, and to provide a method for organizations to be obligated in the reduction in embodied carbon emissions of a product are the three main aims of this carbon labelling schemes (Fig. 1).

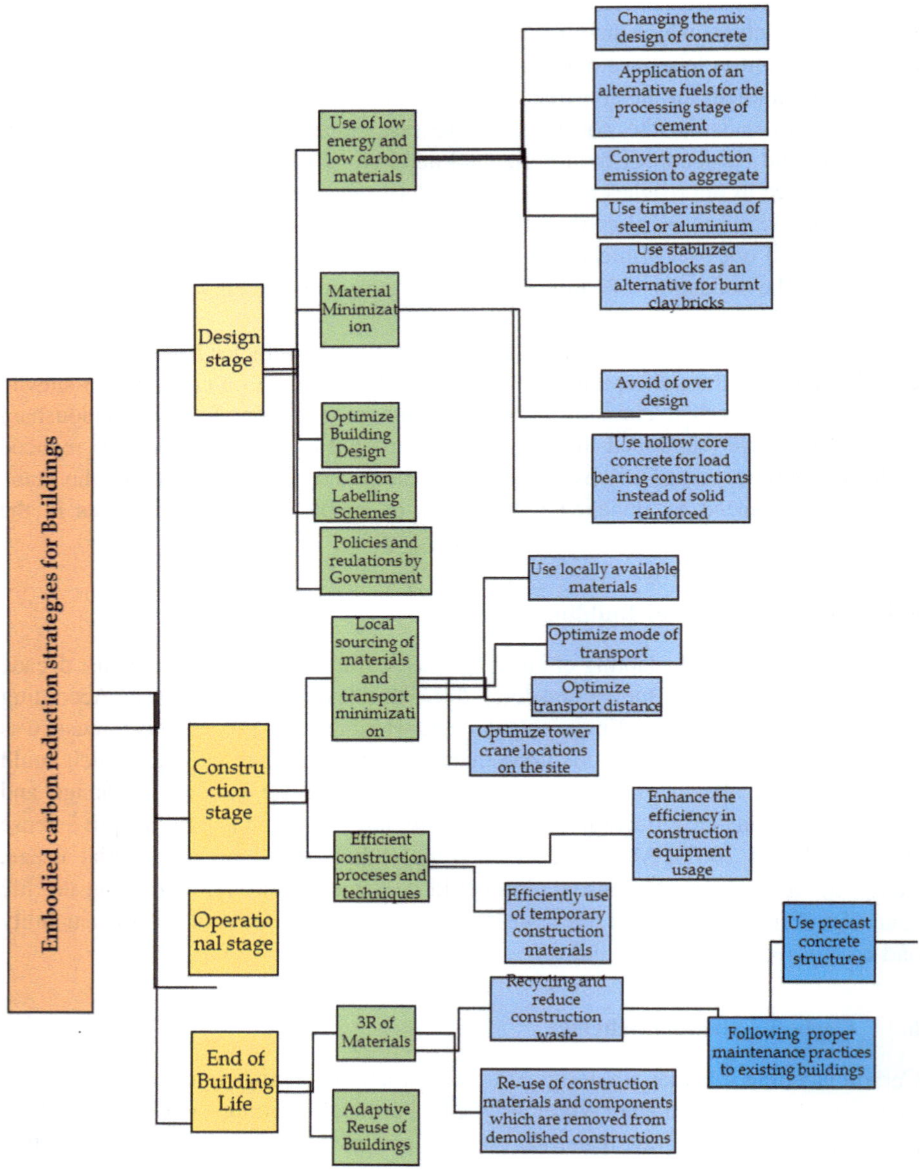

Fig. 1. Carbon reduction strategies for buildings

7 Conclusion

Carbon dioxide, methane, hydro fluorocarbon, per fluorocarbon, nitrous oxide and sulphur hexafluoride are the six main GHG exists in the atmosphere. These GHGs have increased due to activities carried out by humans. Therefore, the carbon footprint is defined as the total amount of GHGs produced directly or indirectly as a result of an activity. The researcher focused on a building's carbon emission and there are two general groups of carbon emission related to buildings; operational emission and embodied emission. According to the research gap, priority was given to embodied carbon in buildings and initial and recurring embodied carbon was identified as the two major categories of embodied carbon.

The use of low energy and low carbon materials, reuse, recycle and reduce of carbon intensive construction materials, materials minimization, optimizing building design, local sourcing of materials and transport minimization, policies and regulations by government, efficient construction process and techniques, adaptive reuse of buildings and carbon labelling schemes are identified as the main global embodied carbon reduction strategies.

Acknowledgement. Authors express the gratitude to each and every individual for their encouragement, values and ideas, assistance and specially their commitment towards this research to make it a success.

Further, authors would like to acknowledge the technical assistance given by Miss L A Siriwardena, Technical Officer (Grade 1), Department of Building Economics, University of Moratuwa.

References

Akbarnezhad A, Xiao J (2017) Estimation and minimization of embodied carbon of buildings: a review. Buildings 7(4):5

Anon (n.d.) Embodied carbon assessment. Circular Ecology. http://www.circularecology.com/embodied-carbon.html. Accessed 07 Feb 2018

Asdrubali F, Baldinelli G, D'Alessandro F, Scrucca F (2015) Life cycle assessment of electricity production from renewable energies: review and results harmonization. Renew Sustain Energy Rev 42:1113–1122

Avetisyan HG, Miller-Hooks E, Melanta S (2012) Decision models to support greenhouse gas emissions reduction from transportation construction projects. J Constr Eng Manag 138 (5):631–641

Basbagill J, Flager F, Lepech M, Fischer M (2013) Application of life-cycle assessment to early stage building design for reduced embodied environmental impacts. Build Environ 60:81–92

Blue Planet Company (n.d.) Economically sustainable carbon capture. http://www.blueplanet-ltd.com/?page_id=54

Boardman B (2008) Carbon labelling: too complex or will it transform our buying? Significance 5(4):168–171

Bradley PE, Kohler N (2007) Methodology for the survival analysis of urban building stocks. Build Res Inf 35(5):529–542

Bribian IZ, Capilla AV, Uson AA (2011) Life cycle assessment of building materials: Comparative analysis of energy and environmental impacts and evaluation of the eco-efficiency improvement potential. Build Environ 46(5):1133–1140

Bullen PA (2007) Adaptive reuse and sustainability of commercial buildings. Facilities 25 (1/2):20–31

Bueren EV, Jong JD (2007) Establishing sustainability: policy successes and failures. Build Res Inf 35(5):543–556

Carmichael DG, Williams EH, Kaboli AS (2012) Minimum operational emissions in earthmoving. In: Construction research congress

Chau C, Hui W, Ng W, Powell G (2012) Assessment of CO_2 emissions reduction in high-rise concrete office buildings using different material use options. Resour Conserv Recycl 61:22–34

Chen T, Burnett J, Chau C (2001) Analysis of embodied energy use in the residential building of Hong Kong. Energy 26(4):323–340

Chen Y, Ng ST (2015) Integrate an embodied GHG emissions assessment model into building environmental assessment tools. Proc Eng 118:318–325

Chou J-S, Yeh K-C (2015) Life cycle carbon dioxide emissions simulation and environmental cost analysis for building construction. J Clean Prod 101:137–147

Crawford RH (2011) Life cycle assessment in the built environment. Spon Press, London

Dakwale VA, Ralegaonkar RV, Mandavgane S (2011) Improving environmental performance of building through increased energy efficiency: a review. Sustain Cities Soc 1(4):211–218

Department of the Environmental and Heritage (2004) Adaptive reuse: preserving our past, building our future, s.l.: s.n

Dhakal S (2010) GHG emissions from urbanization and opportunities for urban carbon mitigation. Curr Opin Environ Sustain 2(4):277–283

Ding GK (2004) The development of a multi-criteria approach for the measurement of sustainable performance for built projects and facilities. University of Technology, Sydney

Dixit MK, Fernandez-Solis JL, Lavy S, Culp CH (2010) Identification of parameters for embodied energy measurement: a literature review. Energy Build 42(8):1238–1247

Dutil Y, Rousse D, Quesada G (2011) Sustainable buildings: an ever evolving target. Sustainability 3(2):443–464

Foraboschi P, Mercanzin M, Trabucco D (2014) Sustainable structural design of tall buildings based on embodied energy. Energy Build 68:254–269

Gajewski A, Siergiejuk J, Szulborski K (2013) Carbon dioxide emission while heating in selected European countries. Energy Build 65:197–204

Gan VJL et al (2017) Sustainability analyses of embodied carbon and construction cost in high-rise buildings using different materials and structural forms. HKIE Trans 24(4):216–227

Giesekam J, Barrett JR, Taylor P (2015) Construction sector views on low carbon building materials. Build Res Inf 44(4):423–444

Gonzalez MJ, Navarro JG (2006) Assessment of the decrease of CO_2 emissions in the construction field through the selection of materials: practical case study of three houses of low environmental impact. Build Environ 41(7):902–909

Green Construction Board (2013) Low carbon routemap for the UK built environment, s.l.: s.n

Gustavsson L, Joelsson A, Sathre R (2010) Life cycle primary energy use and carbon emission of an eight-storey wood-framed apartment building. Energy Build 42(2):230–242

Hakkinen T, Kuittinen M, Ruuska A, Jung N (2015) Reducing embodied carbon during the design process of buildings. J Build Eng 4:1–13

Hammad AWA, Akbarnezhad A, Rey D, Waller ST (2016) A computational method for estimating travel frequencies in site layout planning. J Constr Eng Manag 142(5):04015102

Higuchi T, Morioka M, Yoshioka I, Yokozeki K (2014) Development of a new ecological concrete with CO_2 emissions below zero. Constr Build Mater 67:338–343
Huang W et al (2017) Carbon footprint and carbon emission reduction of urban buildings: a case in Xiamen City, China. Proc Eng 198:1007–1017
Ibn-Mohammed T et al (2013) Operational vs. embodied emissions in buildings – a review of current trends. Energy Build 66:232–245
Imbabi MS, Carrigan C, McKenna S (2012) Trends and developments in green cement and concrete technology. Int J Sustain Built Environ 1(2):194–216
IPCC (2007) Climate change 2007: mitigation of climate change Working Group III contribution to the fourth Assessment Report of the International Panel on climate Change. Cambridge University Press, Cambridge, United Kingdom
Iwaro J, Mwasha A (2013) The impact of sustainable building envelope design on building sustainability using Integrated Performance Model. Int J Sustain Built Environ 2(2):153–171
Junnila S, Horvath A (2003) Life-cycle environmental effects of an office building. J Infrastruct Syst 9:157–166
Kang G et al (2015) Statistical analysis of embodied carbon emission for building construction. Energy Build 105:326–333
Kwok KYG, Kim J, Chong WKO, Ariaratnam ST (2016) Structuring a comprehensive carbon-emission framework for the whole lifecycle of building, operation, and construction. J Architect Eng 22(3):04016006
Lai JH, Yik FW, Man C (2012) Carbon audit: a literature review and an empirical study on a hotel. Facilities 30(9/10):417–431
Lee B, Trcka M, Hensen JL (2011) Embodied energy of building materials and green building rating systems – a case study for industrial halls. Sustain Cities Soc 1(2):67–71
Lewis P et al (2009) Requirements and incentives for reducing construction vehicle emissions and comparison of nonroad diesel engine emissions data sources. J Constr Eng Manag 135:341–351
Lockie S, Berebecki P (2012) Methodology to calculate embodied carbon of materials (RICS). Royal Institute of Chartered Surveyors, United Kingdom
Lopez-Mesa B, Pitarch A, Tomas A, Gallego T (2009) Comparison of environmental impacts of building structures with in situ cast floors and with precast concrete floors. Build Environ 44 (4):699–712
Martos JL, Styles D, Schoenberger H, Zeschmar-Lahl B (2018) Construction and demolition waste best management practice in Europe. Resour Conserv Recycl 136:166–178
Monahan J, Powell J (2011) An embodied carbon and energy analysis of modern methods of construction in housing: a case study using a lifecycle assessment framework. Energy Build 43(1):179–188
Nadoushani ZM, Akbarnezhad A (2015) Effects of structural system on the life cycle carbon footprint of buildings. Energy Build 102:337–346
Nadoushani ZSM, Hammad AWA, Akbarnezhad A (2017) Location optimization of tower crane and allocation of material supply points in a construction site considering operating and rental costs. J Constr Eng Manag 143(1):04016089
Ng W, Chau C (2015) New life of the building materials - recycle, reuse and recovery. Energy Proc 75:2884–2891
Ortiz O, Castells F, Sonnemann G (2009) Sustainability in the construction industry: a review of recent developments based on LCA. Constr Build Mater 23(1):28–39
Ozkaymak M et al (2017) CO_2 emission during the combustion of Orhaneli lignite coal. World J Eng 14(1):27–34
Pomponi F, Moncaster A (2016) Embodied carbon mitigation and reduction in the built environment – what does the evidence say? J Environ Manag 181:687–700

Rai D, Sodagar B, Fieldson R, Hu X (2011) Assessment of CO_2 emissions reduction in a distribution warehouse. Energy 36(4):2271–2277

Ramesh T, Prakash R, Shukla K (2010) Life cycle energy analysis of buildings: an overview. Energy Build 42(10):1592–1600

Rashid AF, Yusoff S (2015) A review of life cycle assessment method for building industry. Renew Sustain Energy Rev 45:244–248

Reddy BV (2009) Sustainable materials for low carbon buildings. Int J Low-Carbon Technol 4 (3):175–181

Saghafi MD, Teshnizi ZSH (2011) Recycling value of building materials in building assessment systems. Energy Build 43(11):3181–3188

Sandanayake M, Zhang G, Setunge S (2016) Environmental emissions at foundation construction stage of buildings – two case studies. Build Environ 95:189–198

Serrano AR, Alvarez SP (2016) Life cycle assessment in building: a case study on the energy and emissions impact related to the choice of housing typologies and construction process in Spain. Sustainability 8(3):287

Shi X, Yang W (2013) Performance-driven architectural design and optimization technique from a perspective of architects. Autom Constr 32:125–135

Thongkamsuk P, Sudasna K, Tondee T (2017) Waste generated in high-rise buildings construction: a current situation in Thailand. Energy Proc 138:411–416

Thormark C (2002) A low energy building in a life cycle – its embodied energy, energy need for operation and recycling potential. Build Environ 37(4):429–435

Upton B, Miner R, Spinney M, Heath LS (2008) The greenhouse gas and energy impacts of using wood instead of alternatives in residential construction in the United States. Biomass Bioenergy 32(1):1–10

Weight D, Connaughton J, Jones C (n.d.) Information sheet for construction clients and designers: cutting embodied carbon in construction projects. http://www.wrap.org.uk/sites/files/wrap/FINAL%20PRO095-009%20Embodied%20Carbon%20Annex.pdf. Accessed 29 Apr 2018

Wheating NC (2017) Embodied carbon: a framework for prioritizing and reducing emissions in the building industry. s.l.: s.n

Wilkinson SJ, James K, Reed R (2009) Using building adaptation to deliver sustainability in Australia. Struct Surv 27(1):46–61

Wu P, Feng Y, Pienaar J, Xia B (2015) A review of benchmarking in carbon labelling schemes for building materials. J Clean Prod 109:108–117

Wu P, Xia B, Pienaar J, Zhao X (2014) The past, present and future of carbon labelling for construction materials – a review. Build Environ 77:160–168

Yeo D, Gabbai RD (2011) Sustainable design of reinforced concrete structures through embodied energy optimization. Energy Build 43(8):2028–2033

Yu D, Tan H, Ruan Y (2011) A future bamboo-structure residential building prototype in China: life cycle assessment of energy use and carbon emission. Energy Build 43(10):2638–2646

Zhong Y, Wu P (2015) Economic sustainability, environmental sustainability and constructability indicators related to concrete- and steel-projects. J Clean Prod 108:748–756

Sustainable Building and Infrastructure

Lessons to Learn and Challenges to Address in Resilient Historic Towns

M. De Fino(✉), E. Cantatore, A. Scioti, and F. Fatiguso

Polytechnic of Bari, Via Orabona 4, 70125 Bari, Italy
mariella.defino@poliba.it

Abstract. The concept of resilient cities, including the environmental, social and economic dimensions of the urban sustainability against both slow and fast moving changes over long time horizons, is very distinctive when applied to historic towns. For built heritage, beyond the conventional goals, namely wellbeing and safety of citizens, functionality and reliability of services and infrastructures, effectiveness of management and development strategies, as well as availability of economic resources under disruptions or threads, a further aspect should be considered. It concerns the preservation and promotion of the original value and identity of the places by the capability to prevent, adjust and overcome the obsolescence at different levels – physical, technological, normative and functional. In this regard, two approaches are required: on the one hand, the critical understanding of the inherent qualities that made the historic towns resilient to past transformations – lessons to learn; on the other hand, the identification of strategies that could enhance such a resilient behaviour in the future against an exceptionally precipitous evolution of the outer alteration factors – challenges to address. In the light of the above-mentioned issues, the paper is going to develop some insights on the aspects that most influence the resilience of the historic towns, with specific attention toward the Mediterranean climate of South Italy. Firstly, some performance-based categories are identified, connected to the inherent resilience of the built heritage – e.g. relationships with the surroundings, employment of resources, natural regulation of microclimate, aggregation schemes and uses – as well as to the requirements of contemporary resilience – e.g. safety, accessibility, fruition, and efficiency. Thus, some qualitative and quantitative indicators are introduced and discussed, based on the validation by representative case studies, as a support toward the development of a resilience metrics for historic towns.

Keywords: Urban resilience · Historic towns · Built heritage conservation

1 Introduction

The concept of resilient cities, including the environmental, social and economic dimensions of urban sustainability over long time horizons, is very distinctive when applied to historic towns. In reality, the urban resilience of the built heritage certainly involves the requirements set by several international organizations, such as GFDRR, UNISDR, UN-HABITAT, in terms of its capability to guarantee wellbeing and safety of citizens, functionality and reliability of services and infrastructures, effectiveness of

R. Dissanayake and P. Mendis (Eds.): ICSBE 2018, LNCE 44, pp. 311–322, 2020.
https://doi.org/10.1007/978-981-13-9749-3_29

management and development strategies, as well as availability of economic resources under disruptions or threads. Nevertheless, it also implies the opportunity to safeguard the integrity and identity of the places themselves, as tangible and intangible evidence of culture, tradition and history of the communities. In this regard, with reference to the resilience of the historic towns, various studies and researches refer to ancient sites as heuristic models of the centenary ability to resist, adapt, mitigate and recover against both slow and fast moving changes, as discussed by (Correia et al. 2014), (Hassler and Kohler 2014), Langenbach (2016), Garcia and Vale (2017). However, the same scientific and technical literature highlights how such an inherent resilience is nowadays facing an unprecedented acceleration of landscape and climate changes, as well as cultural, economic and relational modifications of the societies.

Consequently, a challenging declination arises about the conservation-transformation relationship that typically features the building refurbishment and requalification processes. On the one hand, the identification and preservation of the original qualities that made the historic towns capable of showing resilient behavior in the past - lessons to learn – should be pursued. On the other hand, the development of strategies that might extend such a behavior in the future against an exceptionally precipitous evolution of the outer alteration factors – challenges to address – is highly desirable. In the light of the above-mentioned issues, the recognition of the aspects that most influence the resilience of the historic towns is paramount, in order to point out present values and envisaged goals. Yet, such a recognition should preliminarily identify the investigation boundaries, so that the multi-disciplines and multi-scales character of the urban resilience might not "*collapse into meaninglessness*", as underlined by Vale (2014).

In this regard, the present study is going to focus mainly on the physical spaces and structures of the historic towns (streets, squares, buildings, infrastructures …), thus neglecting the issues related to management policies, involvement of citizens and availability of funds. Moreover, it will concern the slow risk coming from obsolescence, instead of sudden threads, such as flooding, landslide and external attacks. More specifically, the paper is going to classify some performance-based categories of resilience, in relation to both the original identity of the built environment toward its conservation and the required contemporary quality in view of transforming, adapting and improving the existing heritage. The categories, based on the critical review of the state-of-the-art, are herein referred to the Mediterranean are of South Italy. Furthermore, some of them are discussed and validated by assessment of a representative case study, where qualitative and quantitative indicators are proposed, as a support toward the development of a resilience metrics for historic towns.

2 General Framework

As outlined in the introduction, the first research phase is concerned with the identification of a general framework of performance-based categories addressing the concept of resilience of Mediterranean historic towns. In detail, based on the state-of-the-art-review, as well as on the authors' experience on the topic, as in De Fino *et al.* (2017). Fatiguso *et al.* (2017), it seems suitable to refer to such an identification of two technical-scientific areas. The former relates to the "inner" resilience, namely the characteristics and values developed by traditional settlements and architectures in order to interact with local environment, resources and communities. The latter concerns the "outer" resilience, which comprises, in the authors' view, all the issues and aspects deriving from external obsolescence factors, in terms of physical, normative, technological and functional decrease of building performances.

According to the above-described classification, a set of requirements are proposed below. They are linked to the three dimensions – technical-environmental (TE), socio-cultural (SC) and socio-economic (SE) - that are commonly used for urban resilience. It is worth mentioning that the requirements are not meant to provide with a closed taxonomy. On the contrary, they should pave the way for an open discussion on facets and implications, even conflicting, toward the development of robust decision-making support tools for the preservation and revitalization of resilient historic towns.

Particularly, the "inner" resilience, as partially addressed by Sharma (2017), Currà and D'Amico (2014), Choudhary (2016), Ray and Shaw (2018), Qian and Li (2017), is herein classified in:

RI.1. Relationship between the urban settlement and the surroundings

RI.1.1. Selection and adaptation to the geo-morphological and hydro-geographic characteristics of the site - TE
RI.1.2. Arrangement and exposure to climatic agents (wind, sun …) - TE

RI.2. Bioclimatic response in summertime

RI.2.1. Use of passive solutions for natural ventilation - TE
RI.2.2. Use of passive solutions for thermal regulation - TE
RI.2.3. Use of passive solutions for shading - TE

RI.3. Management of resources

RI.3.1. Use of local materials and construction techniques - SC
RI.3.2. Water management - TE
RI.3.3. Storage of resources - SE

RI.4. Use models

RI.4.1. Shared aggregation spaces - SC
RI.4.2. Common services and facilities - SE
RI.4.3. Typological and functional layout - SE

In parallel, the "outer" resilience related to the physical, normative, technological and functional obsolescence, as referred to some extent by Martínez-Molina *et al.* (2016), Webb (2017), Marchi et al. (2018), Maio et al. (2018), Srinurak et al. (2016), Micelli and Pellegrini (2018) and Zagroba (2016), is herein classified in:

RE.1. State of conservation (physical obsolescence)

RE.1.1. Environmental quality of outdoor spaces - SC
RE.1.2. Structural stability - SC
RE.1.3. Protection from dampness - SC
RE.1.4. Surface integrity - SC

RE.2. Compliance with codes (normative obsolescence)

RE.2.1. Seismic vulnerability - TE
RE.2.2. Energy efficiency - TE
RE.2.3. Accessibility - SC
RE.2.4. Habitability of indoor and outdoor spaces - TE
RE.2.5. Privacy - SC

RE.3. Energy infrastructures (technological obsolescence)

RE.3.1. Availability of energy grids - TA
RE.3.2. Employment of RES - TA

RE.4. Fruition (functional obsolescence)

RE.4.1. Connection with the modern city - SE
RE.4.2. Typological and functional flexibility - SE
RE.4.3 Distribution of services and infrastructures - SE

In order to validate the above-described framework, the following sections introduce, as an illustrative case study, the historic town of Bitonto in South Italy. For that, some "inner" resilience requirements are assessed by quantitative and qualitative indicators. Specifically, a multi-level approach is applied, where the assessment is developed at the building component scale – herein the roof due to its relevance – and thus reported at the building scale and extended at the district scale.

3 Case Study

The historic centre of Bitonto shows some typological and technical solutions of the so-called *genius loci* – namely the distinctive vernacular identity of a place – making the town highly representative of the coastal and sub-coastal cities of the Apulia Region. In general, it is a medieval settlement, developed under the Norman, Swabian and Angevin dynasties, with additions and modifications during the Spanish domination in the XVI century. It is characterized by compact urban arrangement with very high housing density and use of local calcarenitic limestone as construction material. The main building typologies are the "tower-house" featured by a simple plan, with only one room per floor, two, three or four floors and inner staircase and the "palace-house", typically coming from the aggregation of two/three tower-houses and, therefore, featured by several rooms per floors on several floors.

In detail, with reference to some of the above-described resilience requirements, the settlement is naturally shaped following the close Balice watercourse (RI.1.1), it is arranged through narrow streets providing solar shading on the building facades and refreshing wind flows in summer (RI.1.2.), it is featured by inner and outer courtyards that could enhance social interaction and sharing (RI.4.1). Nonetheless, the historic town has been undergoing, for the last century, a progressive process of transformation, in order to overcome physical, normative, technological and functional obsolescence, by structural strengthening (RE.1.2), provision of energy grids (RE.3.1), functional changes from residential to commercial and entertainment use for improving the overall touristic attractiveness (RE.4.2). Among the occurred and ongoing transformations, the most relevant seem to be those related to the roofs. In fact, on the one hand, the roofs have undergone several renovation either/or replacement works throughout the years due to static failures and water infiltrations, thus potentially losing their "inner" resilience features (RI.2.2–RI.3.1). Furthermore, they are crucial in relation with "outer" resilience performances, such as structural, environmental and energy quality (RE.2.1–RE.2.2), due to their exposure to external aggressive agents (wind, rain, pollution, direct solar radiation …). Consequently, for such building components, the research has developed the assessment of some illustrative "inner" resilience requirements (RI.2.2–RI.3.1) and, thus, it has drawn in the conclusions some insights on the relationship among these requirements and some "outer" resilience ones (RE.2.1–RE.2.2).

3.1 Assessment at the Building Scale

As far as the resilience requirement RI.3.1 is concerned - use of local materials and construction techniques – the assessment is qualitative: it corresponds to the degree of compatibility of the current building component with the original roof solution, in terms of both structural system and external finishing layer. On the contrary, as far as the resilience requirement RI.2.2 is concerned - use of passive solutions for thermal regulation – the assessment is quantitative: it corresponds to the indoor surface temperature of the roof in summer (T_{in}), as calculated by theoretical dynamic simulation, given two main parameters of the current building system:

- Periodic thermal transmittance ($Y_{I,E}$) calculated according to normative standard EN ISO 13786:2017 "Thermal performance of building components - Dynamic thermal characteristics - Calculation methods". It is an indicator of the thermal inertia, deriving from the attenuation factor (fa) and the steady-state thermal transmittance U;
- Albedo (a) assessed according to common values for categories of materials. It is the measure of the diffuse reflection of solar radiation out of the total solar radiation received by the external finishing layer.

Both the assessment criteria – degree of compatibility with the original solution for RI.3.1 and indoor surface temperature for RI.2.2 - have been applied to the case study, after the detailed identification of the current roof components, based on documentary research and onsite inspection. Specifically, n.6 scenarios have been surveyed, corresponding to several combinations of structural system and external finishing layer:

Comb1.a and ***Comb 1.b.*** They both correspond to horizontal slabs made out of steel reinforced concrete beams and hollow brick blocks. They were typically used in the 60's to replace the original timber slabs, because of their poor conditions and due to missing local government control, or to cover a new built upper floor. They are generally covered by waterproofing bituminous membranes, when inaccessible (Comb1.a). Otherwise, they were finished by traditional stone tiles (Comb1.b);
Comb2.a e ***Comb2.b.*** They both correspond to horizontal slabs made out of steel beams and hollow brick blocks. They generally result from replacement work of the early XX century and can be found in those areas with medium-low state of conservation, where limited intervention have been carried out in the last decades. Consequently, they might be considered a historic transformation according to the local construction techniques, consistent with the tradition featuring that period at regional levels. Similar to the previous combinations, they might be covered by waterproofing bituminous membranes, when inaccessible (Comb2.a). Otherwise, they are finished with local clay tiles (Comb2.b);
Comb3.a e ***Comb3.b.*** They both correspond to masonry vaults made out of regional calcareous stone structure and upper filling of mortar mixture and stone fragments. They typically feature in the noble palaces, which still have residential functions. Again, they might be covered by waterproofing bituminous membranes, when inaccessible (Comb3.a). Otherwise, they are finished with stone tiles (Comb3.b);

As previously explained, following the above-described identification, the requirements R1.3.1 and R2.2 have been assessed. Particularly, the former has been further classified into R1.3.1a and R1.3.1b, corresponding respectively to the use of local techniques for the structural component and the use of local materials for the finishing layer. For the assessment, each requirement has been described by a score scale, ranging from 0 to 1, according to the roof configuration.

In detail, for the requirement RI.3.1a, the following scores are proposed:

- "0" for horizontal slabs made out of steel reinforced concrete beams and hollow bricks (Comb1.a-b) due to their inconsistency with the construction and technical characteristics of the original structures, namely timber slabs and stone vaults;
- "0.5" for horizontal slabs made out of steel beams and hollow brick blocks (Comb2.a-b), which still witness the indigenous construction tradition, although resulting from a subsequent alteration of the original features;
- "1" for stone masonry vaults (Comb3.a-b) as representative of the very primary identity of the local built environment places.

Similarly, for the requirement RI.3.1b, the following scores are proposed:

- "0" for waterproofing bituminous membranes;
- "0.5" for clay tiles as representative of a subsequent historic transformation;
- "1" for original stone tiles.

Finally, the same combinations have been assessed against the requirement R.I.2.2. For the purpose, the thermal properties of the structural layers (Table 1) and the optical properties of the finishing layers (Table 2) have been identified, based on common values reported in literature.

Table 1. Thermal properties

	U(W/(m^{2}K)	fa	$Y_{i,e}$	φ(h)
Comb1a-b	1,63	0.29	0.471	9.25
Comb2a-b	1,91	0.35	0.662	8
Comb3a-b	2,07	0.136	0.282	10.6

Table 2. Optical properties

	α	ε
Bituminous membrane	0.1	0.95
Clay tiles	0.4	0.9
Stone stiles	0.6	0.95

Therefore, the above-reported parameters have been implemented into software-based dynamic simulations of the energy and environmental behavior of typical building unit, here three-storey tower-houses, covered by each roof combinations. The simulations, run by Design Builder©, enabled the calculation of the indoor surface temperature in summertime, based on the climatic data of the meteorological station at the near international airport of Bari Palese.

Figure 1 reports some representative results of the roof intrados and extrados mean temperatures, as calculated for a representative hot day – 4th August – with 24.5 °C and 34.5 °C average and maximum daily temperature. Thus, based on the internal mean surface temperatures (Table 3), the requirement R.I.2.2 has been also described by a score scale, ranging from 0 to 1, according the temperature ranges.

In detail, for the requirement RI.2.2, the following scores are proposed:

- "0" corresponding to the range 30°–32 °C. The score is associated with all the combinations featured by bituminous waterproofing membranes as the finishing layer (Comb1.a, Comb2.a and Comb3.a). The results show particularly how the worst scenario corresponds to the combination of horizontal slabs made out of steel reinforced concrete beams and hollow brick blocks (Comb1.a), where both poor thermal properties of the structural component and poor optical properties of the finishing layer occur.
- "0.5" corresponding to the range 28°–30 °C. The score is associated with the horizontal slabs covered by clay tiles (Comb1.b e Comb2.b), where the poor thermal inertia of the structural components are somehow compensated by the medium solar reflectance of the finishing layer.

"1" corresponding to the range 26°–28 °C. The score is associated with stone vaults and stone tiles (Comb3.b), where the density of the massive structural component provides with good thermal inertia and the light color of the finishing reduces the surface temperature peaks over the daytime.

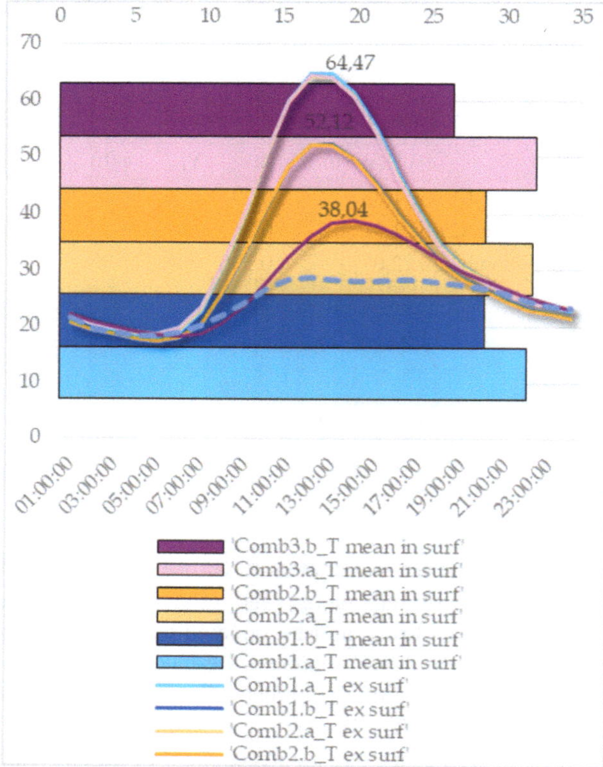

Fig. 1. Roof intrados and extrados mean temperatures on 4th August

Table 3. Internal mean surface temperatures

Comb	1.a	1.b	2.a	2.b	3.a	3.b
T_{mean} [°C]	31,30	28,51	31,70	28,53	31,92	26,41
T env	24,45					

3.2 Assessment at the District Scale

Based on the assessment at the building scale, the entire historic town has been mapped (Fig. 2) to calculate the incidence of the surface of each roof solution out of the total and, thus, derive for each requirement a synthetic score at the district scale as weighted average of the scores achieved by the six combinations (Table 4).

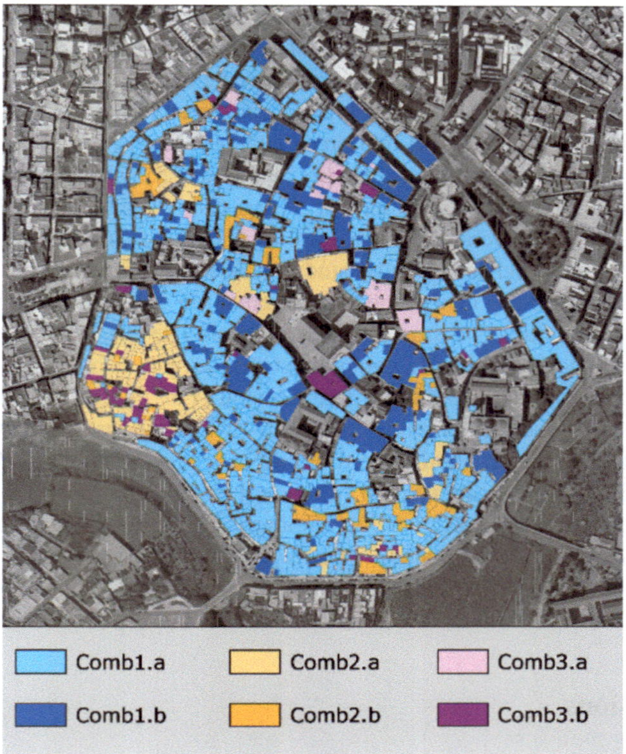

Fig. 2. Distribution of the combinations

Table 4. Assessment scores for the historic town

	% extent	RI.3.1a	RI.3.1b	RI.2.2
Comb1a	61%	0	0	0
Comb1b	17,9%	0	0,5	0,5
Comb2a	9,8%	0,5	0	0
Comb2b	5,3%	0,5	0,5	0,5
Comb3a	2,6%	1	0	0
Comb3b	3,3%	1	1	1
Total		**0,135**	**0,15**	**0,15**

3.3 Discussion of Results

At the construction component scale, as related to the local building typologies and climatic conditions, it should be noted that all the solutions preserving original roof materials and techniques - herein the stone masonry vaults covered by stone tiles (Comb.3.b) – show the most satisfying indoor comfort. Nevertheless, some further systems, including horizontal slabs made out of steel reinforced concrete beams and hollow bricks and horizontal slabs made out of steel reinforced concrete beams and hollow brick blocks, perform relatively better if combined with "semi-traditional" clay tiles, instead of modern waterproofing bituminous membranes (Comb.1.b and Comb.2.b). However, the worst combination from the conservative point of view, namely the horizontal slabs made out of steel reinforced concrete beams and hollow brick blocks and covered by external bituminous membrane (Comb.1.a), is also the poorest in terms of environmental performances. Moreover, at the district scale, the high incidence of this solution affects the overall assessment of the historic town, resulting in a quite low score for all the three indicators under study. Consequently, in view of future requalification and retrofitting policies, the strict correlation between recovery of original materials and techniques and improvement of quality levels and the opportunity to apply the assessment routine as comparative guide and control tool for selecting suitable alternatives should be taken into account. For instance, if all the external finishing layers of that era currently made out of bituminous membranes would be covered by clay tiles, great improvements could be achieved, in terms of compatibility (RI.3.b = 0.53), as well as attenuation of thermal peaks in summertime and improvement of indoor comfort (RI.2.2 = 0.53). Similarly, the replacement of all the existing horizontal slabs made out of concrete beams and hollow brick blocks with structural timber beams and borders and finishing stone tiles would imply a further increase of the district overall score (RI.3.a = 0.92, RI.3.b and R2.2 = 0.85).

4 Conclusions

The paper set out to identify in Mediterranean historic towns – some categories of "inner" resilience – characters and values that the traditional settlements and architectures have developed in relation to the local environment, resources and communities – and some categories of "outer" resilience – issues and aspects of the historic built environments coming from external obsolescence factors. Thus, a score-based assessment of some relevant requirements has been proposed, with specific focus on the "use of passive solutions for thermal regulation" and the "use of local materials and construction techniques" in the roof components. The outcome shows how the construction culture and practice are very well compatible with the desirable capability of protection from atmospheric agents, particularly the direct solar radiation in summertime, and mitigation of indoor discomfort. Such a compatibility has been lost in some modern solutions, where the employment of industrial building products and processes, beyond the missing opportunity to transmit a traditional knowledge to the future generations, has caused a performance drop, in terms of micro-environmental and micro-climatic quality. Nonetheless, although not investigated by the present paper, it should be outlined for further studies that the recovery of traditional solutions seems to be very promising even for some requirements of "outer" resilience, related to seismic protection and energy efficiency among others. In fact, the steel reinforced concrete solutions with outer finishing membrane might negatively affect the vulnerability to earthquakes due to their weight and rigidity, as well as the heat conductivity on the inside due to the high solar absorption and thermal gradient of the outside. The development of a resilience metrics, to which the present paper gives a limited contribution, is highly desirable in order to understand and manage the multiple implication connected with the institutional actions and the technical measures. Finally, it should be highlighted that the safeguard and transmission of the "lessons" about inherent resilience of these historic towns is paramount and vital. For this purpose, the "challenges" against the obsolescence according to modern quality level, although equally relevant, might be addressed by the compatible integration of traditional solutions with cutting edge products and systems, as an alternative to the replacement or alteration of the existing structures.

References

Choudhary P (2016) Vernacular built environments in India: an indigenous approach for resilience. In: Urban disasters and resilience in Asia. Elsevier, pp 269–286. https://doi.org/10.1016/b978-0-12-802169-9.00017-3

Correia M, Dipasquale L, Mecca S (2014) VERSUS: heritage for tomorrow. Correia M, Dipasquale L, Mecca S (eds) Firenze University Press. ISBN 9788866557418 (print)

Currà E, D'Amico A (2014) The role of urban built heritage in qualify and quantify resilience. Specific issues in Mediterranean City. Proc Econ Finance 18:181–189. https://doi.org/10.1016/S2212-5671(14)00929-0

Fatiguso F et al (2017) Resilience of historic built environments: inherent qualities and potential strategies. Proc. Eng. https://doi.org/10.1016/j.proeng.2017.04.262

De Fino M et al (2017) Methodological framework for assessment of energy behavior of historic towns in Mediterranean climate. Energy Build 144:87–103. https://doi.org/10.1016/j.enbuild.2017.03.029

Garcia EJ, Vale B (2017) Unravelling Sustainability and Resilience in the Built Environment. Routledge, Abingdon ISBN 9781138644021 (hardcover)

Hassler U, Kohler N (2014) Resilience in the built environment. Build Res Inf 42(2):119–129. https://doi.org/10.1080/09613218.2014.873593

Langenbach R (2016) What we learn from vernacular construction. In: Nonconventional and Vernacular Construction Materials. Elsevier, pp 3–36. https://doi.org/10.1016/b978-0-08-100038-0.00001-9

Maio R, Ferreira TM, Vicente R (2018) A critical discussion on the earthquake risk mitigation of urban cultural heritage assets. Int J Disaster Risk Reduction 27:239–247. https://doi.org/10.1016/j.ijdrr.2017.10.010

Marchi M et al (2018) Environmental policies for GHG emissions reduction and energy transition in the medieval historic centre of Siena (Italy): the role of solar energy. J Clean Prod 185:829–840. https://doi.org/10.1016/j.jclepro.2018.03.068

Martínez-Molina A et al (2016) Energy efficiency and thermal comfort in historic buildings: a review. Renew Sustain Energy Rev 61:70–85. https://doi.org/10.1016/j.rser.2016.03.018

Micelli E, Pellegrini P (2018) Wasting heritage. The slow abandonment of the Italian Historic Centers. J Cult Heritage 31:180–188. https://doi.org/10.1016/j.culher.2017.11.011

Qian Z, Li H (2017) Urban morphology and local citizens in China's historic neighborhoods: a case study of the Stele Forest Neighborhood in Xi'an. Cities 71:97–109. https://doi.org/10.1016/j.cities.2017.07.018

Ray B, Shaw R (2018) Changing built form and implications on urban resilience: loss of climate responsive and socially interactive spaces. Proc Eng 212:117–124. https://doi.org/10.1016/j.proeng.2018.01.016

Sharma AK (2017) Historic city–a case of resilient built environment. Proc Eng 180:1103–1109. https://doi.org/10.1016/j.proeng.2017.04.270
Srinurak N et al (2016) Analysis of urban morphology and accessibility character to provide evacuation route in historic area. Proc-Soc Behav Sci 216:460–469. https://doi.org/10.1016/j.sbspro.2015.12.061
Vale LJ (2014) The politics of resilient cities: whose resilience and whose city? Build Res Inf 42 (2):191–201. https://doi.org/10.1080/09613218.2014.850602
Webb AL (2017) Energy retrofits in historic and traditional buildings: a review of problems and methods. Renew Sustain Energy Rev 77:748–759. https://doi.org/10.1016/j.rser.2017.01.145
Zagroba M (2016) Issues of the revitalization of historic centres in small towns in Warmia. Proc Eng 161:221–225. https://doi.org/10.1016/j.proeng.2016.08.537

Pedestrian Wind Comfort Study Using Computational Fluid Dynamic (CFD) Simulation

Sarah Fernando[1(✉)], Shiromal Fernando[1], and Priyan Mendis[2]

[1] Civil & Structural Engineering Consultants (Pvt) Ltd., No. 01, 33rd Lane, Bagathale Road, Colombo 03, Sri Lanka
malharifdo@gmail.com

[2] University of Melbourne, Melbourne, Australia

Abstract. Many high-rise buildings are presently being designed and constructed in the central business district of Colombo. These buildings will change the wind patterns and directions which may create an uncomfortable environment at the pedestrian level. Urban authorities in many countries grant a building permit for a high-rise building, only if the wind comfort study proves that the wind environment for the pedestrian remains comfortable and safe. Pedestrian wind comfort and safety studies have been conducted in the past by wind tunnel tests. However, due to the high cost involved, pedestrian wind comfort studies are not generally carried out. With a significant improvement in computer resources in the recent past, Computational Fluid Dynamic (CFD) simulations can be used as an effective alternative method to simulate the pedestrian wind comfort around high rise buildings. This paper presents a validation of pedestrian wind comfort results obtained from a wind tunnel test, using CFD simulation.

Keywords: High-rise buildings · Wind comfort and safety · Computational Fluid Dynamics (CFD) · Wind environment

1 Introduction

Outdoor pedestrian wind environment plays a major role in providing comfort and safety to the pedestrians around high rise buildings. High-rise buildings in urban areas can often introduce uncomfortable or even dangerous wind environments at the pedestrian level due to the form of complex wind patterns. The death of two old ladies has been reported due to a disastrous fall which was caused by high wind speed at the base of a tall building (Lawson and Penwarden 1975). Hence, it is necessary to evaluate the condition of the pedestrian wind environment in terms of comfortability for high rise buildings.

High-rise buildings are particularly influential to complex wind patterns due to the building's complex geometrical features (Ex: sharp/leading edges, trailing edges, complex corner configurations, slender columns etc.) that could result in the behaviour of the wind around the structure. Today, urban authorities in many countries grant a building permit for a high-rise building only if the wind comfort study proves that the

R. Dissanayake and P. Mendis (Eds.): ICSBE 2018, LNCE 44, pp. 323–339, 2020.
https://doi.org/10.1007/978-981-13-9749-3_30

wind environment for the pedestrian remains comfortable and safe (Blocken and Carmeliet 2008). Pedestrian wind comfort studies can be either be performed using wind tunnel tests or using Computational Fluid Dynamic (CFD) simulations. In the past, wind tunnel tests have been conducted intensively as a tool for evaluating pedestrian wind comfort conditions around buildings. Wind tunnel tests are expensive in simulation process and in maintenance. Moreover, only point measurements can be obtained by wind tunnel experiments (Kim et al. 2013). However, with a significant improvement in computer resources in the recent past, Computational Fluid Dynamic (CFD) simulations can be used as an effective alternative method to simulate the pedestrian wind comfort and safety around high rise buildings. CFD studies are most often used due to its ability of full-scale modeling and it facilitates the obtaining of a variety of results from single demonstration. In fact, CFD simulations can be used to evaluate physics that cannot be achieved practically in the wind tunnel test.

This paper presents the CFD evaluation of the pedestrian wind comfort and safety in a high-rise apartment building. The validation process was carried out with the experimental results and the numerical simulation results were obtained using the commercially available software Solidworks Flow Simulation. Wind mitigation measures and improvements are evaluated for the proposed development premises according to the results.

2 Wind Criteria

2.1 Criteria for Comfort and Safety

Many researchers have proposed a variety of criteria to assess the pedestrian level wind climate condition. According to William et al. (1990), either a fixed wind speed is selected and evaluated where the percentage of time that this wind speed can occur varies depending on the activities planned for the area or a fixed frequency of occurrence is selected with acceptable wind speed varying depending upon the anticipated pedestrian activity. Situations where winds are very strong and it directly affects the people's balance, Gust speeds are more important. However, mean wind speeds might be just as important an indicator of comfort in areas such as outdoor pedestrian usage areas. Therefore, it is desirable for the criteria to capture the effects of both mean speeds and gust speeds (ASCE 2003).

Durgin (1995) and Lawson (1978) proposed a way of achieving this, which is to express the criteria in terms of mean speeds, but then to compare not just with the mean speed, but also compare with a parameter called "Gust Equivalent Mean Speed" (UGEM).

$$U_{GEM} = \frac{\widehat{U}}{G} \tag{1}$$

Where $\widehat{U}$ is the peak gust speed and G is a representative fixed gust factor.

For the purpose of feasibility, the pedestrian level wind micro-climate criteria that is described in Table 1 was employed in this study. Wind conditions acceptable for sitting (having lowest wind speeds) are appropriate for outdoor amenity areas and seating areas in open air cafes/parks. Furthermore, a small increase of wind speeds will reduce the humidity of amenity spaces which will be more beneficial. Hence, a mix of sitting and standing conditions for such kind of places would be more appropriate. Standing conditions are preferred at main entrances, drop off areas and bus stops, where pedestrians are about to linger. Strolling conditions are appropriate if pedestrians are expected to move at a more leisurely place (Eg: shop windows, park, Children's play area or landscape area). Moreover, walking conditions are appropriate for side walks and pathways, when pedestrians are not expected to move for a purpose. It should be noted that wind speeds consistently above this level are likely to be a nuisance to pedestrians.

Table 1. Wind comfort criteria used for the study (RWDI 2017)

Comfort category	U_{GEM}/(m/s)	Description
Sitting	≤ 2.38	Calm or Light breezes desired for outdoor restaurants and seating areas where one can read a paper without having it blown away
Standing	$\leq$ **3.88**	Gentle breeze acceptable for main building entrances and bus stops
Strolling	$\leq$ **4.72**	Moderate winds that would be appropriate for window shopping and strolling along a downtown street, plaza or park
Walking	$\leq$ **5.56**	Relatively high speeds that can be tolerated if one's objective is to walk, run or cycle without lingering
Uncomfortable	$\geq$ **5.56**	Strong winds of this magnitude are considered a nuisance for most activities, and wind mitigation is typically recommended

3 Analysis of Wind Data

Wind statistics recorded at the Katunayake International Airport between 1998 and 2017 were observed between May and October, which includes the southwest monsoon and the November to April Period, which includes the northeast monsoon. Figure 1 illustrates the distribution of wind frequency and directionality for those two periods of the year.

During the period of May to October, prevailing winds are from west through south-west directions. During the period of November to April, prevailing winds are from north-east direction.

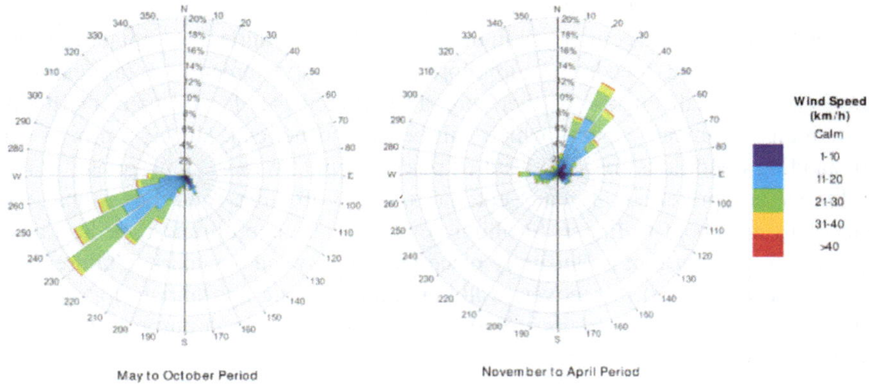

Fig. 1. Seasonal wind roses for Colombo (data from Katunayake International Airport (1998–2017)

4 Wind Behavior Around Structures

The effect of the pedestrian wind microclimate may depend on the building shape, size, orientation and interaction with neighbouring building obstacles. Tall buildings tend to intercept stronger winds at higher elevations and redirect them to the ground level. Figure 2 shows the down washing effect of wind, which is deflected down to the street level, creating swirling turbulence at the pedestrian level. This effect is often the main cause for wind accelerations around large buildings at the pedestrian level.

When oblique winds are deflected down by a building, an increment of the wind activity can be expected around the building corners at the pedestrian level as shown in Fig. 3. (RWDI 2017). This is known as the corner acceleration effect of wind due to building geometry and the behaviour of the wind.

Fig. 2. Down washing effect

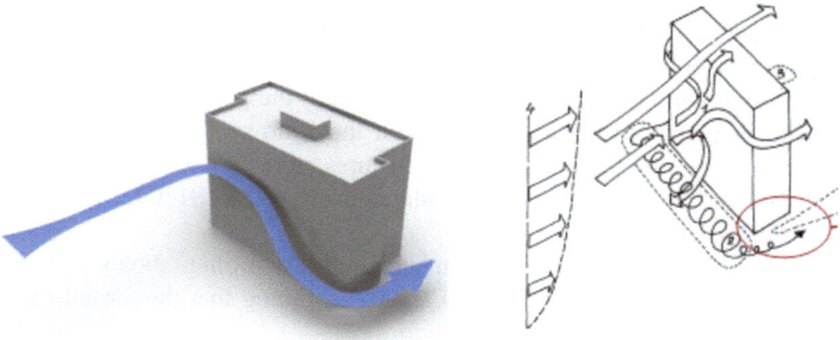

Fig. 3. Corner acceleration effect

5 Case Study

5.1 Model

The Proposed Development is a 36-storey residential development comprising of a tower, a top, and a 4-level podium. The development will be 136 m high above the ground level after completion. Podium level at level 5 provides common facilities and above the podium level, all the other floors are apartments.

5.2 CFD Simulation

In order to evaluate the wind condition related to the pedestrian wind comfort, a CFD simulation was conducted by using the commercially available software Solidworks Flow Simulation. A full-scale solid model of the proposed development was modelled (see Fig. 4) for the purpose of conduction a CFD simulations.

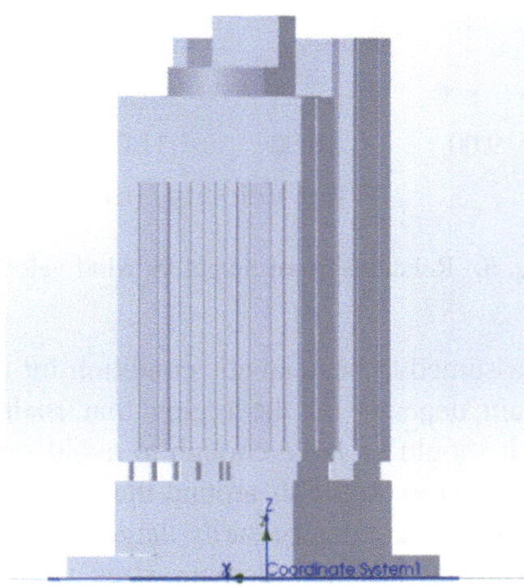

Fig. 4. Solid model of the building

Hence, no scaling was involved in length and time scales of the results. Model setup has been selected for optimizing the computational time whilst keeping the numerical errors to be at a minimum.

Wind profile within the atmosphere extends from the ground level to few hundred meters. The shape, speed and the direction of the wind profile changes with various factors such as terrain type, topography features, direction etc. However, for this analysis, the wind profile which was calculated using AZ/NZS1107.2:2002 (as shown in Fig. 5) was used.

Similar to the wind profile, the turbulence intensity profile (shown in Fig. 6) was used as mentioned in the AS/NZS 1170.2:2002 assuming that the terrain category of the region as 2 and the region as B.

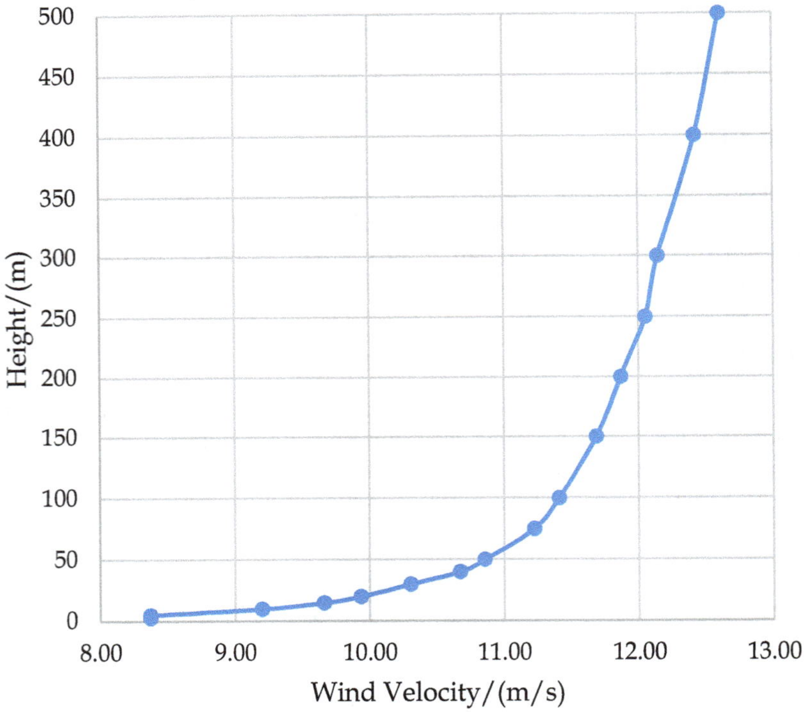

Fig. 5. Relationship of height vs wind velocity

Wall conditions are assumed to be adiabatic condition for this analysis. The size of the computational domain depends on the region that shall be represented by the simulation (see Fig. 7). It should be large enough to avoid reflecting of fluid streams, which may cause abnormal pressure fields around the model.

However, it is also true that unnecessarily large fluid domain demand higher computational cost. Therefore, arbitrary domain sizes were used for this analysis, which are large enough to give a satisfactory level of accurate results with reasonable computational cost.

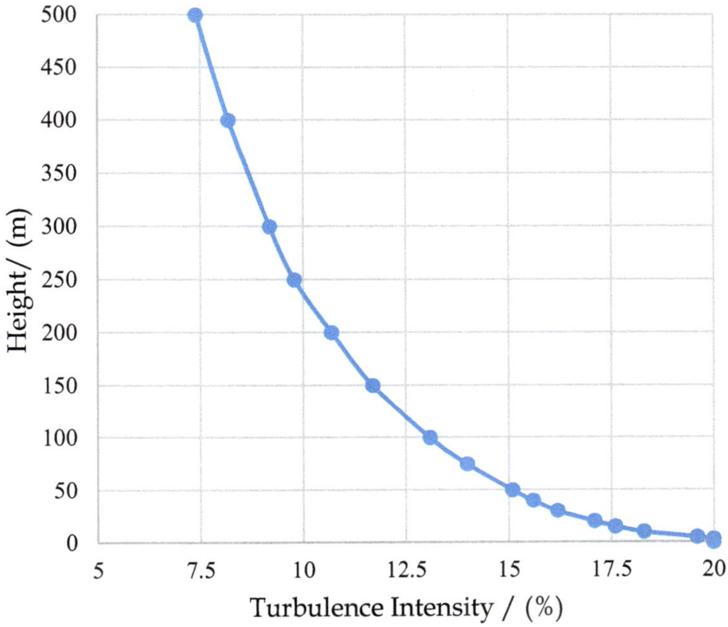

Fig. 6. Relationship of height vs turbulence intensity

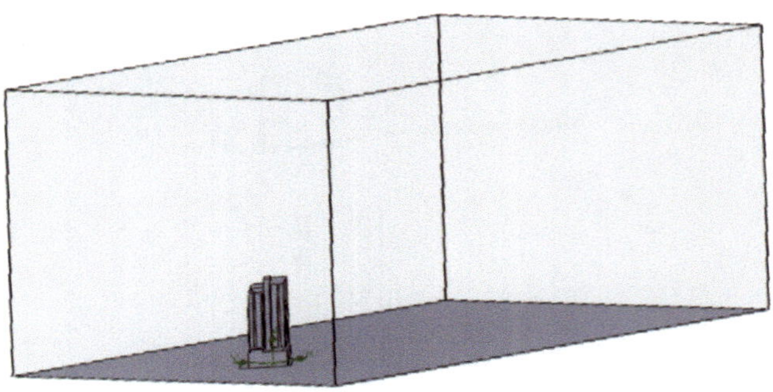

Fig. 7. Computational domain of the study

The mesh has been done via SOLIDWORKS and the model setup consists of two mesh types; one for the global domain and the other for near building surface. This model setup was selected for optimizing the computational time whilst keeping the numerical errors to a minimum.

5.3 Definition of Wind Direction

0° wind direction has been chosen to coincide with the geographical North as shown in Fig. 8. Major wind directions were selected as North-East and South-West for the pedestrian wind comfort analysis.

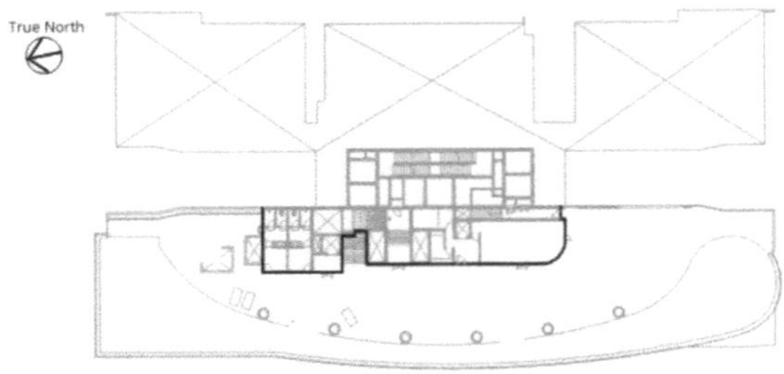

Fig. 8. Geographical north of the building (plan view)

6 Results

Results were obtained using CFD Simulation at three levels (Ground Level, Podium Level and Roof Terrace Level as shown in Fig. 9) to assess the condition where wind directly affects the comfortability of people.

Maximum wind velocity values (from North-East and South-West Direction) at each point obtained from the wind tunnel test (which were shown in Figs. 10, 11 and 12) were compared with the CFD simulation results.

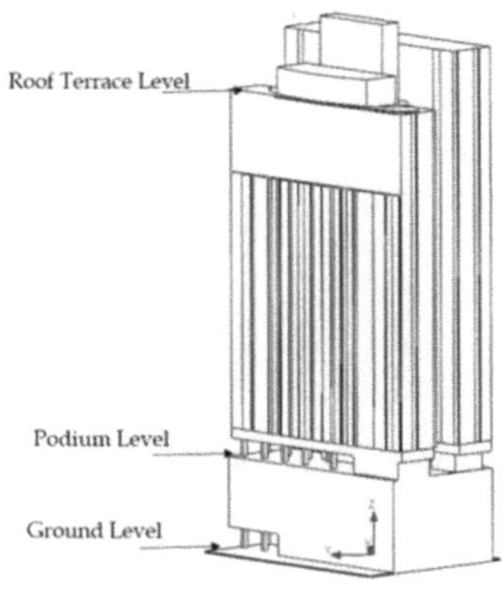

Fig. 9. Considered floor levels of the apartment building

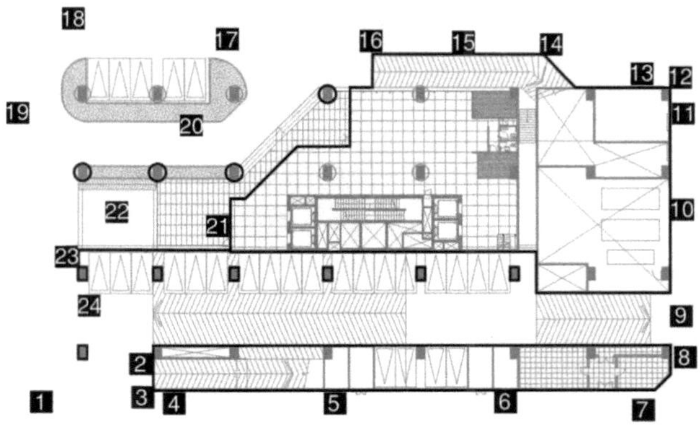

Fig. 10. Ground floor level-locations

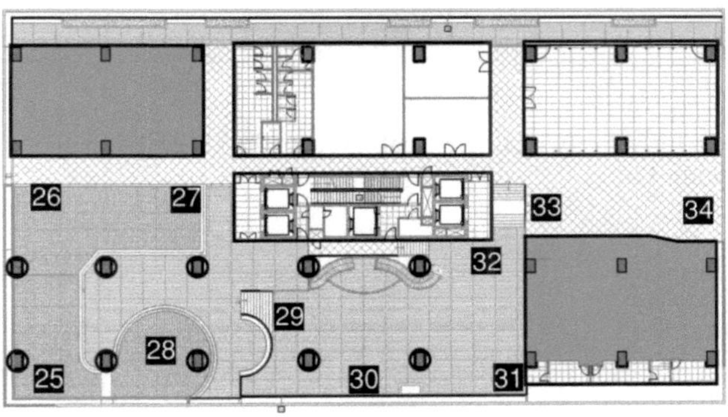

Fig. 11. Podium level-locations

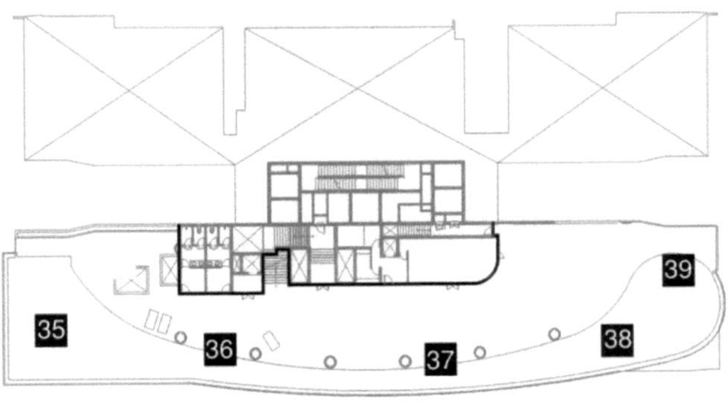

Fig. 12. Roof terrace level-locations

Figures 13, 14 and 15 shows the velocity contours at the ground level, podium level and roof terrace level respectively for the wind direction of North-East.

Figures 16, 17 and 18 shows the velocity contours at the ground level, podium level and roof terrace level respectively for the wind direction of South-West.

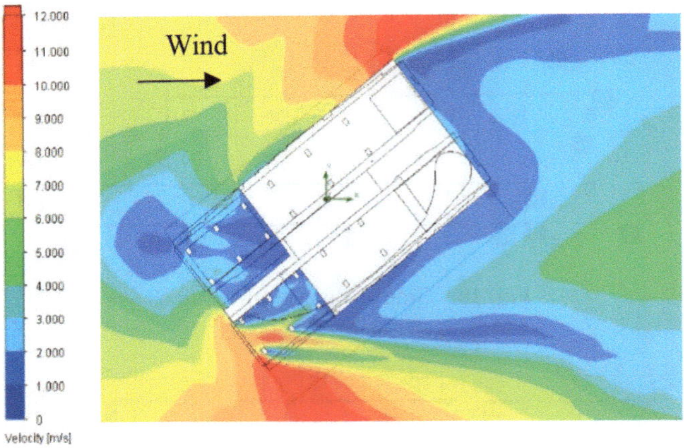

Fig. 13. Contours for velocity in a horizontal level at a height of 1.5 m above ground level for wind direction-North East

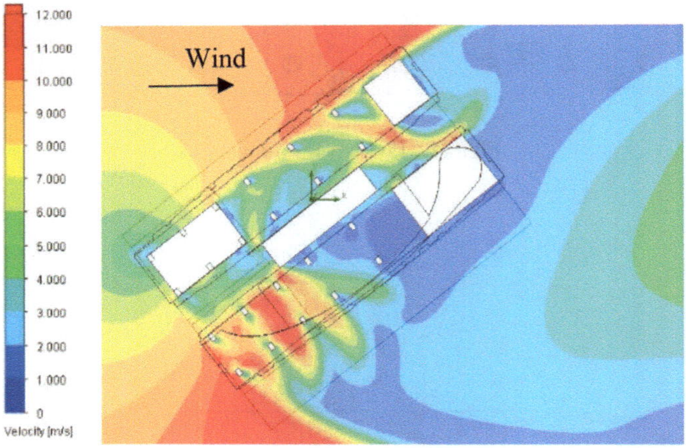

Fig. 14. Contours for velocity in a horizontal level at a height of 1.5 m above podium level for wind direction-North East

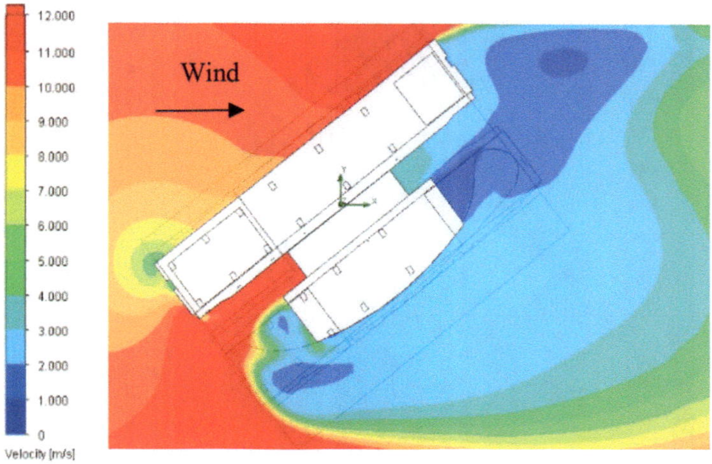

Fig. 15. Contours for velocity in a horizontal level at a height of 1.5 m above roof terrace level for wind direction-North East

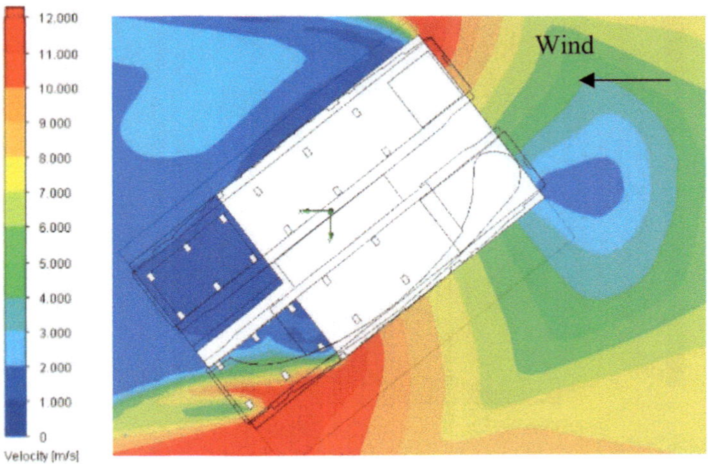

Fig. 16. Contours for velocity in a horizontal level at a height of 1.5 m above ground level for wind direction-South West

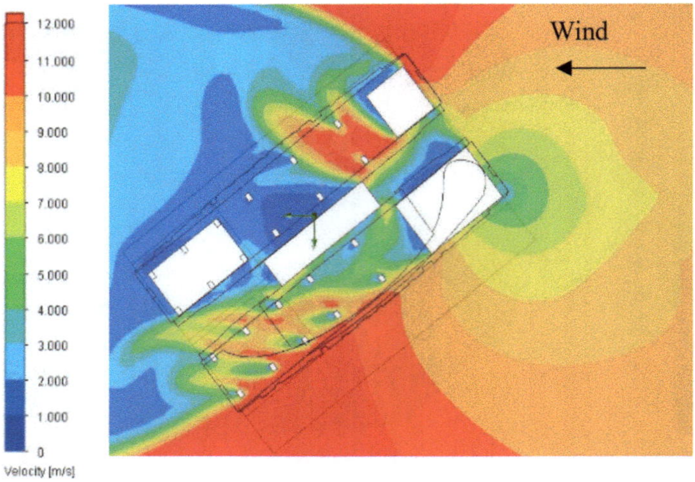

Fig. 17. Contours for velocity in a horizontal level at a height of 1.5 m above podium level for wind direction-South West

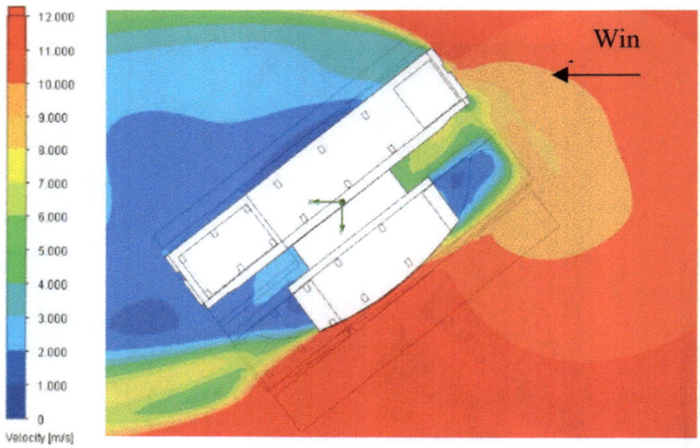

Fig. 18. Contours for velocity in a horizontal level at a height of 1.5 m above roof terrace level for wind direction-South West

Table 2 shows the results of wind condition of each locations, considering the results obtained from both wind tunnel and CFD simulations for the two main directions (North-East and South-West).

Table 2. Summary of the results of pedestrian wind comfort conditions

Location		Wind tunnel/ (m/s)	CFD Simulation/ (m/s)	Condition	Remarks
Ground level	1	7.8	9.6	Uncomfortable	Discrepancies of the results may due to absence of the surrounding of the CFD model, since it may directly affect to the wind behaviour around the structure
	2	5.8	6.9	Uncomfortable	
	3	6.9	7.3	Uncomfortable	
	4	6.1	6.5	Uncomfortable	
	5	4.4	7.5	Strolling-Uncomfortable	
	6	3.3	5.4	Standing-Walking	
	7	4.2	3.5	Strolling-Standing	
	8	3.8	3.0	Standing	
	9	5.0	5.6	Walking	
	10	3.8	4.3	Strolling	
	11	5.3	5.5	Walking	
	12	6.1	7.6	Uncomfortable	
	13	8.3	10.3	Uncomfortable	
	14	5.3	6.8	Walking-Uncomfortable	
	15	5.3	5.0	Walking	
	16	4.7	4.9	Strolling-Walking	
	17	4.2	3.9	Strolling	
	18	3.8	2.0	Standing-Sitting	
	19	3.6	2.1	Standing-Sitting	
	20	3.8	2.0	Standing/Sitting	
Podium level	21	2.5	1.6	Sitting	Discrepancies of the results may due to absence of the surrounding of the CFD model, since it may directly affect to the wind behaviour around the structure
	22	5.0	2.0	Walking - Sitting	
	23	3.3	3.6	Standing	
	24	5.8	6.3	Uncomfortable	
	25	6.4	7.1	Uncomfortable	
	26	5.0	5.5	Walking	
	27	4.7	5.0	Walking	
	28	5.0	10.5	Walking-Uncomfortable	
	29	5.0	5.6	Walking	
	30	4.7	4.9	Strolling/Walking	
	31	3.8	1.5	Standing-Sitting	
	32	3.8	3.0	Standing	
	33	4.2	4.7	Strolling	
	34	3.8	3.7	Standing	

(*continued*)

Table 2. (*continued*)

Location		Wind tunnel/ (m/s)	CFD Simulation/ (m/s)	Condition	Remarks
Roof terrace		6.8	7.3	Uncomfortable	
	36	7.8	8.8	Uncomfortable	
	37	5.3	5.5	Walking	
	38	4.7	3.9	Strolling	
	39	4.4	2.0	Strolling-Sitting	

7 Wind Comfort Assessment

The discussion of the wind comfort at the three levels is based on results obtained using wind tunnel data and CFD simulations.

7.1 Ground Level

Locations represented by 1 to 27 are expected to have wind conditions ranging from suitable to sitting to uncomfortable according to the results obtained in Sect. 6. Uncomfortable wind conditions are formed in the corners of the building due to corner acceleration effect of the wind flow. Mitigation measures are required to achieve desired calmer conditions at the ground level.

Locations represented by 18, 19 and 24 (Main Entrance) shows a calmer wind condition throughout the year. However, lowest wind condition may lead to create discomfort to the pedestrians (due to formation stagnation condition). Such areas need to be considered to avoid that kind of discomfort where pedestrians are likely to stay.

7.2 Podium Level

The wind condition at the podium level ranges from standing to uncomfortable conditions during both monsoons. The amenity area along the podium terrace includes the swimming pool, Children's pool and built-in seating areas. Wind condition of the swimming pool areas (28 and 31) are likely to have walking to uncomfortable conditions according to the results, due to the podium which is exposed to the wind. Wind flow may lead to accelerate through confined spaces as well as corner areas of the building. Open sky seating areas (36 and 37) are predicted to have higher wind conditions which would require mitigation measures to maintain desired (sitting) wind conditions.

7.3 Roof Terrace Level

Roof level terrace areas represented by 38 to 42 and wind condition along the terrace ranges from sitting to uncomfortable according to the simulation. Mitigations measures are required to achieve the desired wind conditions if any seating areas are proposed along this terrace.

8 Mitigation Strategies

In areas where wind condition is unacceptable for the intended pedestrian use or unsafe and will be accessible to pedestrians, it is advisable to develop wind control strategies.

8.1 Mitigation at Ground Level

Wind control measures are recommended along ground areas to eliminate the uncomfortable walking conditions that were described in Sect. 7.1. The addition of hard or soft landscaping elements are advisable as shown in Fig. 9. However, ever-green trees having 2 to 4 m height with low level landscaping elements will diffuse the energy of high wind.

In order to diffuse high wind speeds, porous wind screens can also be used in ground areas. To achieve sitting wind localised sheltering and 1.5 m–2 m planting or screening elements can be used.

It is recommended, that the areas that indicating stagnant conditions, landscaping elements and planting trees should be avoided or minimized in order to unblock the air flow entering such places.

8.2 Mitigation at Podium Level

Seating areas and swimming pool along the podium level, need to be considered when adding high wind speed mitigation measures. The addition of 1.5 m to 2 m height columnar plantations along the balustrade would be significant in breaking up any oncoming flow and provide a desired wind condition in such places. In addition to that, low-level landscaping elements along the periphery of the pool area could be more helpful in achieving calm conditions. Furthermore, low level landscaping such as shrubs in planters, porous screens or sculptures can be used within the podium level to confuse the wind and break high wind speeds.

8.3 Mitigation at Roof Terrace Level

Mitigations that are mentioned Sect. 8.2 are applicable to the roof terrace level. In addition to that, increasing the height of the balustrade from 1 m to 2 m will be beneficial in providing shelters for the areas which are nearby (immediately behind) the balustrade.

9 Concluding Remarks

CFD has been applied to evaluate the pedestrian wind comfort in specific locations in a High-rise Apartment Building. The CFD Simulations were conducted based on the two seasonal monsoons that occur in Sri Lanka (North-East monsoon and South-West Monsoon). Results obtained from CFD simulation have been compared with the results obtained from wind tunnel test for verification. Pedestrian comfort was assessed for the ground level, podium level and the roof terrace level of the development. The wind microclimate in these levels is predicted to range from sitting to uncomfortable during the windiest season.

Overall, the pedestrian comfort conditions around the selected levels is acceptable for its intended use and any windier than desired locations are proposed to be mitigated by the suggested mitigation measures.

References

Lawson TV, Penwarden AD (1975) The effect s of wind on people in the vicinity of buildings. In: 4th international conference on wind effects on buildings and structures, Heathrow

Janssen W, Blocken B, Van Hoof T (2013) Use of CFD simulations to improve the pedestrian wind comfort around a high-rise building in a complex urban area. In: Proceeding of BS2013: 13th conference of international building performance simulation association, pp 1918–1925

Williams CJ, Soligo MJ, Cote J (1992) A Discussion of the components for a comprehensive pedestrian level comfort criterion. J Wind Eng Ind Aerodyn 44–41:2389–2390

Fadl MS, Karadelis J (2013) CFD simulation for wind comfort and safety in urban area: a case study of Coventry University central campus. Int J Archit Eng Constr 2(2):1–13

Blocken B, Carmeliet J (2008) Pedestrian wind conditions at outdoor platforms in a high-rise apartment building: generic sub configuration validation wind comfort assessment and uncertainty issues. Wind Struct 11(1):51–70

ASCE Aerodynamics Committee (2003) Outdoor human comfort and its assessment. State of the art report, Task Committee on Outdoor Human Comfort, American Society of Civil Engineers, Boston, VA, USA

Stathopoulos T (2006) Pedestrian level winds and outdoor human comfort. J Wind Eng Ind Aerodyn 94:769–780

William CJ, Hunter MA, Waechter WF (1990) Criteria for assessing the pedestrian wind environment. J Wind Eng Ind Aerodyn 36:811–815

Montazeri H, Blocken B, Janssen WD, Hooff TV (2012) CFD analysis of wind comfort on high-rise building balconies: validation and application. In: The seventh international colloquium on bluff body aerodynamics and applications, Shanghai, China, pp 1674–1681

Kim H, Kim T, Leigh S (2013) Assessment of pedestrian wind environment of high-rise complex using CFD Simulation. In: Sustainable procurement in urban regeneration and renovation Northern Europe and North-West Russia, pp 1–8

AS/NZS-1170.2 (2002) Australian/New Zealand Standard, Structural Design Action-Part 2: Wind Actions, Standards Australia & Standards New Zealand

RWDI (2017) Pedestrian level wind micro-climate assessment, Apartment complex, Colombo, Sri Lanka, 170 1325

Mohotti D, Danushka K, Mendis P (2015) Wind design of slender tall building: CFD approach. In: 6th international conference on structural engineering and construction management, Kandy, Sri Lanka, pp 194–203

Urban Design Terms of Reference (2014) Pedestrian comfort and safety studies. Planning and Building Department, Development and Design Division, City of Mississauga

Mendis P, Ngo T, Hariots N, Hira A, Samali B, Cheung J (2007) Wind loading on tall building. EJSE special issue: loading on structures, pp 41–54

Performance of Paving Materials in Outdoor Landscaping

G. H. Galabada(✉) and R. U. Halwatura

University of Moratuwa, Moratuwa, Sri Lanka
hyasasiri@yahoo.com

Abstract. Many developments today, have made the human lifestyle more comfortable but seem to be concerned only about the technical and economic viability rather than the environment or the user comfort. Hence, these developments are causing severe environmental issues such as global warming, heat island effect, losing the natural appearance etc. The selection of construction materials considering their properties is vital with respect to sustainable outdoor development. Materials such as clay bricks, asphalts, concrete etc.; are used and since recent times, concrete block pavements have become an attractive engineering and economical material in the construction of walkways, worldwide. However, there are many drawbacks with respect to user comfort via temperature effect in the use of these materials. Therefore, the study of temperature effects of materials, which are used for outdoor developments is important. Accordingly, this research was done to evaluate and compare the thermal performance of existing paving materials used for outdoor development in Sri Lanka. Separate sections were paved with different paving materials and same colour was applied on the surface to avoid the colour effect. The results obtained were evaluated and the thermal performances were compared. The results indicated that the material and its surface colour of surface influenced thermal performance.

Keywords: Outdoor development · Paving materials · Thermal comfort · Surface colour

1 Introduction

Continually increasing urbanization and careless development of infrastructure and open spaces with hard surfaces creates a major impact on the urban climate (Doulos et al. 2004; Babić et al. 2012). Most of the outdoor open spaces adversely affects the behavior of users and the aesthetic aspects of built environment and causes the deterioration of the visual image of most urban areas (Mendoza et al. 2012). One of the most dominant surface elements contributing to the city is artificial pavements exposed to solar radiation because there is no evaporative cooling effect over the surface. They generate long wave radiation than a vegetated surface. They create not only high surface temperatures but also influence the near surface air temperature (Chudnovsky et al. 2004; Kleerekoper et al. 2012; Spagnolo and de Dear 2003). (Rehan 2013) recommended to use locally sourced paving materials manufactured with sustainable

R. Dissanayake and P. Mendis (Eds.): ICSBE 2018, LNCE 44, pp. 340–352, 2020.
https://doi.org/10.1007/978-981-13-9749-3_31

practices to reduce life-cycle environmental impacts (Rehan 2013). Therefore, sustainability in the outdoor development is one of the most important strategies for urban design.

On the other hand, well planned and well connected infrastructures will lead to increased walking activities while limiting the demand for motorized transport. The environmental benefit of limiting motorized transportation are the reduction of fossil energy consumption and greenhouse gas (GHG) emissions (van der Zee 1990). Creating a pedestrian-friendly environment could promote walking as a mode of sustainable transportation and safe recreation, thus encouraging citizens to be physically active (Mendoza et al. 2012), Chiesura 2004, (Kelly et al. 2011). Therefore, pedestrian pavements and recreational areas are designed to be technically, economically and aesthetically suitable for users and developers should involve a pedestrian need hierarchy in outdoor development. Therefore, the design process must apply comprehensive life cycle environmental data to identify suitable construction solutions and urban management strategies that contribute to minimizing environmental impacts (Grimmond et al. 2010; Rahaman and Ohmori n.d.). Construction materials play a very important role in the thermal balance in the urban environment (Santamouris et al. 2011; Scudo and Dessì 2006). The materials' thermal balance is determined mainly by their reflectivity of solar radiation and their emissivity of long wave radiation during daytime (Babić et al. 2012). (Doulos et al. 2004) has determined that the materials thermal balance by their reflectivity of solar radiation and their emissivity of long wave radiation during day time and has reported that the mean daily surface temperature differences are mainly caused by the different albedo value. Further they have mentioned that the physical characteristics of the material that affected their albedo are the colour, surface texture and the construction material. The rough and dark coloured surfaces tend to absorb more solar radiation than smooth light coloured and flat surfaces and therefore the rough and dark coloured surfaces are warmer than the light coloured (Doulos et al. 2004). They have defined the materials light in colour and having smooth surface as 'cold' materials and those having rough and dark coloured materials as 'warm' materials.

The use of proper materials to reduce the temperature effect and improve the thermal characteristics of the outdoor spaces has had extended interest during the resent years. Researchers are trying to examine the impact of materials' thermal performance on the urban temperature (Doulos et al. 2004; Erell et al. 2014).

The urban heat island phenomenon (UHI) is one of the most important occurrences affecting the climate and environment quality and negative influence on the standard of living and health of city residents as well as increasing energy consumption (Babić et al. 2012; Chudnovsky et al. 2004; Kleerekoper et al. 2012). Most of the urban areas where large proportions consist of dry impervious pavements or buildings increase surface temperature than many rural areas, where plant cover and evaporation of soil moisture may be moderate (Santamouris et al. 2012; Kleerekoper et al. 2012).

The urban heat island phenomenon is mainly related to the high density of buildings and urban structures that absorb solar radiation, the use of highly absorbent materials, the lack of green spaces and the absence of vegetation (Chudnovsky et al. 2004; Santamouris et al. 2012). Since, it is very important to identify the thermal behavior of the urban surface, researchers recommend to use cool pavements wherever possible to reduce the urban heat island effect, particularly in the urban environment of cities with hot climate. However, warm materials are used in urban areas instead of cold material due to economic and aesthetic reasons. As a result, the temperature in the urban environment is raised and the demand of cooling energy in the building is increased (Doulos et al. 2004; Rehan 2013; Santamouris et al. 2012).

Heat island phenomenon increases temperatures in urban areas, increases the energy consumption for cooling purposes and affects the global environmental quality of cities. Therefore, the hidden approach of the lowering top layer air temperature in a city gives the benefit of reduced air conditioning loads in buildings; and improved thermal comfort for pedestrians in outdoor urban spaces. To mitigate the heating of such surfaces that result in elevated air temperature and contribute to the urban heat island phenomenon, numerous researchers have suggested the use of paving materials which have a high albedo as much as possible. These studies have directed recommendations and guidelines for urban planners and architects on practical implementation of cool paving which are presenting a high solar reflectivity and emissivity as an effective mitigation technique when applied to buildings and open spaces (Santamouris et al. 2012; Dareeju et al. 2013).

Although many experimental data are available to evaluate the performance of cool materials in isolated buildings (Jaffal et al. 2012; Halwatura 2013) very limited information is available concerning the mitigation potential of the reflective materials when applied in open urban areas (Santamouris et al. 2011).

Therefore, this paper represents the results of field experiment done to compare the temperature behavior on commonly used paving materials in Sri Lanka for outdoor landscaping.

2 Objective of the Research

To compare the actual performance of existing paving materials used for outdoor development in Sri Lanka.

3 Materials and Method

The research was conducted in University of Moratuwa premises, one of the main universities in Sri Lanka. The climate of the area is characterized as hot and dry during the observations, which is similar to its climate for most parts of the year as well as most city areas in Sri Lanka. The climatic data of the area shows that February and March were comparatively the hottest months of the year, with the daily average temperature ranging from 27–28 °C and the daily maximum temperature ranging from 30 °C to 33 °C. The daily average relative humidity is around 70–80%, and the average wind speed is 5–6 km/hr during the period of measurement [10]. Figure 1 shows the locations of the four observation areas, which were paved with three different paving materials and the fourth left natural ground. The selected paving materials were cement concrete blocks, burnt clay bricks and asphalts. An areas of 2 m × 2 m was paved with selected paving materials while keeping the same sub-base. Temperature measurements were taken as two sets of observations. The materials were tested with their native colour and then it was tested after painting them with the same colour to avoid the Albedo effect. The colour applied was beige. Figure 2 shows the four observation areas with three different paving materials after painting while Fig. 3 shows the change in colour on the paving areas.

Fig. 1. Paving areas with their native colour.

For all observation areas, the temperatures were recorded on the top surface and point which is 300 mm above from the top surface simultaneously at 10 min intervals. The temperatures were collected using a data logger. The ground surface and the ambient temperatures were also recorded at the same time intervals. Observations were recorded for four days and 24 h for each day.

Fig. 2. Coloured paving areas and data logger

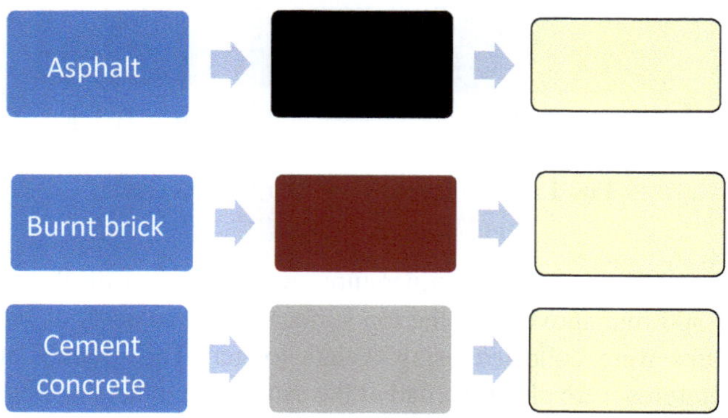

Fig. 3. Change in colour on paving areas.

4 Results and Discussion

The average surface temperature of the studied objects, the ambient temperature and the natural ground temperature were evaluated to compare temperature variation of manmade hard surface with natural surface. Results obtained from temperature measurements were analyzed to compare the temperature variation for each paving material.

The temperature variations on the paved area for 24 h in the 1st case of this study were analyzed first. (i.e. the areas paved with selected paving materials with their native colour). The results observed are shown in Fig. 4. From 00.00 h to 7.00 h all the surfaces show nearly same temperature with variation less than 0.8 °C. After 7.00 h the temperatures increased rapidly for all three paved areas. The maximum average ambient temperature of these day was 32.71 °C at 12.00 h. The maximum temperature of the asphalt top surface was 52.78 °C at 13.00 h, which is 20.07 °C higher than the ambient temperature at the time. At the same time, the temperature at a height of 300 mm above the surface reached 33.6 °C, showing a difference between the ambient temperature and the temperature at a height of 300 mm above of 0.89 °C. The temperature difference between top surface and the 300 mm above was 19.18 °C.

The maximum temperature on the top surface of burnt brick paved area was 47.56 °C at 14.00 h. At this time the ambient temperature was 31.33 °C. The temperature difference between top surface and the ambient temperature was 16.23 °C. At the same time, the temperature at a height of 300 mm above the surface reached 32.74 °C. This shows temperature difference of 1.4 °C between ambient temperature and the temperature at 300 mm above and 14.82 °C between the top surface and 300 mm above.

In the case of cement block paved area the maximum temperature on the top surface of was 49.79 °C, which is 18.46 °C higher than the ambient temperature at 14.00 h. At the same time, the temperature at a height of 300 mm above the top surface reached 31.99 °C, showing a difference of 0.66 °C between ambient temperature and the temperature at a height of 300. The temperature difference between top surface and the 300 mm above was 17.80 °C.

All most all the cases the top surface temperature remains above the ambient throughout the day, while this effect was become null after 20.00 h at a height of 300 mm above the paved surface.

The maximum temperature difference between surface and the ambient temperature was shown on dark asphalt, while the second maximum was given by cement concrete. The temperature difference between 300 mm above and the ambient temperature shows a significant difference in the day time for dark asphalt and burnt brick. However, the same difference was not that significant for cement paving. Hence, it's clear that the darker paving materials give more contribution to the urban heat island than the light colour surfaces.

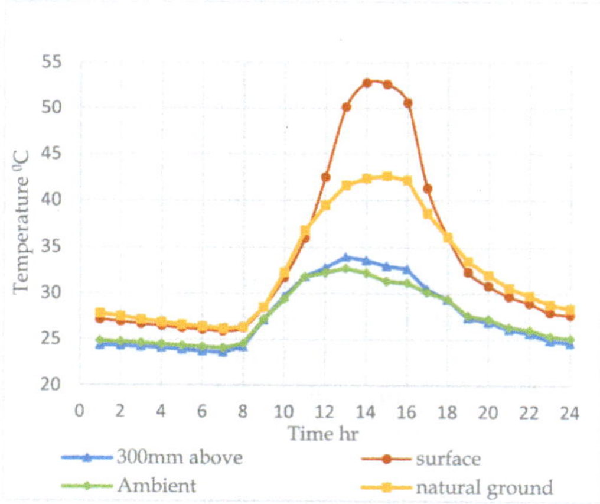

a: temperature variations on the asphalt paved area

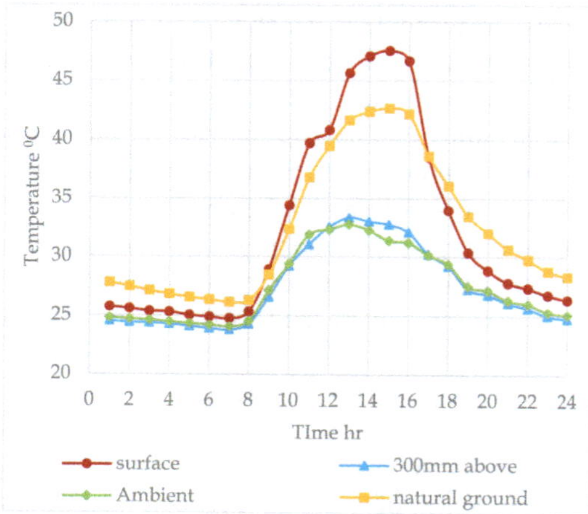

b: temperature variations on the burnt brick paved area

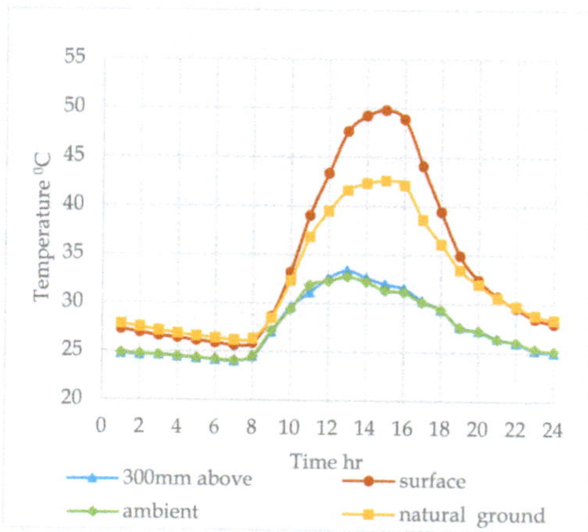

c: temperature variations on the cement concrete area

Fig. 4. The temperature variations on paved areas with their native colour

Then the same testing was repeated for the painted paved areas. (i.e. the areas paved with same selected paving materials by painting the surface). Figure 5 shows the temperature variation for painted areas. The temperature differences from 01.00 h to 7.00 h were marginal and increased slowly, however there was slight temperature differences between different materials. After 7.00 h the temperatures increased rapidly for all three paved areas. The maximum average ambient temperature of this day was 34.43 °C at 12.00 h. The maximum temperature on the asphalt top surface was recorded as 51.73 °C showing a difference of 17.3 °C with the ambient temperature. The temperature difference between the ambient and surface temperature was 3.27 °C less than the temperature difference between ambient temperature and surface temperature on their native colour. At the same time, the temperature at a height of 300 mm above the top surface reached 38.88 °C, showing a difference between the ambient temperature and the temperature at a height of 300 mm above the top surface of 4.45 °C. The temperature difference between top surface and the 300 mm above was 12.85 °C. This shows 6.33 °C lesser value than that of first case.

The maximum temperature on the top surface of burnt brick paved area was 43.99 °C at 13.00 h. showing a 10.11 °C difference between the ambient temperatures at that time. This was 6.12 °C less than the temperature difference between ambient temperature and surface temperature on their native colour. At the same time, the temperature at a height of 300 mm above the top surface reached 38.03 °C, showing a difference between the ambient temperature and the temperature at a height of 300 mm above the top surface of 4.15 °C. The temperature difference between top surface and the 300 mm above was 5.96 °C. This shows 8.87 °C lesser value than that of first case.

In the case of the cement block paved area, the maximum temperature on the top surface was 42.97 °C at 13.00 h showing a difference between the ambient temperature of 9.1 °C. At the same time, the temperature at a height of 300 mm above the top surface reached 33.43 °C, showing very little difference between the ambient temperature and the temperature at a height of 300 mm above of 0.45 °C. The temperature difference between top surface and the 300 mm above was 9.54 °C again after 20.00 h, the temperatures of the top surface and at a height of 300 mm above the top surface cooled down nearer to ambient temperature.

After applying the same colour on the surfaces, the difference in temperature with ambient temperature decreased by a considerable amount. The difference in temperatures between surface and 300 mm above were also decreased by a significant amount. This shows that the colour is an important factor, which influences the thermal behavior of paving surface and the urban heat island.

Figure 6 shows the comparison of temperature variation on the top surface of selected paving areas with their own colour, natural ground and ambient temperature while Fig. 7 shows the temperature variation for paving after painting to the same colour. In the first case almost all the surfaces showed a minor temperature variations during morning (i.e. 00.00 h to 7.00 h). Asphalt, burnt brick and cement concrete block paving surfaces heated up quicker than ground. During 13.00–14.00 h asphalt surface showed the highest temperature with a difference of 10.45 °C with natural ground. Cement blocks and brick surfaces took the second and third places with temperature differences of 6.88 °C and 4.75 °C respectively.

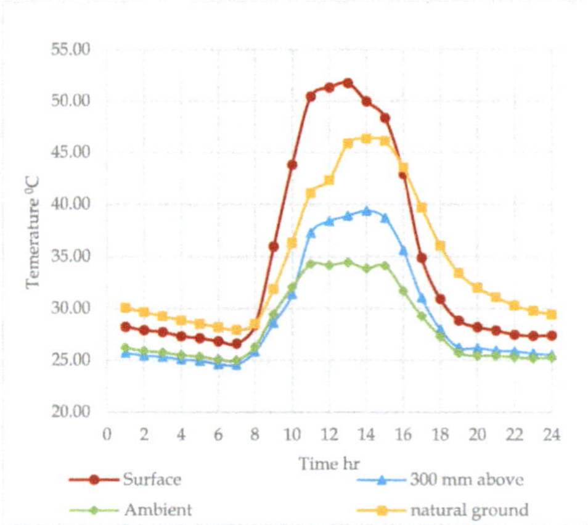

a: temperature variations on the Asphalt paved area

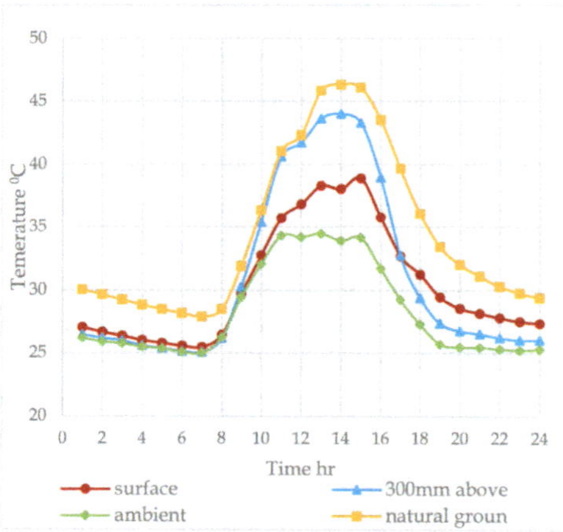

b: temperature variations on the burnt brick paved area

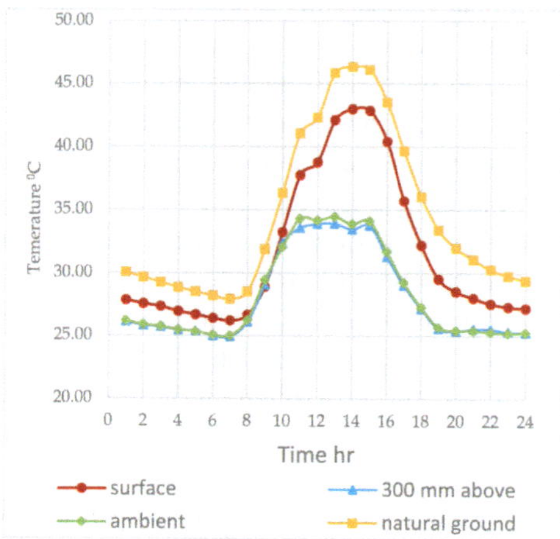

c: temperature variations on the cement concrete area

Fig. 5. The temperature variations on paved areas after applying colour

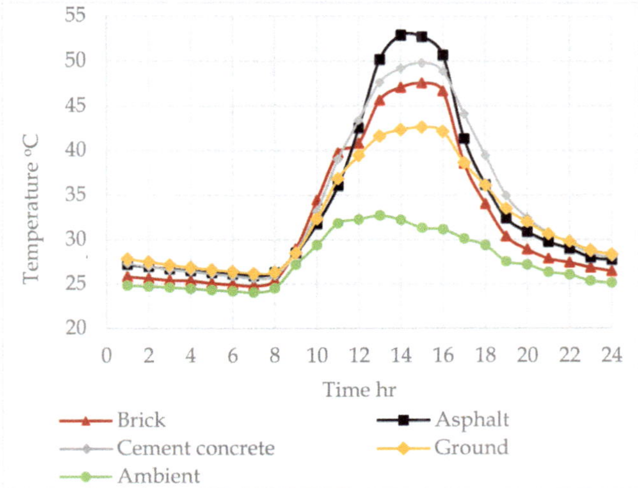

Fig. 6. Comparison of temperature variation on the top surface with native colour

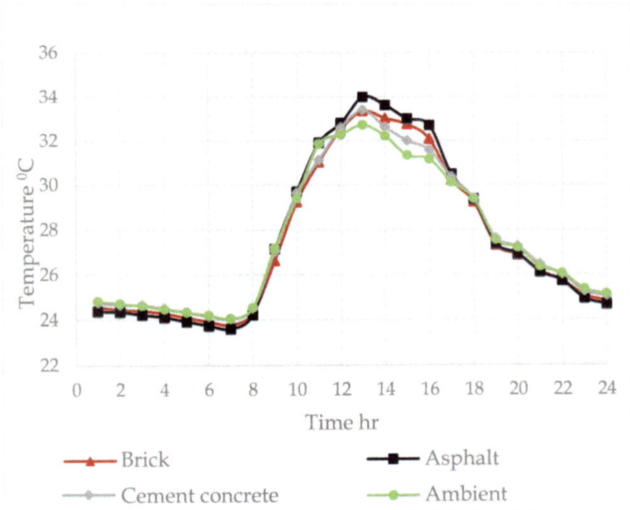

Fig. 7. Comparison of temperature variation on the top surface with coloured surface

In the second case, temperature variation on the top surfaces were less during early morning (i.e. 00 h to 07 h). At peak temperature during day time (at 13–14 h), the asphalt surface showed the highest temperature difference of 5.87 °C with natural ground. However, it was less when compared to the temperature difference between natural ground and the asphalt top surface with their native colour, which was 10.45 °C.

Figures 8 and 9 show the comparison of temperature variation on a 300 mm above the surface of all types of paving areas and the ambient temperature. For both cases,

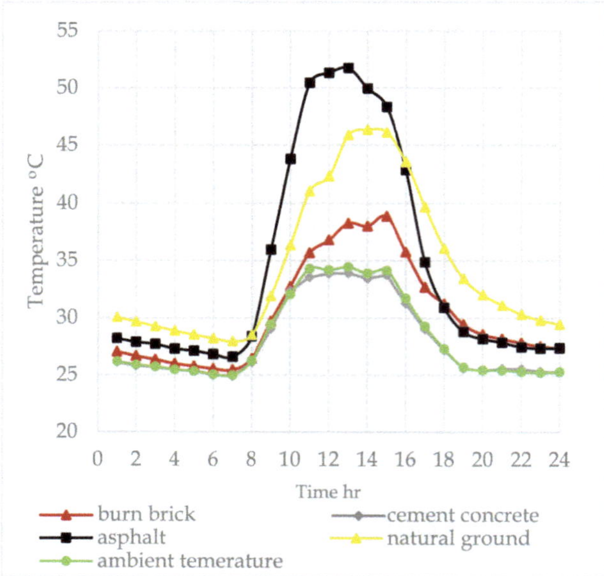

Fig. 8. Comparison of temperature variation of 300 mm above top surface with native coloured surface

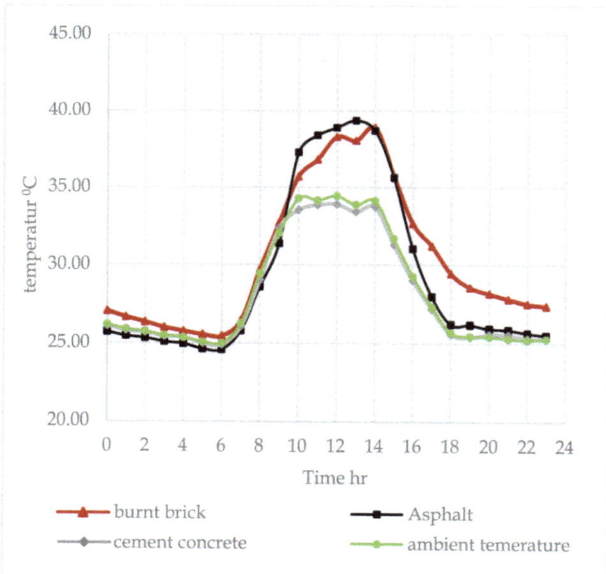

Fig. 9. Comparison of temperature variation of 300 mm above top surface with coloured surface

though almost all the temperatures were a similar pattern during 00.00 h to 7.00 h and 20.00 h to 00.00 h, the temperature between 7.00 h to 20.00 h significantly differed than that of the ambient temperature. In the first case, during the peak time at 13.00–14.00 h, the asphalt surface showed the highest temperature with a difference of 27.41 °C than the ambient temperature and that of the coloured asphalt was 14.21 °C.

5 Summary and Conclusions

Nowadays, much development is seen with many models and techniques for designing pedestrian ways or sidewalks around the world. However, in Sri Lanka, the transport authorities still think of managing only the motorized vehicles. Therefore, the most vulnerable road users, the pedestrians, are neglected with regard to their safety as well as convenience. This research work is an attempt to raise awareness in transportation planners and researchers of the existing paving materials and their thermal properties, with regard to their comfort. From the present study, it can be concluded that manmade hard surfaces heat more than natural ground. Further, paved areas exposed to direct solar radiation elevated the temperature of the atmosphere above them throughout the day. This indicates that such impervious artificial surfaces can be regarded as "heat sources" to the urban environment and the usage of impervious pavements should be reduced. The knowledge of thermal behavior of various urban surfaces and landscape components is an important tool for planners and designers. If utilized properly, it can lead to climatic rehabilitation in urban areas and an improvement of the outdoor thermal environment.

References

Babić S, Deluka-Tibljaš A, Cuculić M, Šurdonja S (2012) Analysis of pavement surface heating in urban areas. Građevinar 64:125–132

Chiesura A (2004) The role of urban parks for the sustainable city. Landsc Urban Plan 68 (1):129–138

Chudnovsky A, Ben-Dor E, Saaroni H (2004) Diurnal thermal behavior of selected urban objects using remote sensing measurements. Energy Build 36:1063–1074. https://doi.org/10.1016/j.enbuild.2004.01.052

Dareeju B, Meegahage JN, Halwatura RU (2013) Influence of grass cover on flat reinforced concrete slabs in a tropical climate

Doulos L, Santamouris M, Livada I (2004) Passive cooling of outdoor urban spaces. The role of materials. Sol Energy 77:231–249. https://doi.org/10.1016/j.solener.2004.04.005

Erell E, Pearlmutter D, Boneh D, Kutiel PB (2014) Effect of high-albedo materials on pedestrian heat stress in urban street canyons. In: Urban climate, ICUC8: the 8th international conference on urban climate and the 10th symposium on the urban environment, vol 10, pp 367–386. https://doi.org/10.1016/j.uclim.2013.10.005

Grimmond CSB, Roth M, Oke TR, Au YC, Best M, Betts R, Carmichael G, Cleugh H, Dabberdt W, Emmanuel R, Freitas E, Fortuniak K, Hanna S, Klein P, Kalkstein LS, Liu CH, Nickson A, Pearlmutter D, Sailor D, Voogt J (2010) Climate and more sustainable cities: climate information for improved planning and management of cities (producers/capabilities perspective). In: Procedia environmental sciences, world climate conference, vol 3, no 1, pp 247–274. https://doi.org/10.1016/j.proenv.2010.09.016

Halwatura RU (2013) Effect of turf roof slabs on indoor thermal performance in tropical climates: a life cycle cost approach [WWW Document]. J Constr Eng. https://doi.org/10.1155/2013/845158

Jaffal I, Ouldboukhitine S-E, Belarbi R (2012) A comprehensive study of the impact of green roofs on building energy performance. Renew Energy 43:157–164. https://doi.org/10.1016/j.renene.2011.12.004

Kelly CE, Tight MR, Hodgson FC, Page MW (2011) A comparison of three methods for assessing the walkability of the pedestrian environment. J Transp Geogr 19:1500–1508. https://doi.org/10.1016/j.jtrangeo.2010.08.001 Special section on Alternative Travel futures

Kleerekoper L, van Esch M, Salcedo TB (2012) How to make a city climate-proof, addressing the urban heat island effect. Resour Conserv Recycl Clim Proofing Cities 64:30–38. https://doi.org/10.1016/j.resconrec.2011.06.004

Mendoza J-MF, Oliver-Solà J, Gabarrell X, Rieradevall J, Josa A (2012) Planning strategies for promoting environmentally suitable pedestrian pavements in cities. Transp Res Part D: Transp Environ 17:442–450. https://doi.org/10.1016/j.trd.2012.05.008

Rahaman KR, Ohmori N, Harata N Evaluation of the roadside walkway environment of Dhaka city. Proc East Asia Soc Transp Stud 5:16

Rehan RM (2013) Sustainable streetscape as an effective tool in sustainable urban design. HBRC J 9:173–186. https://doi.org/10.1016/j.hbrcj.2013.03.001

Santamouris M, Gaitani N, Spanou A, Saliari M, Giannopoulou K, Vasilakopoulou K, Kardomateas T (2012) Using cool paving materials to improve microclimate of urban areas – design realization and results of the Flisvos project. Build Environ 53:128–136. https://doi.org/10.1016/j.buildenv.2012.01.022

Santamouris M, Synnefa A, Karlessi T (2011) Using advanced cool materials in the urban built environment to mitigate heat islands and improve thermal comfort conditions. Sol Energy Prog Sol Energy 2(85):3085–3102. https://doi.org/10.1016/j.solener.2010.12.023

Scudo G, Dessì V (2006) Thermal comfort in urban space renewal. In: Proceeding 23rd PLEA

Spagnolo J, de Dear R (2003) A field study of thermal comfort in outdoor and semi-outdoor environments in subtropical Sydney Australia. Build Environ 38:721–738. https://doi.org/10.1016/S0360-1323(02)00209-3

van der Zee D (1990) The complex relationship between landscape and recreation. Landscape Ecol 4:225–236

Determining Criticality of Performance Indicators for a Construction Company

K. K. Tripathi[1(✉)], K. N. Jha[2], and A. K. Jain[2]

[1] B. S. Anangpuria Institute of Technology and Management, Faridabad 121004, India
kktripathi05@gmail.com
[2] Indian Institute of Technology Delhi, Hauz Khas, New Delhi 110016, India

Abstract. Because of stiff competition in the construction business, it is necessary for construction companies to measure their performance on a regular interval for long-term survival. Therefore, it is imperative for construction companies to determine the parameters critical to the measurement of their performance. This study identifies the performance indicators, financial as well as non-financial, for construction companies to measure their performance. In a previous study, the authors identified a total of 20 performance attributes and applied principal component analysis which extracted six components called performance indicators. These performance indicators are profitability and asset management; key stakeholders' satisfaction; time and cost predictability; environment, health, and safety; quality consciousness; and employee turnover. The present study utilized structural equation modelling (SEM) to determine the criticality of these performance indicators for which a questionnaire survey and structured interview approach were adopted. A total of 106 responses were collected from 90 different construction companies operating in various parts of India. The findings indicate that 'quality consciousness' is the most critical performance indicator whereas the 'employee turnover' is the least critical performance indicator in measuring the performance of a construction company. The performance indicators obtained from this study may help construction companies to measure and compare their performances with their competitors and setting out a strategy to remain competitive in the market.

Keywords: Performance attributes · Performance indicators · Questionnaire survey · Principal component analysis · Construction company

1 Introduction

Performance measurement has been considered fundamental to management planning and control by both academicians and practitioners and hence gained more attention from many researchers (Tsolas 2011). According to Kagioglu *et al.* (2001), it is defined as the process of evaluating how successfully the companies or individuals have been achieving their strategies and objectives. Performance measurement helps the construction companies to concentrate on their long-term strategic view, objectives, and optimization of their operations in the companies.

R. Dissanayake and P. Mendis (Eds.): ICSBE 2018, LNCE 44, pp. 353–364, 2020.
https://doi.org/10.1007/978-981-13-9749-3_32

To measure the performance of the construction companies, the companies must first identify the appropriate key performance indicators (KPIs). According to Cox *et al.* (2003), "performance indicators can be defined either by quantitative measures ($/unit) or qualitative measures such as worker behaviour on the job". It helps top-level management to watch the performance of the corporate or department in one place. The identification of KPIs is the commencement in developing a correct framework for measurement of a construction company's performance (Lin *et al.* 2011).

Traditionally, performances of the companies were measured in financial terms such as turnover, profit, sales per employee, return on investment, etc. which has been criticized (Love and Holt 2000). It has also been found that financial parameters measure only the short-term performance of the company and ignores the issues related to the long-term sustainability of the company (Kaplan and Norton 1996). The traditional financial measures of performance do not reflect the need of stakeholders, fail to provide information about what customers are getting against what they really wanted and do not identify the performance of the competitors (Isik 2009). Hence, construction companies should use the performance indicators, both financial as well as non-financial to evaluate and compare their performances with others which will enhance the effectiveness and efficiency of the company.

Performance measurement of any construction company in terms of their failure or success is very complex because of the participation of various stakeholders like the clients, project management consultants, contractors, and public etc. For a given company, the objectives of all the stakeholders need not be the same. For one stakeholder, the parameter for measuring the performance of a company may be high profit while for the other stakeholder, it can be client satisfaction. Hence defining the performance in terms of failure or success without defining the objectives of the various stakeholders involved has no meaning (Tripathi and Jha 2018a).

In the present study, the factors based on which the performance of a construction company can be measured, have been described. Researchers have identified various performance indicators for measuring the performance of a construction company. But their research works are basically project specific and in the context of developed countries which may not be applicable in the performance measurement of construction companies in other countries. Very few of them have focused on the performance indicators of construction companies working in developing countries like India (Tripathi and Jha 2018a). Moreover, the performance of projects and the construction companies were measured mainly based on cost, time, and quality achieved on those projects whereas, the performance relative to these three parameters is influenced by some other parameters like environment, health, and safety (EHS), productivity of the employees, and key stakeholders' satisfaction etc. This study attempts to identify all such parameters which measure the performance of a construction company.

2 Literature Review

The purpose of the current study is to determine those indicators that are critical for the measurement of construction company's performance. Very limited studies are available in the literature in this area. These are briefly discussed below.

Mbugua *et al.* (1999) developed a framework which will develop the information and hence give improved construction performance grounded on financial (liquidity, profit, potency etc.) and non-financial (leadership, customer's satisfaction, impact on society, learning and growth, quality etc.) measures. Cox *et al.* (2003) established a correlation between quantitative and qualitative performance measurement parameters to identify the most extensively used parameter. They found that quality control, timely completion, cost, safety, etc. are the extremely vital indicators for the construction profession. But they neither incorporated company level indicators like the company's financial standing, market condition, relationship with external agencies etc. nor evaluated its efficiency.

Bassioni *et al.* (2004, 2005) proposed a theoretical framework employing a balanced scorecard and business excellence model to evaluate the business performance of construction companies. They divided the framework into performance driving factors (resource management, leadership, strategic management, capital management, risk management, work culture, etc.) and performance results factors (people, partnership and supplier results, project results, customer and society results, organizational business results, etc.). Elyamany *et al.* (2007) developed a performance measurement model based on the financial ratios along with economic and trade factors and concluded that the company with higher financial performance index had better performance. Balatbat *et al.* (2010) measured the performance of Australian publicly listed company with other listed companies in Australia using profitability ratio, market performance, and equity analysis and performance ratio.

Luu *et al.* (2008) performed the strengths, weaknesses, opportunities, and threat (SWOT) analysis to judge the strategic performance of huge contractors in Vietnam. The KPIs known were from four perspectives: money, customer, internal method, and learning and growth. Chan (2009) developed a scientific performance measuring framework for the Malaysian industry to watch their progress towards achieving the goals commenced within the industry master plan 2006–2015. The performance measures known were identical to the ones known by Luu et al. (2008). Tsolas (2011) integrated the data envelopment analysis (DEA) and the ratio analysis while developing a framework to gauge the performance of the construction companies with regards to profit and effectiveness of the construction companies listed on the Athens Exchange.

Ali *et al.* (2013) used 47 indicators for performance evaluation of construction companies in Saudi Arabia via relative importance index (RII). Profitability, cash flow, service and work quality, financial stability, market share, growth, customer satisfaction, safety, business efficiency, and effectiveness of planning were the highest 10 KPIs. Hassaan *et al.* (2013) applied the analytical hierarchy process (AHP) along with a fuzzy set theory for performance evaluation of contractors on 34 quantitative and qualitative criteria during an award of tender in Egypt. According to the respondents, the financial stability and past performance were the foremost significant measures for contractor selection.

The literature review reveals that various frameworks/models were developed by researchers to gauge the performance of construction companies. Most of the researchers administered the research focused on a specific country, however, none of them have administered the research for India. In most of the studies, only contractors were considered by the researchers, but this study has thought about the opinions of other stakeholders such as the clients and project management consultants along with the contractors. This study also attempts to include all possible performance evaluation parameters pertinent to a construction company. Hence, authors have attempted to fill the gaps in the previous studies with the help of the current study.

3 Research Method

The research method contains a total of four steps. The first two steps were carried out by the authors in their previous study and the remaining two steps are pertaining to the present study. These are described in detail in the subsequent steps.

Step 1: Identification of Performance Attributes and Data Collection
Twenty attributes were extracted from the literature to measure the performance of a construction company. Table 1 shows the list of these attributes along with their sources. Based on the attributes identified, a questionnaire was designed. A five-point Likert scale was used to measure the degree of importance of these attributes in measuring the performance of a construction company. In the Likert scale, the degree of importance of performance attributes ranges from 1 (very low importance) to 5 (very high importance).

A total of 106 completed questionnaires were received from 90 different construction companies. Out of 106 questionnaires, 29 questionnaires were collected via email and 77 questionnaires via personal meeting. The average experience of the respondents was 20 years whereas the average experience of the companies participated in the survey was 21 years.

Table 1. List of performance attributes and their sources

Sl. no.	Performance attributes with id	Sources
1	Size of the organization (PA-1)	Mbugua et al. (1999), Chan (2009)
2	The productivity of employees (PA-2)	Mbugua et al. (1999), Cox et al. (2003), Chan (2009)
3	Good track record of timely completion of the projects (PA-3)	Luu et al. (2008), Skibniewski and Ghosh (2009)
4	Health and safety consciousness (PA-4)	Cox et al. (2003), Chan (2009), Luu et al. (2008)
5	Customer satisfaction in terms of product and services (PA-5)	Rimbalova and Vilcekova (2013), Menches and Hanna (2006)
6	Client satisfaction in terms of product and services (PA-6)	Chan (2009), Skibniewski and Ghosh (2009), Nemcova-Zunana (2009)
7	Cost performance of projects (PA-7)	Menches and Hanna (2006), Bassioni et al. (2004)

(*continued*)

Table 1. (*continued*)

Sl. no.	Performance attributes with id	Sources
8	Impact on society (PA-8)	Mbugua et al. (1999), Nemcova-Zunana (2009)
9	Impact on the environment (PA-9)	Rimbalova and Vilcekova (2013), Nemcova-Zunana (2009)
10	Optimum liquidity ratio (PA-10)	Elyamany et al. (2007), Balatbat et al. (2010)
11	Higher profitability ratio (PA-11)	Skibniewski and Ghosh (2009), Horta et al. (2010)
12	The higher annual growth rate of the organization (PA-12)	Chan (2009), Luu et al. (2008), Horta et al. (2010), Yu et al. (2007)
13	Predictability of cost in construction (PA-13)	Rimbalova and Vilcekova (2013), Chan (2009)
14	Predictability of time in construction (PA-14)	Nemcova-Zunana (2009), Bassioni et al. (2004)
15	Rework or defect rectification (PA-15)	Menches and Hanna (2006), Kagioglou et al. (2001)
16	Adopting learning and growth culture in the organization (PA-16)	Mbugua et al. (1999), Chan (2009), Luu et al. (2008)
17	Higher wages of the employees (PA-17)	Rimbalova and Vilcekova (2013), Nemcova-Zunana (2009)
18	Employee turnover (PA-18)	Chan (2009), Rimbalova and Vilcekova (2013)
19	A good relationship with the client (PA-19)	Mbugua et al. (1999), Menches and Hanna (2006)
20	Annual construction demand/market share (PA-20)	Chan (2009), Luu et al. (2008), Yu et al. (2007)

Step 2: Identification of Performance Indicators

Performance attributes identified in Step 1 were grouped into six components, called performance indicators when subjected to principal component analysis. The output of the principal component analysis indicated that the Kaiser Meyer Olkin (KMO) value was 0.793 (>0.05) indicating sample adequacy for principal component analysis. The result of the principal component analysis is shown in Table 2. 1. These indicators are profitability and asset management (PI-1), key stakeholders' satisfaction (PI-2), time and cost predictability (PI-3), environment, health, and safety (PI-4), quality consciousness (PI-5), and employee turnover (PI-6) as shown in Table 2 (Adapted from Tripathi and Jha 2018a). The total variance explained by all the indicators accounts for 67.448%.

Table 2. Results of the principal component analysis

Sl. no.	Performance indicators with variance (%)	Performance attributes with factor loading
1	PI-1 Profitability and asset management (13.893%)	PA-12 The higher annual growth rate (0.801)
		PA-11 Higher profitability ratio (0.742)
		PA-10 Optimum liquidity ratio (0.637)
		PA-2 The productivity of employees (0.609)
2	PI-2 Key stakeholders' satisfaction (12.688%)	PA-5 Customer satisfaction (0.830)
		PA-6 Client satisfaction (0.816)
3	PI-3 Time and cost predictability (11.698%)	PA-13 Predictability of cost in construction (0.876)
		PA-14 Predictability of time in construction (0.862)
4	PI-4 Environment, health, and safety (11.161%)	PA-9 Impact on the environment (0.806)
		PA-4 Health and safety consciousness (0.734)
5	PI-5 Quality consciousness (10.581%)	PA-15 Rework or defect rectification (0.769)
		PA-16 Adopting learning and growth culture (0.719)
		PA-20 Annual construction demand/market share (0.525)
6	PI-6 Employee turnover (7.427%)	PA-18 Employee turnover (0.673)

Reliability Test

To ensure the internal consistency within the attributes grouped under components and reliability of the data, Cronbach's alpha ($C\alpha$) reliability test was performed using statistical package for social science (SPSS) version 21. The $C\alpha$ value ranges between 0 and 1. A higher $C\alpha$ value points towards the greater internal consistency and vice versa. As per Tripathi and Jha 2018b, a $C\alpha$ value greater than 0.7 is acceptable. In this study, the $C\alpha$ value is 0.844, which shows a good overall internal consistency among the attributes. Hence, the attributes grouped under the individual performance indicators were considered reliable for further analysis (Chan *et al.* 2014).

Step 3: Development of a Hypothesized Model

After grouping of the performance attributes, a hypothesized SEM model was proposed to test the relationship between performance indicators and performance of a company as shown in Fig. 1.

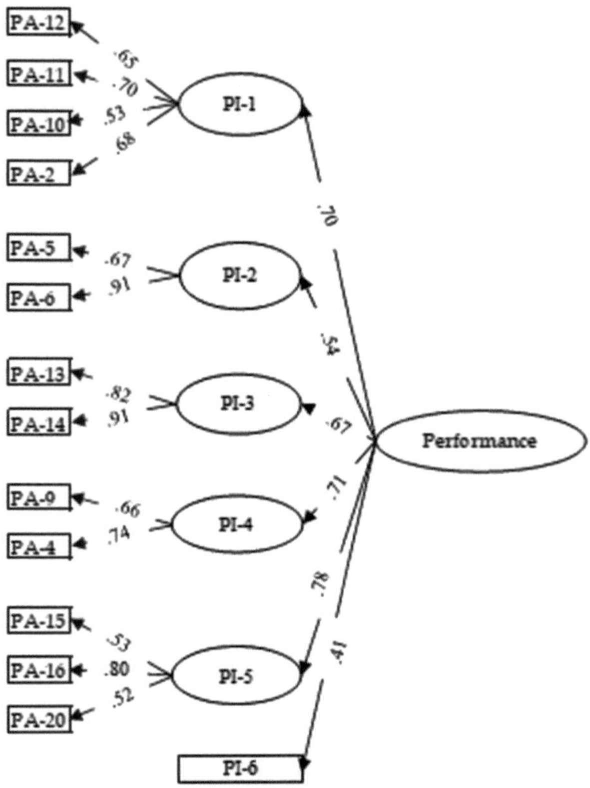

Fig. 1. Hypothesized and final model

Based on the model proposed, the hypothesis that performance indicators measure the construction company's performance was tested as follows:

1. Null hypothesis (H0): Path coefficient between performance indicators and the construction company's performance is not significantly different from zero.
2. Alternate hypothesis (Ha): Performance indicators measure the performance of a construction company.

Step 4: Validation of the Hypothesized Model

The adequacy of the SEM model is established by assessing the results of the covariance structural analysis, which is shown by the various goodness-of-fit (GOF) indices suggested by different researchers. If the model is not adequate to interpret, it needs to be revised. Different GOF indices evaluate the adequacy of a model from different aspects. From the many GOF indices available within the SEM literature, following GOF indices were proposed in this study for validating the hypothesized model (Tripathi and Jha 2018b).

(1) The ratio of chi-square ($\chi2$) to the degree of freedom (df), (2) Incremental fit index (IFI), (3) Tucker-Lewis index (TLI), (4) Comparative fit index (CFI), and (5) The root-mean-square error of approximation (RMSEA).

The recommended level of these indices is given in Table 3 (Tripathi and Jha 2018b). The hypothesized model was analysed using an analysis of moment structure (AMOS) version 21.

Table 3. GOF measures (Adapted from Tripathi and Jha 2018b)

Sl. no.	The goodness of fit (GOF) indices	Recommended level of GOF indices	Values of GOF indices obtained in the final model
1	Chi-square/degree of freedom ($\chi2$/df)	1 to 2	1.171
2	Incremental fit index (IFI)	0 (no fit) −1 (perfect fit)	0.969
3	Tucker-Lewis index (TLI)	0 (no fit) −1 (perfect fit)	0.952
4	Comparative fit index (CFI)	0 (no fit) −1 (perfect fit)	0.967
5	Root mean square error of approximation (RMSEA)	<0.05 (very good) - 0.1 (threshold)	0.041

Table 3 shows the GOF indices of the hypothesized SEM model. The values of GOF indices indicate that the hypothesized SEM model was adequate to explain the interrelationships between performance indicators and the construction company's performance. Hence, the hypothesized SEM model was not revised and was acceptable for interpretation.

All the standardized path coefficients shown on the arrow lines in the Fig. 1 were positive and statistically significant which indicate that there is a relationship between the construction company's performance and performance indicators. The larger path coefficient indicates that the attributes or indicators are more critical for measuring the construction company's performance. Accordingly, 'quality consciousness' emerged as the most critical performance indicator whereas the 'employee turnover' emerged as the least important performance indicator. The hypothesis H1, which assumes that the performance indicators measure the performance of the construction company is found to be supported because the path coefficients were significant at the 0.05 significance level.

4 Result and Discussion

The results of the study reveal that the quality consciousness (PI-5) is the most critical performance indicator with a path coefficient of 0.78. Quality consciousness is an important area for good quality construction and its integration in the company. Nowadays, clients are more worried about the quality rather than the cost. Also, correcting a work of poor quality involves an investment of time and money and creates an obstruction in the construction. Hence, quality consciousness is considered as a significant indicator for the measurement of the performance of a construction company. Construction companies can increase their market by enhancing client satisfaction with a superior product.

Environment, health, and safety (PI-4) is the second most critical performance indicator with a path coefficient of 0.71. It must be an environmental responsibility for both an individual as well as the company. Much of the earth's resources are controlled by the construction companies which diminish environmental nuisance. In addition, the cost of construction can also be reduced by environment-friendly construction using energy-efficient construction materials and techniques. Inefficient health and safety management on construction sites results in the loss of life and wastage of construction materials and hence increases the associated costs. Therefore, construction companies must be held responsible for the society as well as the environment because they interfere in various areas of our life.

Profitability and asset management (PI-1) received a path coefficient of 0.70 and was placed in the third position. Asset management can be defined as the technique of managing assets for the companies through cash, bonds, and stocks. A company's performance is gauged by the profitability and assets of the company. Profitability is one of the most significant parameters for measuring the performance of a construction company (Yu *et al.* 2007).

Time and cost predictability (PI-3) received a path coefficient of 0.67 and was placed in the fourth position. Predictability is very important for any construction company because it psychologically influences client. If clients are confident that their project will be handed over by the company within the cost and within the scheduled time, then it is most likely that they can deal with the company. The lack of predictability in the company negatively influences the clients and reduces the chances of doing business with the company. Therefore, all companies must possess predictability otherwise its number of clients will decrease affecting its overall profits.

Key stakeholders' satisfaction (PI-2) received a path coefficient of 0.54 and was placed in the fifth position. Key stakeholders' satisfaction is a significant indicator for measuring the construction company performance. Client satisfaction is found to be fundamental to the success of the business. Construction companies always attempt to maximize the Key stakeholders' satisfaction through good quality of products and services rendered by them. They manage their resources in such a way that their client's objectives are achieved to have repeat business.

Employees turnover (PI-6) was placed at the last position with a very low path coefficient of 0.41. It is expressed as the percentage of the staff leaving the company for one year. For any company, the objective should be to have employee turnover as low as possible to maintain a consistent workforce and to grow more skilled staff. High employee turnover results in higher cost to the company because additional time and resources are required to fill the vacant position and to train a new employee. Low employee turnover permits the company to concentrate on its business rather than preparing the new employee, adapting them to the company's needs. Therefore, low employee turnover is an indicator of the company performing well in the industry.

The above results show that the respondents have given more weight to the performance indicators 'quality consciousness' and 'environment, health, and safety' when compared to the performance indicator 'profitability and asset management'. It indicates that the construction companies apprehend that the traditional financial parameters are no longer a comprehensive measure of company performance. The insufficiency of the traditional financial performance indicators has drawn the attention of construction companies in non-financial performance indicators. Several construction companies have a long-term strategy that focuses on the satisfaction of the stakeholders, good quality of the work, time and cost predictability etc. Hence, the company performance from the perspective of their key stakeholders has become a priority for construction companies.

The results of this study recommend that these indicators could be considered as a yardstick for the Indian construction companies while measuring their performance with their competitors. Most of these indicators are the prerequisite of the award of the tender. Clients prefer the above criteria to be satisfied by the construction companies to decrease the risk of budget and time overruns, low-quality work and workmanship and so forth. Construction companies can enhance their performance by meeting these criteria and increase the chances of getting increased business despite the competition in the market.

5 Conclusion

This study tries to find performance indicators that can measure the performance of the construction companies executing real estate projects. This study utilised a questionnaire survey and an SEM approach to analyse the causal relationship between performance indicators and company's performance. The results enable a construction company to understand and formulate a strategy to perform well with their competitors. The SEM supports the hypothesized positive relationships between performance indicators and the performance of a construction company with very strong path coefficients at the significance level of 0.05.

The SEM model reveals that the quality consciousness (PI-5) is the most critical indicator for the measurement of construction company's performance followed by environment, health, and safety (PI-4), profitability and asset management (PI-1), time and cost predictability (PI-3), key stakeholders' satisfaction (PI-2), and employees turnover (PI-6). Due to rapid changes and challenges in the construction business, the traditional parameters of performance measurement may not be enough in developing adequate results for the various stakeholders. It has been found that high financial performance is essential but not enough for satisfactory performance of a company. All these indicators contribute to enhancing the performance of a construction company leading to sustainable growth. The probability of getting the increased business of a construction company will increase if proper attention is drawn towards improvement in these areas. The clients select a construction company which fulfils the above criteria to ensure enhanced value for money and overall satisfactory results in construction.

The scope of this study was limited to the construction companies associated only with the real estate business operating in India. Hence, the result of this study may or may not be applicable to the construction companies in other countries. However, despite various limitations, this study may be beneficial for neighboring countries like Bangladesh, Nepal, Sri Lanka, Pakistan, Afghanistan etc. and other developing countries because of resemblance in working conditions.

References

Ali HAEM, Al-Sulaihi IA, Al-Gahtani KS (2013) Indicators for measuring performance of building construction companies in Kingdom of Saudi Arabia. J King Saud Univ - Eng Sci 25:125–134

Balatbat MCA, Lin C, Carmichael DG (2010) Comparative performance of publicly listed construction companies: Australian evidence. Constr Manag Econ 28(9):919–932

Bassioni HA, Price ADF, Hassan TM (2004) Performance measurement in construction. J Manag Eng ASCE 20(2):42–50

Bassioni HA, Price ADF, Hassan TM (2005) Building a conceptual framework for measuring business performance in construction: an empirical evaluation. Constr Manag Econ 23 (5):495–507

Chan TK (2009) Measuring performance of Malaysian construction industry. Constr Manag Econ 27(12):1231–1244

Chan IYS, Liu AMM, Richard Fellows R (2014) Role of leadership in fostering an innovation. Constr Manag Econ 30(6):1–7

Cox RF, Issa RRA, Ahrens D (2003) Management's perception of key performance indicators for construction. J Constr Eng Manag ASCE 129(2):142–151

Elyamany A, Basha I, Zayed T (2007) Performance evaluating model for construction companies: egyptian case study. J Constr Eng Manag ASCE 133(8):574–581

Hassaan HS, Fors MN, Shehata MS (2013) Fuzzy decision model for construction contractor's selection in Egypt: tender Phase. In: Proceedings of the international conference on industrial engineering and engineering management, institute of electrical and electronics engineers (IEEE), Bangkok, Thailand

Horta IM, Camanho AS, Da Costa JM (2010) Performance assessment of construction companies integrating key performance indicators and data envelopment analysis. J Constr Eng Manag ASCE 136(5):581–594

Isik Z (2009) A conceptual performance measurement framework for construction industry. Ph. D. thesis, The Graduate School of Natural and Applied Science of Middle East Technical University

Kagioglou M, Cooper R, Aouad G (2001) Performance management in construction: a conceptual framework. Constr Manag Econ 19(1):85–95

Kaplan RS, Norton DP (1996) The balanced scorecard. Harvard Business School Press, Boston

Lin G, Shen GQ, Sun M, Kelly J (2011) Identification of key performance indicators for measuring the performance of value management studies in construction. J Constr Eng Manag ASCE 137(9):698–706

Love PED, Holt GD (2000) Construction business performance measurement: the stakeholder perspective measurement alternative. Bus Process Manag J 6(5):408–416

Luu T-V, Kim S-Y, Cao H-L, Park Y-M (2008) Performance measurement of construction firms in developing countries. Constr Manag Econ 26(4):373–386

Mbugua LM, Harris P, Holt GD, Olomolaiye PO (1999) Framework for determining critical success factors influencing construction business performance. In: Hughes W (ed) 15th Annual ARCOM conference of association for research in construction management held in Liverpool, John Moores University, 15–17 September 1999, vol 1, pp 255–264

Menches CL, Hanna AS (2006) Quantitative measurement of successful performance from the project manager's perspective. J Constr Eng Manag ASCE 132(12):1284–1293

Nemcova-Zuzana (2009) Application of key performance indicators in the construction industry: a comparison of the United Kingdom with the Czech Republic. The European Council of Construction Economists, 6 p

Rimbalova J, Vilcekova S (2013) The proposal of key performance indicators in facility management and determination the Weights of Significance. J Civil Eng 8(2):73–84

Skibniewski MJ, Ghosh S (2009) Determination of key performance indicators with enterprise resource planning systems in engineering construction firms. J Constr Eng Manag ASCE 135 (10):965–978

Tripathi KK, Jha KN (2018a) An empirical study on performance measurement factors for construction organizations. KSCE J Civil Eng 22(4):1152–1166

Tripathi KK, Jha KN (2018b) Determining success factors for a construction organization: a structural equation modeling approach. J Manag Eng 34(1):1–15

Tsolas IE (2011) Modelling profitability and effectiveness of Greek-listed construction firms: an integrated DEA and ratio analysis. Constr Manag Econ 29(8):795–807

Yu I, Kim K, Jung Y, Chin S (2007) Comparable performance measurement system for construction companies. J Manag Eng ASCE 23(3):131–139

Seepage in Iranamadu Dam and Its Influence on the Stability

T. Venuja and L. C. Kurukulasuriya(✉)

Faculty of Engineering, University of Peradeniya, Peradeniya, Sri Lanka
chank@pdn.ac.lk

Abstract. Earth dams have played an important role in agriculture since ancient times. Failure of side slopes caused due to seepage or drawdown is one of the major considerations in the design of earth dams as the failure of a slope is catastrophic. The Iranamadu dam which was raised at different times showed a significant increase of seepage when the reservoir level increased beyond a certain elevation. In this research, the seepage problem of the Iranamadu dam was analysed using Geo-Studio (SEEP/W) software to estimate the seepage rate with the determination of the phreatic surface and to investigate the influence of the anisotropy of hydraulic conductivity of core material on seepage. For these analyses, volumetric water content function to be used in the unsaturated zone above the phreatic surface was selected from the van Genuchten model. Both steady and transient state seepage were carried out to compare the seepage rates with field measurements and the effect of anisotropy of hydraulic conductivity using the observed reservoir level history spanning over 170 days. Furthermore, the stability of the dam was assessed under seepage conditions using the SLOPE/W software. SEEP/W results show a good agreement in seepage rates with field measurements under both seepage condition when the reservoir level is above a certain elevation. Within a small range of reservoir level, some deviation of the predicted seepage rate is observed. Additionally, when the phreatic surface passes through the core section, a drop-in the seepage rate was observed due to the low permeability of the clay core. The seepage rate increased with k_y/k_x ratio and the Iranamadu dam seepage field measurements showed a good agreement when k_y/k_x ratio is 1. Therefore, the behaviour of the Iranamadu dam with respect to the seepage flow, shows a near isotropic hydraulic behaviour. The SLOPE/W results conclude that when the reservoir water level increases, the stability of the downstream slope reduced beyond the recommended value but the stability of the upstream slope was found to be satisfying the safety criteria under both steady and transient seepage conditions.

Keywords: Earth dam · Seepage · Steady state · Transient state · Slope stability

1 Introduction

Dams are constructed to meet various functional requirements such as flood control, hydroelectric power generation, navigation, recreation, irrigation and as a source of water. Most of the large embankment dams in the world were built during the middle

R. Dissanayake and P. Mendis (Eds.): ICSBE 2018, LNCE 44, pp. 365–377, 2020.
https://doi.org/10.1007/978-981-13-9749-3_33

decades of the twentieth century and can be classified into two main categories; earth-fill and rock-fill dams. These dams represent about 85% of all dams built in the world (Elshemy et al. 2002). These embankment dams are more common than other type of dams because of using ordinary construction methods utilizing cheap raw soil materials and subsurface materials (Abhilasha and Antony Balan 2015). Furthermore, the self-weight of these earth dams can be effectively used to prevent sliding and overturning.

Failure of an earth dam may occur due to different reasons, such as failure of upstream or downstream slope under earthquake conditions, seepage through the dam, piping through dam body or under the foundation, and drawdown (Athani et al. 2015; Kirra et al. 2015a). One of the main causes of failure is seepage in their body and foundation. Failure due to seepage has occurred in more than 35% of all earth dam and failures and structural failures account to about 25% of dam failures (Elshemy et al. 2002). Hence, seepage analysis is an important aspect in the design of an earth dam. Normally, seepage flow occurs along the path that connects through the pores within soil particles. This flow most likely is laminar that complies with Darcy's law. Darcy's law originally applied for the saturated domain, but can also be used in the unsaturated domain (Andreea 2016).

This seepage can give rise to stability problems under high pressure head that causes reduced shear strength. All earth dams will have some seepage and it is unrealistic not to expect this (Andreea 2016). Therefore, seepage must be controlled to prevent leeching out of embankment or foundation material. The ground conditions and the geological features of the dam site greatly influence the amount of seepage and its relevant effects.

1.1 Research Area

The Iranamadu dam is one of the largest earth dams in Sri Lanka. Figure 1 presents the cross section of the Iranamadu dam comprising of a central puddle clay core, original shoulder fill, shoulder fills placed in 1951 and 1975 and the embankment foundation.

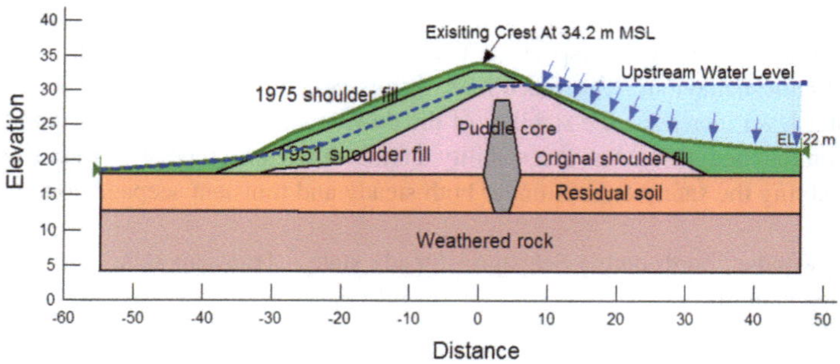

Fig. 1. Iranamadu dam cross section

In this research, seepage problem of the Iranamadu dam was analysed using Geo-Studio (SEEP/W) software to estimate the seepage rate under both steady and transient state conditions and, the influence of anisotropy of hydraulic conductivity of core material on seepage was also examined. In addition, stability of upstream and downstream slopes of the Iranamadu embankment was also investigated under both steady and transient state seepage conditions using SLOPE/W limit equilibrium software.

1.2 Research Objectives

- Establish historical data on the seepage and reservoir water level measurements over a continuous period of time.
- Compare field seepage data and results from numerical analysis under steady and transient states using SEEP/W software.
- Study the effect of soil anisotropy on seepage through the dam by numerical modelling for transient state.
- Investigate the stability of the dam incorporating the effect of seepage.

2 Literature Review

Seepage analysis can be divided into two stages; steady state analysis and transient analysis. Steady state flow analysis is where the boundary conditions inside and outside of the ground do not change with time. Therefore, the inflow is always equal to the outflow within the analytical range. Transient analysis is used when the boundary conditions inside or outside of the ground changes with time. Transient analysis can display different inflow and outflow with time, even when the same boundary conditions as those for the steady state analysis are applied. The main differences between transient and steady state analyses are that the boundary conditions change as time passes and the fact that the transient analysis requires the definition of the volumetric water content function (Andreea 2016). In the slope stability analysis, several limit equilibrium methods have been developed to determine the critical slip surface and the corresponding minimum factor of safety.

Recently, the Finite Element Method (FEM) and the Limit Equilibrium Method (LEM) are used for the seepage and stability analyses of embankment dams respectively. Several authors, namely Uromeihy and Barzegari (2007), Kamanbedast and Delvari (2012), Hasani et al. (2013) and Kirra et al. (2015a, b) had performed seepage and/or stability analysis of an embankment dam. Abo (2001) used FEM to determine the free surface position, quantity of seepage, pore water pressure distribution, total head distribution and the effect of anisotropy of the material in AL-Adheem zoned earth dam, Iraq.

GeoStudio has been used for seepage and slope stability analyses of earth dams by many researchers. Kamanbedast and Delvari (2012) used the GeoStudio (SEEP/W) software to investigate seepage condition and flow of water through the soil in the Maroon dam, Iran. Hasani et al. (2013) also studied the seepage analysis using SEEP/W software in the Ilam earth dam, Iran for four mesh sizes in order to assess the effect of meshing on accuracy of results and slope stability analysis have been carried out under different conditions and methods using SLOPE/W software. Kirra et al.

(2015b) also used GeoStudio software to analyse the seepage and stability of the Mandali earth fill dam, Iraq.

3 Materials and Methods

Based on the geotechnical investigation details of the Iranamadu dam, soil parameters were identified for each layer and which are tabulated in Table 1.

Table 1. Soil parameters used in the software analysis (Report of design development of remedial works Iranamadu embankment 2013)

Layer name	Soil type	c′ (kPa)	ϕ′ (°)	Unit weight (kN/m^3)	Permeability (m/s)
Weathered rock	Silty clayey sand	0	45	21	1×10^{-7}
Residual soil	Silty sand	0	35	21	1×10^{-5}
Puddle clay core	Sandy clay with gravels	5	20	16	1×10^{-7}
Original shoulder fill	Clayey gravel	0	30	20	1×10^{-4}
1951 Shoulder fill	Clayey sand	0	30	20	1×10^{-7}
1975 Shoulder fill	Silty sand	0	30	20	1×10^{-5}

3.1 Field Measurements

Based on the Iranamadu dam field records collected from the Department of Irrigation, Kilinochchi, a seepage location has been identified downstream in the side of the dam embankment at Ch 0+1544 m and seepage discharges were measured from 26th December 2014 to 13th June 2015 totalling 170 days under different reservoir levels at a location which is 41 m below the dam crest. Also, a temporary filter of 6.0 m × 3.2 m has been built at that measuring location and the seepage discharge was collected into a measuring cylinder through a pipeline from where the rate of seepage was calculated. Then the upstream reservoir water levels were recorded with the time period. Based on the field measurements, upstream reservoir water levels were plotted with time and seepage rates were plotted with different reservoir water levels as shown in Figs. 2 and 3, respectively.

In Fig. 2, upstream reservoir water level has reduced from 30.33 m to 27.71 m during the period of 170 days.

As expected, seepage rate has decreased with the reservoir water level as shown in Fig. 3. There is a uniform reduction of the seepage rate in zone I from 30.33 m to 29.0 m of the reservoir water level. After that, a sudden reduction can be seen in zone II between 29.0 m to 28.75 m of the reservoir water level. In zone III, the seepage rate is seen reducing uniformly with the reduction of the reservoir water level.

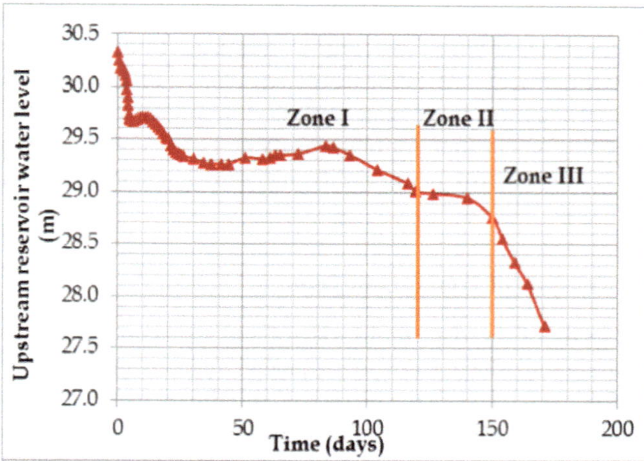

Fig. 2. Variation of upstream reservoir water level with time

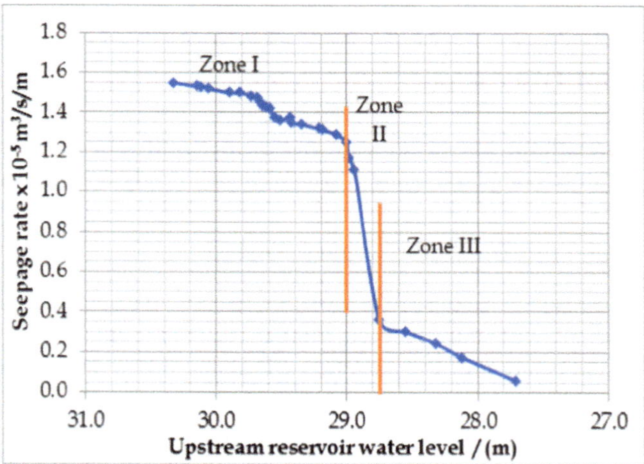

Fig. 3. Variation of Seepage rate with upstream reservoir water level

3.2 Seepage Analysis of Iranamadu Dam

Seepage analysis was carried out using SEEP/W software assuming isotropic hydraulic characteristics. The saturated/unsaturated hydraulic model was selected to accommodate seepage flow above the phreatic surface under unsaturated conditions. Therefore, the hydraulic conductivity function and the volumetric water content function were defined for both analyses. To estimate the hydraulic conductivity function, van Genuchten model was used as the estimation method and sample function option was used to estimate the volume water content function.

In the steady state analysis, the seepage rate was calculated by considering different upstream reservoir levels. In transient analysis, a steady state analysis at maximum reservoir water level of 30.33 m was taken as the parent analysis. In this analysis, the variation of reservoir level with time (Fig. 2) was used to define the input hydraulic head function. Furthermore, the influence of anisotropy of hydraulic conductivity of core material on seepage was investigated by varying soil anisotropy ratios (k_y/k_x) from 1 to 3 and by keeping k_x constant for both steady and transient state analyses, where k_x and k_y are the coefficients of permeability measured in the horizontal and vertical direction respectively.

3.3 Stability Analysis of Iranamadu Dam

The stability analysis of the dam under steady and transient seepage conditions was carried out by assigning shear strength parameters (Table 1) for the selected Mohr-Coulomb material model and using Bishop's simplified method in SLOPE/W. The stability was analysed for both upstream and downstream slopes of the Iranamadu dam embankment. Then, the results of the analysis were compared with the safety regulation criteria set by the British Dam Society (BDS 1994), according to which the factor of safety should be more than 1.2 under rapid drawdown and 1.3 under steady state seepage conditions.

4 Results and Discussion

Seepage rates through the Iranamadu dam and the determination of the position of the phreatic surface were computed using SEEP/W under steady and transient states.

4.1 Steady and Transient State Seepage Analysis

In the steady state seepage analysis, the variation in reservoir upstream water level was assumed as constant with time in each drop of 0.15 m from 30.33 m to 27.71 m. Figure 4 shows the position of the phreatic surface and flow fields when the reservoir water level was at 30.33 m.

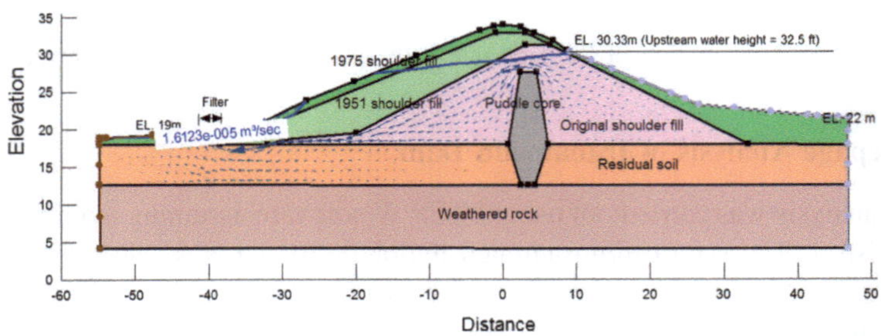

Fig. 4. Phreatic line and flow fields for the reservoir water level at 30.33 m

As indicated in Fig. 4, the seepage rate at the filter location is computed to be 1.612×10^{-5} $m^3/s/m$ (1.393 $m^3/day/m$) for the reservoir water level at 30.33 m. The seepage rates thus computed for different reservoir water levels are plotted in Fig. 5 along with the field measurements.

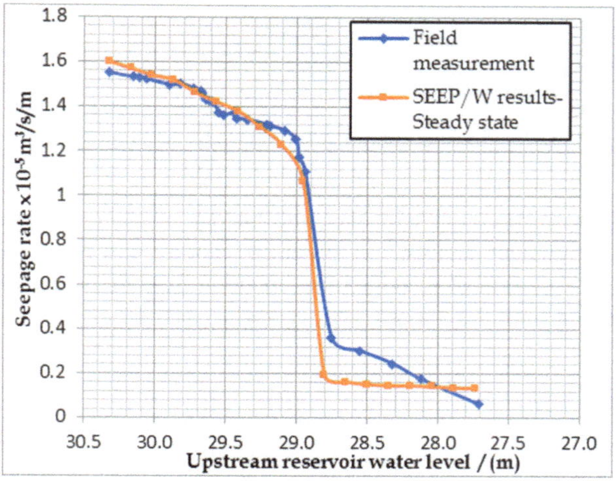

Fig. 5. Variation of seepage rates with different upstream reservoir water levels

Based on Fig. 5, it can be seen that SEEP/W software results and field measurements closely match for reservoir water levels varying from 28.8 m to 30.33 m. The significant drop in seepage rate for reservoir water levels below 29.0 m for both field and SEEP/W results is due to the change of the phreatic surface position from passing through the original shoulder fill to the clay core of the dam. Figures 6 and 7 show the change of phreatic surface for reservoir water level at 29.0 m and 28.8 m, respectively.

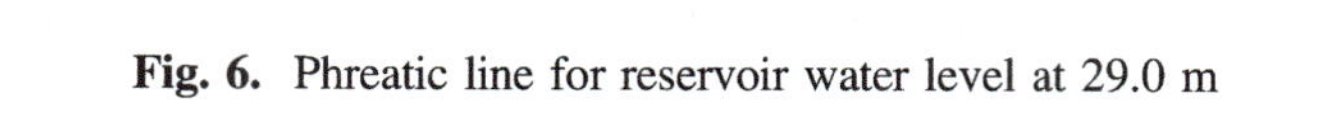

Fig. 6. Phreatic line for reservoir water level at 29.0 m

When the reservoir water level further decreases below 28.8 m, the software results showed a greater difference with the seepage rate calculated from field measurements. It may be due to the presence of non-homogeneous soil in clay core section. However, in the analysis, the clay core was taken as a low permeability homogenous section. Therefore, when the phreatic surface reaches the clay core section, there is no significant change in the seepage rate when reservoir water level reduces other than that due to changes in the water head.

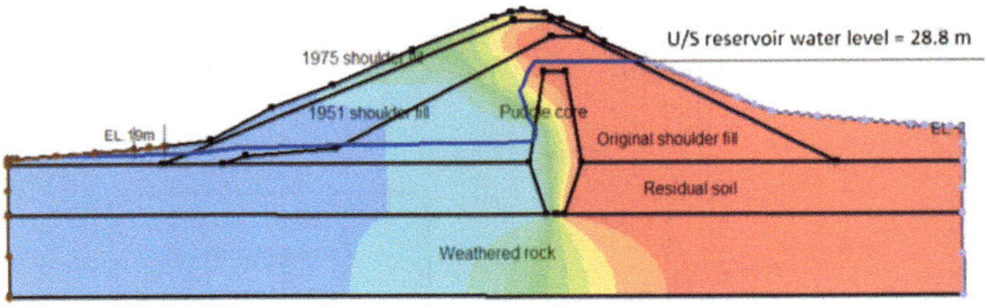

Fig. 7. Phreatic line for reservoir water level at 28.8 m

Seepage rate corresponding with the time history of reservoir level variation was calculated using transient analysis in order to capture the real time behaviour of seepage flow. The variation of seepage rates obtained from the above analysis and field measurements along with the variation of reservoir water level with time are shown in Fig. 8.

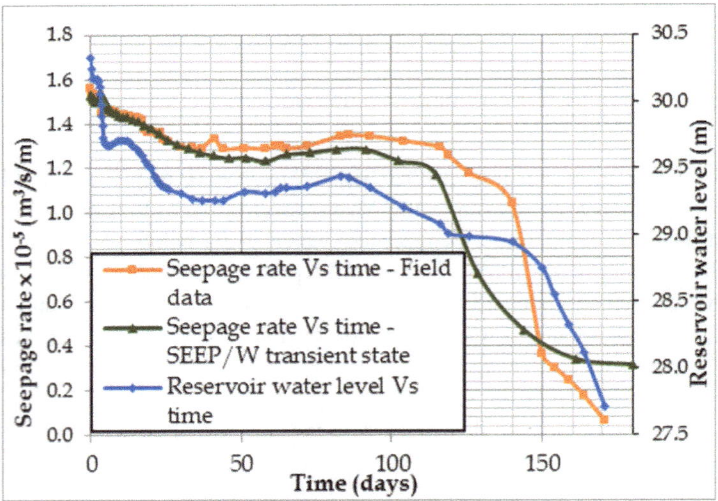

Fig. 8. Variation of seepage rate and reservoir water level with time

Based on Fig. 8, the seepage rates and reservoir water level reduce with time. The variation of seepage rates in field and SEEP/W software results are almost same up to 115 days when the reservoir water level reached 29.1 m. As the water level further decreased, the software results showed a significant variation from field measurements due to the presence of non-homogeneous clay core. After 115 days, the phreatic surface passes through the puddle clay core. Hence, there is a significant difference in field and SEEP/W results. Figure 9 shows a comparison of phreatic surfaces in 115 and 144 days. Further, it was noticed that the upstream water levels are at 29.1 m and 28.8 m after 115 and 144 days, respectively. Therefore, both steady state and transient seepage analysis show almost similar variation due to the change in phreatic surface.

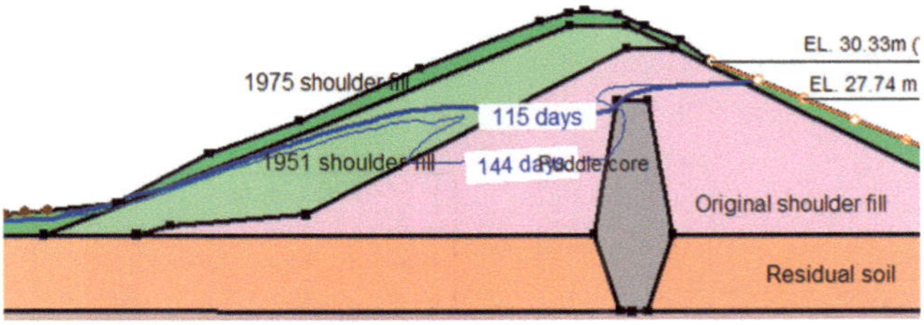

Fig. 9. Comparison of phreatic surface at reservoir level of 29.1 and 28.8 m after 115 and 144 days respectively

Figure 10 shows a comparison of observed seepage rate under steady and transient state analyses with field measurements at different reservoir water levels.

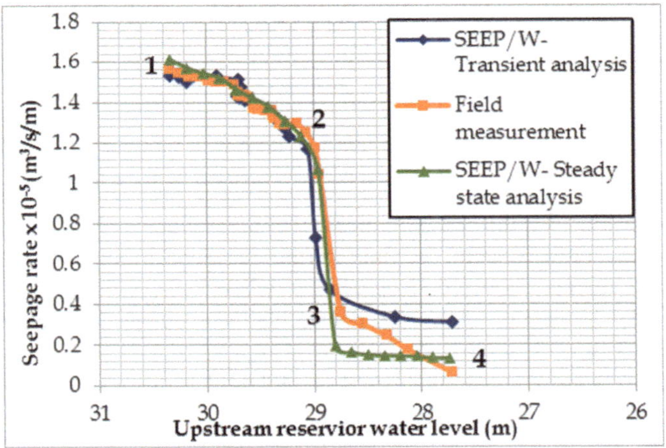

Fig. 10. Comparison of observed seepage rate with results of steady and transient state analyses

In Fig. 10, the transient and steady state analyses match with field data for the phases from 1 to 2 and 2 to 3. For the phase between 3 and 4, a deviation of the seepage rate from each line is observed. The deviation of the transient analysis is greater than that in the steady state analysis, especially at lower reservoir elevations, indicating the over prediction of seepage rate in the transient analysis due to the effect of drawdown phase of reservoir water level. When reservoir water level draws down suddenly, the pore water pressure will reduce and more pores will drain out. Therefore, a greater seepage rate is observed in transient analysis.

4.2 Effect of Anisotropy of Hydraulic Conductivity on Seepage

The effect of anisotropy of hydraulic conductivity on seepage was examined under steady and transient states using SEEP/W software. Hydraulic conductivity ratio, k_y/k_x was changed by keeping k_x constant and seepage rates were calculated with time under k_y/k_x ratios of 1, 2 and 3. Figure 11 shows the variation of seepage rate with time for different k_y/k_x ratio.

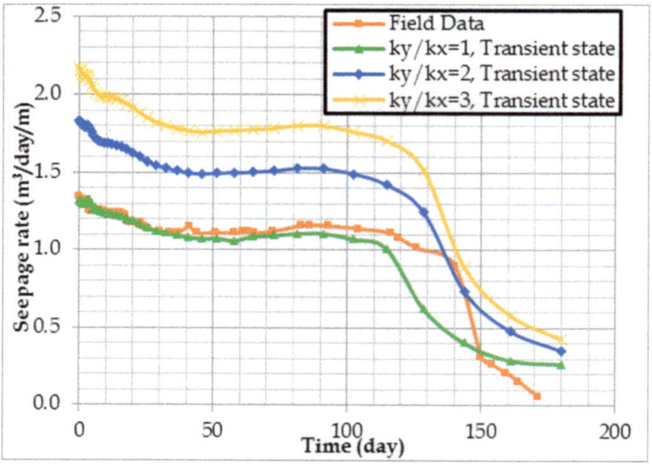

Fig. 11. Variation of seepage rate in transient state with time for different k_y/k_x ratio

As shown in Fig. 11, the results indicate that the seepage rate is increased with the increase of k_y/k_x ratio. Except for time period between 115 to 150 days, the field seepage measurements show good agreement when k_y/k_x ratio is 1. Therefore, the behaviour of the Iranamadu dam with respect to the seepage flow rate shows a near isotropic behaviour.

In the steady state seepage analysis, four different water levels were selected and seepage rates were calculated for different k_y/k_x ratio. The calculated seepage rates are given in Table 2.

Table 2. Seepage rates under steady state for different reservoir water levels

Reservoir water level (m)	Seepage rate × 10^{-5} $m^3/s/m$			
	$k_y/k_x = 1$	$k_y/k_x = 2$	$k_y/k_x = 3$	$k_y/k_x = 5$
30.33	1.610	2.204	2.583	2.931
29.10	1.230	1.564	2.067	2.436
28.75	0.164	0.214	0.260	0.325
27.70	0.135	0.194	0.237	0.282

Based on the results of the SEEP/W analysis, the rate of seepage increases when increasing the upstream water level and k_y/k_x ratio. This agrees with Seepage modelling with SEEP/W (2012), which stated that the seepage rate increases with anisotropy ratio (k_y/k_x).

4.3 Slope Stability Analysis Using SLOPE/W Software

The slope stability analysis was carried out corresponding to the transient seepage analysis. The variation of minimum factor of safety corresponding with the reservoir water level with time for downstream (D/S) and upstream (U/S) slopes are shown in Figs. 12 and 13, respectively.

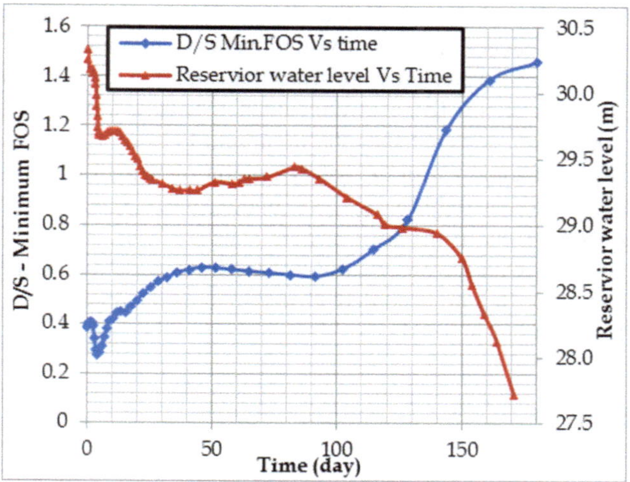

Fig. 12. Variation of downstream minimum factor of safety and reservoir water level with time in transient state analysis

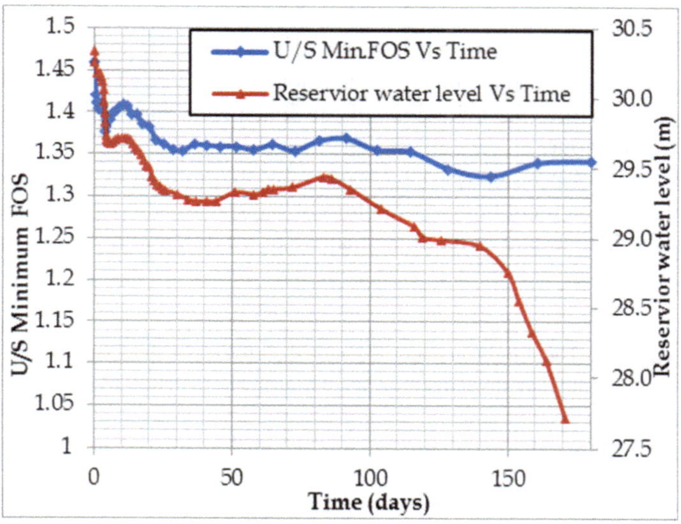

Fig. 13. Variation of upstream minimum factor of safety and reservoir water level with time in transient state analysis

Based on Fig. 12, the Iranamadu dam downstream slope does not satisfy the minimum FOS recommended up to 144 days. Thus, it infers that when the reservoir water level increases above 28.85 m, there is a possibility of a slip in the downstream slope. This is due to the passage of the phreatic surface through the critical slip surface.

As per the Fig. 13, the minimum factor of safety values for upstream slope varies from 1.46 to 1.32 when the reservoir water level is reduced, due to the removal of the balancing hydraulic forces acting on the upstream side embankment and the near undrained condition prevailing within the clay core. Also, after 140 days, when the reservoir water level reduces significantly, the upstream minimum FOS increases due to the dissipation of pore water pressure from the saturated zone within the embankment. Souliyavong et al. (2012) also concluded the same results. As a result, the Iranamdu dam upstream slope satisfies the minimum FOS of 1.3 under transient state seepage condition as mentioned in the safety criteria of British Dam Safety (1994).

Furthermore, the slope stability analysis under steady state condition also gives similar variation of results to that of transient state analysis. The results of stability analyses compare well with the observations of the Iranamadu dam slopes, where there was no upstream slope failure recorded in the Iranamadu dam while a downstream slope failure had occurred in 1983.

5 Conclusions

Based on the results obtained from the seepage and stability analyses, the following conclusions can be made.

- In both steady and transient state seepage condition, SEEP/W software results show a good agreement in seepage rates with those obtained from field measurements.
- When the phreatic surface is intercepted by the puddle core section, the sudden drop-in seepage rate is observed due to the low permeability of puddle core section.
- The seepage rate increases with increase of k_y/k_x ratio by keeping $k_x$constant and the Iranamadu dam seepage field measurements show a good agreement when k_y/k_x ratio is 1. Therefore, the behaviour of the Iranamadu dam shows a near isotropic hydraulic behaviour.
- When the reservoir water level increases, the minimum FOS of the downstream slope reduces which could lead to failure of the Iranamadu dam and the possibility of upstream slope failure of the Iranamadu dam for different reservoir water levels is highly unlikely.

Acknowledgements. The Engineers and Technical Officers of the Department of Irrigation, Kilinochchi are gratefully acknowledged for the permission granted to use the data related to the Iranamadu dam.

References

Abhilasha PS, Antony Balan TG (2015) Numerical analysis of seepage in Embankment dams. J Mech Civil Eng 13–23. https://pdfs.semanticscholar.org/2ef9/5e1b61aa0133de1f5b6c05d7fd767fbc2694.pdf

Abo AA (2001) Calculating seepage through earth dams by finite element method. College of Engineering, University of Baghdad, Iraq

Andreea C (2016) Unsaturated slope stability and seepage analysis of a dam. Energy Procedia 85:93–98

Athani SS, Shivamath, Solanki CH, Dodagoudar GR (2015) Seepage and stability analysis of earth dam using finite element method. Aquat Procedia 4:876–883

BDS (1994) The British dam society at the institution of civil engineers, Great George Street, London

Elshemy M, Nasr RI, Bahloul MM, Rashwan IM (2002) The effect of blockages through earth dams on the seepage characteristics. Faculty of Engineering, Tanta University, Egypt

Hasani H, Mamizadeh J, Karimi H (2013) Stability of slope and seepage analysis in earth fills dams using numerical models (case study: Ilam dam - Iran). World Appl. Sci. J. 21(9):1398–1402

Kamanbedast A, Delvari A (2012) Analysis of earth dam, seepage and stability using ANSYS and GeoStudio software. World Appl. Sci. J. 17(9):1087–1094

Kirra MS, Shahien M, Elshelmy M, Zeidan BA (2015a) Seepage and slope stability analysis of Mandali earth dam, Iraq: case study. In: International conference on advance in structural and geotechnical engineering

Kirra MS, Zeidan BA, Shahien M, Elshelmy M (2015b) Seepage analysis of Walter F. Geoge Dam, USA: a case study. In: International conference on advance in structural and geotechnical engineering, pp 1–13

Report of design development of remedial works Iranamadu tank embankment (2013)

Seepage modelling with SEEP/W (2012). GEO-SLOPE International Ltd

Souliyavong T, Gallage C, Egodawatta P, Maher B (2012) Factors affecting the stability analysis of earth dam slopes subjected to reservoir drawdown. In: Second international conference on geotechnique, construction materials and environment

Uromeihy A, Barzegari G (2007) Evaluation and treatment of seepage problems at Capar-Abad Dam, Iran. Eng Geol 91:219–228

Application of Principle of Network Connectivity in Creation of Sustainable Urban Form: Case of Negombo Water Fronts

D. S. N. Jayakody(✉)

Urban Development Authority, Gampaha District Office, Minuwangoda, Sri Lanka
sauri.jayakody626@gmail.com

Abstract. City is a vital organism that entitled with the diverse planning attempts. Blue green infrastructure planning is the one of attempt which has been practised recently. Mainly the Blue Infrastructure Planning involves with the four major core principles of Green-Grey Integration, Network connectivity, multifunctionality and social inclusion. The principle of network connectivity has been considered in the research and it involves with the both structural and functional connections between blue green spaces in order to create added values from an interlinked system. Numerous quantative approaches have been developed to measure and map the connectivity. Measures of structural connectivity are based on the concepts of presence or absence or the size, form and shape of corridors. Initially, connectivity was calculated using the various indices of distance, frequency and density. The research finding has been adopted in Negombo Development Plan 2018–2030 which is formulated by Urban Development Authority. It adheres with the strategic planning approach with the use of diverse methods of multy criteria analysis techniques, field surveys, observation and key formal interviews to integrate the blue green spaces of the city with the purpose of enhancing the sectoral tourism, recreation and self-sufficient affluent city economy connecting the all blue spaces of urban waterfronts. The research attempts to find how the network connectivity influence to reinforce a sustainable urban form in the city and activate the networks of green and blue spaces with full spectrum of benefits to denizens and the journeyrs at all spatial scales that enable to open a positive platform to merge with the emerging urban challenges of the city and ultimately it influences to show the creation of sustainable urban form in Negombo with the integration of blue-green networks and considerable positive outcomes.

Keywords: Blue-green infrastructure · Network connectivity · Self sufficient city · Sustainable urban form

1 Introduction

Blue green infrastructure planning associates with the major core principles of green grey integration, network connectivity, multifunctionality and social inclusion that contribute to create a green space networks and multiple function and services. The

R. Dissanayake and P. Mendis (Eds.): ICSBE 2018, LNCE 44, pp. 378–393, 2020.
https://doi.org/10.1007/978-981-13-9749-3_34

principle of network connectivity is an innovative approach in the creation of sustainable city form to face the major urban challenges and mitigate the risks of the cities. more than half the world's population live in cities are rising rapidly (UNEP 2011).

The major factors of Increase of population density, development pressure lead to densification of cities. Many cities are situated in the coastal or delta areas where disaster risk increases. This is firstly a result of climate change, including rising sea levels, increased river discharge and extreme precipitation. Secondly, this results from land use changes such as the increase of sealed surfaces and buildings and from subsidence.

Frequent flooding and the consequent damage are a threat to a healthy and vital urban environment. Furthermore, heat waves become more frequent because of climate change. This may lead to health problems in cities becoming urban heat islands.

Well planned and designed cities adhere with the green infrastructure, including water, greenery and soil. This can contribute to climate change adaptation and promote and support healthy urban living. Effective application requires design principles that address the direct relationship between green infrastructure and impacts of climate change is import but also the quality of the ecosystem (Kemperman and Timmermans 2014).

Three categories of health aspects of water and ecosystem services support human health as directing exposure to water contributing to medical health, encouraging healthy living by creating possibilities to exercise and enhancing the aesthetical aspects of water contributing to mental health. Further the principle affects to generate the economic benefits for the city that can be derived from the aquatic ecosystem services (Fig. 1).

Fig. 1. Design principle of blue green infrastructure on city level

The European Commission has adopted a Green Infrastructure Strategy (COM 2013) and states that green infrastructure (GI) can make a significant contribution to effective implementation of all policies where some or all of the desired objectives can be achieved in whole or in part through nature-based solutions (Merk 2012).

There is usually a high return on GI investments and overall reviews of restoration projects typically show cost-benefit ratios in the range of 3 to 75. Well planned and designed green infrastructure, including water and soil, can contribute to climate change adaptation and at the same time promote and support healthy urban living (Worpole 2007).

1.1 Objectives of the Research

The key objectives of this research open to assess how the functional use of urban blue and green infrastructure both at the street and city level can contribute to a liveable, healthy and sustainable city. Liveability can be improved with a variety of ecosystems services with the creation of connectivity and integration of blue green infrastructure and develop innovative combinations of existing solutions with added value, using state of the art expertise, tools and models to climate change mitigation, climate change adaptation and resource efficiency and giving rise to recognition of the central role that green space networks.

Successful green-blue infrastructure planning naturally requires an integration of different scales of planning and application. Accordingly, green-blue infrastructure planning will likely be for a whole urban area or catchment, but will need to consider possible interventions at all scales of local, neighbourhood and city scales their cumulative impact (Fig. 2).

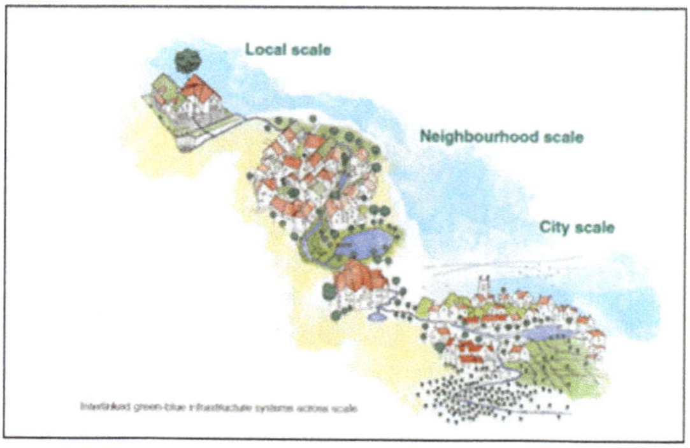

Fig. 2. Interlinked green blue infrastructure systems across scale

1.2 Study Area

City of Negombo was selected as the study area. It is situated in Negombo Municipal Council Area in Gampaha District, Western Province and the west coast of Sri Lanka. It is located 35 km away from Colombo, the commercial capital of Sri Lanka and 7 km away from the Katunayake Bandaranayake International Airport and free trade zone. It is bounded, North by Maha Oya, South by Diya honda Ela, East from Roma Ela and the west from lagoon and Indian Ocean.

The total extent of Negombo Municipal Council area is 2800 ha and it is within the boundary of Negombo Divisional Secretariat Division with 39 Grama Niladari Divisions. The total population of Negombo is 175000. Negombo features a tropical rainforest climate under the Koppen climate classification. Mainly the city receives rainfall from the South-western monsoons from May to August and October to January. During the remaining months there is a little precipitation due to Convective rains. The average annual precipitation is about 2400 mm. The average temperature varies 24 to 30 °C, and there are high humidity levels from February to April (Fig. 3).

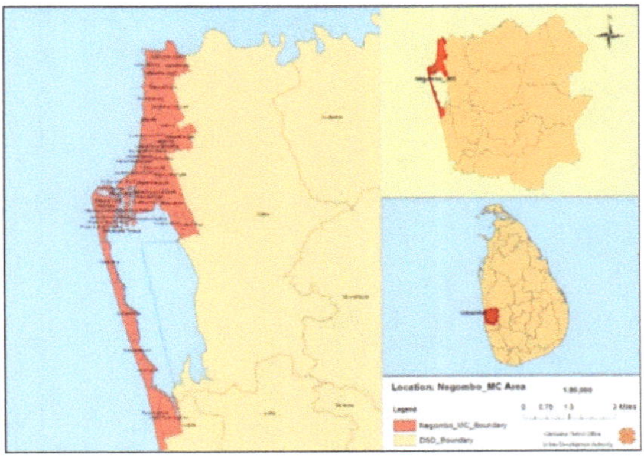

Fig. 3. Location of study area

Negombo currently act as the one of major commercial, transport center, an educational Facilitator and health service provider for the Region that is function with both rail and road network together in Gampaha District.

Negombo has an identifiable blue green network that blend with the Lagoon, sea, Hamilton Canal, Ma oya river, 24 km long golden sandy beach and 18 number of mangrove species. This contribute to strengthen the self-sufficient economy of the city that is created through three majors of tourism, fishery and commerce. Simply 45% of economy is based on fishery, 35% of economy is based on tourism and 25% of economy is based on commerce and other service based on the blue green network.

This affect to Attract 350000 tourists annually and they come to Negombo for leisure activities and this lead to become the major tourist destination. Negombo has an active fishery industry which has a plenty of fish production. It gets the third place in Sri Lanka and Contributes 16% for the national economy by fish production. Which is 80% of marine fish and 20% of lagoon fish. Shrimp and prawn's production highly contribute for its economy (Fig. 4).

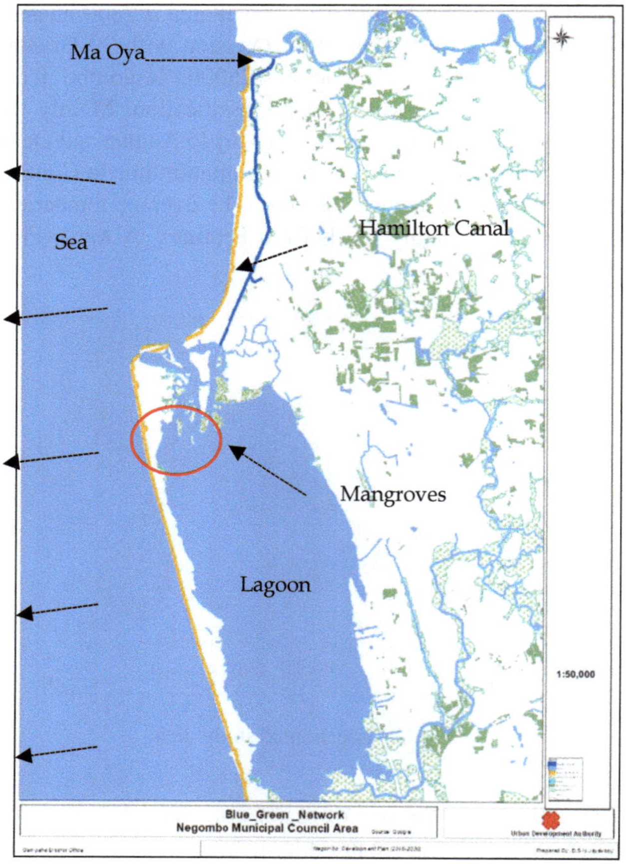

Fig. 4. Blue green network of Negombo

2 Methods of the Research

The research involves with the Multi criteria analysis of development pressure, environmental, sensitivity, liveability, potential and density and GIS mapping, field survey and key interviews. 39 number of Grama Niladari Divisions of Negombo MC area were selected and surveyed for the research.

3 Analysis and Synthesis

The value of green spaces in Negombo Municipality area were rediscovered with the use of multi criteria analysis of development pressure, sight scene hotspot, sensitivity, liveability, potential and density analysis. The development pressure analysis proved that the development is spread towards the coastal belt and sensitive blue green spaces with urban trend phenomena (Figs. 5 and 6).

Core Values	Weight	Parameters	Values		Score
Land use	20%	Land use distribution	Built-up		5
			Vegetation		3
			Hydro		2
			Other		1
Accessibility	20 %	Tendency to accessibility of transport network	Major Roads (A Roads)		5
			Minor Roads (B Roads)		4
			Other Roads (C Roads,		3
			jeep/Cart track)		2
Population density	20%	Density		<700	1
				700 – 2500	2
				2500 - 5000	3
				5000 – 10000	4
				10000<	5
Population Growth rate	20%	Rate		(-6.4) – (-2.2)	1
				(-2.2) – (-0.69)	2
				(-0.69) – (-0.16)	3
				(-0.16) – (1.34)	4
				1.34 – 5.57	5
Local Connectivity	10%	Hierarchical level of town	Level 1 town centre		
Buildings	10%				

Fig. 5. Criteria of development pressure analysis

The sensitivity analysis enhanced that the existing blue green spaces of Negombo lagoon, Ma oya, Hamilton Canal and mangroves species are the most sensitive elements in the city and the it is the most important blue green infrastructure for the entire eco system of Negombo and the inhabitants who are dependent on this infrastructure (Fig. 7).

Thee green corridors were provided to access and integrate the beach parks as a single promenade. The proposed public outdoor recreation space plan of Negombo has created a positive linkage among the Ma oya, lagoon and beach strips with Liner parks, mangrove parks, and active pedestrian friendly cycling and walking paths.

The network connectivity integrated with the proposed Water transport along the Hamilton Canal to create an active blue space up to Ma oya environment. The proposed strategic action opened a platform for the conservation of blue green spaces and activation of the spaces in to the healthy and economically viable context of Negombo with promoting the tourism and fishery activities (Auttred 2015) (Fig. 8).

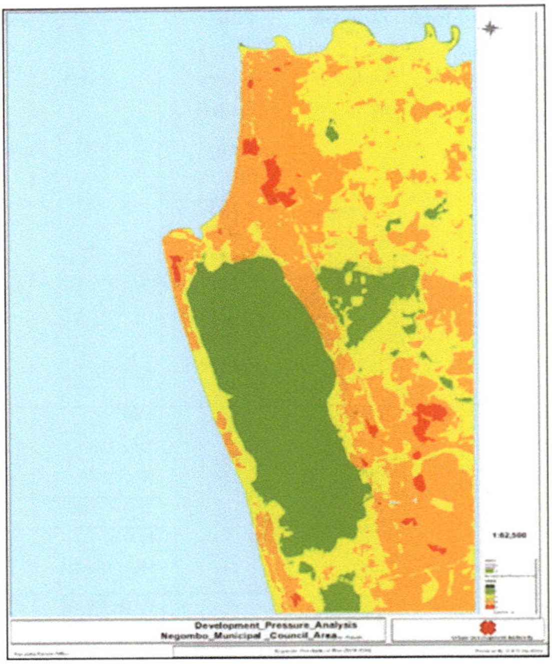

Fig. 6. Development pressure analysis

Core Values	Weight	Features	Parameter	Score
Hydro system	35%	Maha Oya	45m	5
		Sea	100m	5
		Lagoon	100m	5
		Hamilton and Dutch Canals	6.5m	5
		Water/Tributaries/Lakes		5
		Water retention area (Marsh / Wetland)		4
		Inundation Area		3
		Sea Level Rise		3
Vegetation cover	35%	Mangrove/marsh	100m	5
		Scrub		4
		Paddy		4
		Agriculture		3
		Cultivation areas (Coconut / Rubber)		3
		Home garden		2
Geomorphological and relief features	10%	Beach	60m	4
		Soil- Red Yellow podzolic		1
		Alluvial Soils of variable texture and drain		5
		Bog and half – dog soils of variable texture and drain		4
		Latosols and regosolt on alled red and yellow		2
		Regosols on recent beach and dune sand		3
Human-Use resources	20%	Culturally important locations / Heritage sites		4
		Religious places		3
		Parks and Play Ground		3

Fig. 7. Criteria of environmental sensitivity analysis

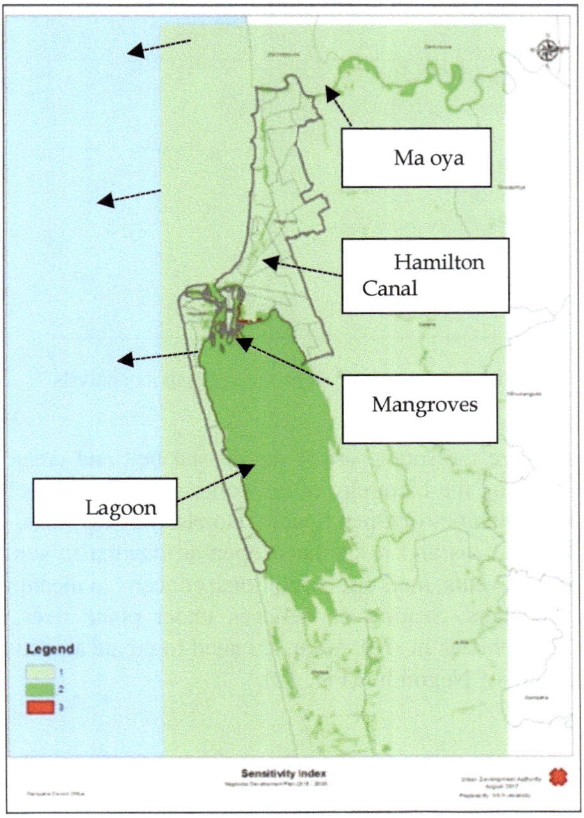

Fig. 8. Environmental sensitivity analysis

Further the field survey and key interview elaborated that the main functions, economy and other services of Negombo are oriented with the existing green spaces which provide the aesthetic value and recreational facilities. There is a growing awareness that green infrastructure can be more beneficial to society than merely serving aesthetics and recreation. For example, the municipality of Negombo is interested in making better use of green infrastructure in the search of measures that can help to achieve a healthy city.

Town planners want to take the capacity of green elements for regulating temperature, air quality, water storage and drainage and noise reduction into account. In order to be able to do so, there is a need for translation of scientific knowledge on functionality of green infrastructure into practical design principles. It also requires practical guidelines for how to integrate these principles into the design of multi-functional green infrastructure with the principle of blue green network connectivity.

The principle was adapted with the use of sight scene hotspot analysis of the south part of beach promenade of Negombo. The five major hotspots were identified and the hotspots were converted on to central, mini and minor beach parks (Fig. 9).

Criteria	Value
Beach Width	20
Sea Depth	20
Contour	5
Accessibility	15
Soil	5
Distance between Lagoon and Sea	15
Landuse	5
Nodes Buffer	15

Fig. 9. Criteria of sight scene hotspots analysis

In order to optimize the south part of the coastal belt and create an active beach promenade with creating the hierarchy of networks among the beach parks. The provision of the typical amenities of Open Spaces, Hotels, playing areas, strolling, cycling, inline-skating, beer gardens and restaurants, open-air-cinema in summer, sunbathing, and occasional public events, markets, exhibitions, concerts, a meeting place, Chatting, restrooms, picnic shelters, Seating on benches under plane trees, clean and well-maintained, safe, no vehicles in sight were activated to create an active open space for the users and journers of Negombo (Fig. 10).

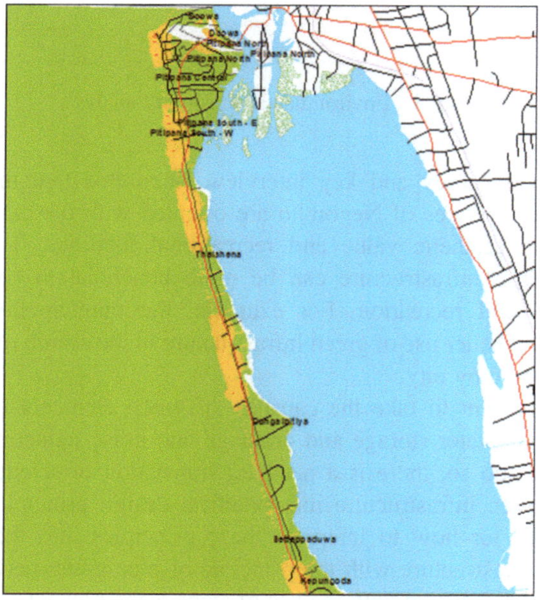

Fig. 10. Sight scene hotspots analysis

Thee green corridors were provided to access and integrate the beach parks as a single promenade. The proposed public outdoor recreation space plan of Negombo has created a positive linkage among the Ma oya, lagoon and beach strips with Liner parks, mangrove parks, and active pedestrian friendly cycling and walking paths.

The network connectivity integrated with the proposed Water transport along the Hamilton Canal to create an active blue space up to Ma oya environment. The proposed strategic action opened a platform for the conservation of blue green spaces and activation of the spaces in to the healthy and economically viable context of Negombo with promoting the tourism and fishery activities (Fig. 11).

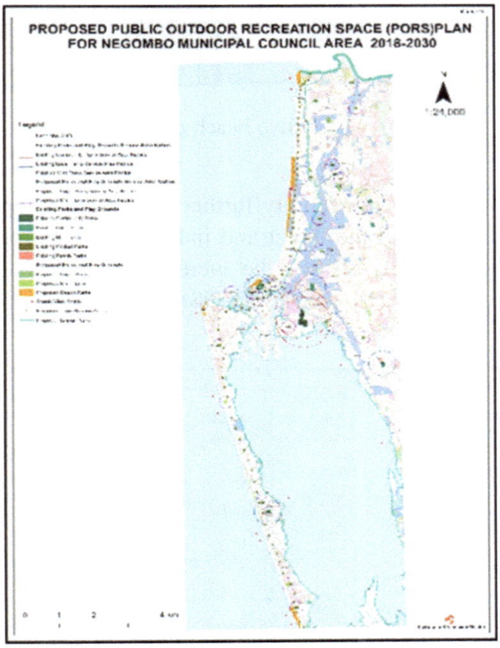

Fig. 11. Proposed PORS plan

4 Findings and Results

The research emphasises the key findings. This attempts to optimize the blue green infrastructure. The identified gap of blue green infrastructure and city scape were integrated with the principle of network connectivity and it affects to interact between two major components of Blue green infrastructure and the city fabric.

Further the application of principle of network connectivity influences to create a vibrant active, accessible, comfort sociability and a reputed image for the city (Fig. 12).

Fig. 12. Active beach promenade

The principle of network connectivity further influenced to upgrade the liveability and potential of Negombo and the direct and indirect impacts for city economy were interlinked and the tourist arrivals will be increased up to 300000 with the tourist prediction scenarios (Baudry and Merriam 1998) (Figs. 13 and 14).

Livability Analysis		
Index	**Criteria**	**Value**
Population Index	Population Growth Rate	30%
	Population Prediction	
Human Index	Buildings	10%
	Landuse	
	Service Nodes Buffer	
	Hospital Buffer	
	School Buffers	
Recreation Index	Mini Parks	10%
	Local Parks	
	Community Parks	
Environment Index	Soil	10%
	Wetland	
	Hamilton Canal	
	Beach Buffer	
	Water Buffer	
	Mangroves	
	Scrubs_Paddy_Coconut	
Accessibility Index	Nodes Buffer	20%
	Line Connectivity	
	Highway Interchanges	
	Railway	
	Road Network	
Restriction Index	Airport - Height -Limitation	20%
	Obstacles – Restricted - Zone	

Fig. 13. Criteria of liveability analysis

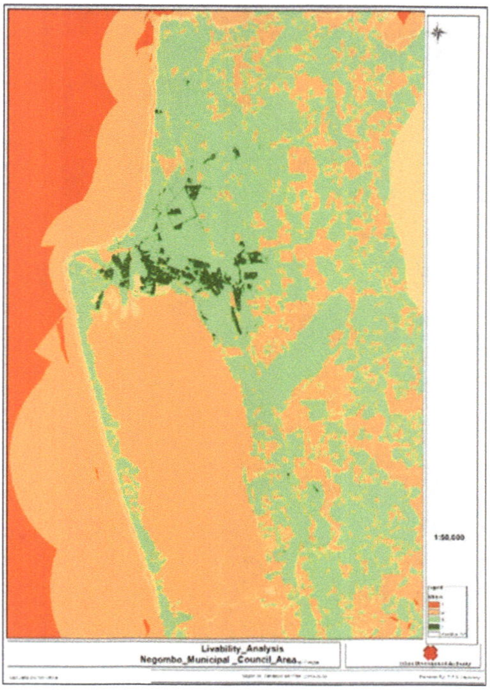

Fig. 14. Livability analysis of Negombo

The entire area is emerged as more potential in terms of hotels, tourism, fishery, residential, commercial and other uses and many projects were initiated to accommodate the area (Figs. 15 and 16).

Index	Score
Development Index	20
Population Index	10
Restriction Index	10
Human Index	20
Recreation Index	10
Environment Index	10
Accessibility Index	20

Fig. 15. Criteria of potential analysis

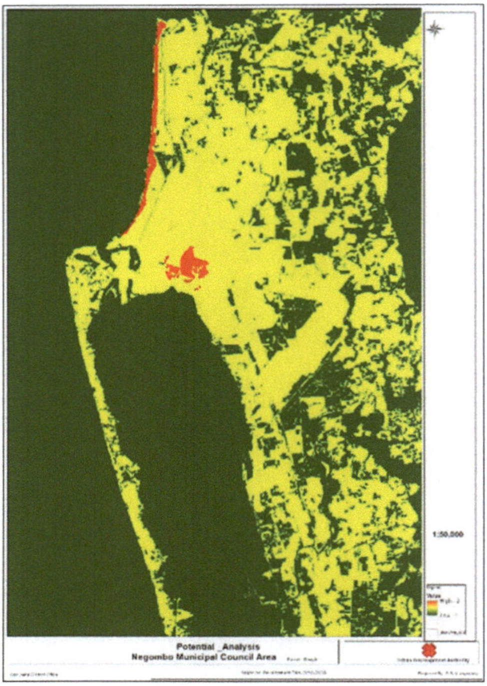

Fig. 16. Potential analysis of Negombo

Land value of the area was risen with the integration of blue green infrastructure and it became the major influence on city economy and the coastal areas will be converted as the potential tourism areas and active open spaces and the significant hotel development, demand and competition for the space will be originated on the ground of the application of blue green network connectivity (Worpole 2007) (Fig. 17).

Further the application of principle of Connectivity encourages the structural and functional upgrading between green spaces to create an added value from inter linked systems and a sustainable, healthy form of the city with self-sufficient economy (Taylor 2006).

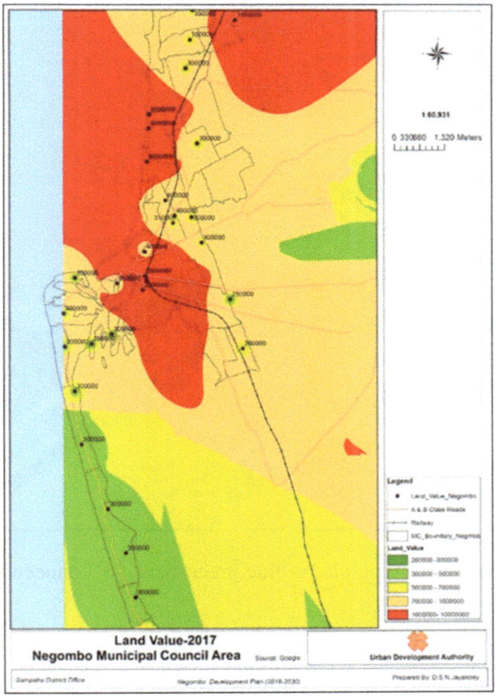

Fig. 17. Land value of Negombo

5 Conclusion and Recommendations

This research interprets the how far the application of the design principle of blue green network connectivity is worth for the sustainable city form with addressing the urban challenges of climate resilience, healthy and liveable urban environment with sustainable self-sufficient city economy. A healthy and liveable urban environment contributes to the strengthening of the socio-economic climate in cities and improve a variety of ecosystems services.

Research opens a platform for the future researchers to apply the modern design principals for the cities.

The urban environment in which people live, work and play affect for their health with the characteristics of a healthy a clean and safe environment, sufficient green space, nature and water, healthy and sustainable homes, attractive and varied public spaces, wide range of public services.

The strategic approach followed in the research interlinks to harness the potential of blue green spaces to make a major contribution to sustainable development by enhancing social cohesion, supporting the economy, and adapting to a changing climate, and highlights the importance of green infrastructure solutions in cities. The design principle can adhere to the urban challenges of protecting bio diversity, promoting green economy, increasing the social cohesion and minimising the disparities (Fig. 18).

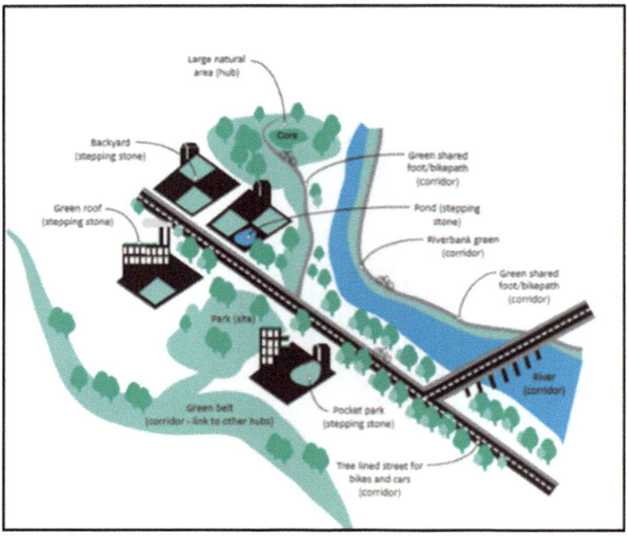

Fig. 18. Application of principle of blue green network connectivity in Negombo

Meanwhile the physical arrangement of the city is converted with the use of blue green infrastructure and the colours, building designs and blue green oriented regulations, conservation buffers, and green blue corridors will be adapted in the city and green walkable street areas, user friendly open spaces are activated with the multiple benefits for the users and journeyrs of the city.

Acknowledgements. Author wishes to thank the Plnr. Janak Ranaweera, Director of Western Province of Urban Development Authority, Plnr J. Somasiri, Deputy Director of the Gampaha District Office and the staff and the Plnr Samantha Kumara, Deputy Director of Matale District Office and A.M.C. Samanthilaka, U. D. Hettiarachchi, Town planners of Kelaniya and Biyagama Development Plans in Gampaha District office guidance, supervision and sponsorship to become the success of the research study. Sumithra Wijesinghe, City Planner of Negombo MC supporting for the field surveys and the process of data collection and Plnr, Nileesha Senavirathna encouraging to launch the research and those who support to carry out the research.

References

UNEP (2011) Green Jobs towards a green economy, pathways to sustain able development and poverty eradication. Kenya

UNEP (2013) Measuring progress towards an inclusive green economy. Kenya

Merk O (2012) Financing green urban infrastructure. Paris

Dunn AD (2010) Sitting Green intra legal and policy solution to alleviate urban poverty and promote a healthy communities. Boston College

Kemperman A, Timmermans H (2014) Green spaces in the living environment and social contacts of the aging population, Landscape and Urban Planning

Worpole K (2007) The social value of public spaces. Joseph Rowntree Foundation, York

Taylor PD (2006) Landscape connectivity. Cambridge
Baudry J, Merriam G (1998) Connectivity and connectedness, functional versus structural patterns in landscapes connectivity
Auttred AG (2015) The spatial and temporal components of functional connectivity in fragmented landscapes
Department of Environment, Land, Water and Planning (2017) Planning a Green-Blue City, Australia
Gehrels H, van der Meulen S, Schasfoort F, Bosch P, Brolsma R, van Dinther D, Geerling G, Goossens M, Jacobs C, Kok S, Massop HTL (2016) Designing green and blue infrastructure to support healthy urban living. TO federate

Development of a Post-tentioned Masonry Retaining Wall System

W. W. P. K. Perera(✉)

Department of Civil Engineering, National Engineering Research and Development Centre of Sri Lanka,
No: 2P/17B, Industrial Estate, Ekala, Ja-Ela, Sri Lanka
wasantha@nerdc.lk

Abstract. A Retaining Wall is a commonly used structural element in the field of construction. A retaining wall is a structure constructed to hold back any material (mainly soil) and prevent it from failure. A retaining wall is capable of resisting the material pressure of the material that it is holding back. Random Rubble Masonry (RRM), Reinforced Cement Concrete (RCC) and Gabion Wall (GW) are the most common types of construction materials used in the construction of retaining walls. RRM & GW are gravity type retaining walls and they require significant width from the land. RCC retaining walls require large amount of tor steel reinforcement, concrete and vertical form work, which are costly. Masonry walls such as brick walls and block walls cannot be used as retaining walls as they cannot withstand the tensile stresses developed due to lateral earth pressure. This paper discusses the development of a masonry block wall system which can be used as a retaining wall by post-tensioning eccentrically. The compressive stresses developed due to post-tensioning is used to neutralize the tensile stresses developed due to lateral earth pressure. This system eliminates concrete and vertical formwork and minimizes steel usage. It has advantages such as cost effectiveness, pleasing aesthetic appearance and simplicity.

Keywords: Retaining walls · Post-tensioned masonry · Eccentrically pre-stressing · Embankment · Lateral pressure of soil

1 Introduction

A retaining wall is a structure designed and constructed to resist the lateral soil pressure and hydrostatic pressure if any of a slope that exceeds the angle of repose of the soil. Angle of repose is the maximum slope a soil type can withstand without any external support. Movement of unstable soil masses can cause lot of inconveniences resulting in loss of property and life. The soil movements are known as slips or landslides depending on their magnitude (Fig. 1).

R. Dissanayake and P. Mendis (Eds.): ICSBE 2018, LNCE 44, pp. 394–410, 2020.
https://doi.org/10.1007/978-981-13-9749-3_35

Fig. 1. Failure of retaining wall

Traditional retaining walls can be classified as gravity walls, piling walls, cantilever walls and anchored walls based on their behaviour. Random Rubble Masonry (RRM), Reinforced Concrete (RC) and Gabion walls (GW) are the most common material types used to construct retaining walls. RC retaining walls are costly as it requires high quality vertical formwork. Since reinforced concrete retaining wall are in-situ construction, lack of supervision and lack of skilled supervisors results in construction of a different structure than what is designed. RRM and GW require more space from the land. This is critical in smaller plots of land with higher land value.

2 Literature Review on Post-tensioned Masonry

Masonry is the first man made construction material in the world invented long before other construction materials such as steel and concrete. Masonry walls have advantages such as availability, simplicity, durability, strength and ability to incorporate architectural elements in to its design. But the application of masonry is limited only to compression members as masonry is weak in resisting tensile stresses. However, it has sufficiently large compressive strength. Due to this fact, traditional masonry cannot be used to construct retaining walls as it cannot withstand the tensile stresses developed due to lateral earth pressure. Thus a traditional brick work structure subjected to bending will fail in tension long before its compressive strength is exploited [1]. If these tensile stresses can be neutralized by applying a pre-compression to the masonry elements, there is a possibility to use masonry elements in retaining walls. Post-tensioning technique can be applied to brick masonry to achieve this target and it can also be used to get the maximum benefits from this versatile construction material. It

offers new application for masonry with simple construction technique and competitive cost. Post tensioning will increase cracking resistance, and ductility. Post tensioning will also provide effective connection between wall and binding material to avoid the failure of the joints [2].
Advantages of post-tensioning

- Enhances the flexural tensile resistance of the brickwork.
- Highly cost effective and durable
- Simple technique.
- Aesthetically pleasing.
- Convert the brickwork from a brittle material to a ductile material. (It will bend without cracking/ deflect and recover when the load is removed)
- Increases the robustness, resistance to impact and high improvement in accidental damage resistance.

Disadvantages

- Post-tensioning of high-shrinkage units shall be avoided due to possibility of excessive losses of pre-stress.
- Difficulty in finding anchoring methods.
- High accuracy has to be maintained in positioning and tensioning pre-stressing wires.

3 Research Methodology

The methodology described below is adopted in the development of Post-Tensioned Masonry (PTM) retaining wall system.

- Observe the failure modes of retaining walls and loads on a typical retaining wall section.
- Analysis and design of post-tensioned retaining wall system by first principles considering retained height, effective prestressed force and eccentricity as variables.
- Manufacturing of cement: quarry dust blocks using the hydraulically operated cement: soil block making machine
- Construction of pilot project to validate the theoretical results.

3.1 Failure Modes of Retaining Wall and Loads on a Typical Retaining Wall

3.1.1 Modes of Failure of a Retaining Wall

Failure modes of retaining wall are illustrated in Fig. 2.

3.1.2 Loads on a Typical Retaining Wall

Loads on a typical retaining wall is illustrated in Fig. 3 and mainly consist of active soil pressure, hydrostatic pressure, passive soil pressure, surcharge, weight of concrete sections, weights of soil mass on the footing and upward soil pressure.

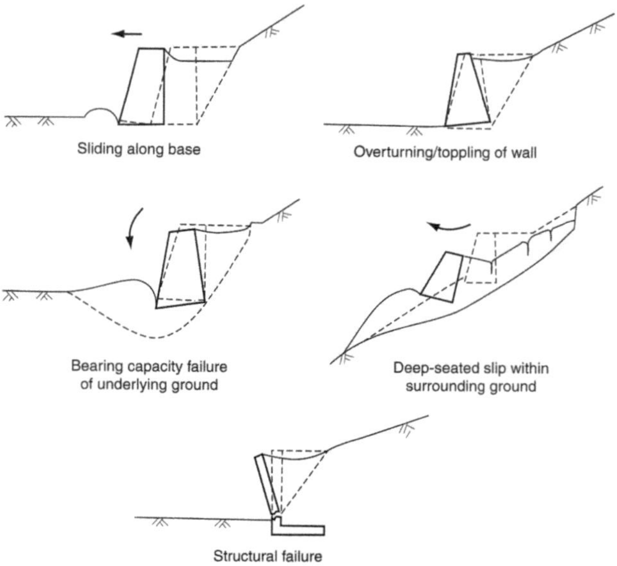

Fig. 2. Modes of failure of a retaining wall

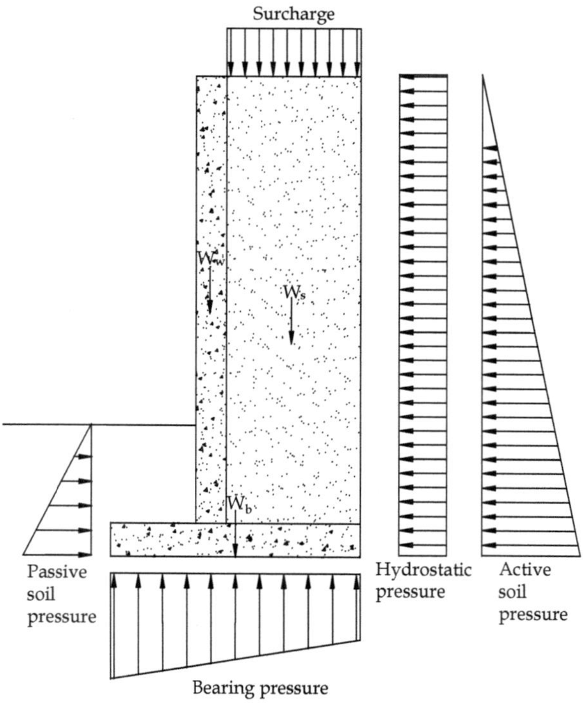

Fig. 3. Loads on a typical retaining wall

3.2 Analysis of Post-tensioned Retaining Wall Section

For the analysis of the retaining wall system with first principles, the section of the retaining wall shown in Fig. 4 is selected.

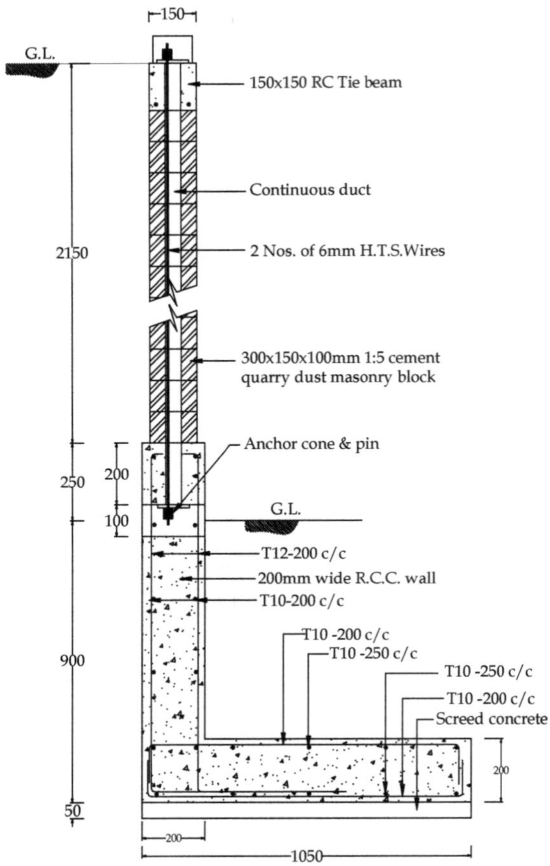

Fig. 4. Section of the retaining wall used to analysis

A retaining wall needs to be checked for sliding, overturning, bearing and bending. The following parameters were selected as variables for the design of Post-Tensioned Masonry (PTM) retaining wall.

Density of insitu concrete:	24 kN/m^3
Grade of insitu concrete:	25 N/mm^2
Density of soil:	18 kN/m^3
Angle of internal friction:	30
Allowable bearing pressure:	125 kN/m^2

Active earth pressure coefficient:	0.33
Passive earth pressure coefficient:	3.0
Surcharge:	Nill
HTS wire diameter:	5 mm
Jacking force:	24 kN
Wire spacing:	150 mm
Total losses of prestress:	30%
Design bending moment:	23.83 kN/m

3.3 Construction of Pilot Project

3.3.1 Selection of Site for the Pilot Project

A site at Dalupitiya Clinical Building premises belonging to the Mahara Pradeshiya Sabha was selected to construct the pilot project of the research (Fig. 5). The necessary approval was obtained from the Mahara Pradeshiya Sabha. The height of the soil embankment to be retained at the site was 2400 mm. The expected length of the retaining wall is 6.7 m. The section of the PTM retaining wall illustrated in Fig. 4 is selected.

Fig. 5. Site selected for the pilot project

3.3.2 Selection of Masonry Block Type for Casting

The available bricks and blocks are first considered for construction of PTM retaining wall. However it was found that the characteristic direct compressive strength of these bricks and blocks are in the range of 1–2 N/mm^2 and are not suitable for PTM retaining wall. A potential axial compressive strength could be obtained from the cement: quarry dust mixed interlocking blocks manufactured using the NERDC developed hydraulic block making machine. A series of sample tests were carried out to find the axial compressive strength (Fig. 6) of masonry blocks produced. The results indicate an axial compressive strength in the range of 7–8 N/mm^2 is possible with a mix proportion of 1:5 cement: quarry dust.

Fig. 6. Direct compression test

3.3.3 Testing of Flexural Tensile Strength of Masonry Blocks

According to the literature the characteristic flexural tensile strength of masonry is assumed as zero. Since a masonry block with high compressive strength is used in this research, a test is carried out to find out the actual characteristic flexural tensile strength of masonry blocks produced. The test results show an average characteristic flexural tensile strength of 2.73 N/mm^2 and this value is used in analysis under this research (Fig. 7).

Fig. 7. Flexural tensile strength test

The following values base on the test results are also used for the calculation.

Density of block: 17 kN/m^3
Compressive strength of blocks, f_k: 7.9 N/mm^2

Characteristic flexural compressive strength is equal to 1.5 times the characteristic direct compressive strength [2].

Material safety factor:	1.2
Design flexural com. strength, f_f:	9.87 N/mm^2
Flexural tensile strength, f_k:	2.73 N/mm^2
Self weight of the block work:	0.03 N/mm^2

3.3.4 Results of Analysis of Stresses on the Post-tensioned Masonry Retaining Wall

Results of Analysis of different type of tresses acting on the masonry retaining wall is illustrated in Fig. 8.

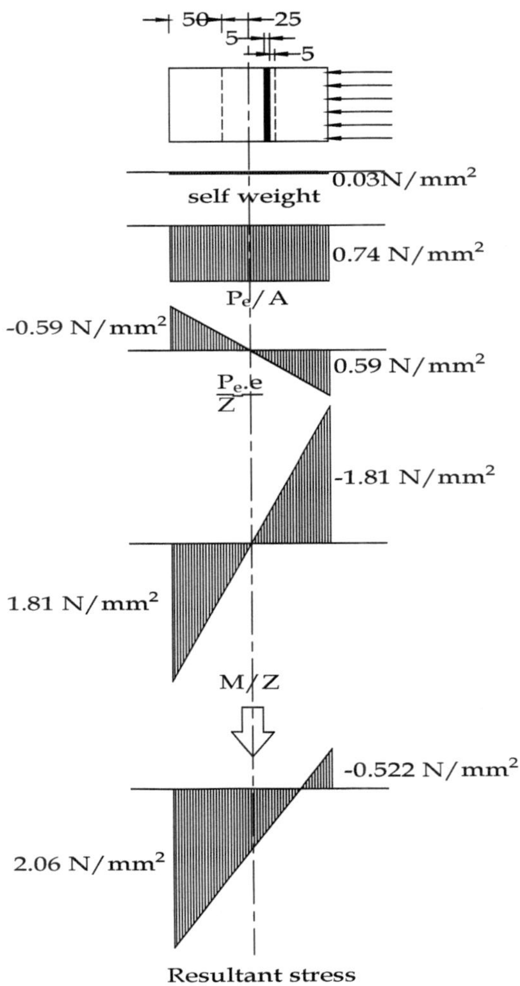

Fig. 8. Stresses on the post-tensioned masonry retaining wall

3.3.5 Casting of Masonry Blocks

Blocks are manufactured using the hydraulically operated block making machine with the cement: quarry dust ratio of 1:5 (Fig. 9). The machine is originally designed and manufactured for cement: soil block making. Since the cement: quarry dust mix is less compressible compared to cement: soil mix, the volume of mix used to manufacture the blocks has to be carefully selected.

Fig. 9. Casting of masonry blocks

3.3.6 Curing of Masonry Blocks

Blocks are kept under a polythene cover and spray water frequently for 14 days for curing. Blocks are stored another 14 day to allow for shrinkage (Fig. 10).

Fig. 10. Curing of masonry blocks

3.3.7 Preparation of Site for Construction

The existing embankment of approximate width of 1.0 m to 1.2 m is excavated up the boundary using an excavator. Then the foundation trench was excavated (Fig. 11) as per the structural detail drawing.

Fig. 11. Excavation of embankment and foundation trench

3.3.8 Construction of Foundation

The insitu concrete foundation base and wall up to the ground level was constructed with necessary reinforcement arrangements placed. Vertical and horizontal ducts which are required to insert High Tensile Steel wires are kept in the exact required positions. PVC pipes and electrical conduit pipes are used to keep ducts in insitu concrete (Fig. 12). Completed RCC foundation and RCC wall is shown in Fig. 13.

Fig. 12. Positioning of vertical & horizontal ducts

Fig. 13. Completed RCC foundation and wall up to ground level

3.3.9 Construction of Masonry Wall

Interlocking masonry block wall is constructed using normal construction procedure as shown in Fig. 14. Cement slurry is used to paste the interlocking blocks. A reinforced concrete tie beam is constructed over the masonry block wall (Fig. 15). Block wall is allowed a period of curing of 14 days.

Fig. 14. Construction of masonry wall

Fig. 15. Casting of tie beam

3.3.10 Inserting the H.T.S. Wires

High Tensile Steel (HTS) wires cut to the exact measurement are inserted through the duct from the top and fixed against the RCC wall using the anchor cone and the pin fixed permanently at the bottom.

3.3.11 Grouting the Ducts

The ducts into which the H.T.S. wires are inserted are filled with cement grout to avoid corrosion. Grouting is done before the anchor cone and pin is placed to the jacking end as shown in Fig. 16.

Fig. 16. Grouting the ducts

3.3.12 Post-tensioning of the Wall

H.T.S. wires of 5 mm diameter with characteristic tensile strength 1670 N/mm^2 is used in post-tensioning the masonry wall. Anchor cones and pins which can accommodate 2 H.T.S. wires are used to hold the wires in position in the fixed end as well as the jacking end. 10 mm thick mild steel bearing plates of size 75 mm × 75 mm are placed in between the cone and the concrete to facilitate smooth transfer of stresses. The wires are then are stressed to 24 kN each using the hydraulic jack system and the jacking force is transferred to the permanent anchor cone and pin as shown in Fig. 17.

Fig. 17. Stressing the H.T.S. wires

The basic stress-strain relationship

$$E = \frac{Fj/A}{e/l}$$

is used to calculate the elongation relevant to the jacking force with following values.

E = Young's modulus (205 kN/mm^2)
F_j = Jacking Force (24 kN)
A = Cross sectional area of the tendon ($\pi \times 2.5^2$)
l = Initial length (2.35 m)
e = Elongation (12 mm)

After the H.T.S. wires are stressed, the anchor cone and the pin are fixed permanently. The steel bearing plates, anchor cones and pins at both the jacking end and the fixed end are covered with the 1: 2 cement sand mortar to prevent corrosion as shown in Fig. 18.

Fig. 18. Covering of anchor cones and pins

3.3.13 Back Filling of Foundation and the Embankment

The excavated trench was backfilled with good quality excavated soil and compacted in 300 mm thick layers with a soil compactor as shown in Fig. 19.

Fig. 19. Backfilling the foundation

Dial gauges were positioned as shown in Fig. 20 and the gap between the retaining wall and the embankment was filled with selected excavated soil. The short term deflection of the P.S.C. wall panels were measured in 300 mm intervals.

Fig. 20. Monitoring the deflection

4 Results

4.1 Short Term Deflection of the Wall

Readings of two dial gauges were recorded and the average of these readings is tabulated in Table 1.

Table 1. Deflection against fill height

Fill height (mm)	Defection (mm)
0	0
300	0.1
600	0.5
600	1.0
1200	1.3
1500	1.7
1800	2.0
2100	2.2

The verticality of the wall is measured with a plumb bob and recorded to check the long term deflection or rotation. The distances to the adjacent building from the RCC part of the retaining wall is measured with a steel tape and recorded to check the long term sliding.

4.2 Results of Cost Compression

The cost of RRM, RC and Pre-Stressed Concrete (PSC) retaining walls of two different heights were calculated and compared with the PTM retaining wall with the similar height and tabulated in Table 2. The structural detail for RRM retaining wall is obtained from typical RRM sections designed by the Road Development Authority of Sri Lanka and the structural design for RC is done by computer software "Prokon". The cost of a PSC retaining wall is obtained from the previous study of the author on "Development of Pre-stressed Concrete Retaining Wall System [3]. The parameters described in Sect. 3.2 above is considered for design of all Retaining walls.

Table 2. Cost comparison for 3 m wall length

Clear wall height (m)	Cost (Rs.) and wall width (mm)			
	RRM	RC	PSC	PTM
1.8	80,000 1050	115,000 200	95,000 100	80,000 150
3.0	140,000 1350	185,000 300	160,000 175	135,000 150

5 Conclusion

PTM retaining wall system can be used as a competitive alternative to the traditional methods of retaining wall construction.

Though the cost of a PTM retaining wall is similar to that of an RRM retaining wall, it needs only 15% of land width when compared to RRM. Hence it is suitable to replace an RRM retaining wall with a PTM retaining wall in places where the land value is critical.

The cost of a PTM retaining wall is 25%–30% less than that of an RC retaining wall and the width of land requirement is less than that of an RC.

As the masonry blocks used to construct the retaining wall is manufactured using the hydraulic cement: soil block making machine, manufacturing of masonry blocks can be done without additional cost for purchasing machines.

The aesthetic appearance of PSC retaining wall is high compared to traditional retaining walls.

Since manufacturing of masonry blocks and construction of foundation can be done in parallel, the construction time can be reduced as well.

6 Future Research

The height of the post-tensioned masonry retaining wall can be increased with a specially designed masonry block with reduced cross sectional area, increased sectional modulus (Z) and the possibility of increasing the eccentricity (cellular block).

Acknowledgement. The author wishes to acknowledge the support given by the Technical Officer, Mr. A.D.A.R. Premawansa in manufacturing of masonry blocks and construction of pilot project in clinical building premises at Dalupitiya.

References

1. Curtin WG, Shaw G, Beck JK, Howard J. Design of post-tensioned brick work: the brick development association, January 1989
2. Ganz HR (2002) Post-tensioned masonry around the world. In: Post-tensioning institute engineering conference, San Antonio, TX, 5–7 May 2002
3. Modulus. Society of Structural Engineers Sri Lanka **27**(01), 8–13 (2013)

Roads for Sustainable Cities – What Is Wrong with Us?

K. S. Weerasekera(✉)

Department of Civil Engineering, The Open University of Sri Lanka, Colombo, Sri Lanka
ksweee@ou.ac.lk

Abstract. When 'roads for sustainable cities' are to be considered, firstly, the sustainability of roads should be brought into discussion. As an initiative, a road user behaviour study was conducted in a busy suburban area outside South-Eastern city limits of Colombo. The study area consisted of heterogeneous vehicle composition; typically combined with high proportion of motorcycles and three-wheelers which are poor in adhering to road rules that can commonly be seen in most South Asian developing countries. Although heavy investments are utilized in road improvement projects in these countries, the outcomes gained often seems to be unsatisfactory. The intent of this study was to find out the reasons for this unsatisfactory outcomes of the investments spent on road improvement projects and suggest measures to overcome these unsatisfactory outcomes for sustainability. The study was conducted in two parts; (i) road user behaviour study, and (ii) field consultation study. Through the road user behaviour study, the undesirable practices of all road users such as passenger vehicles, motorcycles, three-wheelers, buses, trucks, pedestrians and excessive noise were directly identified. It was seen from the consultation study results that the varied expectations of different categories of road stake holders were different from category to category and biased towards their advantage. Finally, through this study it was observed that the maximum benefits expected from investing on road improvement projects are not fully realized when poor road user practices are adopted by the stake holders'.

Keywords: Poor road practices · Colombo urban roads · Road indiscipline

1 Introduction

This study was conducted on roads just outside Colombo city limits to find out the negative impacts on urban roads due to poor road user practices. It was suggested that due to non-homogeneous vehicle composition; especially with high proportion of motor-cycles and three-wheelers combined with other motorized vehicles may result in the dual problem of poor traffic flows and high rate of road risk on urban roads (Howorth, 1995). This was substantiated from the current study conducted around the relatively busy suburban areas of Nawala and Nugegoda (i.e. outside South-Eastern city limits of Colombo, Sri Lanka). It was observed that the inefficient use of the road systems and poor road user behaviour is a major contributor to traffic problems, even where there is no shortfall in road capacity. Hence, this study justifies the views of

R. Dissanayake and P. Mendis (Eds.): ICSBE 2018, LNCE 44, pp. 411–418, 2020.
https://doi.org/10.1007/978-981-13-9749-3_36

some recent studies conducted by (Edirisinghe 2011; Jayatilake et al. 2017 and Dharmakeerthi and Sathyaprasad 2017) which indicate that heavy investments spent on road improvements are not fully harnessed unless road user behaviour practices are improved.

The area indicated in Fig. 1 was selected as the area of study which is the South-Eastern city limits of Colombo. Figure 2 indicates some typical sections of roads in the study area. During this study, a number of road users who have been accustomed to the area for the last 25 years, consisting of drivers, residents, businessmen, daily office travellers and pedestrians were consulted. Consultations consisted of question and answer sessions posing different types of questions regarding road user behaviour patterns of multitude of road users who are familiar with the study area.

2 Methodology

Currently, road safety is a complicated issue due to the increase of road users over time and the improvement of vehicle engine efficiencies. The problem of accidents has become more severe in highway engineering. Hence, the responsibility of traffic engineers to carry out systematic studies of traffic accidents and to find the main causes for accidents and finally come-up with preventive or remedial measures is important.

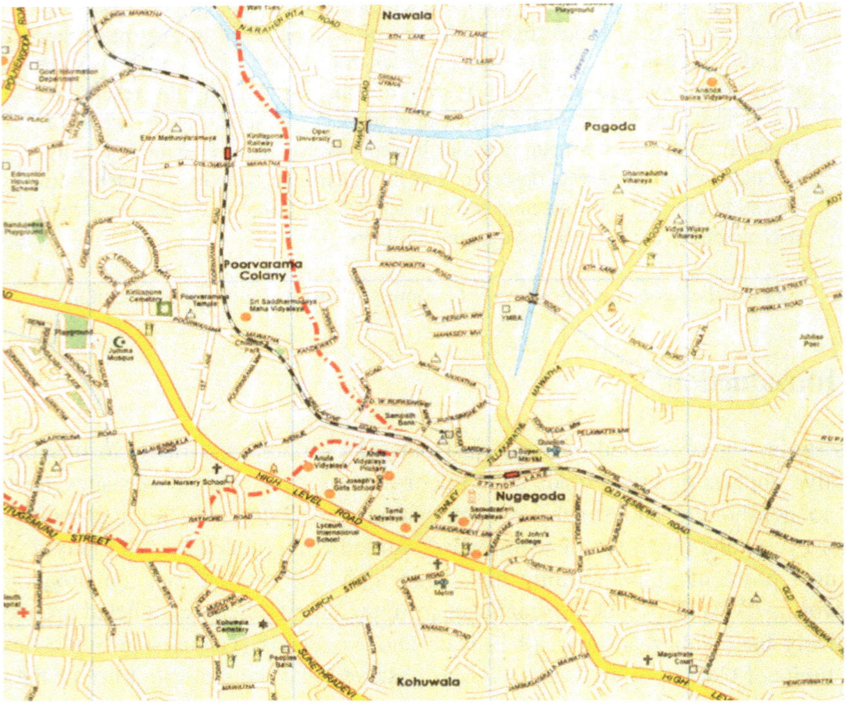

Fig. 1. Road network of the study area (South-Eastern city limits of Colombo)

Fig. 2. Typical road sections of the study area

Adhering to road rules and improved road discipline helps greatly in improving road safety standards. The level of safety of a road system is influenced by the behavioural practices of all road users, not only driver behaviour alone. The other types of road users in addition to standard passenger cars and SUV drivers include motor-cycle riders, three-wheel operators, bus and truck drivers and pedestrians. The traffic composition of Colombo is different from most developed countries due to its extremely heterogeneous vehicle fleet and very high proportion of motor-cycles and three-wheelers. Hence, all road users and their behavioural practices had to be studied carefully to improve road safety and road efficiency under the existing extremely complicated situation. This has observed by Roberts and Simmonds (1995) and Howorth (1995) in their work overseas. The current study tends to observe the local misbehaviour patterns of the local stake holders.

2.1 Road User Behaviour

An observational study was performed to observe and determine road user behaviour operations. Twelve observation locations were selected as indicated in Fig. 1 which is distributed along the road network under study. The observations carried-out were initially focused on studying the behaviours reflecting violations of road rules and other dangerous road behaviour practices. The road user characteristics which were carefully observed were; lane discipline, roadside parking and friction, turning movements, failure to giveaway, irregular lane swapping, bus driver arrogant behaviour and not properly stopping at bus-stops, road rage by large vehicles and the pedestrian movements. Each study location (see Fig. 1) was under observation for the duration one-hour during busy daytime.

2.2 Field Consultations

In addition to the road user behaviour study, a field consultation study was conducted for a total study sample of 500 consultations, which consisted of 100 drivers, 100 local residents, 100 local businessmen, 100 office travellers and 100 pedestrians. Consultations were conducted at random locations in the study area. A questionnaire survey was executed in the field consultation study.

The questionnaire was focused on:

- Sufficiency of surrounding road network
- Adequacy of pedestrian facility
- Adequacy of vehicle parking facilities
- Safety of roads
- Road user behaviour
- Necessity of road widening

3 Results

3.1 Road User Behaviour Study Results

Through field observation studies conducted at 12 randomly selected locations in the study area, different types of adverse road-user behaviour practices of different categories of road-users were identified and are categorized as indicated below.

All Vehicles

- Reluctance to giveaway at un-signalised junctions and small roundabouts
- Reluctance to giveaway at prioritized junctions
- Unauthorized roadside parking
- Unnecessary lane changes and overtaking
- Obstruction to overtaking

Motor-cycles

- Pillion passengers not wearing helmets
- Squeezing through other piled vehicles
- Keeping too close to other vehicles
- Riding on sidewalks
- Excessive speeding every so often
- Cutting across lanes dangerously
- Riding while talking on mobile phones
- Overtaking from the left
- Excessive noise
- Not using indicators during lane changing or turning

Three-wheelers

- Not sticking to lanes
- Sudden lane changes
- Excessive noise
- Travelling on the wrong side
- Cutting across lanes dangerously
- Overtaking from the left
- Keeping too close to other vehicles
- Riding on sidewalks
- Not using indicators during lane changing or turning

Buses

- Blocking other vehicles
- Not always sticking to assigned bus stops
- Overstaying at bus stops
- High speeds
- Travelling on two lanes
- Dangerously sudden stops
- Not using indicators during lane changing or turning
- Excessive use of loud horns

Trucks

- Blocking other vehicles
- Aggressive driving
- High speeds
- Travelling on two lanes
- Not using indicators during lane changing or turning

Pedestrians

- Often not using footpaths
- Jaywalking
- Not using zebra crossings
- Not waiting for green phase for crossing at signalized junctions

Excessive Use of Horns
Excessive use of horns (sometimes very loud) very frequently by Colombo drivers is frustrating. When closely studied, it was found that the reasons are as follows:

- To alert slower vehicles to make way
- To alert pedestrians on the road of approaching vehicles
- To express annoyance
- To facilitate overtaking
- To alert motor-cycles and three-wheelers
- By three-wheel operators to mark their presence

4 Discussion

It is clear from Fig. 3A that drivers are most unhappy about the inadequacy of surrounding road network and closely followed by office travellers. On contrary it is observed that pedestrians are happy about the surrounding road network. It also observe that residents and local businessmen too are generally satisfied with the adequacy of the surrounding road network.

From Fig. 3B it is seen that pedestrians are generally worried about the inadequate pedestrian facilities available. Generally, residents too are not very satisfied about the adequacy of pedestrian facilities that are available currently. On the contrary, drivers are least worried about the availability of pedestrian facilities.

Figure 3C clearly indicates the unhappiness of local businessmen and drivers about the inadequacy of parking facilities. It's also clear that even the local residents expect a higher level of parking facility in the area. It also could be noted that pedestrians and office travellers are not bothered about parking facilities available in the surrounding areas.

From Fig. 3D it is indicated that except pedestrians, all the other categories are generally satisfied about the safety level of these urban roads. Majority of the pedestrians do conclude that the roads are not safe.

Figure 3E indicates that it is mainly drivers and residents who complain about the unsatisfactory level of road user behaviour, but local businessmen and office travellers are not very concerned about road user behaviour. The pedestrians have an opposite view; that they think the road user behaviour is satisfactory from their point of view.

From Fig. 3F it is interesting to note that greater majority of residents and local businessmen are not in favour of widening the roads fearing that their land will get affected. But on contrary majority of all the other categories do agree in widening the roads.

It could be seen from the consultation study results that the expectations of different categories of stake holders' were different from category to category probably biased towards their advantage. For instance, many drivers and travellers preferred road widening to be carried-out, but residents and road side business community were against road widening since they were going to be affected. It also could be seen that pedestrians were unaware of the road capacity situation, but they were worried about the lack of pedestrian facilities in the area. When it comes to adequacy of parking in the area, mainly the business community and drivers were not happy about the inadequacy in parking availability although travellers and pedestrians were least bothered about parking facility in the area. It can be seen that generally all the categories of road users were concerned about road safety, especially pedestrians and drivers who had a high priority concern for improving road safety. Most of the drivers were not satisfied with the behaviour of the road users in general. In this regard, pedestrians were the least bothered.

5 Conclusions

Road safety depends when road user behaviour is combined with vehicle and road characteristics. Hence, road user behaviour of vehicle drivers is due to; lack of understanding of road signs and some carriageway markings, lack of knowledge of priority rules, right of way concept not being properly understood and practiced, reversing to main roads, obstructing the main flow, roadside parking causing side friction, and road raging by bus and heavy vehicle drivers. It could be clearly seen that efficiency of a road depends very much on smooth traffic flow along the roads with; (i) least side friction along mid-blocks, and (ii) orderly maximum throughputs at intersections. The main reasons behind the increase in side-friction and flow in mid-blocks is due to unnecessary lane changing on roads, roadside parking and slowing down due to business activities along the sides of main roads, vehicles cutting across lanes and parking along opposite side of the road and buses overtaking and competing with each other.

The road regulations and conditions that the road users have to adhere are; condition of vehicle (road worthiness), speed limits in urban areas, give-way and priority regulations, adopting correct overtaking practices, buses sticking to assigned bus stops, proper post-accident practices, correct parking practices, sticking to lane discipline and strictly adhering to obligations at pedestrian road crossings.

Addressing all those issues scientifically is vital for the sustainability of roads for sustainable cities. Finally, through this study, it was observed that the maximum benefits expected from investing on road improvement projects are not fully realized when poor road user practices are adopted by the stake holders.

Acknowledgements. Author wishes to acknowledge all who patiently took part in the consultations which consisted of drivers, residents, businessmen, daily office travellers and pedestrians who were familiar with the study area that supported this study with their valuable views and suggestions.

References

Bandara JMSJ (2011) Need for an integrated highway management system: Developments in Sri Lankan context. In: Proceedings of the Joint International Symposium on Social Management Systems (SMSS-2011), pp. 700–703 Colombo, Sri Lanka

Dharmakeerthi UGS, Sathyaprasad IMS (2017) Development of trip attraction rates and parking standards for supermarkets in Kandy area. In: Proceedings of the 5th International Symposium on Advances of Civil and Environmental Engineering Practices for Sustainable Development (ACEPS-2017) March 2017, pp. 198–205. University of Ruhuna, Galle, Sri Lanka

Edirisinghe JE (2011) Driver behaviour at pedestrian crossings. In: Proceedings of the Joint International Symposium on Social Management Systems (SMSS-2011), pp. 696–700 Colombo, Sri Lanka

Howorth N (1995) Road user behaviour and safety in a developing country. In: Proceedings of the 7th World Conference on Transport Research (WCTR), July 1995, pp. 302–306, Sydney, Australia

Jayatilake PASR, Premachandra BMNM, Wickramasinghe WMVSK (2017) Variation of the start-up time of pedestrians with their familiarity to the signal phase. In: Proceedings of the 5th International Symposium on Advances of Civil and Environmental Engineering Practices for Sustainable Development (ACEPS-2017) March 2017, pp. 206–212. University of Ruhuna, Galle, Sri Lanka

Roberts M, Simmonds DC (1995) A strategic modelling approach for urban transport policy development. In: Proceedings of the 7th World Conference on Transport Research (WCTR), July 1995, pp. 285–289, Sydney, Australia

A Sustainable Green Façade Retrofit for An Iconic Built Structure: A Case Study at SAINTGITS

Jacob Thottathil Varghese[1,2(✉)], Sajan Thomas[1], and Joselin Herbert[3]

[1] SAINTGITS College of Engineering, Kottayam, Kerala, India
dr.jacobtvarghese@gmail.com
[2] VIT Vellore, Vellore, Tamil Nadu, India
[3] NICHE, Kumaracoil, Tamil Nadu, India

Abstract. The depleting fossil fuel reserves and the constantly surging levels of atmospheric pollution have made the architects and building owners join hands to foster sustainable initiatives. A modest experimentation in a sustainable domain had been inculcated at SAINTGITS, a much sought academic institution at Kottayam, in the state of Kerala, India. The campus houses multiple high rise buildings and a plenty of other small structures. The iconic structure, the Central Lecture Complex (CLC) exhibits its uniqueness with an arc shaped plinth which houses 3 stories to accommodate several conventional lecture rooms, faculty cabins, sophisticated seminar halls and other essential amenity spaces. It provides natural shade to the SAINTGITS Amphi theatre, structured and maintained at the eastern side. The proposed green façade retrofit on to the East and the West facing walls of CLC has improved the indoor operative temperature and hence, an approximate 15% reduction has been estimated in the building cooling load. An additional 2 m high temporary parapet wall has been recommended on to the existing CLC to provide extended shading hours to the SAINTGITS Amphi, right from 14:15 h onwards. The comprehensive acoustic modelling of the latter has given promising results for the Reverberation Time (RT_{60}), which is on par with the international standards for performance spaces - ISO 3382-1 and ASTM E2235. The analyses inferred that the state-of-the art design of the existing CLC and the proposed green façade retrofits will reduce the entire energy demand by 20% and it certainly will provide guidelines to similar structures at tropical metropolises.

Keywords: Green façade · Sustainable retrofits · SAINTGITS Amphi · Operative temperature · Built structure · Energy analyses

1 Introduction

The industrial revolution has turned the metropolises denser and made way for the evolution of new ones. Massively Built structures, unavailability of free ground space, energy crunch, traffic density, rising city heat and the alarming levels of atmospheric pollution are some of the major concerns that seek urgent intervention (Jacob et al. 2015). High dependence on fossil fuel has made the scientific community to deliberate on alternate sustainable initiatives (Takakura et al. 2000; Jacob et al. 2018). One such

R. Dissanayake and P. Mendis (Eds.): ICSBE 2018, LNCE 44, pp. 419–432, 2020.
https://doi.org/10.1007/978-981-13-9749-3_37

alternative is the implementation of sustainable retrofits on to the existing high rise built structures. These retrofits are trellises of green façades, either draped directly on to the outer fabric of built structures to absorb the direct radiation or erected alongside busy pavements to facilitate the absorption of the atmospheric pollutants (Yang et al. 2008; Jacob et al. 2010, 2015) (Fig. 1).

Fig. 1. A fully grown Green façade trellis (almost 100 m length) alongside the Main Central Road, Kerala state, India.

The former initiative helps to convert 60% of the absorbed radiation into latent heat (Campbell et al. 1998). Hence, the thermal insulation and noise attenuation of the building can have improved (Pal et al. 2000; Wolverton et al. 1989; Jacob et al. 2015). For every 1 °C dip in the operative temperature, the energy load of the respective built structure reduces by 12–15% (Stec et al. 2005; Jacob et al. 2018). Green façades having larger Leaf Area Index (LAI) cools the building more efficiently (Fig. 1) and at least 35% deduction can be expected in the annual cooling load (Wong et al. 2010; Alexandri and Jones 2008)

A modest initiative has been proposed at SAINTGITS, a renowned educational institution in India, to improve the indoor operative temperature. The SAINTGITS group of Institutions, founded in the year 2002, has its lush and verdant campus located at Central Travancore (9.509^{0} N and 76.551^{0} E), Kerala state, India (Fig. 2).

Fig. 2. Areal view of the SAINTGITS group of professional institutions.

2 Methodology

Among the numerous multi-storied structures at SAINTGITS campus, the Centre Lecture Complex (CLC) and the SAINTGITS Amphi are the state-of-the art built structures. Whilst both the structures have arc shaped plinths, the latter is naturally carved as multiple stairs on the hilly terrain (Fig. 3).

Fig. 3. The arc shaped Centre Lecture Complex (CLC) and the SAINTGITS Amphi

At the initial phase, we were successful in preparing and analysing an architectural model which replicated the entire CLC. The simulation results were at par with the previous similar studies (Jacob et al. 2015) and were later validated by conducting real time energy audits.

The next stage of the work estimated the benefits that could be attained, if a suitable green façade was retrofitted on to the East and West facing walls of the CLC building.

The third and the fourth phases contain detailed daylight and shadow analyses, performed exclusively for the entire CLC. Whilst the conclusions from the latter rendered an insight in to the real hours of occupant comfort at the SAINTGITS Amphi, the former had explicitly revealed the ability of CLC's first and the second floors to harness the maximum daylight.

The final phase incorporated the acoustic analyses at SAINTGITS Amphi. The reverberation time was estimated by recreating an architectural model of the entire acoustic space.

3 Results and Discussion

The exclusive analyses, the first of its kind at SAINTGITS, were done for the entire CLC and SAINTGITS Amphi. For the ease of discussing in detail, the entire results are included into five - Subsect. 3.1 whole building's energy load calculation, Subsect. 3.2 attainable dip in operative temperature of a fully air-conditioned & Green façade retrofitted CLC block, Subsect. 3.3 harnessing the natural light source - daylight analyses of the CLC block, Subsect. 3.4 extended occupant comfort at the SAINTGITS Amphi and Subsect. 3.5 estimating the reverberation time at SAINTGITS Amphi.

3.1 Whole Building's Energy Load Calculation

The entire CLC is a full-fledged 3 storied structure with a temporary third floor (Fig. 4).

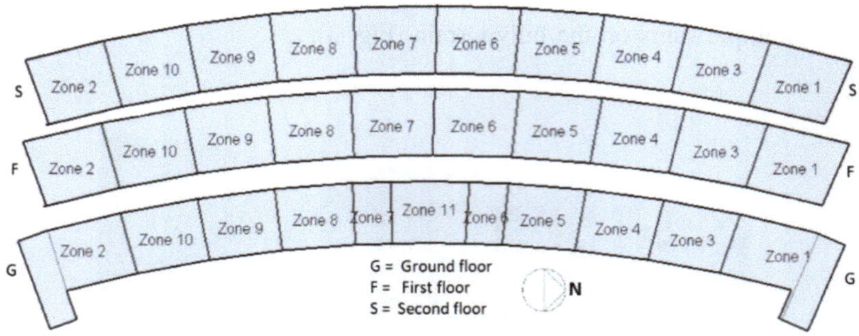

Fig. 4. Spatial arrangement of various zones at the individual levels of the Centre Lecture Complex (CLC).

The ground floor has 11 zones. The first and the second floors share equal number of zones (10 zones per floor) (Fig. 4). All zones in the ground and the topmost floor are earmarked as class rooms (capacity 72 seats in each zone) to accommodate the

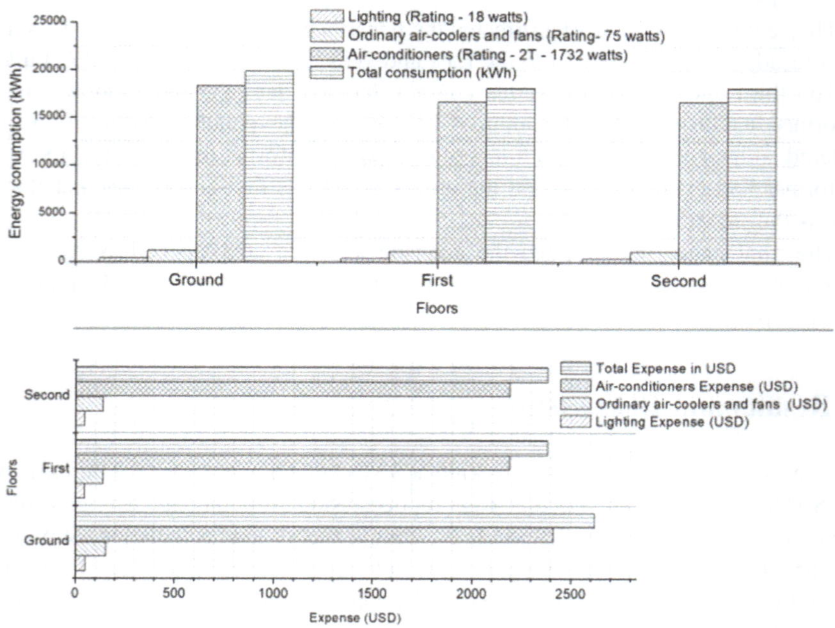

Fig. 5. A stark comparison (floor wise) of the various energy consuming sources and the related monetary outflow in USD – monthly data for the entire CLC block.

sophomores, juniors and senior students. The first floor serves as exclusive zones for two major engineering departments–the Mechanical Engineering and the Electrical Engineering.

Real time energy audits of all the zones were conducted. Monthly analysis of the audit summary revealed that the artificial lighting consumed the least energy units (1071 units) (Fig. 5).

Whilst the energy expended by the ordinary air-coolers and fans was 3348 units per month, a proposed fully air-conditioned CLC's monthly cooling load was approximately 51544 units; 15 times more than the energy expended to operate all ordinary cooling equipment (Fig. 5). Since the first and the second floors share almost the same physical geometry, the monthly power consumption (cooling load alone) of both the floors together is nearly 4400 USD.

Since the combined energy expended by the cooling devices (ordinary air-coolers, fans and air-conditioners) is significantly higher (nearly 10 times) than the artificial lighting load, it is advisable to inculcate reliable and sustainable initiatives to assuage the cooling load.

3.2 Attainable Dip in Operative Temperature of a Fully Air-Conditioned and Green Façade Retrofitted CLC Block

Year after year, as the tropics get warmer, a fully air-conditioned CLC block will rather become essential than luxury. Though it increases the operative cost, to a certain extent, it can be negated by a custom designed green façade retrofits. Earlier experimental and simulation studies on two plant species (*Vernonia elaeagnifolia* and *Passiflora caerulea*) at Vellore Institute of Technology (VIT, Vellore, Tamil Nadu, India) proved that these green drapes can bring down the indoor operative temperature of built structures (Jacob et al. 2011, 2015). Particularly, in summer, the ambient air temperature near the vegetated façade dropped by 1 °C to 3 °C (Hoyano 1988).

The entire Centre Lecture Complex (CLC block) with pitched roof was modelled using the most modern and versatile architectural software, the DesignBuilder (Fig. 6).

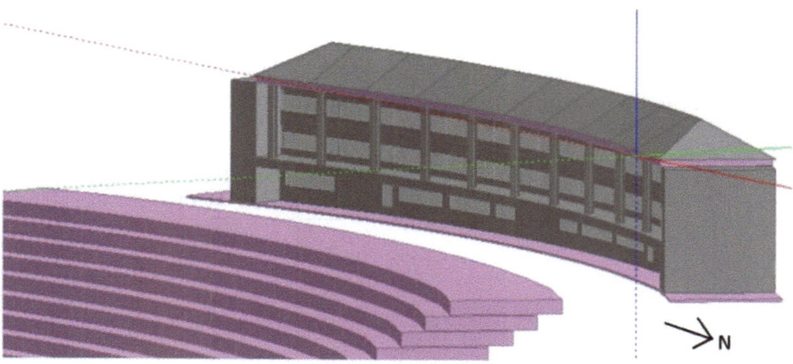

Fig. 6. A comprehensive architectural model of the Centre Lecture Complex (CLC block), modelled in the DesignBuilder.

The energy calculations were done according to the ASHRAE standards by directly plugging in the pre-written energy analyses codes from EnergyPlus, a building energy simulation program, developed by the National Renewable Energy Laboratory (NREL), Colorado, USA.

The East and the West facing walls of the CLC are fully retrofitted by trellises of green façades. The plant species used is a mix of *Vernonia elaeagnifolia* and *Passiflora caerulea* – whilst the former can be draped on high rise structures, the latter is efficient in removing the atmospheric pollutants, hovering at lower heights – roughly up to 4 m (Jacob et al. 2015).

A stark comparison of the average leaf area of the two different plant species shows that a 5 lobbed leaf structure covers nearly twice the area than a single lobbed structure (Fig. 7).

Fig. 7. Comparing the average leaf area of a five lobbed leaf structure (average leaf area = 16 cm^2) and single lobbed *Vernonia elaeagnifolia* (average leaf area = 8 cm^2).

The input parameters of the model are the façade dimensions, building orientation, the weather data file and the U-values. For both the plant species, the U value is taken as 0.192 W/m^2K, which is much closer to the value of straw bale (Jacob 2017; Behzad et al. 2011). The weather data for Trivandrum was used and it was a reasonable approximation for Kottayam.

The operative temperature, which is a direct indication of the indoor occupant comfort, is governed by the air temperature, mean radiant temperature and the air speed - see Eq. (1).

$$\textbf{Operative tempeature} = \frac{\left[\mathbf{t_r} + \left(\mathbf{t_a}\sqrt{\mathbf{10v}}\right)\right]}{\left[\mathbf{1} + \sqrt{\mathbf{10v}}\right]} \tag{1}$$

Where

t_r = air temperature (degree C)
t_a = mean radiant temperature (degree C)
v = air speed (m/s)

Hence, if the air speed (v) is 0.1 m/s, the analytical expression for the operative temperature attains a further simplified form – see Eq. (2).

$$\textbf{Operative tempeature} = \frac{[t_r + t_a]}{2} \tag{2}$$

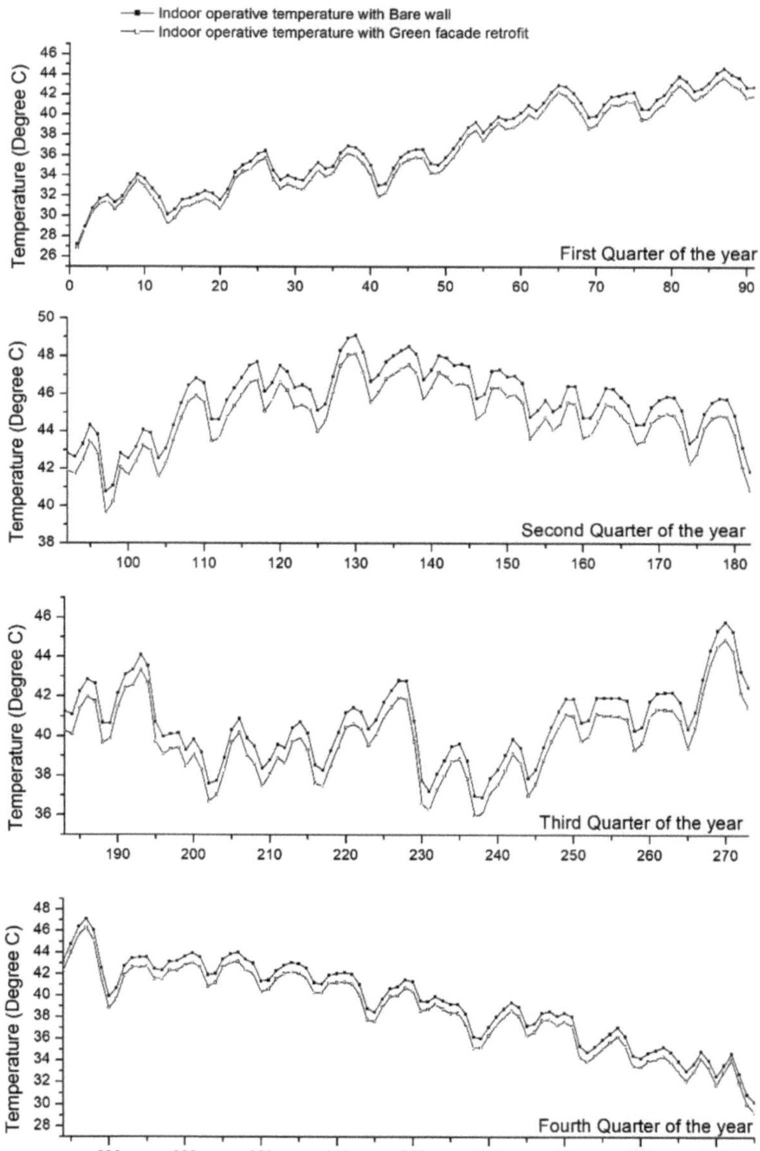

Fig. 8. Indoor operative temperature for a Bare wall vs Green façade retrofitted wall – general trend during the various quarters of the year.

The DesignBuilder model was run for all the days of the year (four quarters) with constant air speed 0.1 m/s. Whilst the operative temperature exhibits mutually opposing trends for first and the last quarters respectively, the trend almost plateaus for both the 2nd and the 3rd quarters (Fig. 8). It is validated by the general perception of the quarterly temperature variation that occurs every year - the first and the last quarters are cooler than the remaining.

It was observed that, throughout the year, with a green façade retrofit on the outer fabric, the average indoor operative temperature dipped by approximately 1 °C than the respective bare wall (Fig. 9).

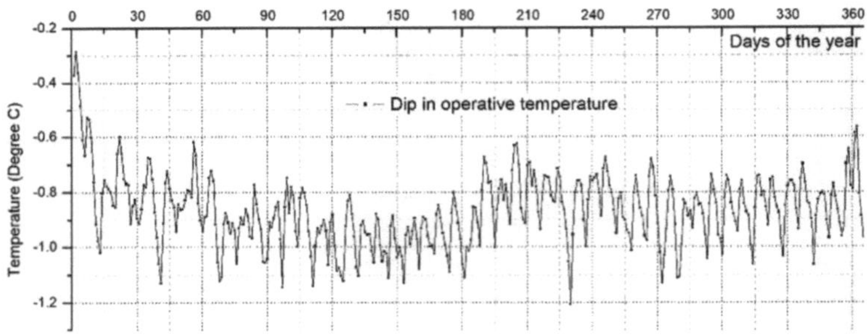

Fig. 9. Dip in the operative temperature experienced due to the green façade retrofit.

To analyse the trend of dip in operative temperature for different values of air speed (v), the data has been plotted for the 15th and the 30th day of every month (Fig. 10).

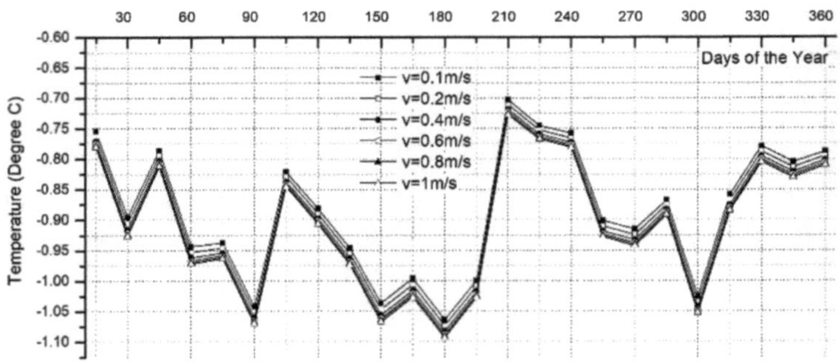

Fig. 10. Pattern of the dip in the operative temperature for different values of air speed.

It is inferred that the increase in air speed improves the air change/hour (ac/h) inside the respective zones and thus, more dip in the operative temperature is experienced. If the air speed is 1 m/s, a maximum dip of 1.1 °C is attainable with a fully grown green façade retrofit.

Based on the earlier similar studies, if the operative temperature drops by more than 1 °C, a modest energy saving of 15% is achievable (Stec et al. 2005; Jacob et al. 2018). Hence, a monthly saving to the tune of 660 USD is reachable with a suitable green façade retrofitted CLC block.

3.3 Harnessing the Natural Light Source – Daylight Analyses of the CLC Block

Though the energy expended for providing artificial lighting is the least among all the other expended categories, it cannot be neglected. The daylight analyses show that the interiors of every zone receive 50% less illuminance than the exteriors (Fig. 11). Hence, throughout the working hours, the existing design of all the zones of the CLC demands artificial lighting. The zones in the ground floor are deprived of natural daylight than the first and the second floors (Fig. 11). The close proximity of the ground floor to a confined veranda is one of the major reasons.

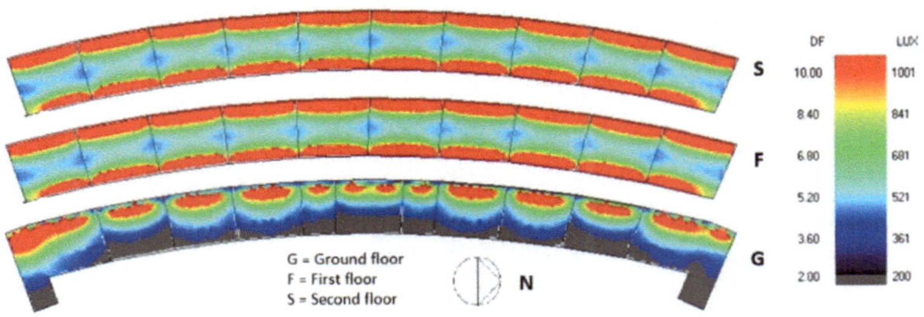

Fig. 11. Daylight analyses of the various zones – arranged floor wise.

By extrapolating the lighting demand, if the artificial lights are assumed to be functioning throughout the day, the monthly lighting load for the entire CLC sums to an approximate amount of 450 USD (31,000 INR). The analyses revealed that by providing sufficient windows at the East facing wall, the artificial lighting load can be reduced by 1/3rd from the existing.

3.4 Extended Occupant Comfort at the SAINTGITS Amphi

The first floor's balcony of the Centre Lecture Complex (CLC) provides extended seating for SAINTGITS Amphi (Fig. 3). During sunny days, the occupant comfort at the latter is challenging – especially during the forenoon hours. Until the afternoon of all the days, the entire amphitheatre receives direct sunlight. The annual sun path diagram explicitly reiterates the fact (Fig. 12).

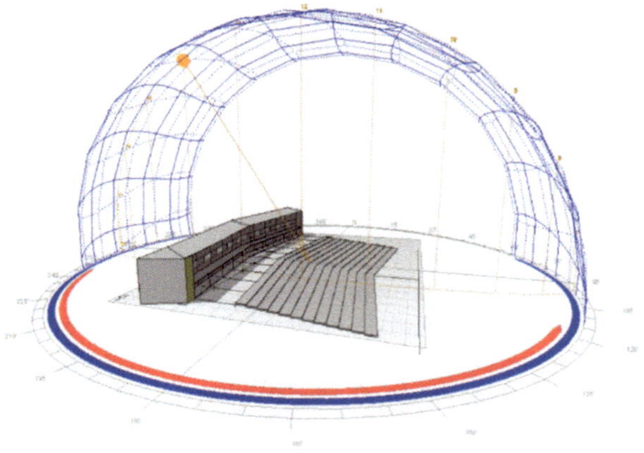

Fig. 12. The annual sun path diagram over the Centre Lecture Complex (CLC Block) and the SAINTGITS Amphi.

Hence, the complete utilization of the amphitheatre could not be claimed during the entire active hours of the day (Fig. 13).

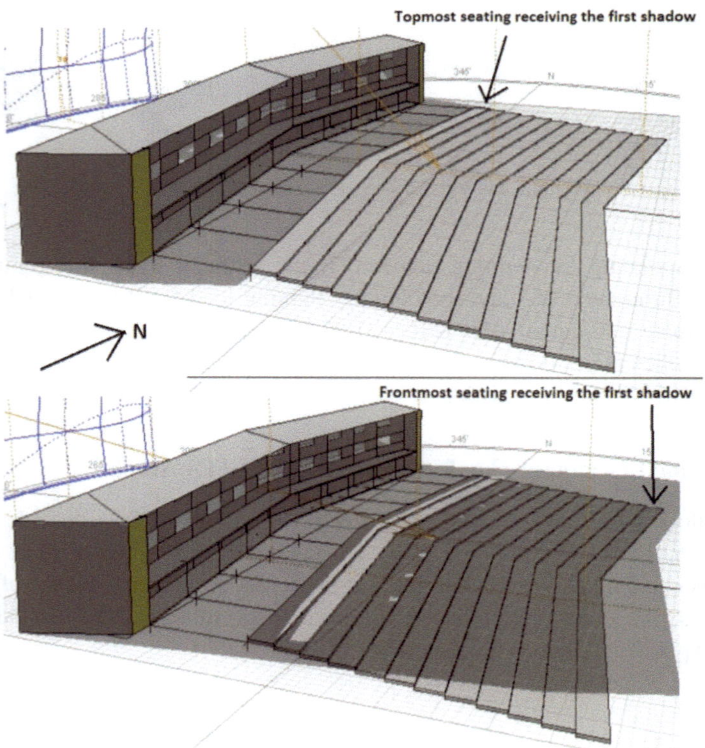

Fig. 13. The extent of shadows cast by the CLC block on SAINTGITS Amphi at 3:10 pm and 5:15 pm.

The shadow details collected for different quarters reveal that the topmost seating at the amphitheatre receives the foremost comfort shadow of CLC at 15:00 h (3.00 pm) - during certain quarters, it even extends to 3:30 pm. The front most row of the amphitheatre receives the shadow, not earlier than 5:15 pm on any of the sunny days (Fig. 13).

The data in Table 1 indicates the respective timings at which the occupants at the amphitheatre receive the benefits of CLC's shadow. The attainable comfort duration also is appended (Table 1).

Table 1. Occupant comfort – the extended duration of attainable benefits at SAINTGITS Amphitheatre.

Month & quarter of the year	Time of receiving the first shadow at the topmost row (pm)	Time of having complete occupant comfort at all the rows of SAINTGITS Amphi (pm)	Time and attainable extended benefits/gains with 2 m high parapet wall above CLC	
			Shadow from (pm)	Gain
Jan (Q1)	3:30	5:15	2:45	45 min
April (Q2)	3:30	5:30	3:00	30 min
July (Q3)	3:30	5:45	3:15	15 min
Oct (Q4)	3:00	5:15	2:15	45 min

If the height of the existing CLC block is increased temporarily by 2 m, either by a parapet wall or vertically arranged temporary corrugated roofing sheets, the duration of the live hours of all the afternoons can be extended by at least 45 min.

While conducting several major events at SAINTGITS, the existing practice is to raise temporary roofs above the amphitheatre. It can comfortably accommodate at least 2500 seats and thus the live and comfortable hours are made available throughout the day.

3.5 Estimating the Reverberation Time at SAINTGITS Amphi

The two major events at SAINTGITS, the cultural festival (NAKSHATRA) and the technical festival (PRATITYA) attract student participants from all parts of the country.

These annual events are conducted at SAINTGITS Amphi, right from the morning hours. Every time, makeshift arrangements are done at the amphitheatre to accommodate the crowd.

An architectural model was created in Autodesk's Ecotect to find the reverberation effects at the amphitheatre (Fig. 14). For evaluating the reverberation effects, the acoustic zone assumed the entire amphitheatre and CLC as a single zone of volume 100187 m^3.

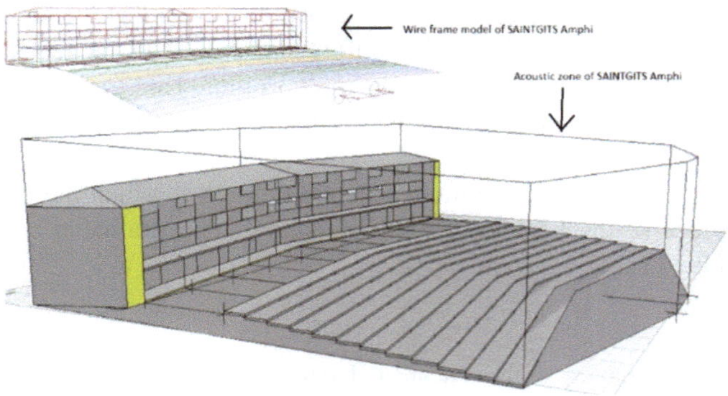

Fig. 14. An Ecotect model of the acoustic zone of SAINTGITS Amphi – the wire frame and rendered view.

The input parameters have been fixed slightly higher than the nominal values. The entire acoustic space is designed to accommodate 2500 seats (persons). Two tower speaker systems having 1000 W each was given as the sound input. The reverberation time (RT_{60} – the time elapsed to shed 60 decibels, measured from the originating time) was calculated by two independent algorithms – (i) the Norris – Eyring (NE) and (ii) the Millington –Settle (MS) (Fig. 15).

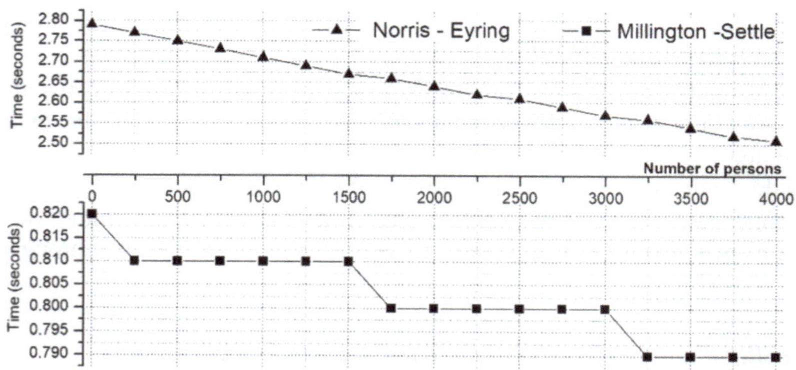

Fig. 15. The Reverberation Time (RT_{60}) measured independently by Norris-Eyring and Millington-Settle algorithms for the SAINTGITS amphi.

The values of RT_{60} were plotted for varying levels of occupancy, independently by each of algorithms (Fig. 15). The Millington-Settle algorithm is widely accepted for its acceptable range of RT_{60} values.

Whilst for the Norris-Eyring algorithm, the plot yielded a steady gradient (RT_{60} between 2.8 s to 2.5 s), the trend observed for Millington-Settle is a sharp gradient followed by a plateau (RT_{60} between 0.82 s to 0.79 s), for 0 to 4000 occupancy (Fig. 15).

Based on the International Standard for performance spaces, the ISO 3382-1 and ASTM E2235, the occupied state is defined as 80%–100% occupancy. Hence, the range of analyses had been limited to the values ranging between 2000 to 2500 occupants respectively for 80% to 100% occupancy. For a performing space to have richer musical sound, the suggested RT_{60} values lie in between 2.5 s–3.5 s. The RT_{60} based on the Norris-Eyring algorithm validates the same (Fig. 15).

4 Conclusion

The five stage scientific analyses of the Centre Lecture Complex (CLC) and the SAINTGITS Amphi are the first of its kind at SAINTGITS campus. The study has revealed the capability of sustainable retrofits on existing built structure. It has explicitly measured the hours of occupant comfort at the amphitheatre and suggested a viable measure to extend the comfort hours. Whilst similar analysis stands promising for several upcoming multi-storied projects at SAINTGITS campus, it revealed the significance of energy analyses for built structures at tropical climes.

References

Alexandri E, Jones P (2008) Temperature decreases in an urban canyon due to green walls and green roofs in diverse climates. Build Environ 43:480–493

Behzad S, Deepak R, Barbara J, Jabuk W, Rosi F (2011) The carbon reduction potential of straw bale housing. Build Res Inf 39(1):1–56

Campbell D, Gaylon S, Norman JM (1998) Introduction to Environmental Biophysics, 2nd edn. Springer, New York

Hoyano A (1988) Climatological uses of plants for solar control and the effects on the thermal environment of a building. Energy Build 11:181–199

Jacob Thottathil V, Balamuralikrishna C, Ghosh S (2010) Use of green facades in sustainable building environments: Quantifying the uptake rates of air pollutants by facades draped with tropical creepers. Proc Int Conf Sustain Built Environ ICSBE, Kandy, Colombo, Sri Lanka 1:363–373

Jacob Thottathil V, Balamuralikrishna C, Ghosh S (2011) Role of green facades in creating sustainable environments: Comparing chemical analyses and SEM images to quantify the atmospheric cleansing by green creepers. In: Proceedings of the International Conference on Ecosystem Conservation and Sustainable Development, ICECSD, Paris, France, vol 77, pp 343–346

Jacob Thottathil V, Ghosh S, Swentank P, Rudranshu S (2015) Evaluating the cleansing efficiency of an extended living façade draped with Vernonia elaeagnifolia. J Green Build 10 (2):157–177

Jacob Thottathil Varghese (2017) Mass and Energy Transfer Process In High Rise Green façade Structures. LAP Lambert Academic Publishing, Beau Bassin. ISBN 978-3-659-51994-9

Jacob Thottathil V, Ghosh S, Anshul SG, Manchanda R, Afsan A, Prathyusha M (2018) The projected thermal attributes and energy saving potential of Green Façade draped multi-storied cuboid buildings at major Indian cities: an ecotect analysis. World J Modell Simul 14(1):12–20

Pal AK, Kumar V, Saxena NC (2000) Noise attenuation by green belts. J Sound Vib 234 (1):149–165

Stec WJ, Van Paassen AHC, Maziarz A (2005) Modelling the double skin facade with plants. Energy Build 37(5):419–427

Takakura T, Kitade S, Goto E (2000) Cooling effect of greenery cover over a building. Energy Build 31:1–6

Wolverton BC, Johnson A, Bounds K (1989) Interior landscape plants for indoor air pollution abatement. Interiorscape 11:37–63

Wong N, Tan A, Tan P, Chiang K, Wong N (2010) Acoustics evaluation of vertical greenery systems for building walls. Build Environ 45:411–420

Yang J, Yu Q, Gong P (2008) Quantifying air pollution removal by green roofs in Chicago. Atmos Environ 42:7266–7273

Embankment Construction on Power Blender Cement Mixed Soft Soil Treated Ground for CKE Bypass

N. Vasanthan[1(✉)], H. Ohashi[1], F. Obara[1], S. D. L. Tilak[2], M. C. W. Dissanayake[2], and A. Karunawardena[3]

[1] Oriental Consultant and Katahira Engineering International JV, Colombo, Sri Lanka
nanthini.ged@gmail.com
[2] Road Development Authority, Ministry of Highway & Road Development, Colombo, Sri Lanka
[3] National Building Research Organization, Colombo, Sri Lanka

Abstract. Power Blender cement mix soft soil improvement method was adopted for the construction of embankment of the 4-lane bypass of Colombo-Katunayake Expressway for the New Bridge Construction over the Kelani River in Sri Lanka. This treatment method is a rapid construction method, first applied in Sri Lanka. The embankment height varies from 2 m to 5 m and peat/organic clay exists for a depth of about 10 m. Soil cement columns are spaced at 1 m × 2 m rectangular shape and the improvement depth is about 10 m. After completion of the ground treatment for the entire area and confirming the quality of the improvement, the embankment was constructed with 2 layers of geotextiles. During the embankment construction, settlement and lateral movement were measured. The expected settlements and lateral deformations were estimated and compared with field measurements. This paper describes the geological condition, embankment construction sequence, settlement estimate and stability analysis, measurement results and the comparison with the respective estimates.

Keywords: Power blender method · Embankment construction · Peat/organic clay · Settlements · Geotextiles · Stability analysis

1 Introduction

The Project consists of a bypass of about 480 m length with 4-lane traffic at the CKE (Colombo - Katunayake Expressway) end section as shown in Fig. 1.

R. Dissanayake and P. Mendis (Eds.): ICSBE 2018, LNCE 44, pp. 433–446, 2020.
https://doi.org/10.1007/978-981-13-9749-3_38

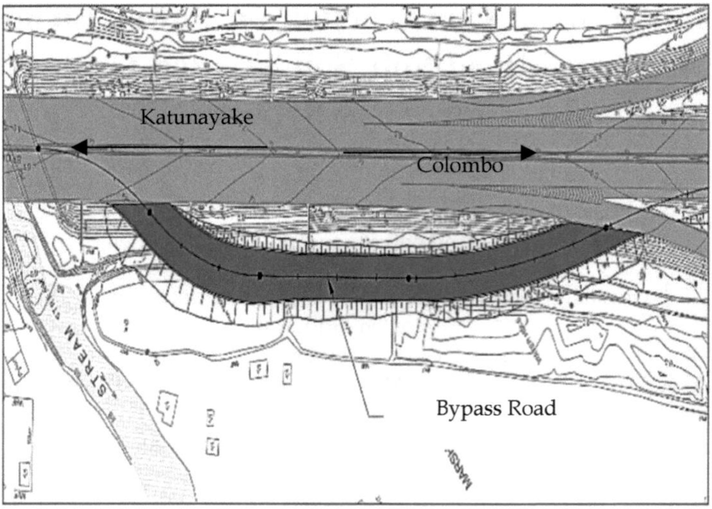

Fig. 1. Bypass section at the CKE end section

The height of the embankment varies from 2 m to 5 m over the soft ground. The existing CKE is running parallel apart to the proposed bypass at about 20 m to 30 m distance. The CKE was constructed on a similar soft ground condition of the proposed bypass using sand compaction piles (SCP) at 1.3 m centre to centre spacing.

Thick fibrous peat and organic clay/silt deposits are distributed over the area with localised swamps. The larger portion of the deposits predominantly consist of a thick layer of very soft to firm fibrous peat and organic clays, and below this layer stiff to hard gravelly clay layers and stiff to very stiff organic silt/clay exist at a deeper level.

The profile indicates some variation of engineering properties of the stratification (Fig. 2). The upper stratum of the bypass area consists of a very soft to firm clay layer with undrain shear strength varying from 10 kPa to 25 kPa. Thick, very soft to soft peat/organic clay/organic silt layers of undrain shear strength of 5 kPa is underlying the above. This layer, if untreated, exhibits significant settlement characteristics of the bypass section. Then a relatively hard clayey layer exists underneath. Soft ground treatment is designed using data from recent geotechnical investigation (12 Nos. of exploratory boreholes) and limited laboratory test results with the experience gained in similar projects in Sri Lanka.

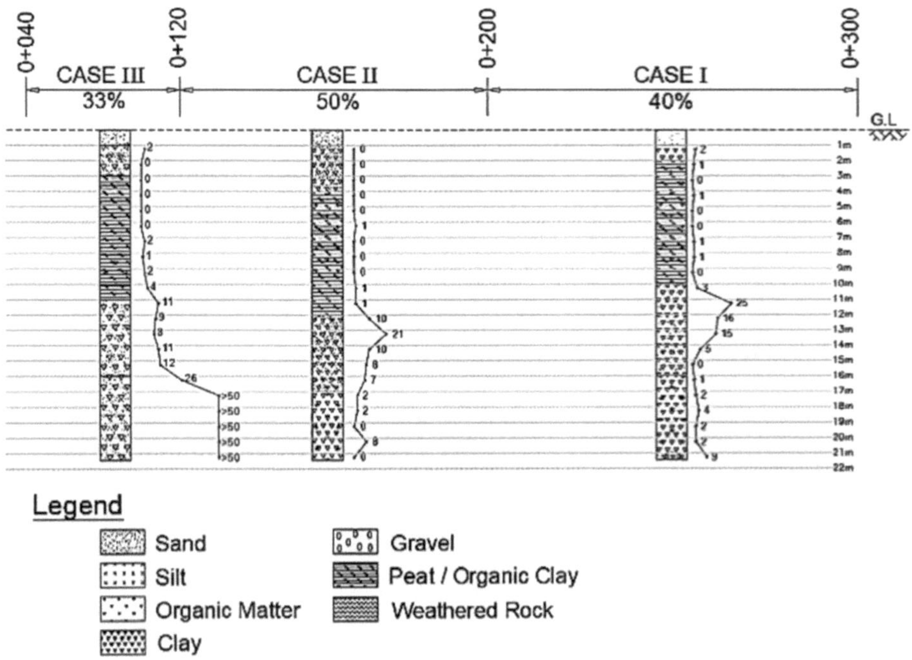

Fig. 2. Subsoil profile with SPT-N value

2 Proposed Ground Treatment

Column Supported Embankment (CSE) using Power Blender Cement Mixing technique is considered (Ohashi et al. 2018) as a technically feasible option for foundation improvement of the bypass due to time constraint in completing the project. Based on subsoil condition and layer thickness recorded from soil investigation, the ground improvement area was divided into three cases; namely Case 1, Case 2 and Case 3, for different mixes and area improvement ratios of 40%, 50%, 33% respectively as shown in Fig. 3.

The load from the embankment is designed to be transferred to Power Blender Cement Mixing columns with two layers of geotextiles of 600 kN/m which is considered in a manner such as to minimize the differential settlements and laid in both directions.

Spacing of the columns is designed to transfer the load effectively to the columns. Soil arching theory helps to load transfer mechanism of column supported embankment. The stresses developed within Power Blender Cement Mixing columns increase with time due to the development of arches above the columns.

Part of the load of the soil wedge below the soil arch is also transferred to the columns by the tension developed within the geotextile layers laid on top of the treated soil. This is called the membrane action of the geotextile layer.

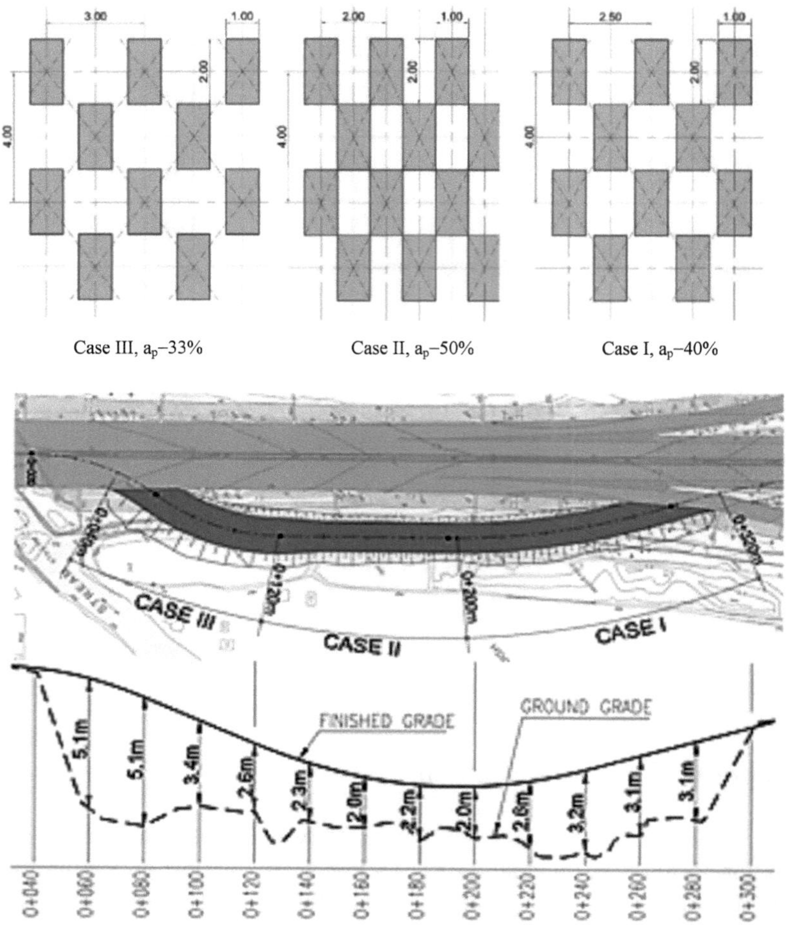

Fig. 3. Improvement sections with different mixes

The load transfer mechanism is a combination of arching within the embankment fill over the Power Blender Cement Mixing columns and membrane action of the geotextile layer.

Figure 4 shows the basic arrangement of a column supported embankment.

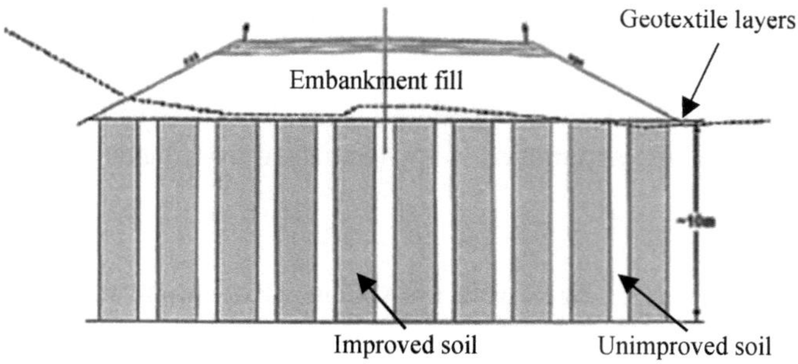

Fig. 4. Basic arrangement of a column supported embankment

3 Methodology

The following procedures have been adopted in the design of CSE with Power Blender Cement Mixing piles for the 480 m long bypass section based on the Federal Highway Administration Design (2013).

- Estimate preliminary column spacing.
- Determine required column load capacity.
- Select preliminary column type based on required column load capacity and site geotechnical requirements.
- Determine capacity of column to satisfy the serviceability design requirements.
- Determine extent of columns required across the embankment width.
- Determine the geotextile requirements based on estimated column spacing and lateral spreading.
- Check the global stability.

4 Settlement Analysis

Road embankments usually experience settlements when they are built on soft compressible foundation soils. Since the area is cement-treated, theoretically, the settlement is relatively very small. However, clayey soil layer exists underneath the treated soil, and may lead to some settlement. Settlements that occur very slowly, over decades or more, may be tolerable by pavements during the service within the scheduled maintenance period. If these settlements occur within a period of a few months or years, they would result in severe pavement damage.

It is, therefore, important to be able to measure the settlement to reasonable accuracy over a period of time to confirm whether the settlement is tolerable by the pavement and contributes to good riding quality by road users.

4.1 Calculation of Stress Increase Due to the Proposed Embankment Construction

Stress increments (σ) due to the proposed embankment construction with varying embankment height (H_e) and traffic load (Q) extracted from Japanese manual are given in Table 1.

Table 1. Stress increment with height of embankment

Case no.	Unit weight, γ (kN/m^3)	H$_e$ (m)	Q (kN/m^2)	Δp (kN/m^2)
1	19	3.2	20	80.8
2	19	2.5	20	67.5
3-1	19	3.4	20	84.6
3-2	19	5.0	10	105.0

4.2 Consolidation Settlement Without Ground Improvement

Consolidation settlement, S_{O}, was calculated using the coefficient of volume compressibility, m_{vO}, stress increase, σ and soft soil thickness, H, as shown in Eq. (1).

$$S_O = m_{v0} \times H \times \sigma \quad (1)$$

The consolidation settlement, S_O estimated without ground improvement is summarized in a Table 2.

Table 2. Consolidation settlement without ground improvement

Case no.	H$_e$ (m)	σ (kN/m^2)	m$_{v0}$ (m^2/kN)	S$_O$ (m)
1	3.2	80.8	0.006	1.53
2	2.5	67.5	0.008	1.36
3-1	3.4	84.6	0.003	0.81
3-2	5.0	105.0	0.002	0.94

4.3 Design Standard Strength of Soft Soil for the Ground Improvement

From the stress increase, σ, area improvement ratio, a_p and the safety factor, F_c, the design strength of the improved ground, q_{uck} is obtained by the Eq. (2).

$$q_{uck} = \sigma / a_p \times F_c \quad (2)$$

For each case, the improvement ratio and the design standard strength are given in Table 3.

Table 3. Improvement ratio and design strength

Case no.	H_e (m)	Δp (kN/m^2)	a_p (%)	F_c	q_{uck} (kN/m^2)	
1	3.2	80.8	40	1.2	242	250
2	2.5	67.5	50	1.2	162	200
3-1	3.4	84.6	33	1.2	308	350
3-2	5.0	105.0	33	1.2	382	400

4.4 Residual Settlement After Ground Improvement

For the calculation of residual settlement after improvement, the following two methods were adopted:

(1) Based on the stress sharing ratio
(2) Considering the arch effect of the embankment fill.

4.4.1 Method Based on the Stress Sharing Ratio

The total residual settlement based on the stress sharing ratio method (S_1) is calculated by the Eqs. (3), (4), (5) and (6).

$$S_1 = \sigma_s/\sigma \times S_0 \tag{3}$$

$$\sigma_s/\sigma = 1/\left((n-1) \times a_p + 1\right) \tag{4}$$

$$n = \sigma_c/\sigma_s \tag{5}$$

$$S_0 = m_{v0} \times E \times \sigma \tag{6}$$

Where S_0 is the settlement of the untreated ground, σ is the average stress (or stress increase), σ_s is the stress on the surrounding soil, σ_c is the stress on the column, a_p is the area improvement ratio and n is the stress sharing ratio (or stress concentration ratio).

A schematic diagram of the method based on the stress sharing ratio is shown in Fig. 5. The "n" value was adopted from Manual for Soft Ground Improvement in Japan.

The "n" value adopted for this project was between 15 to 20.

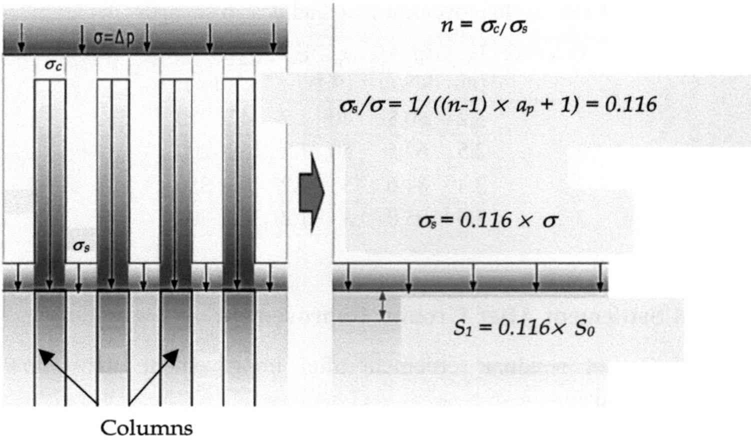

Fig. 5. Schematic diagram of the method based on the stress sharing ratio

For each case, the residual settlement was obtained as given in Table 4. For Case 1 and Case 3, the grounding type (G) was adopted and for Case 2, floating type (F) was implemented owing to a soft ground layer.

Table 4. Total settlement after ground improvement

Case no.	S_0 (m)	a_p (%)	n	β′	S′ (m)	S″ (m)	S_1 (m)
Case 1 G	1.53	40	15	0.152	0.23	–	0.23
			20	0.116	0.18	–	0.18
Case 2 F	1.36	50	15	0.125	0.16	0.07	0.23
			20	0.095	0.12	0.07	0.19
Case 3-1 G	0.81	33	15	0.178	0.14	–	0.14
			20	0.138	0.11	–	0.11
Case 3-2 G	0.94	33	15	0.178	0.17	–	0.17
			20	0.138	0.13	–	0.13

S′- Settlement where column height is equal to soft soil thickness

S″- Settlement where column height is less than soft soil thickness

4.4.2 Method Considering the Arch Effect of the Embankment Material

The total residual settlement after ground improvement is calculated from the deformation coefficient of the composite ground. An unequal settlement between the improved soil and original ground (poor ground) were calculated with consideration of arch action of embankment between each improvement column. The embankment load is transferring between the improved bodies as shown in Fig. 6 due to the arching effect. Soil weight below the arch acts as a load on the unimproved part between the improved bodies.

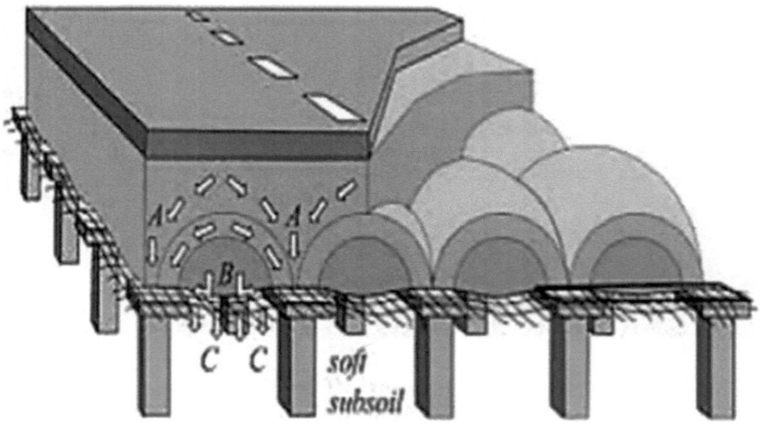

Fig. 6. A schematic diagram of Arching action

The total residual settlement based on the arching effect (S_2) is calculated using Eqs. (7) and (8) by the composite ground which was formed by improved and original soil.

$$S_2 = \left(\sigma / E_{eq} \times H\right) \tag{7}$$

$$\mathrm{E}_{eq} = a_p \times E_c + \left(1 - a_p\right) \times E_s \tag{8}$$

Where E_{eq} is the modulus of elasticity of composite ground, E_c is the modulus of elasticity of improved column and E_s is the modulus of elasticity of unimproved soil.

In order to obtain the amount of settlement of the original ground between each improved column, the consolidation settlement of original ground, S_O is obtained as described in Sect. 4.2 and the amount of settlement of original ground between improved column, S_s is calculated by the ratio of the stress, σ_s acting on the unimproved portion and the average stress, σ acting on the entire improved area using Eq. (9).

$$S_s = S_s \times (\sigma_s / \sigma) \times H \tag{9}$$

The settlement amount of the improved column, S_c was calculated by the ratio of the stress, σ_c acting on the improved portion and the modulus of elasticity of improved column, E_c by Eq. (10).

$$S_c = (\sigma_c / E_c) \times H \tag{10}$$

The differential settlement amount, δ_s between the improved column and original ground was calculated by the Eq. (11).

$$\delta_c = S_s - S_c \tag{11}$$

A summary of total residual settlement and differential settlement are given in Table 5.

Table 5. Differential settlement after ground improvement

Case no.	S_0 (m)	a_p (%)	S_2 (m)	Differential settlement amount		
				S_s (m)	S_c (m)	δ_s (m)
Case 1 G	1.53	40	0.08	0.14	0.06	0.08
Case 2 F	1.36	50	0.14	0.28	0.04	0.24
Case 3-1 G	0.81	33	0.06	0.09	0.05	0.04
Case 3-2 G	0.94	33	0.07	0.08	0.06	0.02

5 Stability of the Embankment Slopes

In these designs the analysis was done with Geostudio-Slope/W software.

The software considers a large number of trial failure surfaces and selects the one with minimum factor of safety. It is taken as the factor of safety of the slope.

In this analysis, the Spencer method was used which covers both circular and non-circular failure surface. It satisfies both force and moment equilibrium conditions.

5.1 Shear Strength Parameters Adopted for the Stability Analysis of the Embankment Slopes

Based on the result of the soil test and experience earned on this area, the shear strength parameters of the original ground were adopted as given in Table 6.

Table 6. The shear strength parameters of the original ground

Soil type	γ (kN/m^3)	Drained		Undrained	
		C' (kN/m^2)	ϕ' (Deg)	C_u (kN/m^2)	ϕ_u (Deg)
Clay	16.0	8.0	22	20	0.0
Peat	12.0	5.0	20	5	0.0
Sand	17.0	0.0	30	–	–
Embankment	19.0	10.0	30	–	–

Shear strength parameters for the improved area were obtained from the design strength of the improved ground, q_{uck} and the area improvement ratio, a_p by the Eq. (12) as it is a composite of improved and original ground and presented in Table 7.

$$C = q_{uck} \times 0.5 \times a_p \tag{12}$$

C is the cohesion for the improved column and the internal friction angle, ϕ for improved column is assumed as 40°.

Table 7. The shear strength parameters of the improved ground

Case no.	Soil parameter for improved ground			
	a_p (%)	q_{uck} (kN/m^2)	C (kN/m^2)	ϕ (deg)
Case 1	40	250	50.00	40
Case 2	50	200	50.00	40
Case 3-1	33	350	57.75	40
Case 3-2	33	400	66.00	40

5.2 Stability Analysis of the Embankment Slopes

Stability analysis of the embankment slope for case 3-2 area is presented in Fig. 7 and the results of all cases are summarized in Table 8. The calculation result was sufficiently secured against the required factor of safety of 1.2.

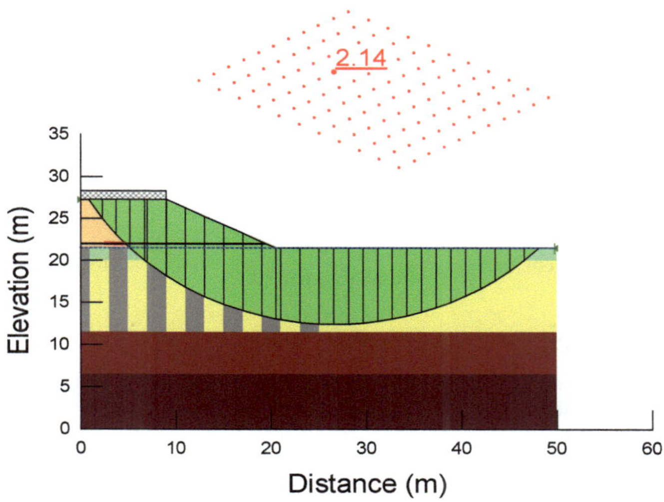

Fig. 7. Stability analysis for Case 3–2 area

Table 8. The results of the stability analysis

Case no.	H_e (m)	Q (kN/m^2)	Factor of safety achieved
Case 1	3.2	20	2.27
Case 2	2.5	20	2.39
Case 3-1	3.4	20	2.51
Case 3-2	5.0	10	2.14

6 Monitoring of Ground Improvement

Settlement plates and horizontal displacement stakes were installed at every 50 m interval to monitor the settlement and horizontal movements respectively as shown in Fig. 8.

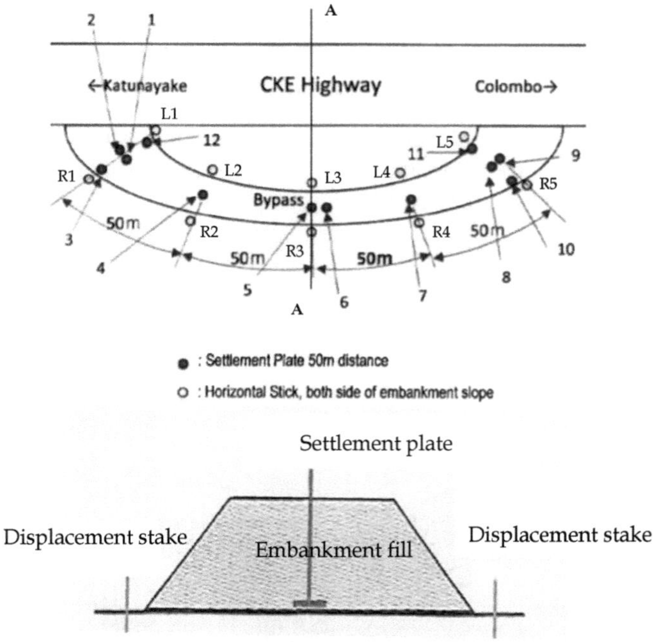

Fig. 8. Plan layout of settlement plates and horizontal displacement stakes and sectional view of A-A

Some of the monitoring results for vertical settlement are given in Table 9, which indicate horizontal displacements to be minimal.

While Point Nos. 2, 6 and 9 are located between improved areas (unimproved portion), the remaining are on improved areas.

Table 9. The results of the vertical settlement

Case	Point no.	Actual settlement until 31st October 2018/(mm)
Case 3	1	31
	2	34
	3	28
	12	11
Case 2	4	33
	5	36
	6	42
Case 1	7	41
	8	59
	9	63
	10	16
	11	23

7 Current Progress of CKE Bypass

Current progress of CKE bypass is shown in Fig. 9 as of October 2018.

Fig. 9. CKE bypass paving work as of October 2018

8 Conclusion

(1) It does not show any consolidation settlement behaviour. Typical total settlement limits ranging between 11 mm to 63 mm.
(2) The settlement behaviour of both the columns and the soils show that there was a significant soil arching in the embankment, and the load acting on the columns are much higher than that acting on the soils between the columns. The measured differential settlements between the columns and the soils are small.
(3) The measurements have proven the effectiveness of this technique in reducing total and differential settlements. The bypass will be open to traffic in early November 2018. The measurement is on-going at the time of writing this paper.
(4) Accordingly, the power Blender method is an efficient method for improving soil in a short period as the required strength can be achieved within curing period, 28-day generally.

Acknowledgement. Authors wish to express their thanks to PMU for their guidance and valuable advice. Also, our thanks extend to Mr. Sato, Mr. Suzuki and Mr. Teruya of SMCC-SANKEN JV who performed the planning, testing and construction.

References

Federal Highway Administration Design: Deep Mixing for Embankment and Foundation Support, Publication No. FHWA-HRT13-046, October 2013

Ohashi H, Jayasekare DA S, Amarasena DGSR, Karunawardena A (2018) Power Blender Cement Mixed Soft Soil Treatment for Colombo Katunayake Expressway (CKE) Bypass, ICSBE 2018. University of Peradeniya, Kandy

Reinforced Concrete Bridge Deck Repair with Titanium NSM

S. L. Platt[1(✉)], K. A. Harries[2], and M. J. McCabe[3]

[1] Architecture and Civil Engineering, University of Bath, Bath BA2 7AY, UK
S.L.Platt@bath.ac.uk
[2] Civil and Environmental Engineering, University of Pittsburgh, Pittsburgh 15260, USA
[3] United States Department of Agriculture Forest Service, Montana, USA

Abstract. The use of titanium bars as a near surface mounted (NSM) repair method for highway bridge deck slabs is investigated in this pilot study. Four full-scale slabs had half of their internal reinforcing steel cut resulting in an approximately 40% loss of slab capacity; these represented 'damaged' slabs. The objective of this study was to evaluate the performance of replacing these two cut bars in order to restore the original capacity of the slab. The four slabs were repaired using NSM titanium bars designed to restore either their capacity or stiffness of the cut bars. One slab from each group was repaired using a full-length straight bar while the other implemented a 'staple' type repair. Both approaches effectively restored the capacity of the slabs although each had significant impacts on the available ductility of the repaired slab. Repairs based on restoring lost bar strength resulted in preferable behavior. Restoring lost stiffness required four times as much titanium and resulted in over-reinforced sections showing little ductility.

Keywords: Concrete · Corrosion · Retrofit · Bond · Ti6al-4V · Serviceability

1 Introduction

Titanium reinforcing bars for concrete rehabilitation have been proposed (Adkins and George 2017) and demonstrated in laboratory tests (Platt and Harries 2018a and 2018b). The primary advantage and motivation for using titanium is its corrosion resistance. An anticipated application for titanium reinforcement is near surface mounted (NSM) reinforcement for strengthening reinforced concrete members in cases where the advantages of titanium may be realized such as in highly corrosive environments, non-magnetic applications, and others in which ductility or high service temperatures may be required.

A single demonstration application is known (Higgins et al. 2015 and 2017) in which NSM titanium 'staples' – straight bars having a 90° bend at either end to affect anchorage – were used in a bridge application.

The fundamental material behaviours of titanium are similar in form to those of steel. Titanium, like steel, exhibits an elastic behaviour to a proportional limit, a definable yield value followed, typically, by some degree of strain hardening, and

R. Dissanayake and P. Mendis (Eds.): ICSBE 2018, LNCE 44, pp. 447–457, 2020.
https://doi.org/10.1007/978-981-13-9749-3_39

displays a great deal of ductility in most cases. This pilot study is part of a large study investigating specific aspects of the use of 6Al-4V titanium (UNS designation R56400) bars having ribbed deformations as a concrete reinforcing material (Platt 2018). When compared to conventional ASTM A615 reinforcing steel, Ti6Al-4V exhibits nearly double the yield strength and half the extensional modulus. Titanium reinforcing bars used in this study were nominally #5 bars, having a nominal diameter and area of 15.9 mm and 200 mm^2, respectively. Results of tension tests compliant with ASTM A370-14 (including Annex A9) in all ways, other than being titanium, are shown in Fig. 1. Nominal (for design) and measured material properties of the bars used in this study are shown in Table 1.

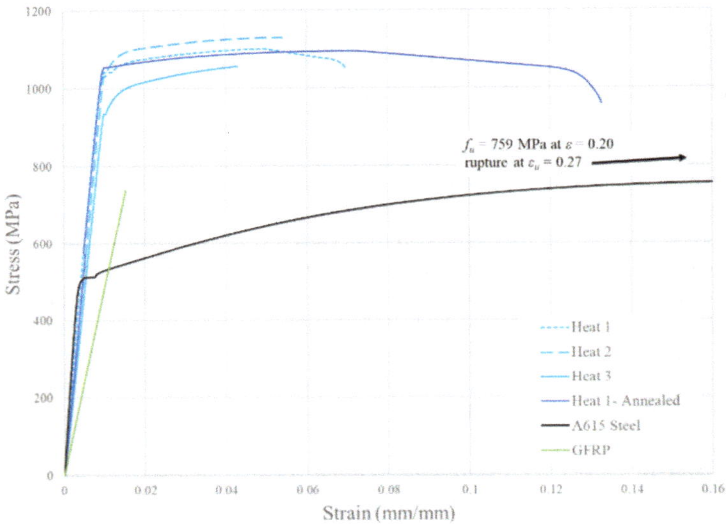

Fig. 1. Representative experimentally determined stress-strain curves.

6Al-4V titanium bars exhibit yield strength approximately twice that of ASTM A615 reinforcing steel and an extensional modulus about 55% of steel. As such, the yield strain of Ti6Al-4V is on the order of 0.008, approximately four times greater than A615 steel. The softer response affects the assumed concrete behaviour and can be a significant disadvantage when considering serviceability; primarily in the form of increased crack widths.

The behaviour of reinforcing steel, particularly under service loads, is a function of axial stiffness, *EA*; thus to directly replace steel with titanium, twice as much titanium is required. Typically, this will not be practical. Alternatively, the design paradigm used for glass fibre-reinforced polymer (GFRP) reinforcing bars [ACI 440.1R], which presents an elastic behaviour to failure with modulus ranging from 20% to 50% that of steel, may be used for titanium-reinforced concrete or for using titanium as a retrofit reinforcement.

Table 1. Geometric and mechanical properties of reinforcing bars reported in this study

	#5 6Al 4V Titanium bars		# 5 ASTM A615 Steel bars	
	Nominal properties	Measured properties (Platt and Harries 2018a)	Nominal properties	Measured properties (McCabe et al. 2014)
Bar area	200 mm^2	236 mm^2	200 mm^2	
Bar diameter	15.9 mm	17.4 mm	15.9 mm	
Density	4430 kg/m^3	4407 kg/m^3	7900 kg/m^3	
Modulus	E_{Ti} = 114 GPa		E_s = 200 GPa	
Yield strength	f_{yTi} = 965 MPa	f_{yTi} = 999 MPa	f_{ys} = 414 MPa	f_{yTi} = 467 MPa
Tensile strength	f_{uTi} = 1100 MPa	f_{uTi} = 1054 MPa	f_{us} > 586 MPa	f_{uTi} = 744 MPa
Elongation at rupture	$\varepsilon_{Tiu} > 0.100$	$\varepsilon_{Tiu} > 0.084$	$\varepsilon_{su} > 0.150$	

2 NSM Titanium Slab Tests

Four steel-reinforced bridge deck slabs, cast in 2013, were available to develop a pilot study to demonstrate the efficacy of titanium NSM retrofit. These slabs were designed based on the AASHTO (2017) prescriptive design method and had four #5 primary reinforcing bars spaced at 140 mm, top and bottom, across the 559 mm slab width. The 'control' specimen used for this study is Slab A, tested in 2013 (McCabe 2013; McCabe et al. 2014). Details of the slab geometry are shown in Fig. 2.

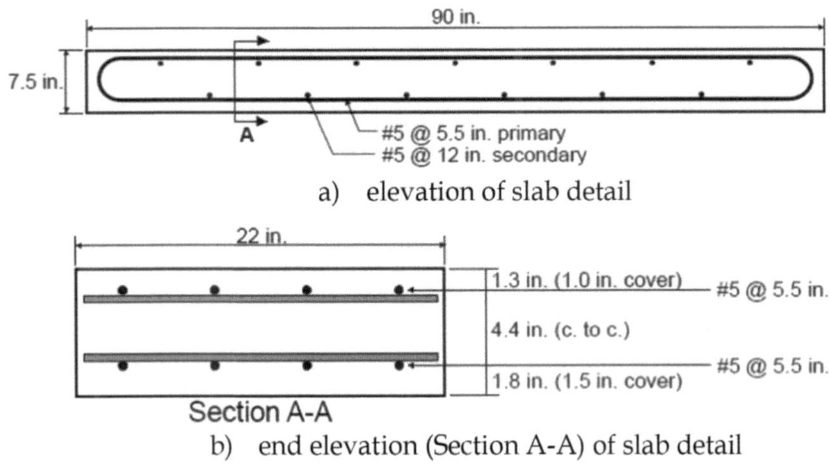

Fig. 2. Details of laboratory control specimen (McCabe 2013) (1 in. = 25 mm)

Each of the four slabs to be retrofit was ‘damaged’ by cutting two of the four existing #5 steel bars at increments along their length sufficient to ensure that the bars no longer contributed to the flexural capacity of the slabs. The slabs were then repaired with one of two types of NSM repairs: a straight NSM bar or an NSM ‘staple’. The staples, as shown in Fig. 3, were short lengths of NSM bar having 90° anchorages at either end. Such a staple may be practical for repairing local damage (as was the case in this study) or in locations were straight bar development is not possible.

Two scenarios are investigated based on repairs that restore either the capacity ($A_s f_{ys} = A_{Ti} f_{yTi}$) or equivalent stiffness ($A_s E_s = A_{Ti} E_{Ti}$) of the cut bars. In either case, #5 titanium deformed bars were embedded within channels cut along the length of the tension face of the slabs. Sikadur 31 (a commercially available two-part structural adhesive) was used to embed the bars. The installations, shown in Fig. 3, were, other than the use of titanium, compliant with the design approach of ACI 440.2R-17.

After allowing the NSM installations to cure for a minimum of 72 h, they were placed in a test frame. Loading and support conditions, shown in Fig. 4a, were identical to those used to test control Slab A (McCabe 2013). The slabs were tested in mid-point flexure over a simple span length (L) of 2135 mm. Primary instrumentation consisted of multiple DEMEC gauges, having a resolution of 8 microstrain and gauge length of 8 in. (203.2 mm), arranged vertically at mid-span (Fig. 4).

Tests of the four NSM titanium retrofit slabs were carried out in a manner (to the extent possible) identical to the control specimen: Slab A reported previously by McCabe et al. (2014).

Load was applied using a 267 kN capacity hydraulic cylinder; load was measured with a precision of 320 N. Load was applied at intervals of approximately 4.45 kN while mid-span displacement was recorded manually at each load interval with a precision of 0.8 mm.

Since there were no test cylinders remaining from the original casting of the slabs and due to undocumented environmental conditions during storage, core samples were taken from the slabs to confirm present *in situ* compressive and tensile strengths. Cores having a diameter of 57 mm (2.25 in.) were removed from the [essentially undamaged] support region of the slabs following testing. The compression and tension strengths obtained from the 57 mm cores (including all corrections) are shown in Table 2. An expected increase in compressive strength is observed. The 1666 day tested strengths reported in Table 2 were used for all subsequent analyses of the slabs. The 132-day strength was used for Slab A. Internal reinforcing steel and NSM titanium material properties are reported in Table 1.

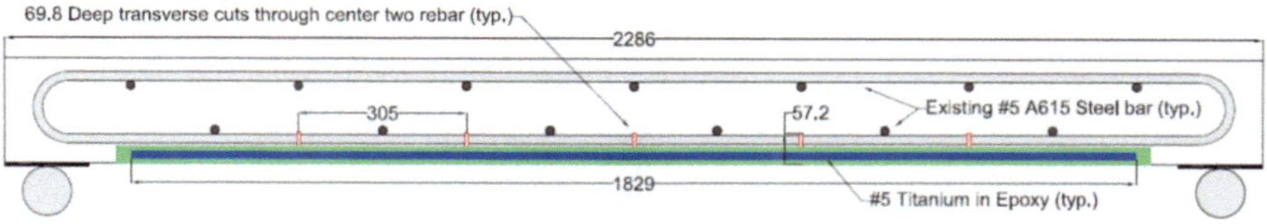

TiNSM-1 and TiNSM-2 elevation with NSM titanium straight bars (units = mm)

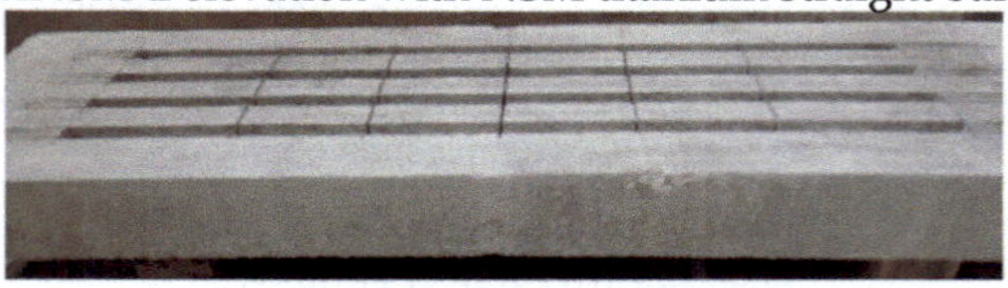

NSM channels on [inverted] slab soffit; transverse cuts through internal reinforcing bars can also be seen.

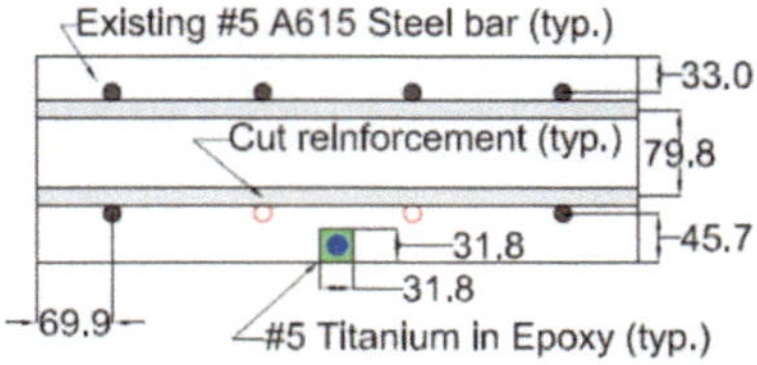

TiNSM-1 section : 1 bar centered in slab soffit

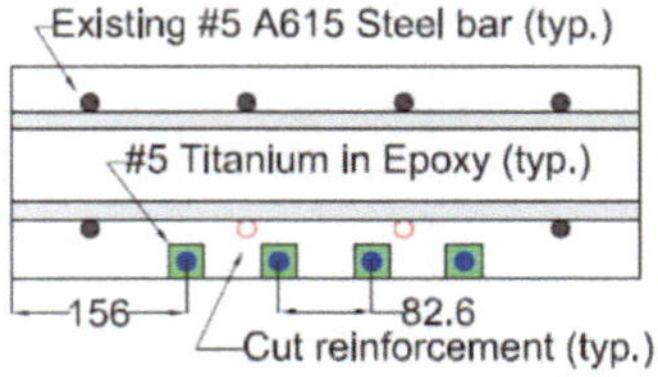

TiNSM-2 section: 4 bars at 83 mm across soffit

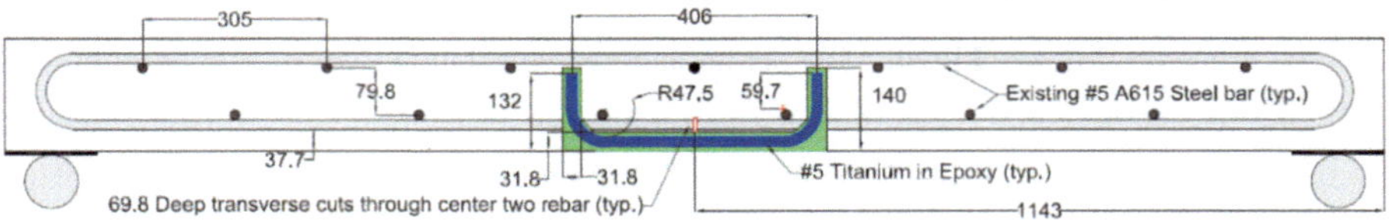

TiNSM-3 and TiNSM-4 elevation with NSM titanium staples (units = mm)

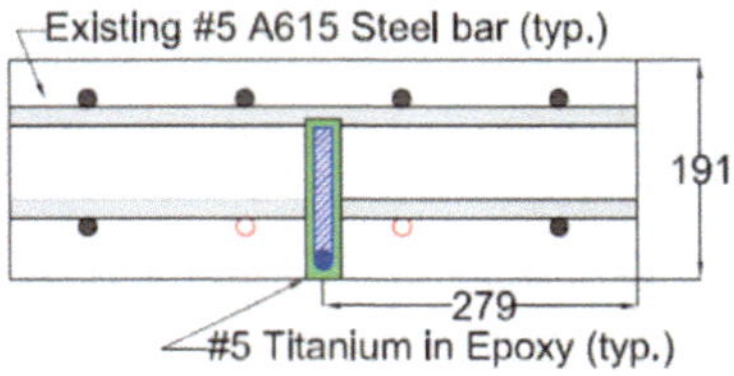

TiNSM-3 section – 1 staple centered in slab soffit

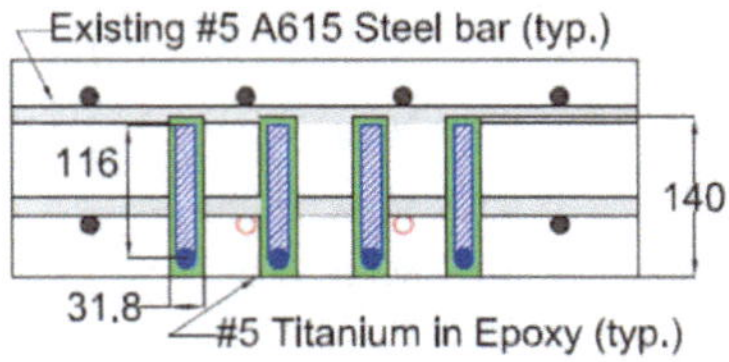

TiNSM-4 section – 4 staples at 83 mm across soffit

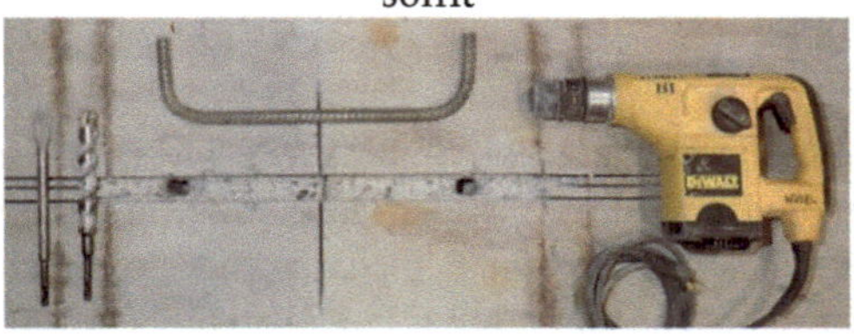

Staple, channel and installation tools

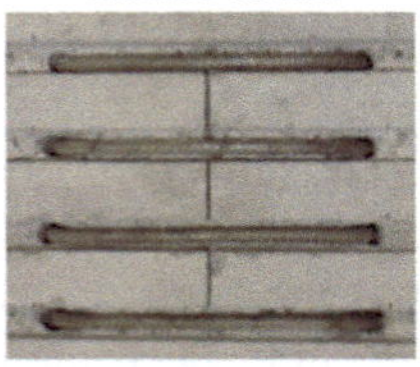

TiNSM4 staples before epoxy embedment

Fig. 3. Repaired cross sections and elevations of slabs based on strength (TiNSM-1 and 3) and Service (TiNSM-2 and 4) criteria (all dimensions in mm).

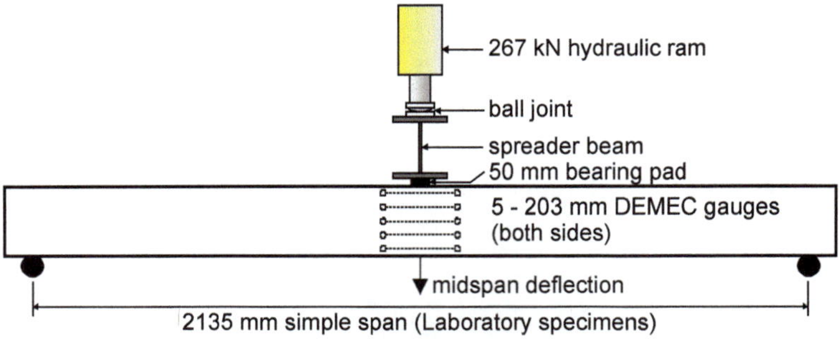

test Set-up (Slab A shown)

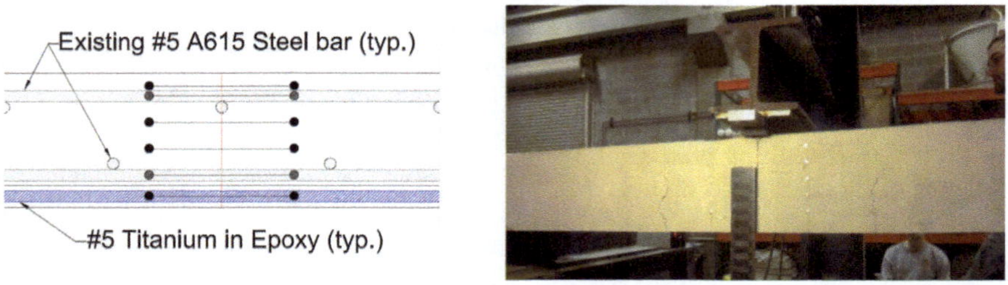

DEMEC and vertical deflection instrumentation (Slab A shown)

Fig. 4. Test set-up and instrumentation

Table 2. Summary of concrete properties

Age	ASTM C39			ASTM C496			ASTM C78		
	Compression test			Split cylinder test			Modulus of rupture		
Days	n	f_c'	COV	n	f_{sp}	COV	n	f_r	COV
28[a]	3	44.8 MPa	0.034	3	$3.12\text{ MPa} = 0.47\sqrt{f_c'}$	0.131	3	$5.45\text{ MPa} = 0.81\sqrt{f_c'}$	0.056
132[a]	3	45.8 MPa	0.084	–	–	–	–	–	–
1666[b]	8	50.7 MPa	0.081	5	$2.98\text{ MPa} = 0.42\sqrt{f_c'}$	0.186	–	–	–

[a] McCabe (2013) [b] Platt (2018)

3 Test Results

The key results for NSM-repaired slabs and the control Slab A are presented in Table 3. Plots of applied moment versus curvature at mid-span are shown in Fig. 5. The reported moments are calculated as: M = PL/4, were P includes the cross-head contribution but neglects the weight of the slab. The curvature is calculated by dividing the difference in strain between the uppermost DEMEC gauge and the gauge located at the existing A615 steel tension reinforcement by the vertical distance separating these gauges. This is consistent with the procedure used for the control Slab A.

The behavior of all slabs was analyzed using the program RESPONSE (Bentz 2000). All material properties used in the RESPONSE models are the measured values

Table 3. Summary of tested slabs

Slab		A	TiNSM-1	TiNSM-2	TiNSM-3	TiNSM-4
Depth of slab	mm	191	216	191	191	191
Design condition			Strength: $A_s f_{ys} + A_{Ti} f_{yTi} \approx$ Slab A	Stiffness: $A_s E_s + A_{Ti} E_{Ti} \approx$ Slab A	Strength: $A_s f_{ys} + A_{Ti} f_{yTi} \approx$ Slab A	Stiffness: $A_s E_s + A_{Ti} E_{Ti}$ $\approx$ Slab A
NSM bars		None	1–#5 straight bar	4–#5 straight bars	1–#5 staple	4–#5 staples
Load at first crack	kN	22.6	23.1	28.2	28.2	32.6
Moment at first crack	kN-m	12.1	22.1	15.0	15.0	17.4
Load at steel bar yield	kN	81.0	81.8	104	72.5	113
Moment at steel bar yield	kN-m	43.1	43.6	55.3	38.7	60.1
Ratio yield capacity to Slab A	–	–	1.01 0.89[a]	1.28	0.90	1.39
Deflection at steel bar yield	mm	9.65	5.59	6.35	5.59	11.2
Curvature at steel bar yield	rad/km	25.6	16.6	21.2	18.4	14.4
Ultimate load	kN	125	179	192	117	130
ultimate moment	kN-m	66.6	95.7	103	62.5	69.6
Ratio ultimate capacity to Slab A	–	–	1.44 1.12[a]	1.55	0.94	1.05
Deflection at ultimate load	mm	–	32.5	15.0	25.4	31.0
Failure mode		Flexural	Flexural	Shear	Flexural	Flexural at end of staples

[a] value normalized to 191 mm slab depth; i.e.: Slab TiNSM-1 ratio multiplied by $(191/216)^2$

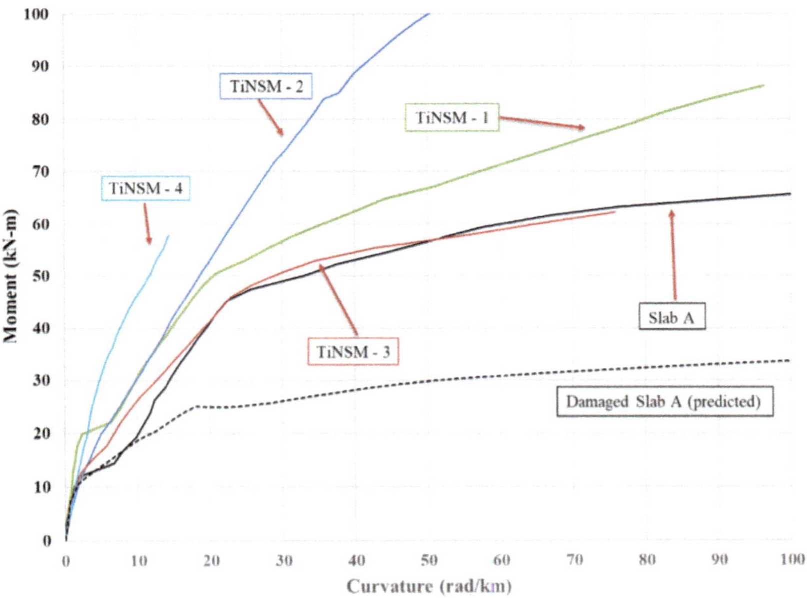

Fig. 5. Summary of Moment-Curvature results from slab flexure tests

given in Tables 1 and 2. Figure 6 shows the predicted moment-curvature responses superimposed with the as tested results.

The predicted moment-curvature responses are for mid-span performance; effects of shear-moment interaction along the relatively short shear spans will result in a marginal reduction of the predicted "pure" moment capacity. In the more heavily reinforced Slabs TiNSM2 and 4, the retrofit flexural capacity exceeds the slab shear capacity and the latter controls the ultimate behavior. RESPONSE-predicted capacities are shown in Table 4.

Table 4. Summary of predicted capacities of NSM-reinforced slabs

	Slab A	Slab A with two bars cut	TiNSM-1	TiNSM-3	TiNSM-2	TiNSM-4
Predicted moment capacity (kNm)	57.3	34.3	84.1	65.6	134	
Predicted peak applied load (kN)	94.8	52.4	137	100	154	
moment capacity accounting for shear (kNm)	50.5	27.9	72.8	53.5	82.3	
Observed moment capacity (kNm)	66.6	–	84.6[a]	62.5	103	69.6
Capacity normalized to Slab A	1.0	–	1.27	0.94	1.55	1.05

[a] value normalized to 191 mm slab depth; i.e.: Slab TiNSM-1 ratio multiplied by $(191/216)^2$

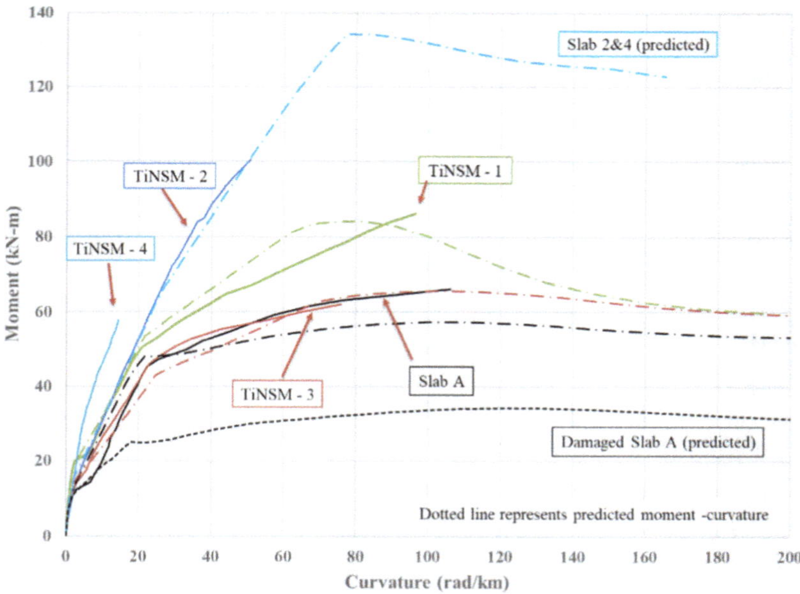

Fig. 6. Summary of moment-curvature results with RESPONSE predicted results

The proposed staple repair was investigated on two slabs having two of the four #5 ASTM A615 reinforcing bars cut only at mid-span (Fig. 4). This effectively dropped the capacity of the slab approximately 40% (Table 4). The development length for a #5 bar is $l_d = 340$ mm. Thus not until 340 mm to either side of mid-span, is the capacity of the cut #5 bars theoretically redeveloped. The staples used were only 406 mm long and thus did not span the entire region of reduced capacity (680 mm). Figure 7 describes the resulting moment capacity envelopes (normalized to the predicted capacity of Slab A). The applied moments at failure are shown by the dashed lines. Clearly Slab A is critical at mid-span, as is TiNSM-3, although there is a relatively small margin at the end of staple. The applied moment of TiNSM-4 is seen to intersect its capacity envelope at the end of the staples, rather than at the increased mid-span capacity. Thus, the staples in this case were too short to develop the predicted capacity of the repaired slab, as reflected in Table 3.

To mitigate the potential weak section at the staple anchorage, the anchorage should be located beyond the point where the cut bars have been fully redeveloped. That is, the staple length should exceed $2l_d$ (of the cut/damaged bars).

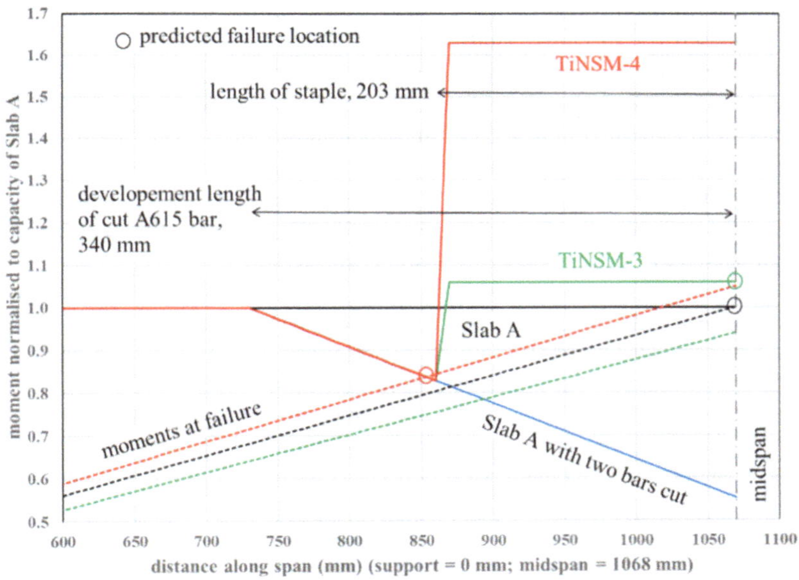

Fig. 7. Generalized capacity envelope of the damaged slab with and without staple repair

4 Conclusion

The application of titanium NSM reinforcement was investigated through a pilot study using four intentionally damaged steel-reinforced bridge deck slabs. Both straight bar repairs and the staple repair based on equivalent stiffness (4 staples) restored the original slab capacity. The repair having the single titanium staple, theoretically sufficient to restore the capacity lost by cutting the internal reinforcement, restored only 94% of the original slab capacity. It is hypothesized that if the staples had been designed to extend beyond the development length of the cut internal steel reinforcing bars, the resulting performance of the NSM repairs would have been improved.

The slabs having only a single titanium bar, developed ductility similar to the undamaged control Slab A although exhibited significant slip at ultimate capacity. The slabs having four bars were both effectively over-reinforced leading to increased capacity but decreased ductility. The slab having four full-length titanium NSM bars, failed upon reaching its shear capacity whereas that having four staples failed 'prematurely' in flexure at the section at the end of the staples. The staples were not long enough to develop the full theoretical capacity of this slab.

Cost of Titanium Bars

Currently the associated costs of implementing titanium as a reinforcement material must be overcome. On a unit weight basis, 6Al-4V titanium bars are approximately 15 times more expensive than A615 steel bars and 5 times more expensive than solid stainless steel bars (Platt 2018; Triantafillou 2012). However, the final in-place cost of titanium bars will be affected by the design paradigm. Platt and Harries (2019) report a preliminary study of the cost of using #5 6Al-4V titanium bars for bridge deck reinforcement (based on an earlier study by Triantafillou (2012) that considered several other bar materials). Depending on the design paradigm used, 6Al-4V reinforcing bars were estimated to cost between $\$430/m^2$ [of bridge deck constructed] and $\$1700/m^2$. A615 bars were estimated to cost $\$117/m^2$ and solid stainless steel bars were $\$294/m^2$. The typical bridge construction unit cost for medium span bridges using conventional A615 reinforcing bars (in 2010) was $\$989/m^2$ (Triantafillou 2012). Therefore, the use of titanium bars represents a significant cost premium although this may be offset by long term savings in maintenance for a non-corrosive deck.

Acknowledgement. The research presented was sponsored by the Perryman Company. The findings and conclusions reported in this paper reflect the opinion of the authors and are not necessarily those of the Perryman Co. Slab construction and testing Slab A was sponsored by the Pennsylvania Department of Transportation. All testing was conducted in the Watkins Haggart Structural Engineering Laboratory at the University of Pittsburgh. The second author acknowledges the support of the Leverhulme Trust (UK).

References

AASHTO (2014) LRFD bridge design specifications (7th edn), American Association of State Highway and Transportation Officials

ACI 440.2R-17 Design and Construction of Externally Bonded FRP Systems, American Concrete Institute

ACI 440.1R-15 (2015) Guide for the Design and Construction of Structural Concrete Reinforced with FRP Bars, American Concrete Institute

Adkins J, George W (2017) Titanium finds a home in civil engineering. In: Concrete International, December 2017, pp 51–55

ASTM A370-14 (2014) Test Methods and Definitions for Mechanical Testing of Steel Products. ASTM International

ASTM A615-16 (2016) Standard Specification for Deformed and Plain Carbon-Steel Bars for Concrete Reinforcement. ASTM International

ASTM C39-16 (2016) Standard Test Method for Compressive Strength of Cylindrical Concrete Specimens, ASTM International

ASTM C496-11 (2011) Standard Test Method for Splitting Tensile Strength of Cylindrical Concrete Specimens, ASTM International

Bentz E (2000) Sectional Analysis of Reinforced Concrete Members, PhD Thesis, Department of Civil Engineering, University of Toronto 2000, pp 310

Higgins C, Knudtsen J, Amneus D, Barker L (2017) Shear and flexural strengthening of reinforced concrete beams with titanium alloy bar. In: Proceedings of the 2nd World Congress on Civil, Structural, and Environmental Engineering (CSEE 2017), Barcelona, Spain, April 2017

Higgins C, Amneus D, Barker L (2015) Methods for Strengthening Reinforced Concrete Bridge Girders Containing Poorly Detailed Flexural Steel Using Near-Surface Mounted Metallics, Report No. *FHWA-OR-RD-16-02*, Oregon Department of Transportation, Salem, OR, pp 138

McCabe M, Harries KA, Sweriduk M (2014) Evaluation of concrete bridge deck rehabilitation by the method of hydrodemolition and latex-modified overlay. In: Proceedings of 15th International Conference on Structural Faults and Repair, London

McCabe M (2013) Structural Evaluation of Slab Rehabilitation by Method of Hydrodemolition and Latex Modified Overlay. (Master's thesis), University of Pittsburgh, PA

Platt S, Harries KA (2019, in revision) Proposed flexural design methodology for titanium reinforcing bars in concrete. Engineering Structures

Platt S (2018a) Development of Titanium Reinforcing Bars for Concrete and Masonry, Doctoral dissertation, University of Pittsburgh

Platt S, Harries KA (2018a) Geometry, Material Properties and Bond Performance of Titanium Reinforcing Bars. Constr Build Mater 187:1253–1266

Platt S, Harries KA (2018b) Study of Galvanic Corrosion Potential of NSM Titanium Reinforcing Bars. Case Stud Constr Mater 9:e00175

Triantafillou L (2012) Cost Comparison of Corrosion Resistant Reinforcing Steel Deployment Considerations. FHWA Office of Infrastructure Research and Development as updated by Wong W (2014)

Physical Behavior of CFRP Retrofitted Reinforced Concrete Slab-Column Connections

R. H. M. Dissanayaka, M. A. L. Silva(✉), L. P. G. Magallagoda, and J. C. P. H. Gamage

Department of Civil Engineering, University of Moratuwa, Moratuwa, Sri Lanka
anulaksilva@gmail.com

Abstract. Carbon Fiber Reinforced Polymer (CFRP) retrofitted application has become widespread in the last decade due to its unique advantages. This paper presents the punching shear strength enhancement of flat slabs which can be achieved with CFRP by embedding it through the slab section and attaching it externally on the tension surface. A common anchorage system that would optimise the use of CFRP has been introduced in CFRP shear reinforcement. The maximum capacity increment in punching shear was 22% and further improvements to achieve even higher load capacities have been discussed. Theoretical expressions have also been compared with the experimental results with recommendations.

Keywords: CFRP · Flat slabs · Punching shear · Reinforced concrete · Internal strengthening · External strengthening

1 Introduction

Strengthening of existing flat slabs might become crucial due to the aging of the structures, changes in the use of the building, exposure to heavy loads, poor detailing at the design stage, early loading of the structure, deterioration and corrosion of rebar as flat slabs are all reasons for punching shear failure.

The key factors which affect the punching shear strength of flat slabs are compressive strength of concrete, ratio of longitudinal reinforcements, dimension of the column and thickness of the slab. Therefore, some of the conventional strengthening methods like concrete jacketing of columns to increase the loading area on the slab, inserting longitudinal steel reinforcements by casting a new concrete layer or introducing Fiber Reinforced Polymer (FRP) and insertion of steel shear reinforcement [3] or FRP [2] have become popular solutions for enhancing punching shear capacity.

Carbon Fiber Reinforced Polymer (CFRP) has become one of the most used retrofitting materials in the construction industry due to the unique combination of characteristics like high tensile strength and low self-weight - thus, a higher strength to weight ratio and ease of installation. However, limited literature is available on the effective use of CFRP to enhance the punching shear capacity.

R. Dissanayake and P. Mendis (Eds.): ICSBE 2018, LNCE 44, pp. 458–469, 2020.
https://doi.org/10.1007/978-981-13-9749-3_40

In recent past, as more attention was given for the use of CFRP materials to enhance the structural performance of strength deficient members [6, 7], different methods of retrofitting using CFRP were found to gain strength in flat slabs. Since punching shear behavior is a type of shear failure which is critical due to its tendency for sudden failure, it has been found that the application of shear reinforcements has the most effective resistance capacity against the shear action. The behavior of flat slab against punching shear was studied in literature using CFRP fabric dowels as shear reinforcement embedded through the slab section near the column where punching shear is critical [1]. It was found out that when the number of shear reinforcement rows around the column increases, the punching shear capacity increases and the mid span deflection at failure decreases. Further, strengthening of flat slabs using FRP rods, bolts and screws as shear reinforcements have been performed [2, 12]. Debonding of FRP rods was observed during the study indicating the insufficiency of area contact to the FRP rod for a proper stress transfer from FRP rod to concrete. Further, the steel screws with proper anchorage provided using tightened nuts gave a higher punching capacity increment than that of given by FRP rods because, FRP rods were not provided with end anchorage at both ends, while steel screws were mechanically anchored with nuts, indicating the importance of proper end anchorage of shear reinforcement.

In this study, taking into account, all the aforementioned drawbacks in previous studies, CFRP fabrics were used in the form of strips to retrofit two alternative arrangements of strengthening by considering the effect of the location of vertical shear reinforcements in terms of the punching shear capacity, mid span deflection and crack patterns. Further, this study proposes a common anchorage system for two separate vertical CFRP shear reinforcements in order to optimize the use of CFRP fabrics.

2 Experimental Details

2.1 Specimen Preparation and Material Properties

Eight slab-column connections were cast with the slab dimensions of 1200 mm × 1200 mm × 100 mm and the stub column dimensions of 100 mm × 100 mm × 150 mm. Reinforcement detailing of each aforementioned specimen is shown in Fig. 1. Concrete and steel material properties are listed in Table 1.

2.2 Strengthening Schemes

Two specimens were kept as control specimens without retrofitting using CFRP and other six were strengthened with CFRP fabric strip widths of 25 mm by embedding them through the slab section and attaching them on to the tension surface of slab as shown in Fig. 2. The strengthening schemes consisted of three different alternative arrangements including two internal embedding of CFRP and one external attachment of CFRP.

In Figs. 2(a) and (b), the two arrangements of CFRP embedment are shown and in each of them, two rows of shear reinforcement were inserted. As shown in the sectional view, two consecutive shear reinforcements of a considering radial section were made

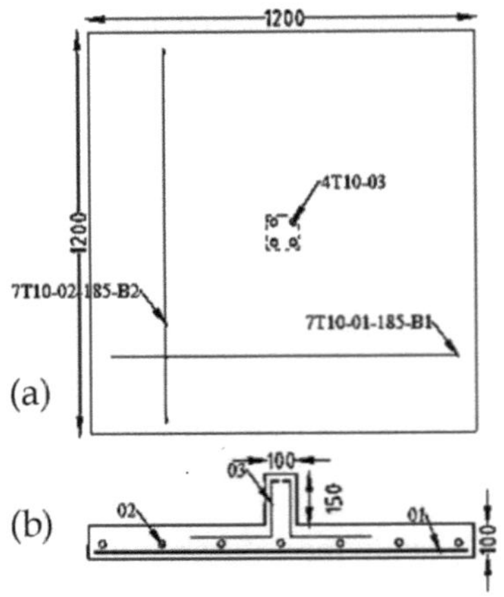

Fig. 1. Steel reinforcement details (a) plan view (b) cross sectional view

Table 1. Material properties of concrete and steel

Properties of concrete	Value
Average compressive strength (N/mm^2)	28.3
water/cement ratio	0.59
Maximum aggregate size (mm)	20
Properties of steel	Value
Yield strength (N/mm^2)	500

of single CFRP strip of 25 mm with separate anchorage provided at the compression face and a common anchorage provided at the tension face. The compression face anchorage length was 200 mm and the tension face common anchorage lengths were different depending on the spacing between two rows of shear reinforcement. The first row of shear embedding was placed at an effective depth (d) away from the column face, which was 70 mm. The second row of shear reinforcements in the type RF-180 and type RF-250 was 110 mm and 180 mm away from the first row of shear reinforcements, respectively. This means approximately, 1.5 d and 2.5 d away from first row of shear reinforcement.

As specified in ACI 318 and Eurocode 2 standards, the critical punching shear perimeter formation of flats slabs is a rectangular perimeter 2, 1.5 and 0.5 times the effective depth away from the column face respectively. Therefore, in both arrangements, CFRP embedment was in the shear critical zone as specified in ACI 318, BS 8110 and Eurocode 2. The arrangement in Fig. 2(c) shows the combination of orthogonal and skewed attachment of CFRP on the tension face of flat slab to identify the contribution of enhancing punching shear capacity.

The surface areas where CFRP sheets needed to be bonded were ground until the low strength cement grout layer got removed, because it is important to maintain a proper bond between CFRP sheet and the concrete substrate.

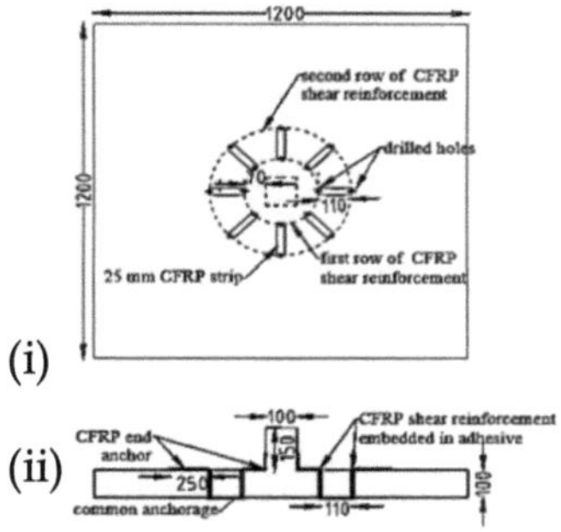

(a) Specimen Type RF-180

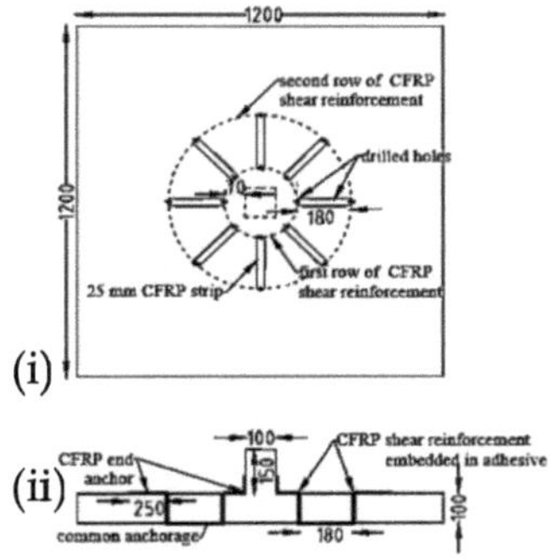

(b) Specimen Type RF-250

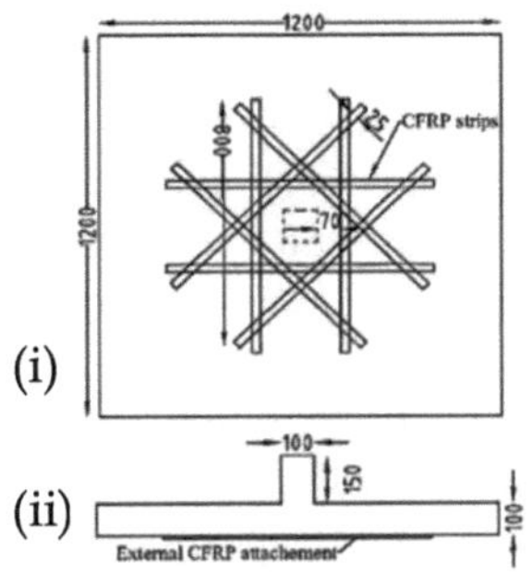

(c) Specimen Type RF-EX

Fig. 2. CFRP applied strengthening scheme (i) plan view (ii) sectional view

The primer was applied on the ground and cleaned concrete surface in order to provide a proper bond between the adhesive and the dry concrete interface. The application of epoxy saturated CFRP was done followed by the application of epoxy adhesive on primer applied concrete surface. The two ends of each CFRP strip were inserted through the drilled holes and attached on the compression surface of the slab. Afterwards, the drilled holes were refilled using a concrete repairing grout ensuring the force transfer between the shear reinforcement and the concrete slab. The properties of CFRP fabric and adhesive are mentioned in Tables 2 and 3.

Table 2. Material properties of concrete and steel

Properties of CFRP fabrics [5]	Value [4]	Properties of CFRP fabric Adhesive	Value
Thickness (mm)	0.166	Tensile strength (MPa)	45
Modulus of elasticity (GPa)	240	Flexural strength (MPa)	60
Yield strength (MPa)	4000	Compressive modulus (GPa)	1.67
Elongation at break	2%		

Table 3. Test Results

Specimen type	Failure load (KN)	Percentage increment of load (%)	Deflection at failure (mm)	Percentage increment of deflection (%)
C1	103	–	13.8	–
C2	98.1	–	11.2	–
RF-180-1	112.815	12.2	17.5	40
RF-180-2	112.815	12.2	15.42	23.36
RF-250-1	98.2	−2.34	18.9	51.2
RF-250-2	107.91	7.2	18.16	45.28
RF-EX-1	117.72	17.08	12.05	−3.6
RF-EX-2	127.53	26.83	10.71	−14.32

2.3 Testing of Specimens

Specimens were tested after the curing period, providing simply supported conditions at four edges of each specimen using steel I sections. A point load was applied on the middle stub column using the hydraulic jack at a rate of 10 kN/min. A steel cap with rubber was used to support the column stub to distribute the load monolithically to the column from the hydraulic jack ensuring the column would not fail prior to the slab. Test setup and instrumentation done, are shown in Fig. 3. The mid span deflection corresponding to the applied load was measured using three dial gauges fixed 75 mm away from the column face while the point load was applied. Crack propagation at the tension and compression surfaces of the specimens were examined throughout the loading procedure and the failure load was recorded.

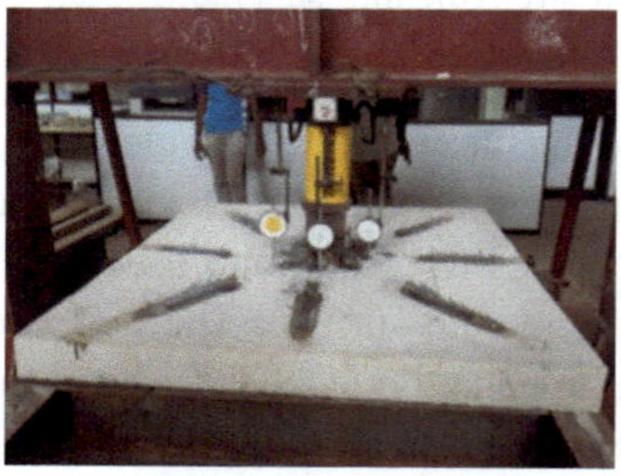

Fig. 3. Instrumentation and test set-up

3 Test Results and Discussion

3.1 Failure Mechanisms

Six flat slab specimens were failed in sudden punching shear failure mode while two failed in flexural failure mode (Fig. 4). The punching shear crack distribution was a conical shape which initiated from the tension surface and ended at compression surface near the column face. In all specimens flexural cracks started appearing before punching shear failure happened.

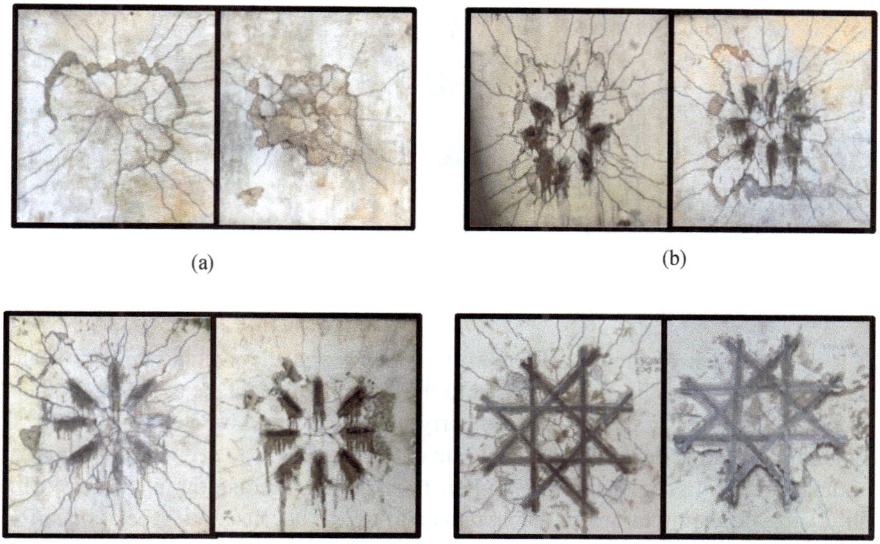

(c) (d)

Fig. 4. Crack patterns on tension surface of tested specimens (a) control specimens (b) type RF-180 (c) type RF-250 (d) type Ex-RF

Figure 4(a) shows the crack pattern on the tension surface of the control sepecimens. Flexural cracks started appearing around the load application of 60 kN and the specimen failed due to punching shear failure, forming a punching shear crack of one to four times the effective depth away from the column face.

Punching shear cracks were between two to five times the effective depth away from the column face in specimen type RF-180 (Fig. 4(b)). In some places, the punching shear crack on tension surface passed through the outer row of CFRP shear reinforcement. The width of the flexural cracks was around 0.3 mm.

Punching shear failure was observed after the appearance of a set of flexural cracks on the tension surface of the type of specimens RF-250 as shown in Fig. 4(c). The punching shear crack formation was in between three to four times the effective depth away from the column perimeter and therefore, the crack passed through the outer row of CFRP shear reinforcement. The number of flexural cracks compared to the type

RF-180 was higher in type RF-250 and large crack widths were observed under the column. The type RF-EX only showed flexural cracks as shown below in Fig. 4(d).

Figure 5 shows the compression surface of RF-250-1 slab, after the collapse. It shows that the crack has passed through the holes. The area has become weak due to the damage occurred during drilling. Truncated cone at the punching shear failure was occurred by pushing the concrete section from this crack.

Fig. 5. Crack patterns on compression surface of tested specimens

3.2 Load Versus Deflection

The load increment of the type RF-180 was 12.2% and the average deflection increment at the failure was 31.6% compared to the control panels. Due to the strengthening of slabs with the CFRP sheets, the load carrying capacity and the deflection of the slab has increased. The deflection increment rate of the type RF-180 seemed to be increased than that of control panels as shown in Fig. 6 indicating a reduction in stiffness. This is contradictory to the literature available related to CFRP because, with the application of CFRP, the structure stiffness should increase. The reason behind this is either the damage done in the slab to install CFRP shear reinforcement might have created micro-cracks resulting in stiffness reduction or the refilling material might not have transferred the stresses from slab to CFRP reinforcement properly.

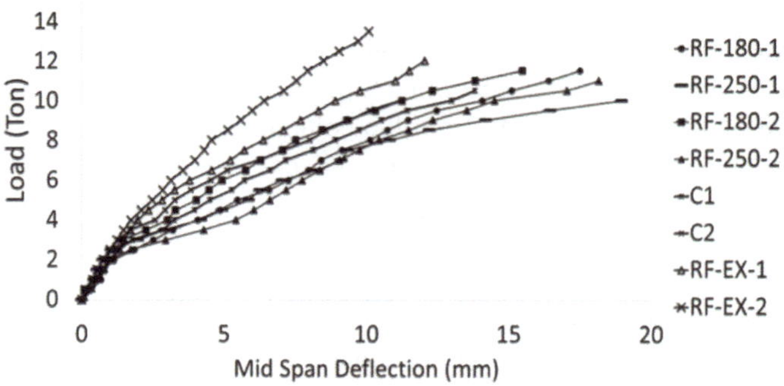

Fig. 6. Load versus mid span deflection of specimens

The type RF-250 has a punching shear capacity increment of 2.5%. The deflection at failure has been increased 48%, compared to the control panels. The deflection increment rates were increased as observed in type RF-180 with respect to the control panels. Due to the retrofitting of the slab, the load carrying capacity and the stiffness of both RF-180 and RF-250 types has been reduced. The reason might be the inadequacy of refilling material of drilled holes to transfer the forces from slab to CFRP shear reinforcement or the damage done to the concrete slab while drilling must have become dominant.

RF-EX panels have achieved the maximum average load bearing capacity of 22% and the deflection at failure was reduced compared to that of control specimens. This shows, the attaching of CFRP at the tension surface has improved punching shear capacity as well as the stiffness of the structure without any damage to the structure.

4 Theoretical Analysis and Comparison

Theoretical analysis was conducted to evaluate the applicability of available analytical models in predicting the strength gain of flat slabs retrofitted with CFRP.

4.1 ACI 318

The punching shear strength of none shear strengthened normal weight concrete can be calculated theoretically using the equations of ACI 318 by ACI Committee [8] given below. The minimum of the following three Q_{up} values denotes the punching shear capacity.

$$Q_{up} = \frac{1}{3} \times \sqrt{f_c'} b_0 d \tag{1}$$

$$Q_{up} = \left(2 + \frac{1}{\beta_c}\right) \times \frac{\sqrt{f_c'}}{12} b_0 d \tag{2}$$

$$Q_{up} = \left(2 + \frac{\alpha_s d}{b_0}\right) \times \frac{\sqrt{f_c'}}{12} b_0 d \tag{3}$$

Where, Q_{up} is the punching shear capacity; b_0 is the critical section perimeter 0.5 times the effective depth away from the face of loading area; d is the effective depth of the slab, α_s is a constant of 40 for interior columns, 30 for edge columns and 20 for corner columns; β_c is the ratio of long side to the short side of the loading area; f_c' is the concrete cylinder compressive strength. Here, all there equations have been derived not considering the effect from flexural reinforcement ratio.

For slabs strengthened against punching shear, the above equations get modified as below by Inacio [11].

$$Q_{up} = 0.5 \times \frac{1}{3} \times \sqrt{f_c'} b_0 d + \frac{A_{sw} f_{yt} d}{s_r} \tag{4}$$

Here, $\mathbf{A}_{sw}$ is the cross sectional area of shear reinforcement $\mathbf{f}_{yt}$ is the tensile strength of shear reinforcement and $\mathbf{s_r}$ is the distance between two consecutive rows of shear reinforcement.

4.2 Eurocode2-2004

The punching shear capacity of none shear strengthened flat slabs has been calculated using Eurocode 2 by British Standard Institution [9] considering the effect of flexural reinforcement and not considering the aspect ratio of loading area as follows.

$$Q_{up} = 0.18K \sqrt[3]{100 \rho f_c'} b_0 d \tag{5}$$

Where, $K = 1 + (200/d)^{1/2}$ and should not be taken more than 2; ρ is the flexural reinforcement ratio and it is limited to 0.02; b_0 is the perimeter of the critical section with round corners at a distance of 2 times the effective depth away from the surface of loading area.

The punching shear strength of retrofitted slabs can be obtained by the equation that has been modified using the simplified critical shear crack theory (CSCT), proposed by Faria [10] considering the concept of equivalent effective depth, d_v.

$$d_v = \frac{d_s A_{sl} + d_f A_{ft} E_f / E_s}{A_{sl} + A_{ft} E_f / E_s} \tag{6}$$

And also by using equivalent reinforcement ratio ρ_{eql} that can be obtain by using,

$$\rho_{eql} = \rho_{sl} + E_f / E_s \rho_{fl} \tag{7}$$

The modified punching shear evaluation equation for retrofitted flat slat slabs is as follows by Inacio et al., (11).

$$Q_{up} = 0.75 \times 0.18K \sqrt[3]{100 \rho f_c'} b_0 d + 1.5 \times \frac{A_{sw} f_{yt} d}{s_r} \tag{8}$$

As shown in Table 4, the non-shear strengthened specimen punching shear capacity prediction by Eurocode 2 is more accurate than ACI 318 prediction. The reason is that the amount of flexural reinforcement has an influence on enhancing the punching shear capacity and the ACI code predicts an underestimated capacity for punching shear.

When it comes to the comparison of test results with theoretical results in CFRP shear strengthened specimens, ACI 318 code predicts more accurate results than Eurocode 2. The contribution of concrete to resist punching shear was estimated as half the contribution when no shear reinforcement exists and the contribution from the shear reinforcement has been taken 100% to enhance the punching shear capacity.

Table 4. Theoretical versus Experimental Results

	Experimental punching shear load (kN) [P_{exp}]	ACI 318-14 Code	Eurocode 2-2004		
		Code predicted load (kN) [P_{ACI}]	P_{exp}/P_{ACI}	Code predicted load (kN) [P_{EC2}]	P_{exp}/P_{EC2}
Type C	100.6	83.96	0.83	95.3	0.95
Type RF-180	112.8	126.5	1.12	147.5	1.31
Type RF-250	103.1	93.6	0.91	117.95	1.14
Type RF-EX	122.63	85.16	1.44	98.3	1.25

Eurocode 2 has overestimated the punching shear capacity. Therefore, it is recommended to reduce the safety factors in the equations.

However, CFRP flexural reinforced specimen experimental results are more compatible with Eurocode 2 prediction values which have been modified using Muttoni equivalent reinforcement ratio and effective depth. The reason is ACI 318 code does not provide sufficient provisions for the amount of flexural reinforcement ratio and therefore, it could not predict accurate results considering the flexural reinforcement ratio increment change.

5 Conclusions

CFRP strengthened flat slab behavior is yet a study area which needs to be explored by performing research studies and analysing them using modified theoretical expressions. Therefore, in this study, the behavior of post installation of CFRP as shear reinforcements was monitored. By observing the ultimate punching shear capacities, the deflection at failure, crack patterns and failure modes, the following conclusions can be made:

- Punching shear crack patterns of non-strengthened flats slabs indicates the punching shear critical perimeter should be approximately at two times the effective depth away from the column face because, the punching shear crack was two to four times the effective depth away from the column face. This observation comples with the Eurocode 2 predictions.
- The maximum punching shear capacity observed in post-installed shear reinforcement strengthening scheme was 12.2% with respect to the non-strengthened panels.
- The type RF-180 obtained a higher punching shear capacity because; the arrangement had two rows of shear reinforcements at the shear critical area while type RF-250 had only one. Therefore, the number of rows of shear reinforcement at the shear critical area is important.
- The punching crack pattern on RF-180 denotes the critical punching shear perimeter location should be closer to 2.5 the effective depth away from column face. This indicates, the application of shear reinforcements in the shear critical area would distribute the shear stress further away from column face.

- Load versus mid span deflection curves indicates stiffness reduction in CFRP shear strengthened slabs. The reasons must be the damage done while drilling the slab to install CFRP shear reinforcements has become server.
- The load carrying capacity and the stiffness has been increased in the specimens where, CFRP was applied externally. The load carrying capacity increment was 22% in average.
- Theoretical predictions calculated using Eurocode 2 and ACI 318 predicted compatible results with the experimental results.
- The punching shear capacity prediction of ACI 318 is more accurate for CFRP shear strengthened specimens because, the proportions provided by concrete and shear reinforcements to resist punching shear are accurate than that of in Eurocode 2.
- Eurocode 2 accounts for the amount of flexural reinforcement ratio in the punching shear capacity. Therefore, it predicts more accurate results when the amount of flexural reinforcement ratio changes than that of ACI 318, using Muttoni modified equivalent effective depth and equivalent flexural reinforcement ratio.

Acknowledgments. Authors would like to extend their gratitude for the technical officers, Mr. D.M.N.L.Dissanayake and Mr. Roshan at Structural Testing Laboratory and Building Material Laboratory; and staff at Structural Testing Laboratory, Department of Civil Engineering, University of Moratuwa.

References

1. Meisami MH, Mostofinejad D, Nakamura H (2014) Strengthening of flat slabs with FRP fan for punching shear. Compos Struct. doi.org/10.1016/j.compstruct.2014.08.041
2. Meisami MH, Mostofinejad D, Nakamura H (2013) Punching shear strengthening of two-way flat slabs using CFRP rods. Compos Struct 99:112–122. https://doi.org/10.1016/j.compstruct.2012.11.028
3. Askar HS (2015) Usage of prestressed vertical bolts for retrofitting. Alexandria Eng J
4. X-wrap lamination adhesive (2017). http://www.x-calibur.us/files/X-Wrap%20Lamination%20Adhesive.pdf. Accessed 21 Jan 2018
5. X-wrap c300 (2017). http://www.x-calibur.us/files/X-Wrap%20C300.pdf. Accessed 21 Jan 2018
6. Wijerathna DMN, Gamage JCPH (2013) Investigation on Effective Use of CFRP laminates for Flexural Performance, In: 4th International Conference on Structural Engineering and Construction Management (ICSCM2013), Kandy, Sri Lanka
7. Wijerathna DMN, Gamage JCPH (2014) A review on Alternative Bonding Techniques to Delay End debonding of CFRP/Concrete composites. In: 5th International Conference on Sustainable Built Environment (ICSBE2014), Kandy, Sri Lanka, pp 144–150
8. ACI Committee 318. (1999). Building code requirements for structural concrete: (ACI 318); and commentary (ACI 318R). Farmington Hills, Mich, American Concrete Institute
9. British Standards Institution (2004) Eurocode 2: design of concrete structures: British standard. London, BSi

10. Faria DMV, Einpaul J, Ramos AP, Fernández Ruiz M, Muttoni A (2014) On the efficiency of flat slabs strengthening against punching using externally bonded fibre reinforced polymers. Construction and Building Materials, Netherlands, vol 73, pp 366–377. https://doi.org/10.1016/j.conbuildmat.2014.09.084
11. Inácio M, Almeida A, Faria D, Lúcio V, Ramos A (2015) Punching of high strength concrete flat slabs without shear reinforcement Elsevier. https://www.sciencedirect.com/science/article/pii/S0141029615005738. Accessed 8 Nov 2018
12. Bartolac M, Damjanović D, Duvnjak I (2015) Punching Strength of Flat Slabs with and without Shear Reinforcement. GRAĐEVINAR 67(8):771–786. https://doi.org/10.14256/JCE.1361.2015

Review on Fire Performance of Cellular Lightweight Concrete

I. R. Upasiri[1(✉)], K. M. C. Konthesingha[2], K. Poologanathan[2], S. M. A. Nanayakkara[3], and B. Nagaratnam[2]

[1] University of Sri Jayewardenepura, 41 Lumbini Avenue, Dehiwala-Mount Lavinia, Sri Lanka
irinduupasiri@gmail.com

[2] Northumbria University Newcastle, Ellison Building ELB117 Northumbria University, Newcastle upon Tyne NE18ST, UK

[3] University of Moratuwa, Katubedda, Moratuwa, Sri Lanka

Abstract. Structural fire damage can be identified as a common accidental disaster throughout the world which cause thousands of deaths, injuries and millions in property damage each year. Fire represents one of the most severe conditions to which structures may be subjected. Generally, structural elements will be exposed to very high temperature (1200 °C) during a fire propagation. Fire safety of a structure is measured in terms of fire resistance, which is the duration that a structural member can exhibit resistance with respect to structural integrity, stability and heat transmission. Concrete generally provides better fire resisting characteristics compared to the other construction materials due to its low thermal conductivity, high heat capacity and slower strength degradation with temperature. Cellular lightweight concrete (CLC) is one of the novel type of concrete which can be identified as a better construction material than conventional concrete due to its numerous advantages. However, limited research work has been carried out to determine the fire performance of CLC. Fire response of structural members depends on the thermal, mechanical and deformation properties of the structural material at elevated temperatures. Even though properties at elevated temperatures for normal weight concrete is available in literature, properties of CLC at elevated temperatures (ambient to 1200 °C) is not thoroughly investigated. Further, CLC fire rating under natural/parametric fire situations and under hydrocarbon fire situations needs to be studied. EN 1992.1.2 provides minimum thickness requirements under standard fire situations for non-loadbearing and load bearing normal weight concrete walls, but for CLC, these values are not available, hence required to be included. Also, parameters and material property limitations related to spalling effect of CLC during fire exposure has not being investigated. Moreover, residual characteristics of CLC walls after fire situations and ability to withstand a second fire situation needs to be assessed.

Keywords: Cellular lightweight concrete · Fire performance · Design guidelines · Insulation · Integrity

R. Dissanayake and P. Mendis (Eds.): ICSBE 2018, LNCE 44, pp. 470–478, 2020.
https://doi.org/10.1007/978-981-13-9749-3_41

1 Introduction

Fire represents one of the most severe condition that structures may be subjected to (Kodur 2014). Structural fire damage causes thousands of deaths, injuries and millions in property damage throughout the world each year (Brushlinsky et al. 2007). According to the Centre of fire statistics data (CTIF 2019), there are nearly 510,000 structural fire situations reported each year, which means once in every 62 s, a structural fire is occurring somewhere in the world. World trade centre fire in 2001, Shanghai fire in 2010 and Grenfell Tower Fire in 2017 are recent massive structural fires reported in the world (En.wikipedia.org 2018). Therefore, appropriate fire safety construction is an important aspect in building construction.

Concrete generally provides the best fire resistant properties compared to other construction materials (Kodur and Raut 2010). This excellent fire resistant characteristics are due to its low thermal conductivity, high heat capacity and slower strength degradation with temperature (Kodur 2014). Concrete is available in various forms and it is often grouped under different categories based on weight (as normal weight and light weight concrete), strength (as normal strength, high strength, and ultrahigh strength concrete), presence of fibres (as plain and fibre-reinforced concrete), and performance (as conventional and high performance concrete) (Kodur 2014).

Lightweight concrete has many advantages over conventional concrete, excellent in acoustic performance, earthquake resistant, good insulation, workability, long-life span due to termite and fire resistance, weather proof and, material savings (Singh 2016) and therefore can be utilised as a better construction material in low rise and multi storey buildings.

Lightweight concrete can be defined as a type of concrete which has low density compared to normal weight concrete. Lightweight concrete has dry density of 300 kg/m^3–1800 kg/m^3, while conventional concrete has a dry density of 2400 kg/m^3. Lightweight concrete can be categorised as lightweight aggregate concrete and cellular lightweight concrete (CLC). Porous lightweight aggregates with low specific gravity such as pumice, scoria and volcanic aggregates are used to produce lightweight aggregate concrete. Due to low specific gravity of aggregates, resultant concrete also has less density. Cellular lightweight concrete does not contain coarse aggregate, thus can be regarded as mortar. It is produced by introducing air or other gas into a cement slurry and fine sand. Cellular concrete can be categorised as aerated concrete and foam concrete. Fine powder of Aluminium is used as air entraining agent to produce aerated concrete. Aluminium powder mixed with the cement slurry, reacts with the calcium hydroxide to produce hydrogen gas, which produces a cellular structure and thus makes the concrete lighter than the conventional concrete. Foam concrete is produced by introducing foam into cement mortar. The foam is created using a foaming agent, mixed with water and air from a foam generator. When the sand in cement mortar is replaced with fly ash, better performance can be obtained in cellular concrete. Since it consumes fly ash which is a waste material from coal fired thermal power plants, cellular lightweight concrete can be identified as a green building material. With the reduction of density, the strength and stiffness of cellular concrete is decreased (Mydin and Wang 2011). CLC is a porous material; since at high temperatures, heat transfer through porous material is influenced by radiation, CLC with its lower thermal conductivity and diffusivity results in better fire resistance. Furthermore, lower density of foam concrete has exhibited better fire performance characteristics (Ramamurthy et al. 2009). However, cellular concrete cannot be used as reinforced concrete as it has a cellular structure and therefore corrosion of steel reinforcement is quite proactive (Singh 2016). In this paper, existing research studies on CLC with respect to fire performance will be assessed and research gaps will be highlighted.

2 Fire Performance of Cellular Lightweight Concrete

Fire resistance of structural members are generally evaluated through standard fire tests (ASTM E119-08b 2008). However, due to high cost and time consumption for fire tests, numerical approaches and finite element modelling has been identified as an effective method of evaluating fire performance of structures (Kodur 2014). These numerical models are developed considering the material properties at elevated temperatures.

2.1 Fire Performance of Cellular Lightweight Concrete with Fire Tests

The simplest way of representing fire is a time temperature curve. ASTM E 119 and ISO 834 are most widely used standard fire curves (Denoël 2007) (see Fig. 1). To determine the fire performance of a structural member, fire tests are carried out. There are full scale fire tests and furnace tests. Full scale fire tests will give exact fire behaviour of the entire structure. However, it is very expensive and time-consuming. Furnace tests are used to evaluate the fire behaviour of structural elements. Furnace temperature needs to be controlled according to the designated fire curve (standard or natural fire curve). Usually, furnaces are equipped with devices to measure temperature, and deformations, and to load test specimens.

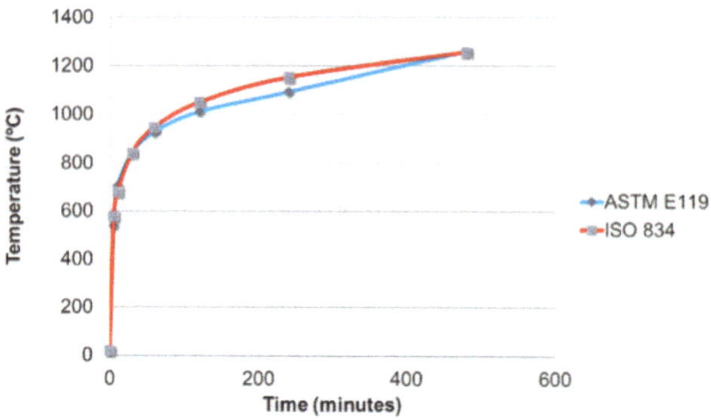

Fig. 1. Time temperature variation of Standard fire curves.

Fire safety of a component is assessed relevant to the standard fire situation. ASTM E 119 considers the inhabit passage of flame or hot gases through the component as a failure criterion while ISO 834 measures it by the ignition of a cotton pad held close to an opening for 10 s. ASTM E 119 has limited the average temperature rise of an unexposed surface to 139 °C while ISO 834 allows the maximum average temperature rise of 140 °C at unexposed surface. Further, ASTM E 119 considers the temperature rise of reinforcement. It is limited to 593 °C. ISO 834 considers the deformation of the member. Downward deformation of flexural member is limited to L/30 where L is the span of the member.

Standard fires are suitable for comparison of fire resistance of materials, but in actual fire situations, ventilation effects, environmental conditions, material composition, boundary conditions and fuel availability of the compartment will affect the intensity of fire. Considering the above effects, more realistic fire curves (natural or parametric fire curves) have been developed. Lie's parametric curve, Swedish curve, Law's maximum temperature equations, Mehaffey's Japanees parametric curves, Ma and Makelainen's parametric curves, Barnett's BFD curves and Euro code parametric curves are such parametric curves developed by various researchers to interpret natural fire situations (Ariyanayagam and Mahendran 2014).

Laurent (2014) has investigated the fire performance of foamed concrete under insulation criteria up to 900 °C temperature and concluded that foamed concrete has better fire performance characteristics compared to normal weight concrete.

For cellular lightweight concrete with different mix proportions, standard fire performance has not been investigated, hence, needs to be determined. Further, with different parametric fire situations and hydrocarbon fire situations research studies on CLC behavior in fire is not available.

Research studies on residual properties of CLC after fire scenarios is not available in the literature where these properties are important to investigate the fire performance of CLC. During fire exposure, structural element will undergo high temperature variation which will induce internal stresses and variation of bonding in CLC. Therefore, though the material is capable of handling fire, after the fire situation, CLC properties might be different from the initial state. Hence, post fire load bearing capacities and capability of handling second fire, needs to be determined.

Breaking up of pieces of concrete when it is exposed to high and rapidly rising temperatures is defined as spalling (Kodur 2014). Spalling can be identified as a special characteristic in concrete (Kodur 2014). Fire-induced spalling is one of the major components to be considered in evaluating the response of concrete members exposed to fire. It occurs due to the build-up pore pressure during heating (Dwaikat and Kodur 2010). Spalling depends on moisture content, permeability, type of fire exposure and tensile strength of concrete (Kodur 2014). Research studies were not done on investigating of spalling of CLC. Spalling might cause a significant effect on fire performance of CLC due to its cellular structure.

2.2 Fire Performance of CLC with Properties at Elevated Temperature

Fire performance of material is influenced by the characteristics of the material, thermal properties, mechanical properties, deformation properties and special characteristics of material in fire. Thermal properties influence the heat transfer in the structural element, whereas mechanical properties influence the strength and stiffness variation. Deformation properties together with mechanical properties influence the extent deformation and strains in the structural member (Dwaikat and Kodur 2010). All these properties vary as a function of temperature and depends on the composition of the material. Hence to predict fire performance, elevated temperature properties need to be determined.

Thermal properties that influence temperature rise and heat distribution in structural members are thermal conductivity, specific heat and mass loss (Kodur 2014). The thermal conductivity of concrete is the rate of heat transferred through a unit thickness of the material per unit temperature difference. Thermal conductivity can be measured using several techniques such as pulse method, hot-guarded plate method (HGP), two-linear parallel probe method, transient hot wire method and laser flash method. Specific heat is the amount of heat per unit mass required to change the temperature of material by one degree. The thermal diffusivity of concrete, often denoted as the ratio between its thermal conductivity and thermal capacity. It is a measure of the rate of heat transported from the exposed surface of concrete to the inside. Flash method is more convenient in measuring thermal conductivity, specific heat and thermal diffusivity of CLC in elevated temperatures. Density of concrete depends primarily upon the type of aggregate. With the increment of temperature density also changes (Kodur 2014). This effect is more predominant in cellular lightweight concrete than conventional concrete since CLC consists more voids. Thermo gravimetric analysis is more suitable to determine the density variation of CLC at elevated temperatures.

Mechanical properties that influence the fire performance of concrete are compressive and tensile strength, modulus of elasticity, and stress strain response at elevated temperatures (Kodur 2014). Thermo-mechanical analysis needs to be done to determine the mechanical properties at elevated temperatures. Both "steady state" and "transient" test methods are available for determining the mechanical properties at elevated temperatures (Kodur 2014). Unlike at ambient temperature, mechanical property tests at high temperature are usually carried out on a wide range of specimen sizes due to a lack of standardized test specifications for undertaking high temperature mechanical property tests (RILEMTC129-MHT 2000).

Deformation properties that affects the fire performance of concrete are thermal expansion and creep (Kodur 2014). Creep, can be defined as the time-dependent plastic deformation of the material. At normal stresses and ambient temperatures, deformations due to creep are not significant. At higher stress levels and at elevated temperatures, however, the rate of deformation caused by creep can be substantial (Kodur 2014). Dialatometric analysis can be done to determine the deformation characteristics at elevated temperatures.

Othuman and Wang (2011) have studied thermal properties of foam concrete at elevated temperatures. They have measured thermal conductivity of foam concrete with various densities up to 600 °C using the hot guarded plate test. Further, they have proposed analytical models to predict thermal conductivity and specific heat of foam concrete at elevated temperatures. In the analytical method, lightweight foamed concrete (LFC) was considered as a mixture of dried LFC and water up to 170 °C. Above 170 °C, LFC was considered as a mix of dried LFC and air pores. Average pore diameter found from the microscopic images used was in the analytical model. Heat transfer of solid material through conduction and heat transfer of air voids through convection and radiation were considered. They have concluded that the models give accurate results predicting the thermal conductivity of foamed concrete at elevated temperatures. Further, they have mentioned that specific heat of LFC can be approximated as a sum of base value of dried LFC with addition of heat required to evaporate free water.

Mydin and Wang (2011) have measured mechanical properties of foamed concrete with various densities in elevated temperatures. They have measured compressive strength, modulus of elasticity, compressive stress-strain relationship and flexural bending strength of foamed concrete with densities 650 kg/m^3 and 1000 kg/m^3. All the tests were conducted increasing the load while keeping the specimen at a constant elevated temperature (100 °C, 200 °C, 300 °C, up to 600 °C) and it was found that the compressive strength of foamed concrete decreases with temperature. However, up to 200 °C, foamed concrete is in a position to maintain 94% of original unheated strength. But at 400 °C. 75% and at 600 °C, only 40% of original strength was retained. Moreover, modulus of elasticity at 200 °C, 400 °C and 600 °C was respectively 75%, 40% and 25% of the original value.

A number of concrete mechanical property predictive models have been proposed for normal-strength concrete and these models have been checked to determine their applicability to foamed concrete. Sayadi et al. (2016) have concluded that the following models as given in Eqs. (1–6) give accurate results for foamed concrete properties in elevated temperatures.

Compressive Strength

$$f_{cT} = f_c \text{ for } T \leq 100\,^\circ\text{C} \tag{1}$$

$$\begin{gathered} f_{cT} = f_c(1.067 - 0.00067\,T) \\ \text{for } 100\,^\circ\text{C} \leq T \leq 400\,^\circ\text{C} \end{gathered} \tag{2}$$

$$f_{cT} = f_c(1.44 - 0.0016\,T) \text{ for } T \leq 400\,^\circ\text{C} \tag{3}$$

f_{cT} – Compressive strength at elevated temperature
f_c – Compressive strength at ambient temperature
T – Temperature

Elastic Modulus

$$E_{cT} = E_c \text{ for } T \leq 60\,^\circ\text{C} \tag{4}$$

$$E_{cT} = \frac{800 - T}{740} E_c \text{ for } 60\,^\circ\text{C} \leq T \leq 800\,^\circ\text{C} \tag{5}$$

E_{cT} – Elastic modulus at elevated temperature
E_c – Elastic modulus at ambient temperature
T – Temperature

Stress-Strain Relationship

$$f'_{cT} = \frac{3\varepsilon_{cT} f_{cT}}{\varepsilon_{0T}[2 + (\frac{\varepsilon_{cT}}{\varepsilon_{0T}})^3} \tag{6}$$

f'_{cT} – Compressive stress at elevated temperature
ε_{cT} – Strain at elevated temperature
ε_{0T} – Strain at maximum stress

Vinith Kumar et al. (2018) have compared thermal behaviour of foamed concrete with clay bricks and concluded that thermal conductivity of foamed concrete varies from 0.021–0.035 W/mK while brickwork has a thermal conductivity of 0.6 W/mK to 1.0 W/mK. Sayadi et al. (2016) has studied the effect of expanded polystyrene (EPS) partials in foamed concrete. Increase of EPS cause reduction in thermal conductivity and higher fire endurance can be obtained with 28% EPS volume in foamed concrete. Kashani et al. (2017) studied the influence of recycled tyre crumbs in cellular lightweight concrete as an insulating constituent. They have concluded that excellent sound and thermal insulation with very low water absorption and total porosity have been achieved by the inclusion of recycled tyre crumb in lightweight cellular concrete. She et al. (2018) proposed a numerical method of predicting thermal behaviour of cellular concrete having different pore shapes.

Even though experimental results for thermal properties of CLC are available up to 600 °C, variation of thermal conductivity, specific heat, thermal expansion with temperature need to be measured up to 1200 °C to model an actual fire situation. Further, density, porosity, strength and elasticity variation with ambient temperature to 1200 °C need to be determined for better prediction in fire performance of CLC. Normally, mechanical properties at elevated temperature are measured keeping the specimen in a constant elevated temperature and increasing the load. However, it will be more realistic if a test method could be developed to increase the temperature while keeping the load constant when determining the elevated temperature mechanical properties.

3 Design Guidelines for Cellular Lightweight Concrete in Fire Situation

EN 1992-1-2 (2004) Design of concrete structures -part 1-2 General rules- Structural fire design describes the principles, requirements and rules for the structural design of buildings exposed to fire. It includes safety requirements, design procedures and design aids for concrete structures under fire situations. The methods described in EN 1992-1-2 are applicable to normal weight concrete up to strength class C90/105 and for lightweight concrete up to strength class LC55/60.

EN 1992-1-2 has tabulated minimum wall thickness requirement for 30, 60, 90, and 120, 180 and 240 min for fire resistant walls for normal weight concrete with siliceous aggregates. It also states that thickness requirement is reduced by 10% when calcareous aggregates are used in normal weight concrete. However, minimum thickness requirement for CLC panels which can be used as non-load bearing walls is not available in the design guidelines.

Further, EN 1992-1-2 states that both load bearing and non-load bearing walls should be limited to wall clear height to thickness ratio of 40 for normal weight concrete. However, applicability of this ratio for CLC walls should be investigated.

Moreover, explosive spalling of concrete can be observed in concrete structures during fire exposure. EN 1992-1-2 has given limitations to the moisture content of normal weight concrete to prevent explosive spalling in the member. However, moisture content limitation for CLC to prevent explosive spalling is not available in design codes of practice and needs to be investigated.

4 Concluding Remarks

Cellular lightweight concrete can be identified as a novel type of concrete which can be used as a construction material for load bearing and non-load bearing walls in low-rise and multi storey buildings. It has numerous advantages against other construction materials. A comprehensive literature survey was done on the fire performance of CLC and the research gaps were identified.

CLC elevated temperature property variations are available up to 600 °C. However, to study the behaviour of CLC in fire situation, CLC elevated temperature properties (ambient to 1200 °C) need to be measured. Properties such as, thermal conductivity, specific heat, thermal expansion, density, porosity, strength and elongation need to be measured in elevated temperatures. Standard fire tests need to be conducted for non-load bearing and load bearing CLC walls to determine the fire rating and spalling of CLC. Moreover, CLC behaviour in hydrocarbon fire situations and fire behaviour under different parametric fire situations needs to be assessed. Post fire residual properties of CLC needs to be measured to predict the adequacy of CLC to withstand in a second fire situation. Properties of CLC at elevated temperatures and fire tests results can be used to update the design guidelines for CLC walls in fire situations. Minimum thickness requirement for non-load bearing and load bearing CLC walls, maximum wall height to thickness ratio and material property limitations to prevent explosive spalling need to be included in relevant design guidelines.

References

Ariyanayagam A, Mahendran M (2014) Development of realistic design fire time-temperature curves for the testing of cold-formed steel wall systems. Front Struct Civil Eng 8(4):427–447

Brushlinsky N, Ahrens M, Sokolov S, Wagner P (2007) World Fire Statistics. Center of fire Statistics

Ctif.org (2019) World Fire Statistics Issue no 22 - 2017–CTIF - International Association of Fire Services for Safer Citizens through Skilled Firefighters. https://www.ctif.org/news/world-fire-statistics-issue-no-22-2017. Accessed 6 Oct 2018

En.wikipedia.org (2018) List of building or structure fires. https://en.wikipedia.org/wiki/List_of_building_or_structure_fires. Accessed 15 Nov 2018

Kashani A, Ngo T, Mendis P, Black J, Hajimohammadi A (2017) A sustainable application of recycled tyre crumbs as insulator in lightweight cellular concrete. J Cleaner Prod 149:925–935

Kodur V (2014) Properties of concrete at elevated temperatures. ISRN Civ Eng 2014:1–15

Laurent C (2014) Investigating the fire resistance of ultra light weight foam concrete. Revista Técnica de la Facultad de Ingeniería Universidad del Zulia 37(1):11–18

Dwaikatand MB, Kodur VKR (2010) Fire induced spalling in high strength concrete beams. Fire Technol 46(1):251–274

Mydin M, Wang Y (2011) Mechanical properties of foamed concrete exposed to high temperatures. Constr Build Mater 26(1):638–654

Othuman M, Wang Y (2011) Elevated-temperature thermal properties of lightweight foamed concrete. Constr Build Mater 25(2):705–716

Ramamurthy K, Kunhanandan Nambiar E, Indu Siva Ranjani G (2009) A classification of studies on properties of foam concrete. Cem Concr Compos 31(6):388–396

RILEMTC129-MHT (2000) Test methods for mechanical properties of concrete at high temperatures, Part 4—tensile strength for service and accident conditions. Mater Struct 33:219–223

Sayadi A, Tapia J, Neitzert T, Clifton G (2016) Effects of expanded polystyrene (EPS) particles on fire resistance, thermal conductivity and compressive strength of foamed concrete. Constr Build Mater 112:716–724

She W, Zhao G, Cai D, Jiang J, Cao X (2018) Numerical study on the effect of pore shapes on the thermal behaviors of cellular concrete. Constr Build Mater 163:113–121

Singh N (2016) Effective uses of light weight concrete. J Civil Eng Environ Technol 3(3):208–211

Standard test methods for fire tests of building construction and materials (2008) ASTM E119-08b, ASTM International, West Conshohocken, PA, USA

The European Union (2004) Eurocode 2: Design of concrete structures - Part 1-2: General rules - Structural fire design

Kodur VR, Raut N (2010) Performance of concrete structures under fire hazard: emerging trends. Indian Concr J 84(2):23–31

Vinith Kumar N, Arunkumar C, Srinivasa Senthil S (2018) Experimental study on mechanical and thermal behavior of foamed concrete. Mater Today: Proc 5(2):8753–8760

Numerical Modelling of Re-cracking Behaviour in Retrofitted Crack Stop Holes in Steel Structures

S. Abeygunasekera[1(✉)], J. C. P. H. Gamage[1], and S. Fawzia[2]

[1] University of Moratuwa, Katubedda, Moratuwa, Sri Lanka
abeygunasekarasampath@gmail.com
[2] Queensland University of Technology, Brisbane, Australia

Abstract. This paper focuses on the delaying of crack initiation process in a crack stop hole (CSH) in steel structures. The cyclic J-Integral (ΔJ) numerical model was adopted to develop this study. Also, the Power law was applied to estimate the required number of cycles for crack initiation at the crack stop hole with the effects of carbon fiber reinforced polymer (CFRP) strengthening CSH improvement techniques. CFRP strengthened CSH indicated a significant improvement in the CSH. This retrofitting technique reported an optimal performance; in the range between 17.7% and 163.6% delay in required cycles for the crack initiation when compared to an un-strengthened CSH.

1 Introduction

Crack stop hole (CSH) technique is a widespread method in the steel industry as a crack repairing method. In this method, a drilled hole is fabricated at the crack tips to crack transfer into a notch. The main purpose of the crack stop hole is the re-distribution of the stresses which reduces stress concentrations at the crack tip. The result is delay in crack propagation and changes in the direction of crack propagation. The crack stop hole technique has been verified as enhancing the fatigue service life of a structure [1, 2]. This technique is a simple and non-expensive method [3]. As a result, this crack stop hole technique is widely applied in the maintenance of metallic structures [4]. This technique is especially used in the aerospace industry.

Due to the removal of material to prepare CSH, the load is redistributed throughout the structure. Hence, the stress is maximum at the edge of the crack stop hole. Therefore, the maximum stress concentration occurs around the CSH. Stress concentration in ductile materials is higher when compared with the yield strength of material. Such behaviour causes the redistribution and deformation of the local plastic stresses. However, stress intensity factor (SIF) could considered as a good indicator representing the stress at the vicinity of the crack tip as it could be used to predict stress and fracture behaviour of the CSH. The simulation of the crack is important for predicting the fatigue life with different techniques which could be used for this purpose. Thus, the J-Integral technique could be considered as one of the best alternatives, as an extension of SIF.

R. Dissanayake and P. Mendis (Eds.): ICSBE 2018, LNCE 44, pp. 479–488, 2020.
https://doi.org/10.1007/978-981-13-9749-3_42

The potential of fatigue damage occurring at the crack stop hole is maximum with respect to the remaining areas, since such a crack stop hole acts as a structural discontinuity. When the CSH is continuously subjected to cyclic loads, it results in the stress concentration increase near the CSH. Due to the increase in the stress concentration, crack initiation restarts from the CSH and this action is called a crack re-cracking on CSH. In fact, only a few numerical investigations are available in the literature to estimate delaying crack incubation and growth at the CSH. According to Tanaka and Mura (1981), the reason for crack initiation originates at the slip bands, without accounting for the effect of cyclic hardening of the material. Also, Bozek et al. (2008) suggested a probabilistic simulation technique to understand the re-cracking behaviour in members. Similarly, Fine and Bhat (2007); Mura and Nakasone (1990); Bobylev et al. (2010) and Xie et al. (2016) have emphasized the ability of delay in crack initiation due to energy barrier techniques.

2 Background

Computerized finite element analysis techniques are more popular due to the practical difficulties of laboratory testing. This study also utilized the J-integral technique in ABAQUS for analysis. The Paris law was used to estimate the rate of crack propagation at the CSH and its regression curve was fitted to obtain the power law approximation. In fact, the rate of crack growth could be estimated accurately by using the results of the power law approximation. When comparing the results of Paris law, the proposed power law approximation showed negligible differences (less than 3%). The Paris law behaviour was based on characteristic of material such as homogeneity, isotropy, and elasticity.

The main problem in CSH is the initiation of re-cracking at the tip of the hole. There are some techniques already applied in this field to understand the delay in this crack propagation. The main objective of this investigation is the analysis of the re-cracking behavior of CSH. In addition, the behaviour of retrofitted CSH was also explored in this context.

Carbon Fiber Reinforce (CFRP) materials show excellent fatigue resistance in strengthened steel structures. Furthermore, CFRP materials exhibit exceptional characteristics; few of them being lightweight, high strength, corrosion resistance, and ease of installation. Field applications also confirmed the usefulness of the CFRP related strengthened steel structures. Therefore, this investigation focuses on the numerical modelling of CFRP strengthened CSH as an alternative solution to delaying re-cracking in a CSH.

3 Methodology

In this method, a single layer of CFRP was attached to the CSH at the bottom surface of the member which is shown in Fig. 1. A tensile load was applied on the bottom surface of the member. Thus, the tendency of crack initiation is high at the bottom surface rather than the upper side of the specimen. Crack stop hole strengthened with CFRP assist in reducing the stress at the edge of the hole. Thus, the dimensions of the CFRP layer were selected as 100 mm length and 40 mm width. Also, uni-directional CFRP material was selected which is commercially denoted as CF 130. Thickness of the CFRP layer was taken as 0.167 mm according to the data sheet provided by the manufacturer. Young's modulus and Poisons ratio of the material was taken as 240 GPa and 0.28 respectively. Interaction mode in the ABAQUS was utilized for modelling of the CFRP layer under vicinity of the crack stop hole in the structural member. Steel plates were considered as the master surface and the CFRP layer was considered as the slave surface.

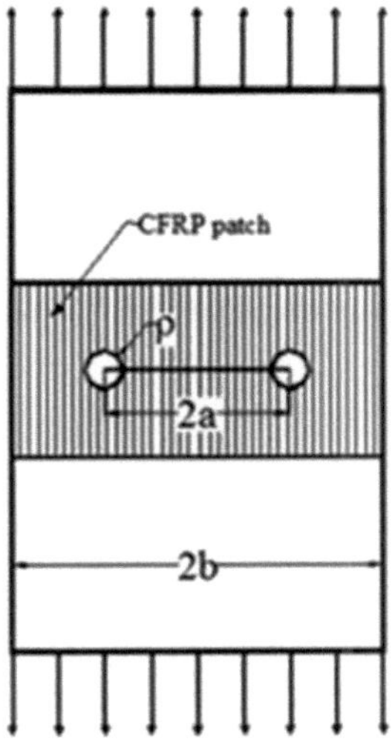

Fig. 1. Schematic view of the CFRP patched with CSH.

CFRP strengthened CSH specimens were kept on two cylindrical shaped, supportive rollers with each having a 25 mm diameter and the span length was fixed at 240 mm between centers of the rollers as shown in Fig. 2. The specimen module was loaded with a low cycle fatigue (LCF) mode using the direct cyclic option in ABAQUS. A Cyclic load was applied on the specimen at the mid-plane using the loading nose and 8 mm radius loading nose was used as shown in Fig. 2. The load was applied at the mid-plane of the specimen. The maximum deflection in the load direction (Y direction) at mid-plane was limited to 5 mm. The magnitude of the frequency was selected according to the ASTM D7774 standards [5]. A constant amplitude load was applied to the model while during the analysis, 5 Hz frequency and 0.05 stress ratio were maintained. Parametric study was conducted by varying the diameter of CSH and alternative bonding techniques of CFRP. The procedure followed is listed below:

1. The geometry of the model using the part modulus of ABAQUS was created.
2. A pre-defined crack profile using a partition cell extrude/sweep edges option of the part modulus was assigned.

3. The material properties, boundary conditions, interaction properties and loads for both the static general step and the cyclic step were utilized in this module. (Static general step for the applying step 1 load, direct cyclic step 2 for fatigue load)
4. Mesh of the model using CPE4R elements was utilized.
5. A special enrichment function called "seams" using contour integral, under interaction modulus were performed to this procedure.
6. The analysis was created and the results were taken from the data file of visualization modulus.

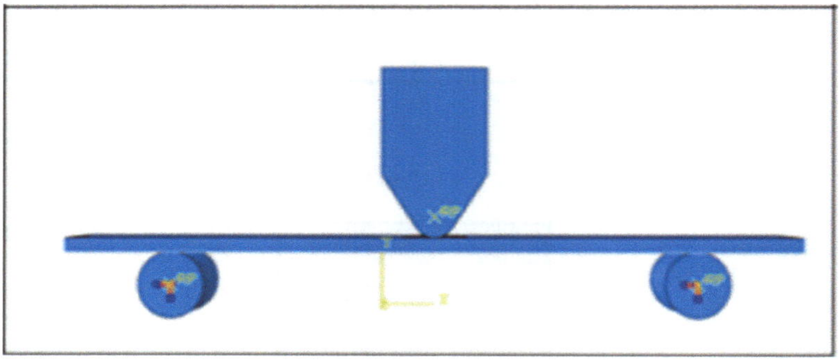

Fig. 2. 3D object with 240 mm span length under the 3 points cyclic flexure load

In this numerical analysis, the diameter of CSH varied from 8 mm to 24 mm. According to the model of the numerical analysis, the minimum and maximum J-integral values were obtained while the cyclic J-integral value was calculated according to Eq. 1. The required number of cycles for crack initiation was calculated by using the power law approximation.

4 Theory

4.1 Theory of Fracture Mechanics

At present, two main fracture mechanics approaches are used for predicting the fatigue life. They are the S-N curve approach method and fracture mechanics approach method. The fracture mechanics approach is utilized in this study where the concept of fracture mechanics was introduced by Leonardo da Vinci several centuries earlier [6] Theory of fracture mechanics is based on the initiation and propagation of the cracks. Since, fracture mechanics are based on the mechanical behaviour of steel structures, the ultimate effect of the fracture is the catastrophic (Unrecoverable) failure of the structure. Therefore, understanding of fracture and its control is important in the real-world application to assure safety. The best way to estimate fatigue life is a prediction of crack initiation. Studies of Griffith (1921), Irwin(1957), and Rice (1968) have greatly contributed to develop the theory on fracture mechanics [7]. Similarly, Irwin in 1948 has extended the Griffith approach to metals [8]. This extension was utilized as energy dissipates due to local plastic flow. During the same period, Orowan [10] independently proposed similar modifications as the Griffith theory [9, 10]. In 1948 Mott extended the Griffith theory considering the rapid propagation of the crack [11]. In 1956, Irwin developed the concept of strain energy release rate [12]. This concept showed a direct relation with the Griffith's theory.

4.2 J-integral Technique

Dr. James Rice in 1968 introduced a path-independent contour integral analysis technique for crack analysis [13]. Rice explained the J-integral method with elastic-plastic fracture mechanics (EPFM). The cyclic J-integral (ΔJ-integral) could be considered as a crack tip parameter. The cyclic J-integral (ΔJ) value can be obtained by using finite element analysis. The result of crack driving force is propagation of a short fatigue crack. In fact, the crack driving force can be alternatively named as a crack tip stress or as the strain energy release rate during the crack growth under elastic-plastic condition. Therefore, the J-integral technique could be viewed as an energy related parameter as well as a stress intensity parameter for non-linear materials. Hutchinson, Rice, and Rosengren have confirmed that the J-integral technique is unique for non-linear materials. Rice has confirmed the basis of EPFM methodology well beyond the validity limits of LEFM. The estimation of cyclic J-integral values from using minimum and maximum values of the J-integral could be explained as follows.

$$\Delta \mathrm{J} = \mathrm{J}_{\min} + \mathrm{J}_{\max} - \left(\mathrm{J}_{\min} + \mathrm{J}_{\max}\right)^{1/2} \tag{1}$$

Where ΔJ is the cyclic J-integral value, J_{min} is the minimum value of the J-integral taken from FEA and J_{max} is the maximum J-integral value available from the same analysis.

4.3 The Power Law

Prediction of crack growth is important for structural elements during their service life, as such predictions helps the estimation of the remaining service life of any structural element. In addition, such a prediction supports to estimate the inspection intervals of structures for maintenance. Therefore, accurate and reliable approximation is required to predict the behaviour of crack propagation in structures. On the other hand, the method of prediction should not be complicated. However, to satisfy all these conditions at the same time is not an easy task as, numerous variables significantly influence the process of crack initiation. The cyclic J-integral technique is based on the theory of fracture mechanics. Dowling and Begley [8] have proposed and implemented the J-integral technique considering the relations of crack growth rate (da/dN), and cyclic J-integral (ΔJ) [14]. The power law approximation is similar to the Paris law. The Power law and fatigue crack growth rate of structures exposed to cyclic loading can be calculated as follows:

$$\mathrm{da/dN} = \mathrm{A}(\Delta \mathrm{J})^{\mathrm{n}} \quad (2)$$

In the Power law application, it is assumed that the cyclic J-integral (ΔJ) includes the sum of elastic and fully plastic solutions. The rate of fatigue crack growth in the Power law depends on constants **A** and **n**. These two parameters are called the **Ramberg-Osgood coefficient and strain hardening index,** respectively. These constants specially represent the tensile data of materials. However, these two parameters do not depend on the geometry and it is based-on material properties. In this analysis, theses power law constants, A and n were considered as 0.95 and 0.47 respectively, using the Paris law results approximation with the Power law. The cyclic J-integral values were obtained from numerical analysis.

5 Model Results

In this study, CFRP strengthened and un-strengthened CSH were modelled to compare the results. All other parameters which critically affected crack behaviour was kept as a constant during the process of numerical analysis. The boundary conditions, properties of materials, loading rates, frequency, stress ratio and aspect ratio were kept as constants throughout the analysis (Fig. 3).

Fig. 3. Visualization of the 3D view under 3 - points flexural cyclic load with an adhesive CFRP strengthen CSH.

The stress concentration at CFRP strengthened CSH contours is shown in Fig. 4.

Fig. 4. Visualization of the bottom view under 3 - points flexural cyclic load with CFRP strengthened CSH.

When the three points cyclic flexural loads are applied on a CFRP strengthened crack stop hole, the change of stress distribution pattern is shown in Fig. 5. In this case, the stress variation around the CSH is not clearly visible as the elastic plastic zone is controlled by the CFRP layer. This is the main reason for the delay in crack initiation at the CSH compared to unstrengthened CSH.

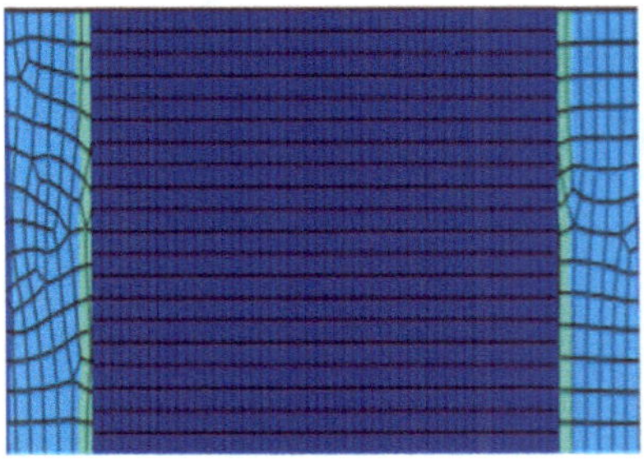

Fig. 5. Stress distributions at CFRP strengthen CSH.

6 Results and Discussion

The analytical results of the CFRP strengthened CSH and un-strengthened CSH were compared as shown in Table 1. It summarizes the maximum and minimum J-integral values together with the variation of different CSH diameters from the range of 8 mm to 24 mm. Interestingly, the number of cycles for crack initiation varied with the hole size. The number of cycles for crack initiation compared with un-strengthened and CFRP layer attached CSH is shown in Fig. 6.

Table 1. J-integral values and cyclic J-integral of CFRP strengthened CSH.

Diameter of the CSH (mm)	J_{Min}	J_{Max}	Cyclic J integral (ΔJ)
8	1.44E−04	8.54E−04	0.0006473207
12	1.24E−04	1.40E−04	0.0001322426
16	1.71E−04	2.25E−04	0.0001998495
20	4.54E−03	8.90E−03	0.0070834286

The variation of the number of cycles for crack initiation with respect to the diameter of the CSH is shown in Fig. 6. The comparison of the un-strengthened CSH and the CFRP strengthened CSH are shown in the same graph. The common behavior of both graphs is sinusoidal wave patterns, which can be observed. According to the investigations conducted by Kormay and Disnig [15], Ack et al. (1988) it was confirmed that theoretical stress concentration factor for a plate with a hole under bending loads were a response to a sinusoidal stress pattern [8, 15]. When the diameter is changed from 8 mm to 24 mm it showed significant delaying of the crack initiation when compared to un-strengthened CSH.

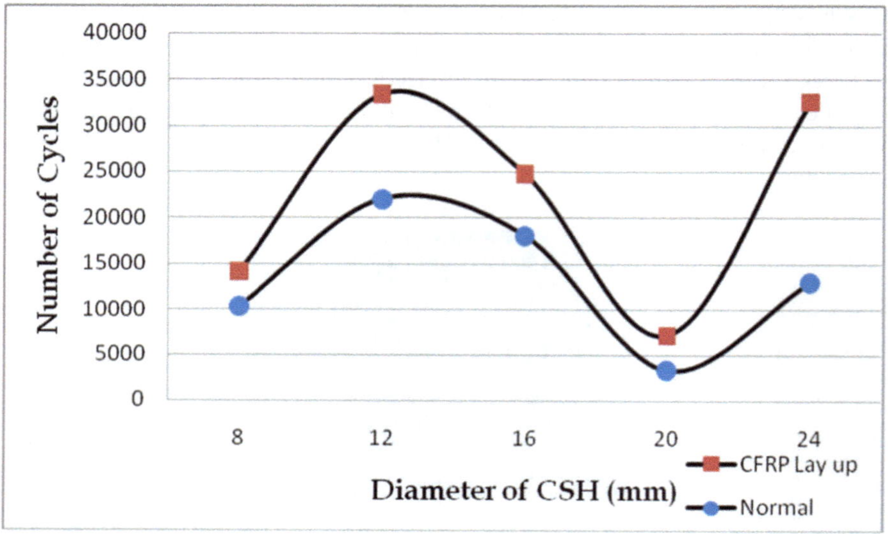

Fig. 6. Comparison of number of cycles for crack initiation with un-strengthened and CFRP strengthened crack stop hole with a hole diameter.

According to this study, the CFRP material exhibited a significant crack initiation delay and its range in variation were recorded as 17.7% to 163.6% with respect to an untreated crack stop hole. This technique is performed with 8 mm to 24 mm range of the diameter of the CSH. However, up to medium size of the CSH diameter, significant improvements were exhibited with respect to the untreated crack stop hole (Table 2).

Table 2. Delaying of crack initiation due to CFRP strengthened

Diameter of CSH (mm)	Percentage of delaying crack initiation compare to un-strengthened CSH (%)
8	162.5
12	90.6
16	163.6
20	17.7
24	50.3

6.1 Discussion

CSH method shows a high potential in successfully delaying crack initiation at the crack stop hole. When the CFRP layer was attached to the bottom surface of the specimen, the effects of the tension load on bottom surface was controlled by the CFR, since the tension plunge shows a high tendency of crack initiation rather than the compression surface (upper side) of the specimen. The CFRP laminate also helps to reduce the stresses at the edge of the CSH. Thus, a number of fiber layers, the module of CFRP material, lay-up angles and bond length could be considered as the main variables which affect the process of crack initiation.

Stress concentration at the CSH is the main reason for crack initiation and it is governed by the residual stress and notch stress. When the CSH diameter is small, the required number of cycles is less. The residual stress is high due to the mechanical stress on the process of manufacturing such as cold forming and rolling. When the hole diameter gradually increases in the notch, the stress decreases and the residual stress, increases due to the process of drilling. According to the net effects of both stresses, the required number of cycles for crack initiation increases as, a comparatively less stress concentration is shown at the CSH. When the hole size was further increased, the required number of cycles for crack initiation decreased again, because residual stress increases due to the drilling process. Net effects of both actions contribute to create high stress concentration at the CSH. When the hole diameter becomes larger, the required number of cycles for crack initiation had drastically increased. This is due to the reduction of notch stress when compared to the residual stress at the CSH. The bottom surface is retrofitted with CFRP and the required number of cycles for crack initiation was increased because the CFRP layer contributed to reduce the stress at the edge of the holes.

7 Conclusions

In this study, CSH was strengthened with externally bonded CFRP sheets which were numerically studied with the required number of cycles and crack initiation for CSH were estimated using J-integral technique. The following conclusions were obtained:

1. Stress concentration at the crack stop hole edges effectively decreased with the CFRP material.
2. The plastic zone of the vicinity of crack stop hole edges can be reduced by the externally bonded CFRP sheets.
3. This retrofitting technique reported the performance; which is in the range of a 17.7% to 163.6% delay in crack initiation compared to the un-strengthened CSH with the diameters changed from 8 mm to 16 mm.

Acknowledgments. The authors gratefully acknowledge the support of the staff in computer laboratory, the Department of Civil Engineering at the University of Moratuwa for their valuable support.

References

1. Wu H, Imad A, Benseddiq N, Castro JTPD, Meggiolaro MA (2010) On the prediction of the residual fatigue life of cracked structures repaired by the stop-hole method. Int J Fatigue 32 (4):670–677
2. Crain JS, Smmons GG, Bennett CR, Barrettgonzalez R, Matamoros AB, Rolfe ST (2010) Development of a technique to improve fatigue lives of crack-stop holes in steel bridges. Transp Res Rec: J Transp Res Board 2200:69–77
3. Wang Y, Pan Q, Wei L, Li B, Wang Y (2014) Fracture toughness and fatigue crack growth analysis of 7050-T7451 alloy thick plate with different thicknesses. J Central South Univ 21 (8):2977–2983
4. Makabe C, Kaito N, Ferdous MS (2014) Method of arresting crack growth for application at a narrow working space. Mech Eng J 1(6):1–12
5. Lin F, Sun JG, Nakamura H, Maeda K (2012) Fatigue crack repair using drilled holes and externally bonded CFRP strips. In: Bridge maintenance, safety, management, resilience and sustainability. Taylor & Francis Group, London
6. ASTM D7774-12 (2013) Standard test method for flexural fatigue properties of plastics, D20.10.24, Ed. West Conshohocken, PA: ASTM International
7. Anderson TL (1991) Fracture mechanics: fundamentals and applications. CRC Press Inc., Florida
8. Rice R (1968) A path independent integral and the approximate analysis of strain concentration by notches and cracks. J Appl Mech 35:379–386
9. Irwin GR (1948) Analysis of stresses and strains near the end of a crack traversing a plate. J Appl Mech 24:361–364
10. Orowan E (1948) Fracture and strength of solids. Rep Progress Phys XII:185
11. Griffith AA (1920) The phenomena of rupture and flow in solids. Philos Trans Ser A 221:163–198
12. Mott NF (1948) Fracture of metals: theoretical considerations. Engineering 165:16–18
13. Irwin GR (1956) Onset of fast crack propagation in high strength steel and aluminum alloys. In: Sagamore research conference proceedings, vol 2, pp 289–305
14. Begley JA, Landes JD (1972) The J-integral as a fracture criterion. ASTM STP 514. American Society for Testing and Materials, Philadelphia, pp 1–20
15. Kormay K, Disnig W (1991) pp 66–69

Comparison of Properties of Masonry Walls Constructed with Commonly Used Bricks and Non-standard Large Bricks

V. L. W. Seneviratne(✉) and D. Nanayakkara

Department of Civil Engineering, University of Moratuwa, Moratuwa, Sri Lanka
vachiralaksitha@gmail.com

Abstract. Brick masonry is a very popular construction material especially among the house builders, because of its better heat insulation properties, compressive strength, high durability, availability and the low cost. In Sri Lanka, brick masonry construction using commonly available burnt clay bricks (normal bricks), in different bond patterns can be seen in practice. However, the English bond is still considered to be the ideal bond pattern with normal bricks. Recently it has been noticed that a burnt clay brick with larger dimensions (non-standard brick) is gaining popularity in the local construction industry due to some advantages such as fast construction and reduction in the required mortar volume due to the reduced number of mortar joints. These larger bricks are used with a header bond pattern to build typical one brick thick walls. Although there are several benefits in using these larger bricks, it is important to determine the strength properties as well, since the strength of brick masonry highly depends on the strengths of brick units, mortar and the bond pattern. If these walls are used as load bearing walls or in-fill walls, the determination of strength properties is essential. The effect on strength properties due to the difference in bond pattern was addressed in this study. Experimental investigation was carried out to compare the strength properties of normal and non-standard brick units and corresponding brickwork in terms of compressive, flexural and shear strengths. Locally available normal brick and non-standard brick samples have been selected and tested according to the relevant standards to assess the strengths of bricks and brickwork. It was observed that the brickwork constructed with non-standard bricks has slightly higher compressive strength compared to brickwork with normal bricks. However flexural strength of masonry constructed with non-standard bricks with a header bond pattern is lower than the flexural strength of normal brick masonry which is constructed with the English bond pattern. Therefore, brickwork constructed with non-standard bricks is not efficient compared to normal brickwork, especially when the walls are subjected to lateral loads. This may be due to the weak bond in the brickwork constructed with non-standard bricks as the header bond pattern was used in them.

Keywords: Compressive strength · Flexural strength · Non-standard brick · Standard bricks

R. Dissanayake and P. Mendis (Eds.): ICSBE 2018, LNCE 44, pp. 489–498, 2020.
https://doi.org/10.1007/978-981-13-9749-3_43

1 Introduction

Considering masonry construction in Sri Lanka, burnt clay bricks are the most commonly used masonry units in the construction industry. Generally, the dimensions of commonly available bricks in the local market vary slightly from the specified dimensions of standard bricks. However, recently it was found that a larger clay brick having dimensions of 225 mm × 150 mm × 75 mm is becoming popular in the construction industry. The properties of brick walls depends on the properties of brick units, mortar and bond patterns. Hence, the investigation on the influence of size difference on the strength of brick units, the effects on strength properties of brickwork due to the difference in bond pattern and the unit size were the main objectives of this study.

Few benefits can be expected when using the large brick units in masonry construction. One benefit is the fast construction due to large unit size. The number of mortar joints would also be reduced due to the larger units. In the construction of one brick thick walls with non-standard large bricks, a better finish could be seen since the header bond pattern is used. Generally, with the normal clay bricks, the English bond is used and due to the uneven sizes of bricks, mortar fillings will be required to maintain the constant wall thickness. Since there are benefits in using non-standard large bricks as explained, it is necessary to investigate the strength characteristics for the effective use of non-standard large size bricks in masonry construction.

2 Objectives

The main objectives of this research study are;

1. Investigation on the strength characteristics of the normal and non-standard large brick units and the corresponding masonry panels.
2. Comparison of the properties of masonry panels constructed with normal and non-standard bricks to ascertain the suitability of the non-standard bricks in masonry construction.

3 Methodology

3.1 Materials Used

3.1.1 Brick Units

In this study, two locally available bricks of different sizes were used. One was a commonly available brick having dimensions of 220 mm × 105 mm × 65 mm while a larger brick having 225 mm × 150 mm × 75 mm was used as the non-standard brick. Recently, this non-standard brick has been widely used in the construction industry. Since the objective of this study is to compare the properties of masonry panels constructed with these two brick types with different sizes, all bricks used for investigation were bought from the same manufacturer ensuring that they are manufactured from the same clay and the same procedure.

Compressive strength of brick units was determined according to the SLS 39: 1978 (Sri Lanka Standards 1978). Figure 1 shows the loading arrangement used.

Fig. 1. Testing for compressive strength of bricks.

A sample of ten bricks was used to determine the compressive strength of bricks. Prior to the compressive load application, bricks were immersed in water at room temperature for 72 h. Compressive load was applied on the bricks using 200T Universal Testing Machine.

3.1.2 Mortar Mix

Mortar mix of 1: 5 (cement: sand) was used for the masonry panels. Ordinary Portland cement and local river sand were used for the mortar. Mortar cubes of 70.7 mm 70.7 mm × 70.7 mm were tested for each batch of mortar as per BS 5628-1: 1992 (British Standard 1992). The mortar strengths were also tested on the same day when masonry panels were tested.

3.2 Compressive Strength of Masonry Panels

Practically, normal size bricks are used for construction of both one brick thick and half brick thick walls. Generally the English bond is used for one brick thick (225 mm) or in other words 9″ thick walls and the stretcher bond is used for half brick or 4½″ thick walls. Using the selected non-standard bricks, 9″ thick walls can be constructed with the header bond and using stretcher bond, 150 mm thick walls can be constructed.

For the comparison of compressive strength of masonry constructed with different sizes of bricks, one brick thick (225 mm) walls were considered. Panel sizes were decided according to BS EN 1052-1: 1999 (British Standard 1999a). Specimen sizes are shown in Table 1. All panels were covered with polythene sheets for three days after construction and then left uncovered until tested. The panels were tested 28 days after construction.

3.2.1 Brickwork Panels Constructed with Normal Bricks

One brick thick masonry panels were constructed with normal bricks using the English bond and the header bond patterns. For the English bond, ten courses high wall panels that were three bricks long, were constructed. For the header bond, wall panels of ten courses high, six headers in a course were constructed. Three masonry panels per each case were constructed. Figure 2 shows these two types of wall panels and their dimensions are given in Table 1.

Table 1. Specimen sizes for compressive strength test

Brick type & bond pattern	Thickness (mm)	Length (mm)	Height (mm)
Normal (English bond)	220	680	740
Normal (Header bond)	220	680	740
Non-standard (Header bond)	225	630	755

Fig. 2. Masonry panels constructed with normal bricks using different bond patterns.

3.2.2 Non-standard Brickwork Panels

One-brick thick masonry wall panels were constructed with non-standard bricks using header bond pattern. The wall panels were nine courses high with a length of four headers. Three specimens were constructed for testing. Figure 3 shows the non-standard brickwork panels and the dimensions are given in Table 1.

Fig. 3. Non-standard brickwork panels

According to the BS EN 1052-1: 1999 (British Standard 1999a) wall panels were tested to determine the compressive strength of brick walls constructed with normal and non-standard bricks.

3.3 Flexural Strength of Brick Masonry

Investigation on flexural strength of masonry was carried out in accordance with BS EN 1052-2: 1999 (British Standard 1999b). According to this standard, the flexural strength of masonry is derived from half-brick thick wall specimens tested under four-point loading. Figure 4 shows the loading arrangement used in the flexural strength test. A further study on one brick thick wall specimens was also considered in the flexural strength test to compare the influence of bond patterns on flexural strength of brickwork. The flexural strength was determined for two directions,

1. Flexural strength when the plane of failure is parallel to the bed joints,
2. Flexural strength when the plane of failure is perpendicular to the bed joints.

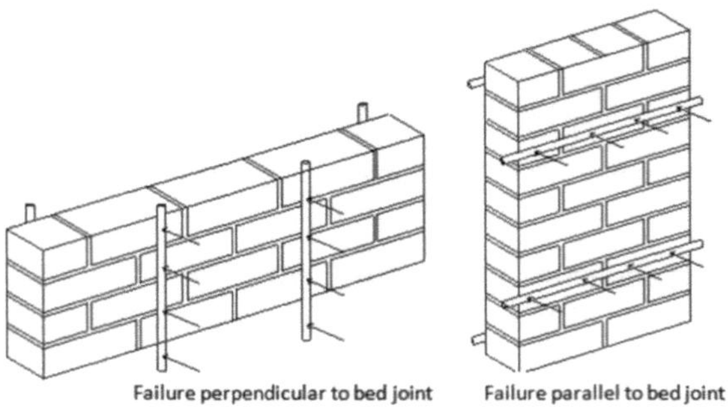

Fig. 4. Layout diagram of flexural strength tests

3.3.1 Flexural Strength of Masonry When Plane of Failure Is Parallel to the Bed Joint

For flexural strength test for failure parallel to the bed joints, normal brick and non-standard brick panels were tested for both half brick and one brick wall thicknesses. According to the standard, the specimen size was two bricks long by ten courses high. Five specimens were tested for each case. Immediately after the construction of wall specimens, they were covered with polythene and pre-compressed by applying a uniformly distributed vertical stress of 3×10^{-3} N/mm^2. All panels were kept undisturbed until testing and panels were tested at an age of 28 days (Fig. 5 and Table 2).

Table 2. Panel sizes for flexural strength test (parallel to bed joint)

Type of brick	Thickness (mm)	Height (mm)	Length (mm)
Normal	105	740	450
Non-standard	150	840	460
Normal	220	740	450
Non-standard	225	840	470

Fig. 5. Loading arrangement for flexural strength test (failure parallel to bed joint)

3.3.2 Failure Perpendicular to the Bed Joint

For flexural strength test (when failure perpendicular to the bed joints), four bricks long by four courses high specimens were tested for both standard and non-standard bricks (Table 3).

Table 3. Specimen sizes for flexural strength test (perpendicular to bed joint)

Type	Thickness (mm)	Height (mm)	Length (mm)
Normal	105	290	910
Non-standard	150	330	930
Normal	220	290	910
Non-standard	225	330	950

Five specimens were tested for each case. Specimens were pre-compressed by applying a uniformly distributed load to give a vertical stress of 3×10^{-3} N/mm^2 and kept undisturbed until testing. Panels were tested at an age of 28 days (Fig. 6).

Fig. 6. Flexural strength test (failure perpendicular to bed joint)

3.4 Initial Shear Strength

Initial shear strength was determined according to BS EN 1052-3: 2002 (British Standards Institute 2002). Six brick masonry triplets were tested for each case, 28 days after casting. Specimens were tested under four point loading at zero pre compression.

4 Results

4.1 Average Unit Compressive Strength

Table 4 gives the compressive strength results obtained for normal and non-standard large brick units.

Table 4. Average unit compressive strengths.

Brick type	Dimensions (mm)			Average mass (kg)	Unit compressive strength (N/mm^2)
	L	W	H		
Normal	212	105	61	2.6	1.67
Non-standard	216	146	78	4.6	1.74

4.2 Average Compressive Strengths of Wall Panels

Compressive strength of brickwork was determined by calculating the mean compressive strength of three masonry specimens. Compressive strength results are given in Table 5.

Table 5. Average compressive strength results of brickwork panels

Brick type	Bond type	Wall thickness (mm)	Average compressive strength (N/mm^2)
Normal	English	225	1.1
Normal	Header	220	1.3
Non-standard	Header	225	1.5

4.3 Shear Strength of Brick Masonry

Average shear strength results obtained for triplets made using normal bricks and non-standard bricks are given in Table 6.

Table 6. Shear strength of brickwork

Brick type	Shear strength (N/mm^2)
Normal	1.16
Non-standard	1.15

4.4 Average Flexural Strength of Wall Panels

Flexural strength of masonry was determined in accordance with BS EN 1052-2: 1999 and the results are presented in Table 7.

Table 7. Flexural strength of masonry panels

Brick type	Wall thickness (mm)	Average panel size (mm)		Average flexural strength (N/mm^2)
		Length	Height	
Plane of failure parallel to the bed joints				
Normal (Stretcher)	105	450	750	0.20
Non-standard (Stretcher)	150	460	920	0.18
Normal (English)	220	450	750	0.33
Non-standard (Header)	220	460	920	0.25
Plane of failure perpendicular to the bed joints				
Normal (Stretcher)	105	890	240	0.72
Non-standard (Stretcher)	150	925	340	0.50
Normal (English)	220	905	300	0.46
Non-standard (Header)	220	940	355	0.35

By analyzing the results obtained from this study, significant differences in results could not be seen for normal and non-standard size bricks. When considering the brick unit strengths, 4% increase can be seen in non-standard brick units compared to normal bricks. It was also found that the compressive strength of non-standard brick masonry is higher than that of normal brickwork. Apart from the higher unit strength of non-standard large bricks, the reduction of the number of vertical mortar joints may be the other reason for this increase in compressive strength. Even though the brickwork with non Standard bricks showed higher compressive strength, flexural and shear strengths of normal brickwork were higher. Figures 7 and 8 clearly show the flexural strength results for one-brick thick and half-brick thick walls constructed with normal bricks and non-standard large bricks.

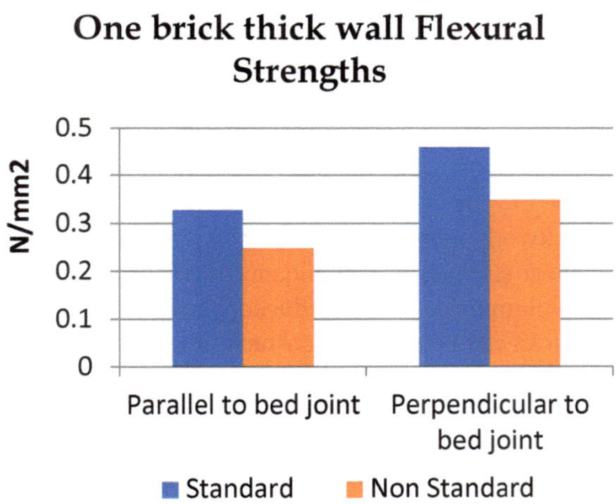

Fig. 7. Flexural strength of one brick thick wall panels

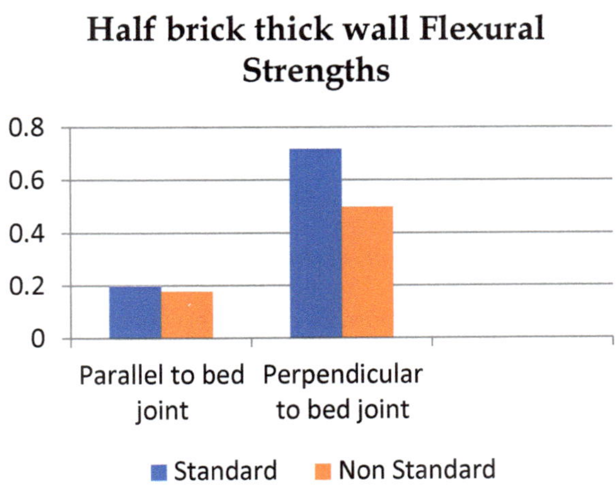

Fig. 8. Flexural strength of half brick thick wall panels

5 Conclusion

According to the test results, when considering the compressive strength of brickwork, the non-standard large bricks can be selected as the better option. However, flexural strength of normal brickwork is higher than that of brickwork made with large bricks. Therefore, non-standard bricks would be ideal for vertical load bearing walls. When walls are subjected to lateral loads, the better option would be normal bricks.

In this study, the main focus was on construction of one-brick thick masonry walls, as the bond patterns were different in walls constructed with two brick sizes. From the results (Table 5) it can be observed that the difference in bond pattern has not caused a significant influence on compressive strength of brickwork. However, a considerable influence of bond pattern on flexural strength can be seen in normal and non-standard brickwork walls.

The English bond is used to construct one-brick thick walls with normal bricks, for which the bricks are laid in two different directions alternatively. However, when header bond pattern is used in non-standard bricks, all bricks are laid in one direction. When bricks are laid only in one direction, resistance to flexure is reduced due to the weak bond pattern. Therefore, the flexural strength of brickwork made with non-standard bricks is low.

Even though brickwork constructed with non-standard bricks does not provide higher flexural and shear strength, it will enhance the speed of construction and reduce the required mortar volume due to the reduced number of mortar joints. Therefore, these non-standard bricks can be used for masonry construction where the walls are not subject to very high lateral loads.

Acknowledgement. The authors wish to acknowledge the financial support provided by the Department of Civil Engineering, University of Moratuwa. Support from all the laboratory technical staff members is also acknowledged.

References

British Standards Institute (1992) Code of practice for use of masonry—Part 1: Structural use of unreinforced masonry. BS 5628-1

British Standards Institute (1999a) Methods of test for masonry—Part 1: Determination of compressive strength. BS EN 1052-1

British Standards Institute (1999b) Methods of test for masonry—Part 2: Determination of flexural strength. BS EN 1052-2

British Standards Institute (2002) Methods of test for masonry—Part 3: Determination of initial shear strength. BS EN 1052-3

Jayasinghe C (2007) Comparative performance of burnt clay bricks and compressed stabilized earth bricks and blocks. Eng J Inst Eng Sri Lanka 40(2):33–40

Sri Lanka Standards Institution (1978) Specifications for common burnt clay building bricks. SLS 39

Sustainable Maintenance of Steel Truss Bridges: A Conceptual Framework

Sudath C. Siriwardane(✉)

Department of Mechanical and Structural Engineering and Materials Science,
University of Stavanger, 4036 Stavanger, Norway
sasc.siriwardane@uis.no

Abstract. This paper proposes a conceptual framework for the essential maintenance of truss bridges based on an accurate remaining fatigue life estimation technique. The framework mainly consists of structural appraisal based critical members and connections identification, remaining fatigue lives estimation of critical connections and time dependent member replacement or strengthening scheme. In the stage of critical member identification, detailed structural appraisal has to be conducted giving priority to visual inspection for condition, FE analysis, and material testing, experimental static and dynamic load testing. Accurate fatigue life estimation techniques of both members and connections consist of measured stress histories and recently developed new fatigue models. A verification of the proposed framework is conducted by comparing the predicted replacement scheme with a sequence of experimental failure and fatigue life of selected test specimens. Finally, the proposed framework was applied to a case study railway bridge to obtain time dependent member replacement scheme and obtained results were compared with previous estimations. Hence, validity and merits of the proposed framework were confirmed.

Keywords: Essential maintenance · Fatigue life · Truss bridges · Member replacement

1 Introduction

Today, many of the structures in the world are getting old and a very large existing stock of civil infrastructures is in need of maintenance, rehabilitation or replacement. Considering the management of structures in terms of maintenance, member replacement had a wide acceptance during this period (Kong and Frangopol 2004). However, there are few studies regarding the member replacement based maintenance guidelines for railway bridges. Most of them are especially based on rating factors against failure (rating of load carrying capacity). Since railway bridges are most liable to fail due to the effect of fatigue, remaining fatigue life based maintenance frameworks illustrates more assurance than rating factor based policies (Network Rail 2001). However, very few studies have only been concentrated to remaining fatigue life based member replacement frameworks. There the remaining fatigue life estimation technique is generally based on a combination of specification load given stress histories, Miner's rule and design fatigue curve. But the experiences from engineering practices have

R. Dissanayake and P. Mendis (Eds.): ICSBE 2018, LNCE 44, pp. 499–509, 2020.
https://doi.org/10.1007/978-981-13-9749-3_44

indicated that fatigue analysis based on specification loads and distribution factors usually underestimates the remaining fatigue life of existing bridges by overestimating the live load stress ranges (Spyrakos et al. 2004). Further, Miner's rule does not properly take account of the loading sequence effect (Suresh 1998; Siriwardane et al. 2008a, b). Therefore, there is a doubt in utilizing the remaining fatigue life based available frameworks/approaches for bridges.

To overcome this problem, this paper proposes a conceptual framework for essential maintenance of truss bridges based on an accurate remaining fatigue life estimation technique. Initially, paper describes the proposed framework in detail. Then, the verification of the proposed framework is confirmed. Finally, the proposed framework is applied to a case study railway bridge to obtain the time dependent member replacement scheme. The obtained results are compared with previous estimations.

2 Proposed Framework

A proposed maintenance framework for truss bridges is discussed in this section. This is a member replacement based essential maintenance framework. It consists of two major parts such as identification of critical members/connections and member replacement/strengthening scheme. The identification procedure of the critical members is dependent on the remaining fatigue life of each member since railway bridges are most liable to fail due to the effect of fatigue. The proposed remaining fatigue life estimation technique is especially dependant on structural appraisal-based stress evaluation and recently developed damage model. Flow chart of the proposed framework is shown in Fig. 1.

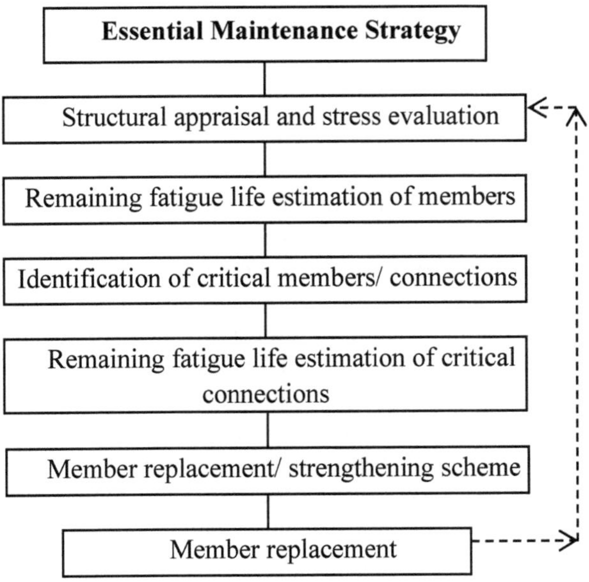

Fig. 1. Flow chart of proposed maintenance framework

2.1 Structural Appraisal and Stress Evaluation

In remaining fatigue life estimation, it is essential to determine the stress ranges generated by the passage of trains over the bridge. Therefore, it is a requirement to know the stress cycles of all the critical members for trains which are included in present and past timetables.

Initially, a condition survey has to be carried out to assess the present geometric condition and damages. Generally, it consists of a detailed visual examination, in-situ measurements of each component of the bridge and non-destructive field examinations. Then, laboratory tests are carried out to determine the current mechanical properties and chemical composition of the bridge materials. Static and dynamic load testing is recommended as the next major step to study the real behavior of the bridge under various load combinations. The obtained results are used to develop a proper analytical model and further assists in evaluating actual dynamic factors of each structural component. Finally, the bridge is subjected to finite element (FE) analysis under test and actual loadings to determine stresses and deflections, as well as variations of stresses under moving loads. Material properties which are obtained through laboratory tests and current geometric properties obtained from condition assessment are applied to the FE model for more realistic outputs. The validation of the FE model has to be done by comparing the results from analysis with those from field-tests. The FE model which gives better comparison to load test results is nominated as “validated analytical model”. Hence, the validated analytical model is used to obtain past and present static stress histories due to passage of trains specified by the owner.

Due to the dynamic effect of moving trains, the actual working stresses should be higher than the analytical static stress. Therefore, dynamic factors are used to multiply the static stress to get the service stresses of each member. Finally, the stress histories have to be converted into stress ranges using the rainflow counting method (Downing and Socie 1982). The described stress evaluation method is briefly summarized as shown in Fig. 2.

2.2 Remaining Fatigue Life Estimation of Members

The recently developed sequential law (Mesmacque et al. 2005; Siriwardane et al. 2007, 2008a, b) is recommended to use for remaining fatigue life estimation, instead of Miner’s rule, which was used as the fatigue theory in previous methods. A detailed description of the damage stress model and the definition of damage indicator, D_i is described in the corresponding paper (Siriwardane et al. 2007, 2008a, b). Here, only the concept of new damage indicator is motioned for comprehension. Suppose a component is subjected to certain stress amplitude or stress range σ_i for n_i number of cycles at load level i and N_i is the fatigue life (failure number of cycles) corresponding to σ_i. Here, the residual life at load level i can be obtained as $(N_i - n_i)$. The stress $\sigma_{(i)eq}$ which corresponds to the failure life $(N_i - n_i)$ is named as i^{th} level damage stress amplitude or stress range (otherwise can be introduced as stress amplitude or stress range relevant to the residual life). Hence, the new damage indicator, D_i is stated as,

$$D_i = \frac{\sigma_{(i)eq} - \sigma_i}{\sigma_u - \sigma_i} \tag{1}$$

where σ_u is the intercept of the Wöhler curve with the ordinate at one-quarter of the first fatigue cycle. Furthermore, it can be stated that, σ_u is the ultimate tensile strength amplitude or range for rotating bending test-based *S-N* curves and it is the ultimate shear strength amplitude or range for torsional fatigue test-based *S-N* curves.

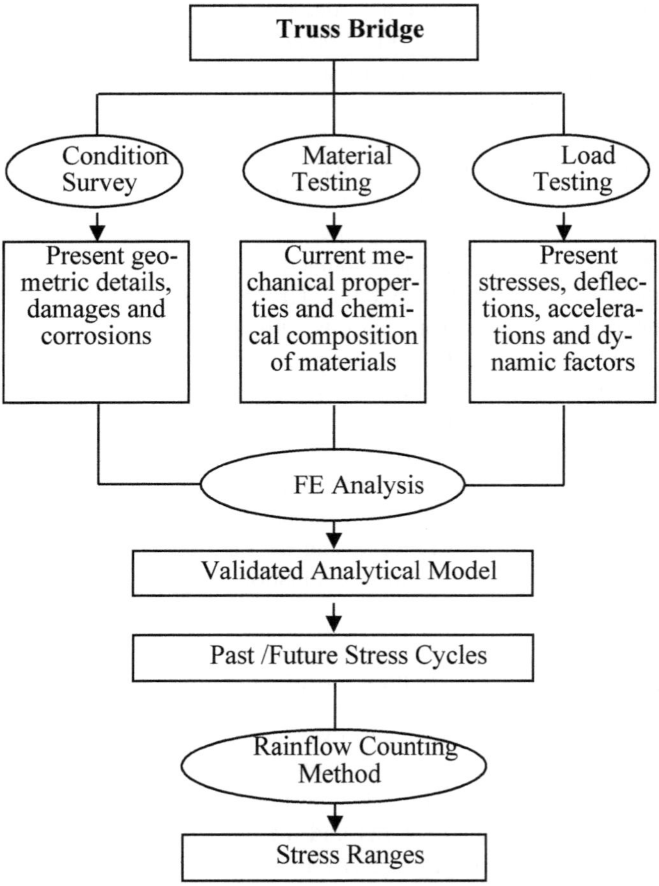

Fig. 2. Flow chart for the past/future stress evaluation method

At the first cycle, the damage stress amplitude or range $\sigma_{(i)eq}$ is equal to applied stress σ_i and corresponding damage indicator becomes $D_i = 0$. According to the proposed methodology, current damage has to be then transformed to the next load level. Therefore, at the last cycle, the damage indicator becomes $D_i = 1$ when $\sigma_{(i)eq}$ is equal to σ_u Therefore, the damage indicator is normalized to one ($D_i = 1$) at the fatigue failure of the material and the same procedure is followed until $D_i = 1$. Here, the defined fatigue failure is the time taken for the occurrence of the first through-thickness crack at the location of maximum stress of the structural component. In the case of railway bridge components, it can probably be taken as the time taken for initiation of crack near a connection (rivet or bolt).

To capture the fatigue damage due to secondary stresses near the riveted connection or discontinuities, detail class (BS 5400 1980) of riveted connection based Wöhler curves are considered for life estimation. But, chosen fatigue curve only describes stress ranges, which are corresponding to more than ten thousands of failure cycles (usually called the partially known Wöhler curve). In the case of the sequential law it is essential to use the Wöhler curve for the full range of the number of cycles. Therefore, the chosen partially known Wöhler curve has to be transferred to fully known Wöhler curve by using Kohout and Vechet Wöhler curve modeling technique (Kohout and Vechet 2001).

2.3 Identification of Critical Members/Connections

The members which have the lowest remaining fatigue life of each member set (set of members which has the same load capacity) is called as the "critical members" in this study. Generally, these members are to be subjected more attention in the member replacement based maintenance. From the previously obtained remaining fatigue lives (Sect. 2.2), one can easily identify these critical members. The connections which are joined to the previously obtained critical members (Sect. 2.2) are termed as "critical connections".

2.4 Remaining Fatigue Life Estimation of Critical Connections

The stress concentration effect in connections between the primary members of bridges was found to be one of main reasons for fatigue damage (Imam et al. 2005; Fisher et al. 1980). Most such connections are subjected to multiaxial fatigue. To capture this effect at connections or discontinuities, detail class (BS 5400 1980, Eurocode 3) of connection based Wöhler curves are considered in the previous life estimation. However, the variation of real rotational fixity, clamping force and geometry at the connection causes changes in the real stress distribution at the connections. Such changes may result in giving over or under predictions to estimated fatigue life (in Sect. 2.2) of the corresponding member (which joined to the connection). As a result, replacement of members based on previously determined remaining lives (Sect. 2.2) may not be an appropriate maintenance procedure. Therefore, replacement of members based on fatigue lives of critical connections is found to be a more appropriate strategy. This section describes the methodology to estimate remaining fatigue life of such connections.

Initially, all the critical connections should be investigated non-destructively to determine their current condition. This is conducted using different types of tests such as X-ray, ultrasonic, magnetic particle, radiographic examinations except visual examination. The connections which do not illustrate significant change from the initial state or condition are not subjected any unexpected stress concentration. Other connections which have been subjected to significant change may have to be assessed for remaining fatigue life sense to determine their degree of criticality.

The remaining fatigue lives of the connections, which are not subjected to significant deviation from the initial condition, is finalized as same as the lowest remaining fatigue life of the member that is joined to the particular connection. The remaining fatigue lives of other connections, where the conditions have been significantly changed, are to be evaluated based on the current geometric condition, secondary stress distribution etc. The latter part of this section describes multiaxial stress-based fatigue life estimation procedure of structural connections.

This procedure is especially based on the newly proposed multiaxial fatigue model. The considered failure mechanism (damage process) is in the mesoscopic scale and detailed descriptions are shown in the authors' previous work (Siriwardane et al. 2008a, b). Initially, the accumulated plastic strain per each stabilized cycle,

$$\varepsilon_s^{pc} = \frac{4}{\sqrt{3}} \frac{2k^* - k_{\max} - k_{\min}}{c} \quad (2)$$

has to be obtained from finite element employed secondary stress analysis of the connection or part of the member (sub model). Where $c = b + 2\eta$. The b and η are the mesoscopic linear hardening modulus and the shear modulus respectively. The k^* is the radius of the smallest hypersphere which contains the entire history of the macroscopic deviatoric stress amplitude of the stabilized cycle. The $k_{\max}$ and $k_{\min}$ are the maximum and the minimum values of mesoscopic yield stresses that can be reached during the loading cycle. Then, the fatigue life is calculated from the new damage indicator,

$$D_i = \frac{\left(\varepsilon_s^{pc}\right)_{(i)eq} - \left(\varepsilon_s^{pc}\right)_i}{\left(\varepsilon_s^{pc}\right)_u - \left(\varepsilon_s^{pc}\right)_i} \quad (3)$$

where $\left(\varepsilon_s^{pc}\right)_i$ is accumulated plastic meso-strain per stabilized cycle of n_i number of cycles at load level i. The N_i is the fatigue life (number of cycles to crack nucleation) corresponding to $\left(\varepsilon_s^{pc}\right)_i$ can be estimated from,

$$N = A\left(\varepsilon_s^{pc}\right)^{-\xi} \quad (4)$$

where ξ and A are material parameters to be determined from fatigue tests. The accumulated plastic meso-strain $\left(\varepsilon_s^{pc}\right)_{iq}$, which corresponds to the failure life $(N_i - n_i)$ is named as i^{th} level damage accumulated plastic meso-strain. The $\left(\varepsilon_s^{pc}\right)_u$ is the accumulated plastic meso-strain which corresponds to one-quarter of first fatigue cycle. According to the proposed methodology, current damage has to be then transformed to the next load level. Similar to the previous sequential law in Sect. 2.2, here too, the damage indicator is normalized to one ($D_i = 1$) at the fatigue failure of the material and the same procedure is followed until $D_i = 1$.

2.5 Member Replacement/Strengthening Scheme

The lowest remaining fatigue life of the connections describes the remaining fatigue life of the bridge. Once the age of the bridge reaches this value (when bridge life becomes zero), it is advisable to replace the corresponding critical member with a new member with longer fatigue life. At the same time, the associated connection is also recommended to be strengthened. After this essential repair, it is recommended to obtain a new sequence for future member replacement by following the same procedure from the beginning. This type of maintenance frameworks extends the service life of the (fatigue capacity) bridge in the safest manner.

3 Verification of Proposed Framework

In this section, the experimental (real) failure life-based member replacement sequence is compared with the proposed framework predicted scheme of five specimen members. The experimental fatigue lives were obtained in the high cycle fatigue regime under variable amplitude combined bending and torsion loading (Marciniak et al. 2008). The comparison between the predicted member replacement scheme and the experimental results-based scheme are shown in Table 1. In addition, it has been compared with two previous frameworks-based member replacement schemes as shown in Table 1. The first scheme is based on load rating factors (AASHTO 1990; Bridge Diagnostics 2002). The second approach is based on remaining fatigue life estimation of members (Caglayan et al. 2009). Here, Miner's rule is considered as the fatigue law (damage model). The Table 1 shows that there is better agreement between the proposed framework-based sequence than the previous frameworks-based sequences. This verification reveals the validity of the proposed maintenance framework in predicting the member replacement scheme in variable amplitude proportional loading conditions.

Table 1. Comparisons of member replacement schemes

Designation of test specimen	Experimental		Previous approach 1 (AASHTO 1990)		Previous approach 2 (Caglayan et al. 2009)		Proposed framework	
	Service life (s)	Sequence	Service life (s)	Sequence	Service life (s)	Sequence	Service life (s)	Sequence
NWL 4	42795	1	27265	1	58243	1	49628	1
KWL 4	90705	2	44325	2	93497	2	78239	2
KWL 2	98948	3	68573	4	202229	5	116975	4
KWL 8	123275	4	158623	5	162519	4	109076	3
NWL 1	130630	5	48329	3	147265	3	123568	5

4 Case Study: Determination of Member Replacement Scheme for a Rail-Way Truss Bridge

The selected bridge is one of the major railway bridges in Sri Lanka spanning 160 m (Fig. 3). It is a six span-riveted bridge with double lane rail tracks having warren type semi through trusses, supported on cylindrical piers. The bridge deck is made of wrought iron and the piers are made of cast iron casings with infilled concrete. The bridge was constructed in 1885. The proposed framework is used to determine the suitable member replacement scheme for the bridge.

Fig. 3. General views and identified corroded locations of the bridge

The condition survey reveals that no visual cracks were observed in any component of the super structure. The obtained values from material testing for elastic modulus, yield strength, ultimate strength in tension, fatigue strength and density are 195 GPa, 240 MPa, 383 MPa, 155 MPa and 7600 kg/m^3 respectively. Static and dynamic load testing were performed to study the real behavior of the bridge under various load combinations. The Bridge deck was analyzed using the finite element (FE) method employed general-purpose package SAP 2000. The FE model validations are performed by comparing analytical results of the FE model and the measurement of the actual bridge. Then, the stress histories were obtained and converted into stress ranges using rainflow counting method. The proposed method (Sect. 2.2) was used to calculate the remaining fatigue life of members. The riveted connections were classified as class Wrought-iron (WI), which is proposed by the UK railway assessment code (Network Rail 2001). The identified critical members which exhibits the lowest remaining life of each member set are shown in Table 2. The corresponding critical connections are also tabulated in Table 2. Designations of the members and connections of the bridge are schematically shown in Fig. 4. The fatigue damage of the critical joints is evaluated based on the state of stress due to release of clamping force while all the rivets of these two connections have no clamping force. The obtained remaining fatigue lives for the all the critical connections are shown in Table 3.

Table 2. Remaining fatigue lives for critical members and corresponding critical connections

Designation of members	Corresponding connections	Remaining fatigue life (years)
TMB3	TB4	323
TMB5	TB6	165
TMB6	TB7	169
CG4, 5	D15-45, D16-46	12
ST 16, 26, 36, 46	D17-47	13
TDT1	TT1, TB2	191
TDT2	TT2, TB3	171
TDT3	TT3, TB4	138
TDT4	TT4, TB5	162

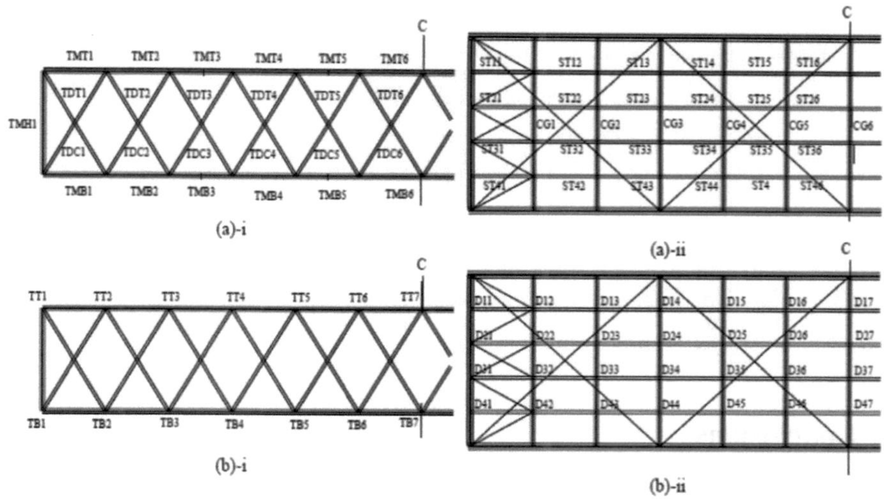

Fig. 4. Designations of (a) Members: i. Main truss girder, ii. Horizontal bridge deck; (b) Connections: i. Main truss girder, ii. Horizontal bridge deck

Table 3. Remaining fatigue lives for critical connections

Bridge component	Designation of critical connection	Corresponding critical member	Remaining fatigue life (years)
Main girder bottom chord	TMB3	TB4	323
Main girder bottom chord	TMB5	TB6	165
Main girder bottom chord	TMB6	TB7	169
Cross girders	CG4, 5	D15-45, D16-46	12
Stringers	ST 16, 26, 36, 46	D17-47	13
Truss diagonal (tension member)	TDT1	TT1, TB2	45
Truss diagonal (tension member)	TDT2	TT2, TB3	171
Truss diagonal (tension member)	TDT3	TT3, TB4	24
Truss diagonal (tension member)	TDT4	TT4, TB5	162

The member replacement sequence and the time of replacement are then determined from the remaining fatigue life of each critical connection. These are shown in Table 4. The determined scheme was also compared with two previous approaches-based member replacement schemes (similar schemes considered in Sect. 3 in Table 1) as shown in Table 4. The comparisons revealed that the proposed framework-based scheme deviates from previous approaches-based schemes.

Table 4. Comparisons of member replacement schemes

Member should be replaced (connection)	Previous approach 1 (AASHTO 1990)			Previous approach 2 (Caglayan et al. 2009)			Proposed framework		
	Rating factor	Replacement years	Sequence	Service life	Replacement years	Sequence	Service life	Replacement years	Sequence
TMB3	3.69	110	4	305	300	5	323	320	6
TMB5	2.70	50	3	156	150	3	165	160	4
TMB6	2.48	35	1	157	150	3	169	160	4
CG4-5	5.63	190	6	20	20	1	12	12	1
ST16-46	4.84	145	5	24	20	1	13	12	1
TDT1	3.39	110	4	179	170	4	45	45	3
TDT2	2.72	50	3	168	170	4	171	170	5
TDT3	2.61	40	2	131	130	2	24	20	2
TDT4	2.55	40	2	132	130	2	162	160	4

5 Conclusions

A remaining service life-based conceptual framework was proposed for essential maintenance of truss bridges. The verification of the proposed conceptual framework was conducted by comparing the predicted replacement scheme with a sequence of experimental failure and fatigue life of selected test specimens. The proposed framework was further utilized to determine the member replacement scheme of a railway truss bridge. It was shown that the proposed framework gives a much safer and sustainable member replacement scheme for railway truss bridges where detailed stress histories are known. Further verifications of the proposed framework are currently under way.

Acknowledgements. The author wishes to express their sincere gratitude to Senior Professor Ranjith Dissanayake, Professor Mitao Ohga, Emeritus Professor M.P Ranaweera and the team of experts who worked in the Sri Lankan Railway Bridge project for their great advice which laid the foundation for this research. The kind support given by the Sri Lanka Railways (SLR) is also appreciated.

References

AASHTO (1990) Guide specification for fatigue evaluation of existing steel bridges, Washington. D.C

Bridge Diagnostics (2002) Load testing, evaluation and rating for railroad flatcar bridge spans over Trinty river, Redding, California, Bureau of Reclamation, Colorado, USA

BS 5400, Part 10 (1980) Code of practice for fatigue, Steel Concrete and Composite Bridges. British Standard Institutions, UK

Caglayan BO, Ozakgul K, Tezer O (2009) Fatigue life evaluation of a through-girder steel railway bridge. Eng Fail Anal 16:765–774

Downing SD, Socie DF (1982) Simple rainflow counting algorithms. Int J Fatigue 4(1):31–40

Fisher JW, Yen BT, Wang D (1980) NCHRP Report No 302: Fatigue and fracture evaluation for rating riveted bridges. Transportation Research Board, National Research Council, Washington, DC

Imam B, Righiniotis TD, Chryssanthopoulos MK (2005) Fatigue assessment of riveted railway bridges. Int J Steel Struct 5(5):485–494

Kong JS, Frangopol DM (2004) Prediction of reliability and cost profiles of deteriorating bridges under time and performance-controlled maintenance. J Struct Eng 130(12):1865–1874

Kohout J, Vechet S (2001) A new function for fatigue curves characterization and its multiple merits. Int J Fatigue 23(2):175–183

Marciniak Z, Rozumek D, Macha E (2008) Fatigue lives of 18G2A and 10HNAP steels under variable amplitude and random non-proportional bending with torsion loading. Int J Fatigue 30(5):800–813

Mesmacque G, Garcia S, Amrouche A, Rubio-Gonzalez C (2005) Sequential law in multiaxial fatigue, a new damage indicator. Int J Fatigue 27(4):461–467

Network rail RT/CE/C/025 (2001) Rail track line code of practice: the structural assessment of under bridges, Railtrack

Spyrakos CC, Raftoyiannis IG, Ermopoulos JCh (2004) Condition assessment and retrofit of historic steel-truss railway bridge. J Const Steel Res 60:1213–1225

Siriwardane SC, Ohga M, Dissanayake R, Taniwaki K (2007) Different approaches for remaining fatigue life estimation of critical members in railway bridges. Int J Steel Struct 7(4):263–276

Siriwardane SC, Ohga M, Dissanayake R, Taniwaki K (2008a) Application of new damage indicator-based sequential law for remaining fatigue life estimation of railway bridges. J Const Steel Res 64(2):228–237

Siriwardane SC, Ohga M, Kaita T, Dissanayake R (2008b) Mesoscopic plastic strain based new fatigue model and its application to a riveted connection. In: Proceeding of international conference on continuum mechanics, nanoscience and nanotechnology, Sri Lanka, pp 77–85

Suresh S (1998) Fatigue of materials. Cambridge University Press, Cambridge

Sustainable Manufacturing

Properties of Portland Blast Furnace Slag Cements Affecting Their Autogenous Shrinkage Behaviors

S. Suntharalingam(✉) and Y. Takahashi

The University of Tokyo, Tokyo, Japan
sharmilane07@gmail.com

Abstract. In Japan, Portland blast furnace slag cements that contain 30–60% of blast furnace slag (BFS) are classified in the slag cement class B in JIS R 5211. Even if slag cements are classified in a category, different autogenous shrinkage behaviors are observed among them. In this study, class B slag cements are collected from eight different plants in Japan. Physical and chemical properties of slag cements are investigated to find the cause of different shrinkage behavior. Physical properties of slag cement are expressed in terms of surface area, mean particle size and particle size distribution. The experimental results showed that the collected slag cements can be differentiated into two different groups based on physical properties. However, different behaviors of autogenous shrinkage cannot be explained, because slag cements, which contain finer particles, have not displayed significant differences in the shrinkage of mortar specimen. Therefore, chemical properties of slag cement could be the main reasons behind the different shrinkage behavior among slag cements. To compare the chemical property of slag cements, the phases of mineral (C_3S, C_2S, etc.) and content of cement and slag mineral (%) in slag cements are analyzed by thermal gravimetric analysis (TGA) and X-ray diffraction (XRD) Rietveld analysis. The results showed that in each plant, different types of BFS are used for producing the slag blended cements, and different quantity levels of slag minerals are found in BFS used in each cement plant. In slag cements classified as Class B in JIS R 5211, different physical and chemical properties are observed. However, the chemical properties of BFS in slag cement mainly influences the shrinkage behavior of slag cement. If the user of BFS cement is concerned about the shrinkage behavior of the concrete, the chemical composition of BFS should be checked to ensure that the concrete has sufficient resistivity against shrinkage.

1 Introduction

Blast furnace slag (BFS) is a by-product of the steel industry that is considered a sustainable alternative to Ordinary Portland cement (OPC). Concrete with slag cement is known for good durability properties and long-term strength development due to its dense microstructure. Japan is one of the major users of blended cements containing BFS and the common slag blended cement is specified as class B in JIS 5211 and it contains blast furnace slag in the range of 30% to 60%.

R. Dissanayake and P. Mendis (Eds.): ICSBE 2018, LNCE 44, pp. 513–525, 2020.
https://doi.org/10.1007/978-981-13-9749-3_45

However, even if supplied class B in JIS 5211 Slag cement concrete is satisfying the demand, different behaviors such as setting time, heat of hydration, shrinkage, etc. in concrete are observed, and some previous research (Sagawa *et al.* 2010; Nito *et al.* 2005; Ekaputri *et al.* 2010) has also reported that commercial cements, classified in slag cement class B with satisfactory existing allowable range of cement properties, are exhibiting different behavior in autogenous shrinkage.

Therefore, slag cements, which are commercially available in Japan, should be differentiated through the physical and chemical properties, to determine the cause of different shrinkage behavior among slag cements; additionally, an investigation should be conducted into what properties of slag cement mainly influence the behavior of shrinkage of concrete.

In this study, the physical properties of slag cements are analyzed through a laser diffraction test; to compare the chemical properties of slag cements, thermal gravimetric analysis (TGA) and X-ray diffraction (XRD) Rietveld analysis are conducted.

2 Properties, Behavior of Compressive Strength and Autogenous Shrinkage of OPC and Slag Cements

2.1 Properties of OPC and Slag Cements

In this experimental study, eight different commercially use slag cements that are available in Japan, are collected and they are coded as **sca**, **scb**, **scc**, **scd**, **sce**, **scf**, **scg**, **sch** and **opc** which is used as a check system. Table 1 shows that, the percentages of BFS and physical properties of OPC and slag cements.

Table 1. Properties of OPC and slag cements.

No	Symbol	Cement type	BFS [%]	Density [kg/m^3]	Surface area [cm^2/g]
1	**opc**	Ordinary Portland cement, JIS R 5210	0	3.15	3430
2	**sca**	Slag cement class B, JIS R 5211	40–45	3.02	3620
3	**scb**		40–45	3.04	3750
4	**scc**		40–45	3.04	3780
5	**scd**		40–45	3.04	3820
6	**sce**		40–45	3.04	4140
7	**scf**		40–45	3.04	4170
8	**scg**		40–45	3.05	3790
9	**scg**		40–45	3.04	3860

2.2 Compressive Strength of OPC and Slag Cements

The compression test results are from each data sheet of cement plant, and the experimental procedure was followed according to the JIS B 5201 for all samples. Cube mortar specimens, with 40 × 40 × 40 mm dimensions, were casted, to test the compressive strength of OPC, and slag cements and fresh mortar were prepared by 1: 3: 0.5 mass ratio between cement, sand and water, respectively. All casted specimens were cured under a constant temperature at 20 °C and a relative humidity of 50%. All specimens were demolded after 24 h, and then immersed into water at a constant temperature of 20 °C. Compressive strength tests were conducted at 3, 7 and 28 days of age.

Figure 1 shows the compressive strength at 3, 7 and 28 days of collected OPC and slag cements. Different compressive strength developments are observed among the OPC and slag cements at 3 and 7 days. However, it is interesting to note that at 28 days, the OPC and Slag cements show an almost equal compressive strength development.

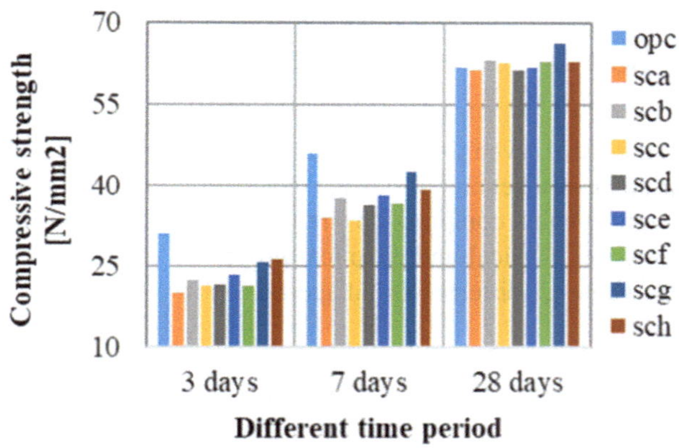

Fig. 1. The compressive strength of OPC and slag cements.

2.3 Autogenous Shrinkage of OPC and Slag Cements

Cylindrical mortar specimens, of ø 5 × 10 cm, were used to measure autogenous shrinkage and fresh mortar (Takahashi and Suntharalingam 2018). These were prepared with 35% of water to binder ratio and sand where the maximum aggregate size and density were 2 mm and 2.63 g/cm^3, respectively, as fine aggregate for 40% volume of fresh mortar. To improve the workability of the fresh mortar, a water-reducing agent of 0.55% cement weight was used (BASF 6500 XD2). Strain gauges of the embedded type (KM-30-120-H2-11) are used to measure the autogenous shrinkages of different OPC and slag cement mortar specimens, under sealed conditions at a constant temperature 20 °C (Ekaputri *et al.* 2010). The results are processed to eliminate the effects of the thermal expansions, through the proposed model of early age heat expansion coefficients by Holt (2001) and temperature histories at the early ages.

The results showed that slag cements had larger autogenous shrinkage than ordinary Portland cement. Slag cements, which are classified in a particular category, show differences in level of autogenous shrinkage behavior, with the results being presented in Fig. 2.

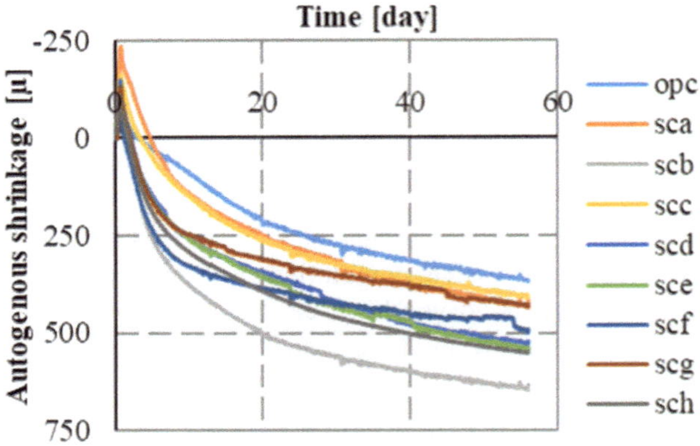

Fig. 2. Autogenous shrinkage of OPC and slag cements.

It is clearly shown that, even if collected slag cements are classified in a category in JIS R 5211, different autogenous shrinkage behaviors among slag cements can be observed, while almost equal compressive strength among OPC and slag cements at 56 days of age. Therefore, physical and chemical properties of slag cements were investigated to find the cause of different shrinkage behavior.

3 Physical Properties of OPC and Slag Cements

The particle size and distributions of OPC and slag cements were measured using a laser diffraction technique, through the SALD-3000 J particle Size Analyzer in the range of particle size from 0.1 μm to 2000 μm. The mean particle size and the relationship between surface area and mean particle size of OPC and Slag cements are shown in Figs. 3 and 4, respectively.

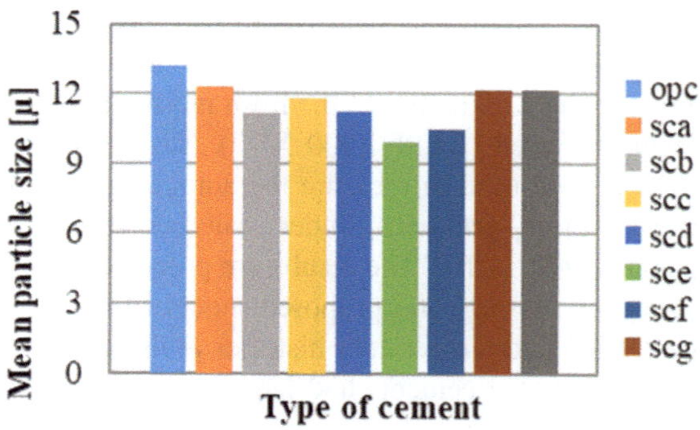

Fig. 3. Mean particle size of OPC and slag cements.

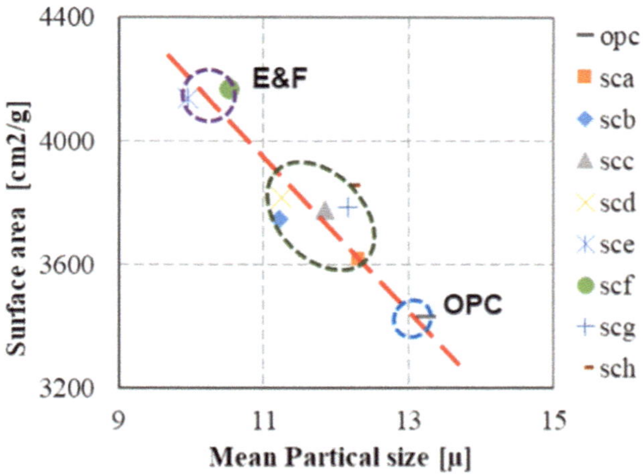

Fig. 4. Relationship between surface area and mean particle size of OPC and slag cements.

The results show that the mean particle size and surface area of the collected slag cement particles are different; when the surface area of slag cement particles with their mean particle size is compared, **sce** and **scf** are significantly deviate from other slag cement, as they have a smaller particle size and a greater surface area among collected slag cement. However, there are no significant differences in the autogenous shrinkage behavior of **sce** and **scf** from other slag cements.

It is evident that the physical properties of Class B slag cement in JIS 5211 are not a major influence factor for showing the significant difference in autogenous shrinkage among slag cements. Therefore, the chemical properties of slag cement could be the main factor. Chemical properties of slag cement are investigated by the thermal gravimetric analysis (TGA) and the X-ray diffraction (XRD) Rietveld analysis for differentiation of collected slag cements.

4 XRD Rietveld and TGA Analysis

The Shimadzu XRD 6100 machine was used to determine the diffraction profile pattern for OPC and slag cements sample in XRD test, and the measurement of XRD was conducted using an internal standard substance with an average particle diameter of 3 μm. Corundum ($\alpha1\ Al_2O_3$) was added by 10% of the sample. In the XRD measurement conditions were as follows: target Cu-K_α, tube voltage 45 kV, tube current 40 mA; range of scanning from 5 to 70°; 2θ step width 0.02. Rietveld analysis is distributed by Siroquant version 3 software. Micro absorption was corrected with the particle diameter of quantitative minerals being 10 μm.

4.1 Quantitative Value of OPC Minerals in Slag Cement

In the quantitative value of cement mineral in slag cement, the following minerals were subjected for Rietveld analysis, C_3S (Belov), C_3S (mono.), C_2S, beta, C_2S, alpha, C_3A, (cubic), C_3A (ortho.), C_4AF, Periclase, Mayenite, Gypsum, Calcite 2, Portlandite, Bassanite of cement mineral and as internal standard substance corundum ($\alpha 1\ Al_2O_3$).

Obtained quantitative values for cement mineral in Rietveld analysis were corrected by the internal standard substance amount and ignition loss according to Eq. (1).

$$W_{CX} = (W_X/W_C) * 10 * (1 + Ig/100) \tag{1}$$

Where

W_{CX} = Corrected quantitative value (%)
Wx = Quantitative value of mineral before correction (%)
W_C = Quantitative value of corundum (%)
Ig = Ignition loss between 30 °C and before starting the crystallization effect of slag materials (%)

In determining ignition loss by the TGA test, mass losses are measured between 30 and 1000 °C and they can be used for Ordinary Portland cement that does not contain BFS. However, when heating the slag cement in the range between 30 and 1000 °C, further losses or increases in weight of slag cement sample nearly above 760 °C temperature were observed; this is because of the slag materials being in the form of amorphous material nearly below 760 °C temperature. However, when continuously heating slag cement above 760 °C, slag materials start to shift into their crystal forms.

Figure 5 shows that slag materials in **scd** start to transform from amorphous phase to crystal form, above 760 °C. Therefore, the quantitative value of each mineral should be corrected with ignition loss, to compare the percentage of each minerals in slag cement before the start crystallization effect, for an accurate analysis. Here, to correct quantitative values of each mineral, ignition loss of slag cements is calculated at a weight loss temperature between 30 °C, and before starting the crystallization effect of slag materials.

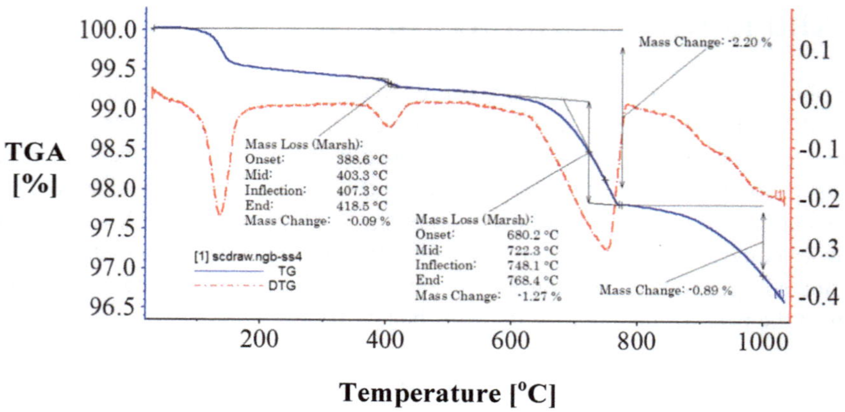

Fig. 5. TGA result of **scd.**

Figure 6 shows that the corrected quantitative value of main cement mineral (C_3S and C_2S) in OPC, which was used to blend with blast furnace slag to produce slag cements in each plant (here, the percentage of each cement minerals were calculated without considering the blast furnace slag portion in slag cement). The result shows that significant differences were observed in percentages of C_3S and C_2S of used OPC for producing slag cement. However, the pozzolanic reaction in slag cement is what mainly decides the long-term behavior of autogenous shrinkage of mortar specimens. Therefore, the content of slag material should be analyzed to understand behavior of autogenous shrinkage.

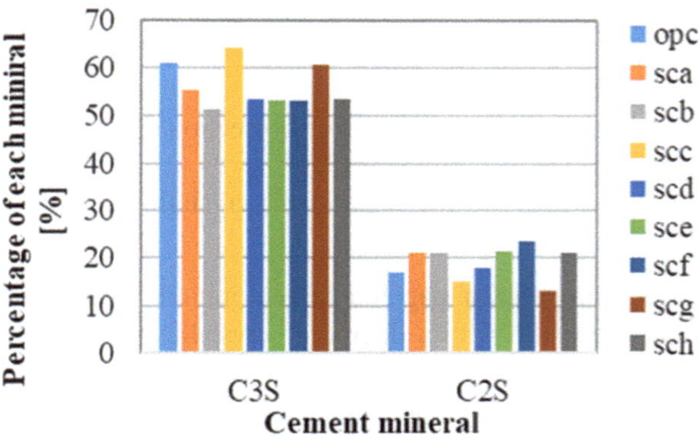

Fig. 6. Corrected quantitative value of main mineral (C_3S and C_2S).

4.2 Quantitative Value of Crystallized BFS Minerals in Slag Cements

BFS fine powder is in amorphous phase in slag cement. Therefore, BFS should be transformed into a crystal, through heat treatment for the XRD test. In a TGA test, it is clearly observed that if heated nearly above 760 °C, the blast furnace slag starts to obtain a crystal structure. Sagawa and Nawa (2006) proposed that, for determining the content of BFS in slag cement, the sample should be heat-treated at 900 °C for 30 min to crystallize slag material; as for the above crystallization conditions, a complete crystallization of the slag, and a minimization the decomposition of the cement mineral, can be achieved.

Due to the aforementioned heating process, there was C_2S (α'L), which was thought to have been generated by decomposition of C_3S in cement mineral and blast furnace slag fine powder changed into Gehlenite ($2CaO{\cdot}Al_2O_3{\cdot}SiO_2$), Akermanite ($2CaO{\cdot}MgO{\cdot}2SiO_2$), Merwinite ($3CaO{\cdot}MgO{\cdot}2SiO_2$), which are represented by abbreviations of C_2AS, C_2MS_2 and C_3MS_2, respectively. The crystal structure of the above minerals was based on data preset in Rietveld analysis software (Siroquant version 3).

The following cement and crystalized slag minerals were the subject of Rietveld analysis: C_3S (Belov), C_3S (mono.), C_2S, beta, C_2S, alpha, C_3A, (cubic), C_4AF, Anhydrite, Wollastonite, Lime of cement mineral, Akermanite, Gehlenite, Merwinite, Mayenite of crystalized slag mineral and as internal standard substance corundum (α 1 Al $_20_3$).

Obtained quantitative values of cement and crystalized slag mineral in Rietveld analysis are corrected by the internal standard substance amount and ignition loss according to Eq. (2).

$$W_{CX} = (W_X/W_C) * 10 * (1 - Ig/100) \tag{2}$$

Where

W_{CX} = Corrected quantitative value (%)
Wx = Quantitative value of mineral before correction (%)
W_C = Quantitative value of corundum (%)
Ig = Ignition loss between starting point of the crystallization of slag materials and end of heat-treatment (%)

The quantitative value of each slag and cement mineral in heat-treated samples (at 900 °C for 30 min) are listed with a degree of crystallization effect in slag material between the starting temperature of crystallization (at nearly 760 °C) and the end of heat treatment. Therefore, to accurately analyze, the quantitative value should be corrected with ignition loss when comparing the percentage of each mineral in slag cement, at the starting point of the crystallization effect. In this heat-treated sample analysis, Ignition losses are calculated for correcting the quantitative value of each mineral weight loss between the starting point of the crystallization of slag materials and end of heat treatment.

Figure 7 shows the corrected quantitative value of crystallized slag minerals (C_2AS, C_2MS_2 and C_3MS_2) at the starting point of slag mineral crystallization (at nearly 760 °C) through XRD Rietveld method in blended slag cements. Among slag cements, several percentage differences are observed in each crystallized slag minerals: **sca** have the higher content of Merwinite and Gehlenite, compared with other slag cements and Akermanite in **sce**. Figure 8 shows the sum of corrected quantitative value of crystallized slag minerals (C_2AS, C_2MS_2 and C_3MS_2) at the starting point of crystallization of slag mineral in slag cements: among these, several percentage changes were observed in total crystallized slag minerals. There are two possibilities for this difference in content of total crystallized slag mineral in slag cements, which are

(a) Mixing percentage of blast furnace slag and
(b) Chemical composition of blast furnace slag (CaO, SiO_2, etc.)

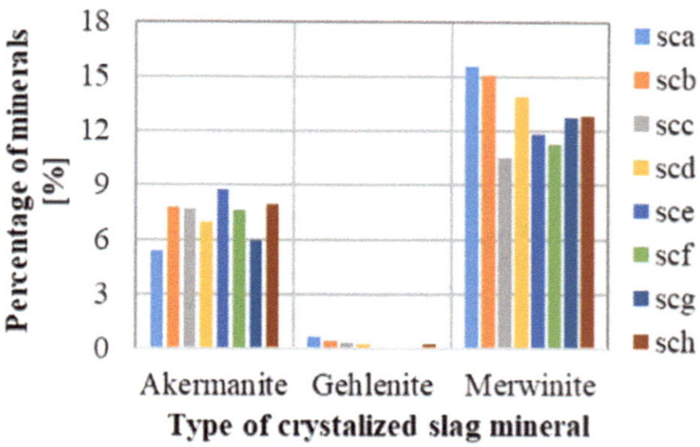

Fig. 7. Content of crystalized slag minerals in slag cements.

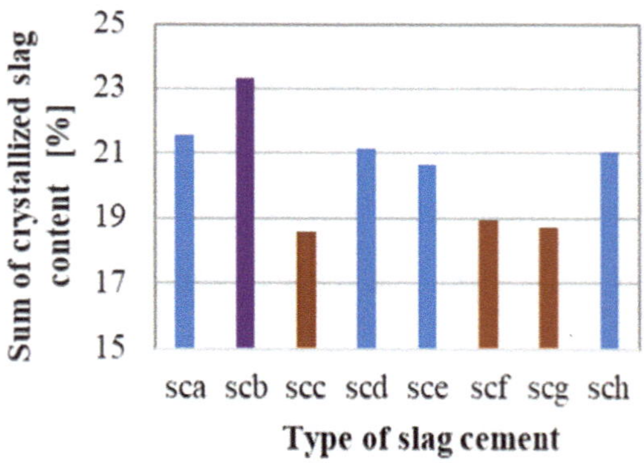

Fig. 8. Content of crystalized slag minerals in slag cements.

The autogenous shrinkages of slag cements at 3, 7, 28 and 56 days are compared with the sum of crystallized slag minerals (Akermanite + Gehlenite + Merwinite), which is shown in Fig. 9. The higher the content of crystallized slag in cement, the greater the autogenous shrinkage. The strong relationship between content of crystallized slag and autogenous shrinkage can be observed at the 56 days period, because of the pozzolanic reaction by slag mineral which will be influenced at a later stage. Therefore, behavior differences in autogenous shrinkage among slag cements are affected by to the level of crystallized slag mineral in heat treated slag cements.

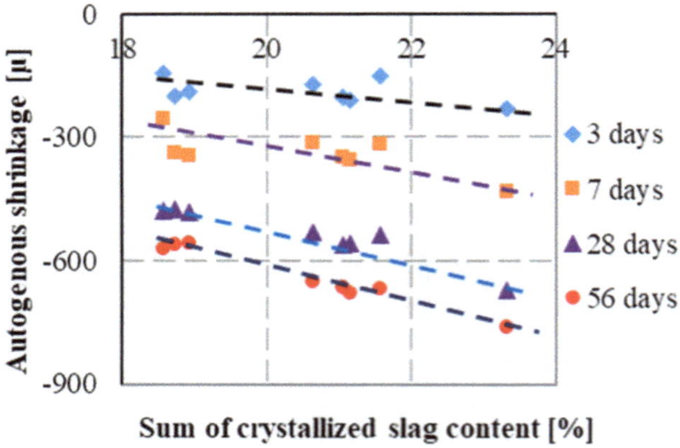

Fig. 9. Relationship between autogenous shrinkage and crystalized slag minerals of slag cements.

Additionally, the types of BFS used for producing the blending slag cements in each manufacturing plant, were analyzed in terms of crystallized slag minerals composition. Compositions of crystallized slag minerals in blended slag cements is shown in Fig. 10, and they are calculated through the percentage of each crystallized slag mineral in total crystallized slag mineral by Eq. (3).

$$P_S = [W_S/\Sigma W_S)] * 100 \tag{3}$$

Where

P_S = The percentage of mineral in blast furnace slag (%)
W_S = Quantitative value of each slag mineral in slag cement (%)
ΣW_S = Sum of Quantitative value of slag mineral in slag cement (%)

From this analysis, different compositions of crystallized slag minerals are observed in slag cements from each plant. **sca** has the higher percentage of Merwinite among crystallized slag minerals, compared to other slag cements. Similar types of slags were used for blending with OPC to produce **scb** and **scd,** and, in **sce** and **scf**, nearly the same type of slag without Gehlenite mineral were observed in heat treated samples.

Figure 11 shows the relationship between crystallized slag minerals of heat-treated slag cement and composition of slag material (CaO, SiO_2, MgO, etc.) (Sagawa and Nawa 2012). The composition of crystallized slag minerals in heat treated samples is dependent on the composition of slag in cement. It is evident that different types of BFS are used in each plant for production of slag cement.

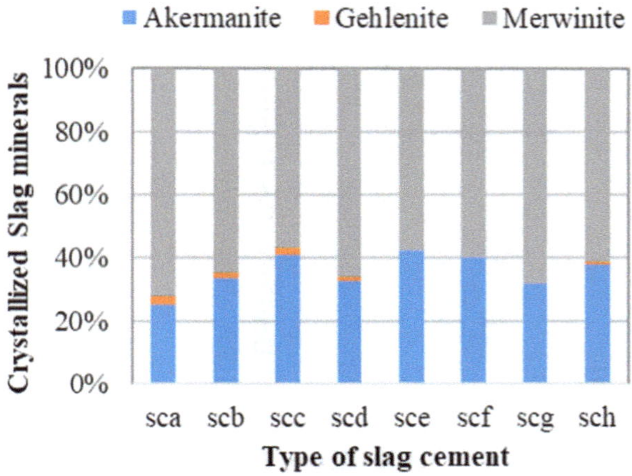

Fig. 10. Types of blast furnace slag used in slag cements.

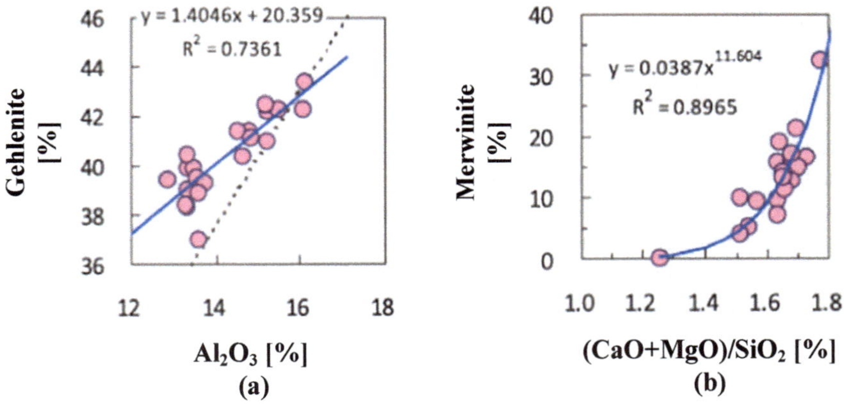

Fig. 11. (a) & (b): The relationship between crystallized slag minerals and composition of slag material (Sagawa and Nawa 2012).

From studies about slag cement chemical properties, different quantities and chemical compositions of slag are observed in collected slag cements (Class B in JIS 5211). These different blast furnace chemical compositions change the Ca/Si ratio and the density of C-S-H product among slag cements, due to the pozzolanic reaction (Sagawa and Nawa 2010), which is shown in Fig. 12. The Ca/Si ratio and the density of C-S-H product mainly influence the autogenous shrinkage behavior of cement, and low densities of C-S-H produces larger shrinkage than high densities of C-S-H product, a fact reported by Maruyama and Kurihara (2016).

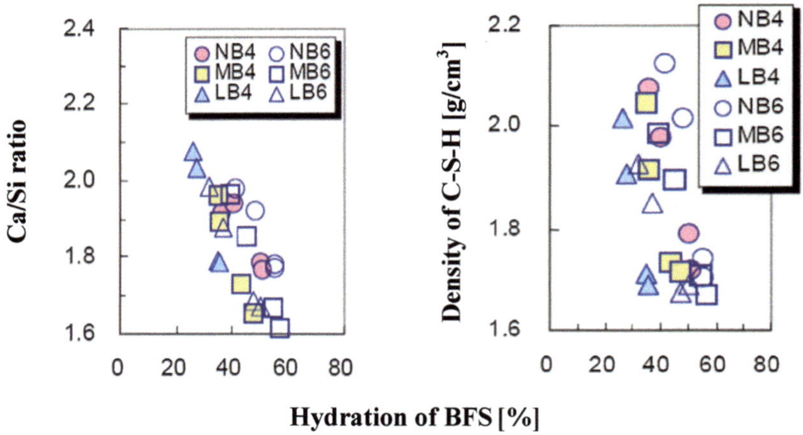

Fig. 12. The relationship between Ca/Si ratio and the density of C-S-H and hydration of BFS (Sagawa et al. 2010).

5 Conclusion

In this study, slag cement (Class B in JIS 5211), which is commercially available in Japan, is differentiated in two different groups based on their physical properties. However, the physical properties of slag cement does not constitute a major influencing factor for the different shrinkage behavior, because slag cements contain finer particles that do not significantly affect autogenous shrinkage behavior. In chemical property analysis, the different quantities and types of BFS that were used to produce slag cements, were studied by TGA and XRD Rietveld analysis; autogenous shrinkage is correlated more with the content of crystallized slag minerals in slag cement at 56 days of hydration. If the user of BFS cement is concerned about the shrinkage behavior of the concrete, the chemical composition of BFS should be checked, to ensure that the concrete has sufficient resistivity against shrinkage.

Acknowledgments. This study was financially supported by JSPS KAKENHI Grant No. 17H01284 and Research scholarship by Asian Development Bank-JSP.

References

Ekaputri JJ, Ishida T, Maekawa K (2010) Autogenous shrinkage of mortars made with different types of slag cement. Proc JCI 32(1):353–358

Holt EE (2001) Early age autogenous shrinkage of concrete. Technical Research Centre of Finland, Espoo, VTT Publications, p 446

Hoshino S, Yamada K, Hirao H (2006) XRD/Rietveld analysis of the hydration and strength development of slag and limestone blended cement. J Adv Concr Technol 14(3):357–367

Maruyama I, Kurihara R (2016) Impact of two types of calcium silicate hydrate on drying shrinkage of Portland cement paste. In: The 70th cement technology convention, vol 1109

Nito N, Hiroshima A, Ohtomo T, Miyazawa S (2005) Property of concrete using portland blast-furnace slag cement b class with different slag grain and a chemical composition. Cement Sci Concr Technol 59:231–238

Sagawa T, Nawa T (2006) Quantitative hydration analysis of blast furnace slag cement by rietveld method. Concr Res Technol 17(3):1–11

Sagawa T, Nawa T (2010) Hydration-microstructural evolution and drying shrinkage of portland cement-blast furnace slag system. J Struct Constr Eng 75(652):1029–1037

Sagawa T, Ishida T, Luan Y, Nawa T (2010) Hydrate composition analysis and micro structure characteristics of Portland cement-blast furnace slag system. Cement Sci Concr Technol 66 (3):311–324

Sagawa T, Nawa T (2012) Quantitative determination of blast-furnace slag content in slag-blended cement by combination method of heat-treatment and XRD/Rietveld analysis. Cement Sci Concr Technol 68

Takahashi Y, Suntharalingam S (2018) Experimental study on the autogenous shrinkage behaviors of different Portland blast furnace slag cements. In: Proceedings of the 8th international conference the Asian Concrete Federation, pp 409–418

Fire Performance of CFRP Strengthened Steel I Beams Cured at Elevated Temperature

E. R. K. Chandrathilaka(✉) and J. C. P. H. Gamage

Department of Civil Engineering, University of Moratuwa, Moratuwa, Sri Lanka
kanishkachandrathilaka@gmail.com

Abstract. Elevated temperature curing increases the glass transition temperature of the polymeric bond between CFRP and steel. Fire performance of an elevated temperature cured and ambient temperature cured CFRP strengthened steel I beam was evaluated using a numerical model. The effects of a range of thicknesses and conductivities of insulation layers were also analyzed. Model results are in good agreement with the theoretical results. Insulation is a compulsory requirement for CFRP strengthened steel I sections. On average, a 35% increase in fire performance was noted at elevated temperature cured CFRP strengthened I beams compared to the ambient cured I beam.

Keywords: Glass transition temperature (T_g) · CFRP/steel bond · Fire · Steel I beams · Insulation

1 Introduction

The use of CFRP (Carbon Fiber Reinforced Polymer) in strengthening of steel structures is rapidly spreading due to its superior properties such as high strength to weight ratio, easy installation process and corrosion resistance. However, one of the main drawbacks of the use of CFRP in strengthening of steel structures is strength degradation of CFRP/steel bond at elevated temperatures. The main cause of the strength degradation is low glass transition temperature in epoxy adhesive (ARELDITE 420 A/B 2018).

The elevated temperature curing has been used to increase the bond performance of CFRP/epoxy/steel bond (Nguyen et al. 2011; Bai et al. 2014; Nguyen et al. 2012). However, the use of the methods mentioned in the literature (Nguyen et al. 2011; Bai et al. 2014, Nguyen et al. 2012) were not practically feasible. Therefore, a better feasible elevated temperature curing method is required. Chandrathilaka et al. (2019) introduced the use of halogen floodlights in elevated temperature curing of CFRP/epoxy/steel bond. The new method was compared with the traditional laboratory experiment procedure with standard oven. It was concluded that the new method was more effective than the standard method.

Gamage et al. (2006) have performed experimental thermo-mechanical analysis to determine the effective thickness of insulation for the concrete/epoxy/CFRP bond. Other than the insulation thickness, thermal conductivity of insulation and type of fire was considered in the analysis. Imran et al. (2018) has predicted the temperature variance of CFRP strengthened steel tubular columns and checked for the effective

R. Dissanayake and P. Mendis (Eds.): ICSBE 2018, LNCE 44, pp. 526–537, 2020.
https://doi.org/10.1007/978-981-13-9749-3_46

thickness for the insulation. In both cases experimental thermal analysis was carried out to validate the results of the numerical thermal analysis. However, performing experimental thermal analysis is much more complex, expensive and time consuming process. Therefore, using of theoretical equations to validate the model will be a good alternative.

Zang and Usmani (2015) have investigated the thermal behaviour of steel I sections in an event of fire numerically and using theoretical formulas. The performance of both analytical methods was significantly similar at lower temperatures. When the temperature exceeds 200 °C, the difference between the results of numerical thermal analysis and theoretical thermal analysis tends to increase.

However, at temperatures less than 100 °C, the results of theoretical and numerical thermal analysis had the same values. Zang et al. (2012) have also used different equations to perform a sensitivity study on their accuracy and predicting applicability to the situation.

In this study, a numerical model was developed to predict the fire performance of CFRP strengthened steel I beam cured at elevated temperature.

2 Numerical Analysis

Transient heat analysis was performed using transient thermal tools in the software (ANSYS mechanical user's guide 2013). The thermal behavior of the CFRP strengthened steel I beams were predicted for a range of environmental temperature variations with the time. A 3-D model was developed to predict the thermal behavior of CFRP strengthen I beam with different insulation layers with varying thermal conductivities and insulation layer thicknesses.

2.1 FE Mesh and Boundary Conditions

Steel I section UC 254 × 254 × 167 was used in the FE analysis to evaluate the fire performance. It was assumed that the CFRP strengthening was done by wrapping the CFRP around a steel column section. A 1 m long steel I section was modelled to evaluate the thermal behaviour of CFRP strengthened steel I section. SOLID70, 8-node brick elements were used in each material. A programme controlled mesh was used in the model with user input maximum mesh size between 25 mm to 50 mm with a minimum mesh size of 1 mm. The aspect ratio was selected to be 1:1 (Fig. 1). The two ends of steel I beam, CFRP layer and insulation were modelled as perfectly insulated as shown in Fig. 2. Heat was introduced on the surface of insulation as a convection (Fig. 2).

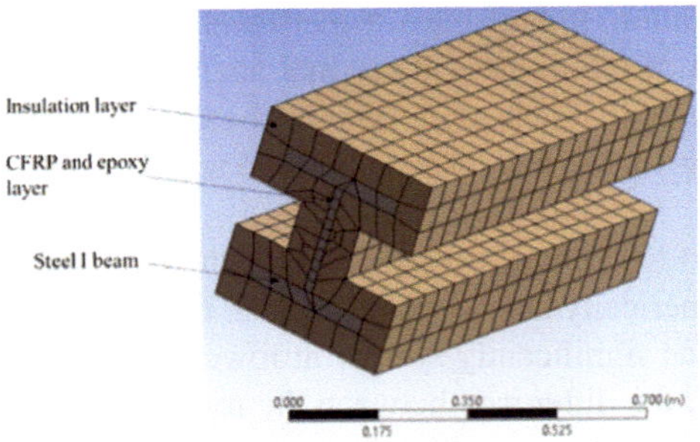

Fig. 1. FE mesh

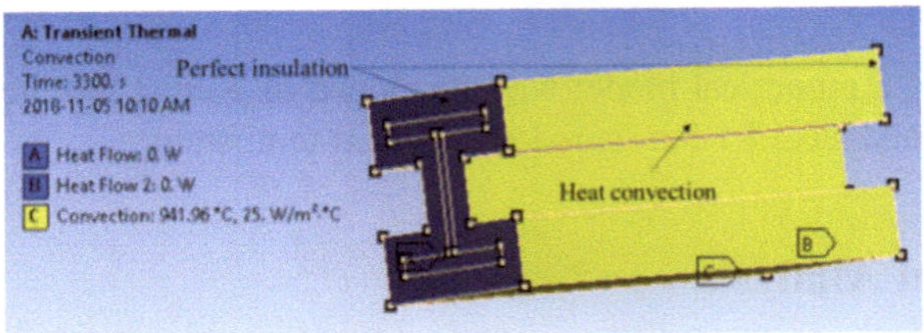

Fig. 2. Boundary conditions

2.2 Material Properties

Thermal properties of the used materials in the FE analysis are listed in Table 1.

Table 1. Material properties (X-Wrap C300, ARELDITE 420 A/B, CAFCO 300, CAFCO 300 AC, CAFCO® BLAZE-SHIELD® II)

Material	Thermal conductivity (W/m.°C)	Specific heat (J/kg.°C)	Density (kg/m^3)
Steel	60.5	511	7800
CFRP (X-Wrap C300)	5	1000	1000
Epoxy adhesive (ARELDITE 420 A/B)	1	1000	1000
Insulation 1 (CAFCO 300)	0.0839	700	240
Insulation 2 (CAFCO 300 AC)	0.078	700	240
Insulation 3 (CAFCO® BLAZE-SHIELD® II)	0.043	700	240

2.3 Analysis Procedure

The solving was done using a computer with 8 GB RAM and i7 processing unit. The transient thermal solver in ANSYS was used. The solving steps were applied as 20 s time steps. Approximately 1 h was taken to complete the solving process. The convection heat was applied as a time dependent temperature profile on the insulation surface of CFRP strengthen I beam. The temperature of convection surface was inserted according to ISO 834-11:2014 as shown in Eq. (1) and Fig. 3.

$$T_f = 345.log_{10}(0.133t + 1) + T_0 \tag{1}$$

T_f = Fire temperature (°C)
t = time (s)
T_0 = Initial temperature (°C)

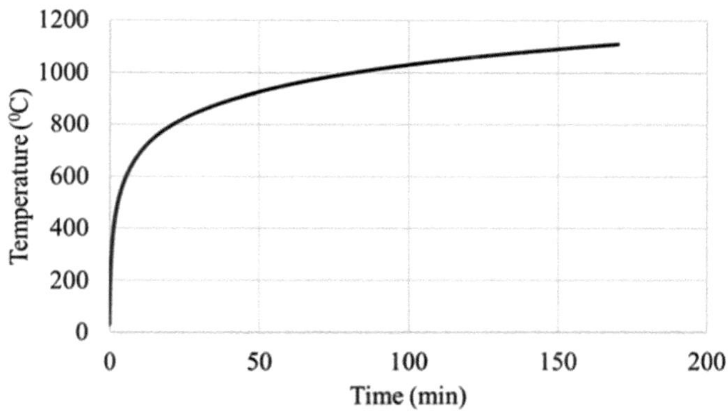

Fig. 3. Standard fire curve (ISO 834-11:2014)

Insulation layer parameters were varied as shown in Table 2.

Table 2. Insulation layer parameters

Series	Thermal conductivity of insulation (W/m.°C)	Thickness of insulation (mm)	Fire rating (mins)	
			Ambient Cured	Elevated temperature cured
Series 1	0.0839	50	18	26
	0.078		19	28
	0.043		31	44
Series 2	0.043	25	17	23
		40	25	36
		50	31	44
		55	33	49
		60	37	53
		75	46	66

3 Theoretical Analysis

Milke (2000) has suggested an equation to predict the fire performance of an insulated structural steel section.

$$\Delta T_s = \frac{k_i}{h}\left[\frac{T_f - T_s}{c_s\left(\frac{W}{D}\right) + 0.5 c_i \rho_i h}\right]\Delta t \quad (2)$$

T_s = Steel temperature (°C)
T_f = Fire temperature (°C)
k_i = Thermal conductivity of insulation material (W/m.°C)
h = Insulation thickness (m)
c_s = Steel specific heat (J/kg.°C)
W = Steel weight per unit length (kg/m)
D = Heat perimeter (m)
c_i = Specific heat of insulation material (J/kg.°C)
ρ_i = Density of insulation material (kg/m^3)
Δt = Time interval (s)
ΔT_s = Steel temperature change (°C)
T_f = Standard fire temperature (°C)

However, the main vulnerability of CFRP strengthened steel sections are the epoxy adhesive layer which connects the CFRP to steel. The glass transition temperature of the bond has a value between 50 °C to 61 °C, varying with the curing type and condition (Chandrathilaka et al. 2019). Therefore, the validity of Eq. (2) depends on the thermal properties of CFRP and adhesive. Zang and Usmani (2015) has shown the similarity of heat conductivity and Ohm's law in electric-circuit theory. The thermal resistance of the insulators was defined as,

$$R_i = \frac{h_i}{k_i} \quad (3)$$

R_i = Thermal resistance
h_i = Insulation thickness (m)
k_i = Thermal conductivity of insulation material (W/m.°C)

However, in cases of low insulation thicknesses and high thermal conductivities the thermal resistance becomes very low, resulting in low temperature variance. The thermal conductivity of insulation material (0.043 W/m.°C–0.0839 W/m.°C) is much lower compared to thermal conductivity of CFRP (5 W/m.°C) and adhesive layers (1 W/m.°C). The insulation thickness varies between 25 mm to 75 mm, while the CFRP and adhesive layer thicknesses were 0.166 mm and 0.5 mm, respectively. Therefore, the accuracy of Eq. (2) should be the main focus in this case, ignoring the CFRP and adhesive layers and considering the steel temperature as bond temperature.

The same steel I beam was used in the theoretical analysis to evaluate the fire performance. It was assumed that the CFRP strengthening was done by wrapping the CFRP around a steel column section as mentioned in above.

4 Results and Validation

Temperature variation in epoxy layer of non-insulated steel I section and standard fire curve is shown in Fig. 4. At the start of the fire, the temperature difference between environment and CFRP strengthen steel beam was about 45%. However, after 100 s, the difference was almost negligible. The temperature of epoxy layer has reached the glass transition temperature of the epoxy layer (50 °C) within 10 s. The fire rating of non-insulated CFRP strengthen steel I section was less than 10 s, which was very weak in fire performance. Therefore, insulation is a compulsory requirement in CFRP strengthened steel I sections in civil engineering applications to maintain required fire rating of buildings.

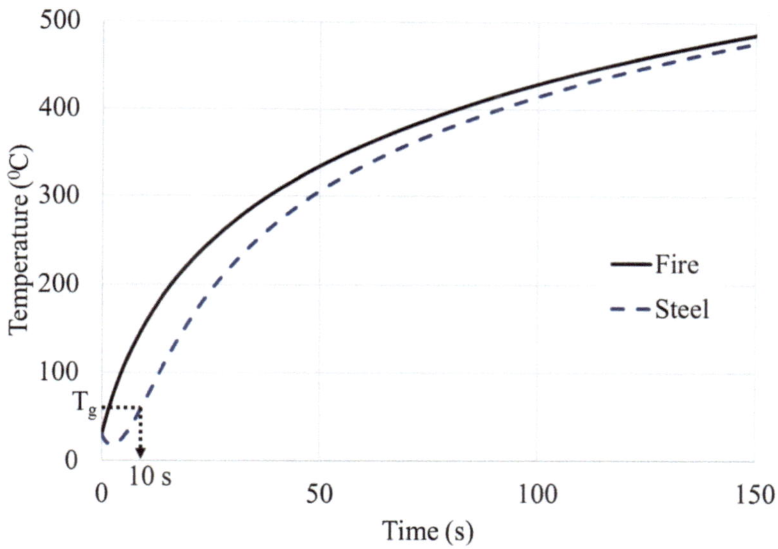

Fig. 4. Temperature variation in epoxy layer of non-insulated CFRP strengthen steel I beam and standard fire with time

Theoretical thermal analysis and numerical thermal analysis results for varying insulation layer thickness and thermal conductivity is shown in Figs. 5 and 6. For the different thermal conductivities of insulation material, the results had an average less than 1% difference between numerical and theoretical analysis. However, for the different insulation layer thicknesses, the variance between theoretical and numerical analysis results were an average of 2%. For the insulation thicknesses of 40 mm, 50 mm and 55 mm, the difference is far more negligible. However, at 25 mm insulation thickness, the difference between numerical and theoretical analysis were as high as 5% with theoretical thermal analysis, showing higher temperature value than the numerical analysis at the same time. When the insulation thicknesses are 60 mm and 75 mm the numerical analysis has over-predicted the temperature on average difference of 3% than the theoretically predicted temperature. However, the overall slight variation between theoretical and numerical results provides evidence for the accuracy of the model. Variation of temperature across an arbitrary cross section of the insulated CFRP strengthened steel I section (with thermal conductivity of 0.043 W/m.°C and insulation layer thickness of 50 mm) is shown in Fig. 7.

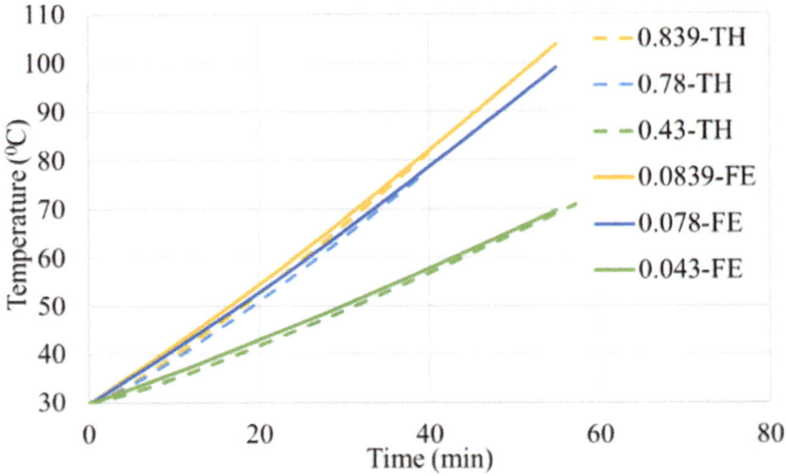

Fig. 5. Validation of FE model with theoretical analysis for different thermal conductivities of insulation materials

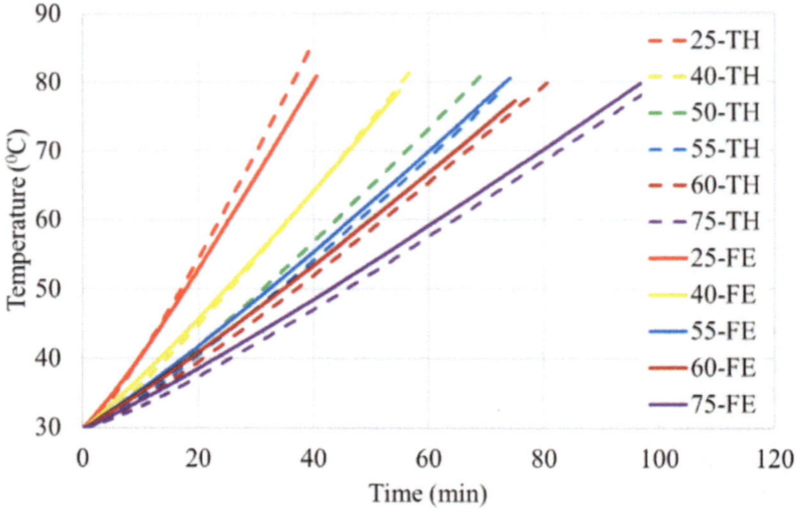

Fig. 6. Validation of FE model with theoretical analysis for different insulation layer thicknesses

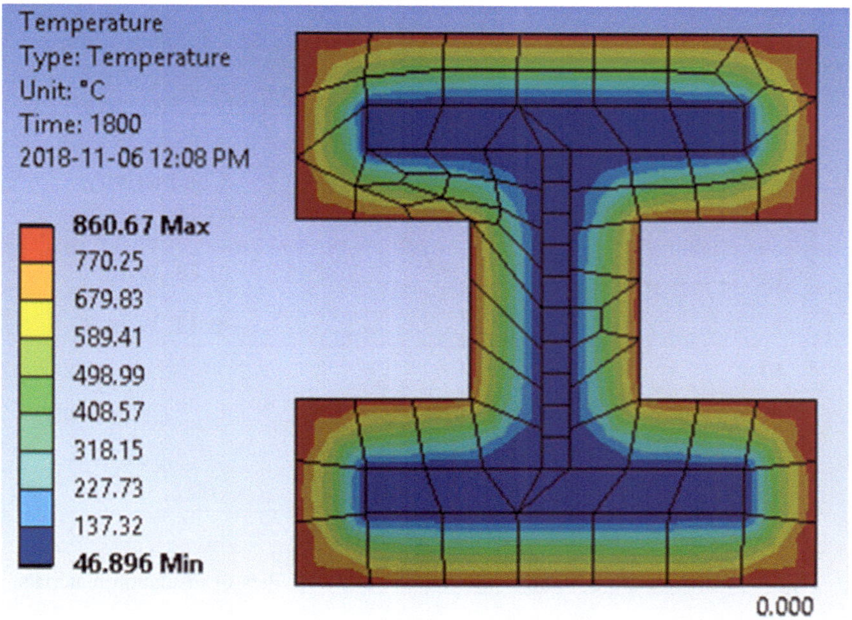

Fig. 7. Variation of temperature across an arbitrary cross section of insulated CFRP strengthened steel I section

5 Fire Rating of CFRP Strengthened Steel I Section

Fire rating of the CFRP strengthened steel I sections were calculated by referring the Tg of CFRP/epoxy/steel bond for elevated temperature curing (61 °C) and ambient curing condition (50 °C) (Chandrathilaka et al. 2019). On average, 35% increase of fire rating was noted in elevated temperature cured CFRP strengthened I sections compared to ambient cured CFRP strengthened I sections as shown in Figs. 8 and 9. However, to achieve 1-hour fire rating there is a requirement of thermal conductivity of 0.043 W/m.°C for the insulation material with 75 mm insulation layer thickness. At 75 mm insulation layer thickness, 20 min increase in fire rating was noted in elevated temperature cured CFRP strengthened steel I sections compared to ambient cured ones. However, use of 0.839 W/m.°C and 0.78 W/m.°C thermal conductivities for the insulation, indicated the same fire rating values as they were having a comparatively similar thermal conductivity while 0.043 W/m.°C thermal conductivity insulation performs 63% better than those two types. This shows that decreasing the thermal conductivity by 45%, fire rating can be increased by more than 65%. However, when the insulation thickness increases by 300%, the increase in fire rating was only 190%. Therefore, the use of low thermal conductive insulation with lesser thickness may increase the fire rating than using high thermal conductivity insulation with high insulation layer thickness. Thermal profiles within the epoxy bond line of insulated beam with 0.043 W/m.°C conductivity and thickness of 50 mm are shown in Fig. 10.

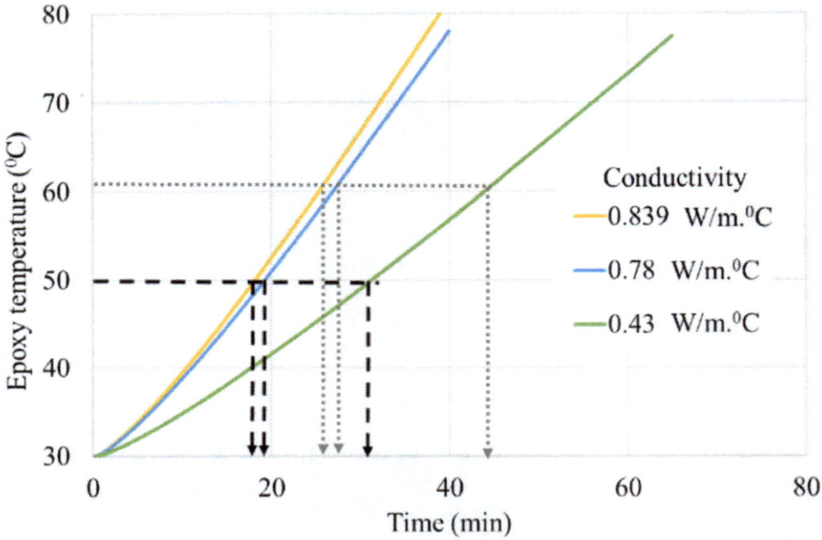

Fig. 8. Variation of fire ratings with thermal conductivities of insulation materials

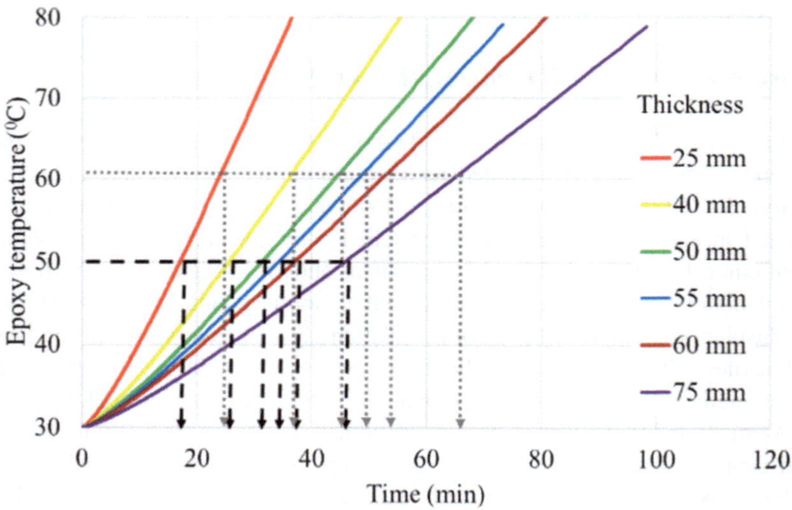

Fig. 9. Variation of fire ratings with insulation thickness

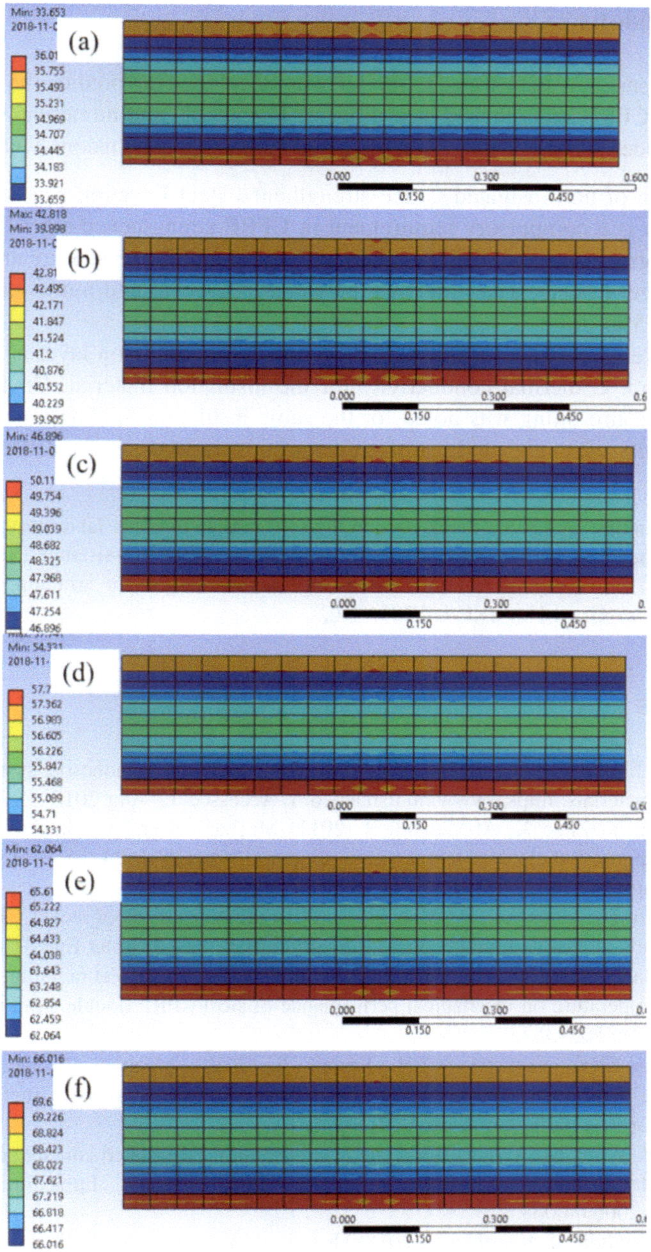

Fig. 10. Thermal profile within the epoxy layer for insulated I beam with conductivity 0.043 W/m.°C and thickness of 50 mm at (a) 10 min (b) 20 min (c) 30 min (d) 40 min (e) 50 min (f) 55 min

6 Conclusions

A three dimensional heat transfer model was developed to predict the thermal performance of CFRP strengthened steel I beam. Theoretical and numerical analysis were compared under standard fire conditions. The following conclusions were made,

1. Fire rating of non-insulated CFRP strengthened steel I section was less than 10 s. Insulation is a compulsory requirement in CFRP strengthened steel I sections.
2. A good agreement was noted between numerical and theoretical fire analysis results for the steel I beam. The difference between theoretical and numerical results was less than 5%.
3. To achieve at least 1-hour fire rating, 75 mm thick insulation layer is required with 0.043 W/m.°C thermal conductivity for the insulation material. On average, 35% increase in fire rating was noted for the same insulation thickness and the thermal conductivity when elevated temperature curing was done after installation of CFRP.

Acknowledgments. A special thank goes to the staff in the computer laboratory of Department of Civil Engineering, University of Moratuwa for their valuable assistance. Senate Research Council of the university of Moratuwa is greatly acknowledged for arranging the necessary financial support (Grant No: SRC/LT/2016/19).

References

ARELDITE 420 A/B (Technical data sheet) Two component epoxy adhesive system, Huntsman advanced materials. https://www.huntsman.com. Accessed 12 Aug 2018

Nguyen T, Bai Y, Zhao X, Al-Mahaidi R (2011) Mechanical characterization of steel/CFRP double strap joints at elevated temperatures. Compos Struct 93:1604–1612. https://doi.org/10.1016/j.compstruct.2011.01.010

Bai Y, Nguyen TC, Zhao XL, Al-Mahaidi R (2014) Enhancement of steel/CFRP adhesively-bonded joints at elevated temperatures through curing. Aust J Struct Eng 15(4):367–376

Nguyen TC, Bai Y, Zhao XL, Al-Mahaidi R (2012) Effects of ultraviolet radiation and associated elevated temperature on mechanical performance of steel/CFRP double strap joints. Compos Struct 94(12):3563–3573

Chandrathilaka ERK, Gamage JCPH, Fawzia S (2019) Mechanical characterization of CFRP/steel bond cured and tested at elevated temperature. Compos Struct 207:471–477. https://doi.org/10.1016/j.compstruct.2018.09.048

Gamage JCPH, Wong MB (2006) Bond characteristics of CFRP plated concrete members under elevated temperatures. Compos Struct 2006(75):199–205. https://doi.org/10.1016/j.compstruct.2006.04.068

Imran M, Mahendran M, Keerthan P (2018) Heat transfer modelling of CFRP strengthened and insulated steel tubular columns. Constr Build Mater 184:278–294. https://doi.org/10.1016/j.conbuildmat.2018.06.205

Zhang C, Usmani A (2015) Heat transfer principles in thermal calculation of structures in fire. Fire Saf J 78:85–95. https://doi.org/10.1016/j.firesaf.2015.08.006

Zhang C, Li G, Wang Y (2012) Sensitivity study on using different formulae for calculating the temperature of insulated steel members in natural. Fire Technol 2012:343–366. https://doi.org/10.1007/s10694-011-0225-x

ANSYS mechanical user's guide (2013) ANSYS, Inc. http://www.ansys.com

X-Wrap C300 (Technical data sheet) High strength carbon fiber fabric for structural strengthening, X-CALIBUR structural systems. https://www.x-calibur.us. Accessed 12 Aug 2018

CAFCO 300 (Technical data sheet) Wet mix spray-applied fire resistive materia. https://isolatek.com. Accessed 12 Oct 2018

CAFCO 300 AC (Technical data sheet) Wet mix spray-applied fire resistive materia. https://isolatek.com. Accessed 12 Oct 2018

CAFCO® BLAZE-SHIELD® II (Technical data sheet) Wet mix spray-applied fire resistive materia. https://isolatek.com. Accessed 12 Oct 2018

ISO 834-11:2014 (2014) ISO. https://www.iso.org/standard/57595.html. Accessed 10 Aug 2018

Milke JA (2000) Analytical methods for determining fire resistance of steel members n.d.: 209–238. Section four Chapter 9

Treatment of Wastewater with High Nitrogen Loading by Moving Bed Biofilm Reactor

H. K. D. T. Karunarathne(✉), L. K. U. De Silva, and K. B. S. N. Jinadasa

Faculty of Engineering, University of Peradeniya, Peradeniya, Sri Lanka
tharukaakarunarathna@gmail.com

Abstract. Wastewater of municipal toilets contains higher nitrogen to carbon ratio than other types of wastewater. The removal of nitrogen from municipal toilet wastewater has become a major concern due to problems caused by discharge of insufficiently treated wastewater, such as eutrophication and toxicity. Moving Bed Biofilm Reactor (MBBR) has been developed in order to facilitate nitrogen removal. Treatment of nitrogen is commonly achieved in separate aerobic and anoxic tanks in typical MBBR systems. However the treatment process using a single MBBR tank as an advantageous method because of the low cost and low complexity, has not been thoroughly studied. This research is focused on the performance of MBBR in nitrogen removal based on simultaneous nitrification and de-nitrification in single aerobic MBBR with continuous operation. A moving-bed biofilm reactor system with an anoxic–aerobic arrangement was set up and the wastewater extracted from the wastewater treatment plant at temple of Tooth Relic, Kandy, was kept in the prepared laboratory set up for the accumulation of microorganisms. The operation was carried out in 3 steps. In the first step, the DO concentration in the reactor was maintained at 5 mg/l and the synthetic wastewater was introduced initially to the reactor at low nitrogen concentration. Then, the nitrogen concentration was increased varying the BOD/N. In the second step BOD/N ratio was maintained at 4.2, which is the BOD/N value of public toilet wastewater and the DO concentration was varied. Volatile Suspended Solids (VSS) of the reactor was kept at around 2200 mg/l. In the third step the DO concentration was maintained at 1.5 mg/l since it was the optimum DO concentration for highest denitrification in the 2nd step. Then the VSS of the reactor was maintained at 2200 mg/l and the BOD/N ratio was varied. Under the operating conditions, 67% of BOD and COD removal and 60% of TN removal could be achieved from the prepared laboratory set up. Further it was observed that the removal of nitrogen increased when the BOD/N ratio was increased.

Keywords: MBBR · Public toilet wastewater · High nitrogen loading · DO concentration

R. Dissanayake and P. Mendis (Eds.): ICSBE 2018, LNCE 44, pp. 538–549, 2020.
https://doi.org/10.1007/978-981-13-9749-3_47

1 Introduction

Many processes are being used for treatment of sewage in Sri Lanka. These conventional methods are not favorable for the treatment of BOD in wastewater and they are not successful when considering the removal of nutrients such as nitrogen and phosphorous.

The moving bed biofilm reactor was developed for the first time when nitrogen removal from wastewater became a major problem. It was first developed in Norway in the late 1980s and in early 1990s. Now, the MBBR has been successfully used for treatment of industrial and municipal wastewater.

The MBBR has several advantages compared to Conventional Activated Sludge (CAS) processes because of simple and reliable operation. There is no clogging. It does not need periodical backwashing because in the MBBR, the biomass grows on bio film carrier which moves freely with wastewater in the MBBR reactor. The process in the MBBR can be aerobic, anaerobic or anoxic. In the aerobic process, the biofilm carrier movement is done through the agitation of the air supplied into the reactor, while in the anoxic and anaerobic processes a mixer inside the reactor is used for agitation (Ødegaard 2006).

In conventional methods of wastewater treatment, nitrogen removal is achieved by nitrification and de-nitrification using two separate reactors with different environmental conditions. However, this research is specified on evaluating the performance of new MBBR setup with simultaneous nitrification and de-nitrification in a single tank. This method is advantageous because of the low cost and the less complexity.

2 Background Information on Biological Wastewater Treatment Using MBBR

The original and mostly used biofilm carrier (K1) is made of high density polyethylene (Density 0.95 g/cm^3) and shaped in a small cylinder with a cross in the inner side of the cylinder and "fins" to the outer side, as illustrated by the Fig. 1. Biomass attached to these carries is used for treatment in MBBR.

Fig. 1. Bio carriers (Ødegaard 2006)

The objectives of biological treatment of wastewater are to oxidize dissolved and particulate biodegradable constituents into acceptable end products, to capture and incorporate suspended and non-settleable colloidal solids into biological floc or biofilm, to transform or remove specific nutrients such as nitrogen and phosphorus. The removal of carbonaceous BOD, the coagulation of non-settleable colloidal solids, and the stabilization of organic matter are accomplished biologically using a variety of microorganisms, principally bacteria. The colloidal and dissolved carbonaceous organic matter is converted into various gases and into cell tissues by microorganisms. These products are then removed by gravity settling (Metcalf et al. 2003).

Organic matter and carbon dioxide are very common sources of cell carbon for microorganisms. Organisms that use organic carbon for the formation of cell tissue are called heterotrophs. Organisms that derive cell carbon from carbon dioxide are called autotrophs. The energy needed for cell growth may be supplied by light or by a chemical oxidation reaction. Nutrients, rather than carbon or energy source, may be at times the limiting material for microbial cell growth. The principal inorganic nutrients needed by microorganisms are N, S, P, K, Mg, Ca, Fe, Na, and Cl.

Nitrogen removal of wastewater is achieved in a treatment process by nitrification and de-nitrification. Nitrification is the term used to describe the two step biological process in which ammonia (NH_4-N) is oxidized into nitrite (NO_2-N) and nitrite is oxidized into nitrate (NO_3-N) under aerobic conditions. The biological reduction of nitrate into nitric oxide, nitrous oxide and nitrogen gas under anoxic condition is termed as de-nitrification (Metcalf et al. 2003).

3 Literature Review

Sidek et al. (2015) stated that MBBR is effective in COD, nitrogen and Total Suspended Solids (TSS) removal than the CAS process, while the CAS process is better in BOD removal than the MBBR. In terms of sludge production, MBBR produces fewer amount compared to that of CAS. For overall performance, MBBR has better constituent removal efficiency compared to CAS process in the experimental basis study.

Most of the literature regarding nitrogen removal from MBBR is based on nitrification and de-nitrification done in two separate reactors (Zinatizadeh and Ghaytooli 2015). Separate reactors or sequences in intermittent aeration systems are required to provide the two different environmental conditions. However, recent studies have revealed that these two important steps can occur in the same reactor (Pochana and Keller 1999).

According to Metcalf et al. (2003) under low DO concentration conditions, de-nitrification can occur in the biomass floc interior, while nitrification is occurring at the floc exterior. Under these conditions, nitrogen removal that occurs in a single aeration tank is referred to as simultaneous nitrification and de-nitrification (SND).

Pochana and Keller (1999) stated that the effectiveness of the de-nitrification process decreases when oxygen concentrations are higher than 0.2 mg O_2/L. DO concentration for nitrification should be higher than 2 mg O_2/L otherwise DO can be the limiting factor. At the same time, 0.2 mg O_2/L is considered as a critical value, at which nitrification no longer occurs (Bliss et al. 1986). However DO concentration around 0.5 mg/L is suitable to achieve a nitrification rate equal to the de-nitrification rate which can lead to Simultaneous Nitrification and De-nitrification (SND).

Wang et al. (2006) stated that SND could be successfully established at DO of about 2 mg/l. About 89.9% of total nitrogen (TN) removal efficiency through SND had been achieved in the system of moving bed biofilm system which was used to treat municipal wastewater.

To accomplish de-nitrification in any process, organic carbon is found to be one of the most essential factors since it is needed as a carbon and energy source for the bacteria (Pochana and Keller 1999). Isaacs et al. (1994) proposed that COD/N ratio for de-nitrification is in the range of 3.5 to 4.5 g COD/g N.

Sumino et al. (2006) examined the effects of C/N ratio and total organic carbon (TOC) loading on nitrogen removal through simultaneous nitrate reduction and anaerobic ammonium oxidation in a single reactor and proved that nitrogen removal ratio was affected by C/N ratio and TOC loading, not by the amount of granular sludge provided.

In the case of a moving bed system, the thickness of the biofilm should be considered. Several Experiments have been performed to determine the effect of the floc size on SND in order to test the hypothesis that Simultaneous Nitrification and De-nitrification (SND) is a physical phenomenon. The underlying physical explanation is that a substantial anoxic mass fraction exists in the center of the biomass flocs resulting from an oxygen diffusion limitation into the flocs. With respect to activated sludge processes, large floc sizes create an anoxic zone inside the flocs leading to de-nitrification in that area (Rosenwinkel and Cornelius 2005).

4 Methodology

4.1 Reactor Configuration

A moving-bed biofilm reactor system with an anoxic–aerobic arrangement, as shown in Fig. 2, was used to treat the synthetic wastewater. The reactor working volume was 4.0 L. And a square shaped Perspex tank of total volume of 8.0 L was used. The biofilm carriers filled 50% by volume of reactor. Influent was fed using an overhead tank and the treated effluent was let to be overflowed through a pipe just above the head level of the reactor into a secondary clarifier. An aeration unit was used to aerate the reactor.

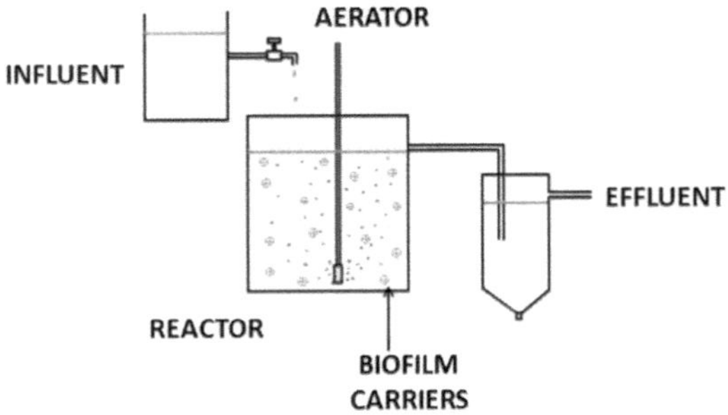

Fig. 2. Schematic diagram of the reactor

4.2 Wastewater Characteristics

Public toilet wastewater was collected from three different locations in the Kandy town area and those were tested to find out the real characteristics of public toilet wastewater. According to the laboratory experiment results, the average BOD/N ratio of public toilet wastewater was 4.2. Synthetic wastewater was prepared and fed into the system according to the results obtained.

4.3 Operation and Procedure

After setting up the model, the operational phase was started. A batch experiment was carried out to find the HRT value for the setup. Finally the HRT of 6 h was maintained in the reactor. Then the wastewater from the wastewater treatment plant at temple of Tooth Relic was kept for accumulation of microorganisms for two weeks.

The testing procedure adopted comprised three major steps. In the first step the synthetic wastewater was introduced initially to the reactor at low Nitrogen concentration and then the Nitrogen concentration was increased varying the BOD/N ratio (Table 1). Four samples were tested for different BOD/N ratios. The BOD concentration was maintained as a constant parameter and Nitrogen concentration was increased using NH_4Cl and $NaHCO_3$ to obtain those four BOD/N ratios. Each sample was treated for 2 weeks. Samples from influent and the effluent were collected and tested to find out the treatment efficiency. The inflow rate, HRT, DO concentration and pH of the reactor during the first step were 11.1 ml/min, 6 h, 5 mg/l and 7 respectively.

Table 1. Sample sets used for step 1

Sample	BOD/N
1	9.5
2	7.5
3	5.5
4	3.5

In the second step of the procedure the BOD/N ratio was maintained at 4.2, which was the BOD/N value for the real public toilet wastewater and the DO concentration was varied. Since the de-nitrification was not efficient in the first step of the procedure, the DO concentration was varied as shown in the Table 2. VSS of the reactor was kept at around 2200 mg/l. Once the influent was treated for 6 h, samples from influent and effluent were taken and were tested for the same parameters as in the first step.

Table 2. DO concentrations used for step 2

Sample no.	DO concentration (mg/l)
1	5
2	4
3	2
4	1.5
5	1
6	0.6
7	0.3

In the third step, the DO concentration was maintained at 1.5 mg/l since it was the optimum DO concentration for the highest de-nitrification in the second step. Then the VSS of the reactor was maintained at 2200 mg/l and the BOD/N ratio was varied as shown in the Table 3.

Table 3. Sample sets used for step 3

Sample no.	BOD/N ratio
1	3
2	4.2
3	5
4	7

4.4 Data Collection

Samples from influent, effluent and from the reactor were collected for each test. Samples of biomass attached carriers also were collected and biomass was separated by vigorously stirring in water. The parameters measured are shown in the Table 4.

Table 4. Parameters measured for three steps

Parameter	Influent	Effluent	Reactor
BOD	√	√	
COD	√	√	
NH_4^+-N	√	√	
NO_3^-N	√	√	
NO_2^-N	√	√	
VSS		√	√

5 Results and Discussion

5.1 Observation for Step 1

During the first step the DO concentration was maintained at 5 mg/l. The initial value of VSS of the reactor was 2200 mg/l. The initial BOD/N ratio was 9.5. Then the BOD/N ratio was reduced by increasing nitrogen concentration in influent. The BOD removal of the system is shown in the Fig. 3. These results have proven the fact that, this system can be used to remove organic matter efficiently. From the initial stage onwards, the efficiency of the system for BOD removal was around 90%.

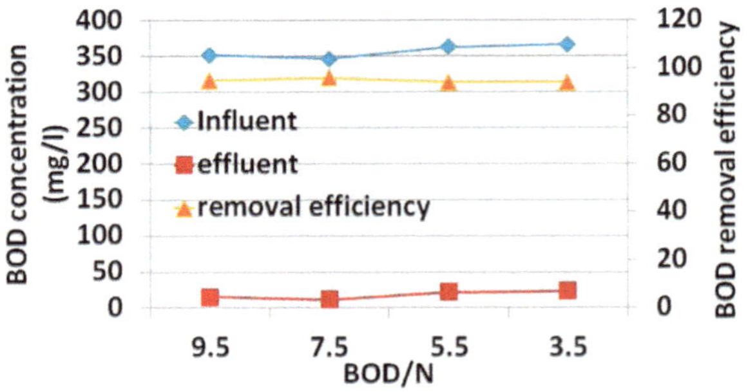

Fig. 3. BOD removal of system in step 1

The COD removal of the system is shown in the Fig. 4. According to the results it has been proved that the COD removal of the system also was efficient. From the initial stage onwards, the efficiency of the system for COD removal was higher than 90%.

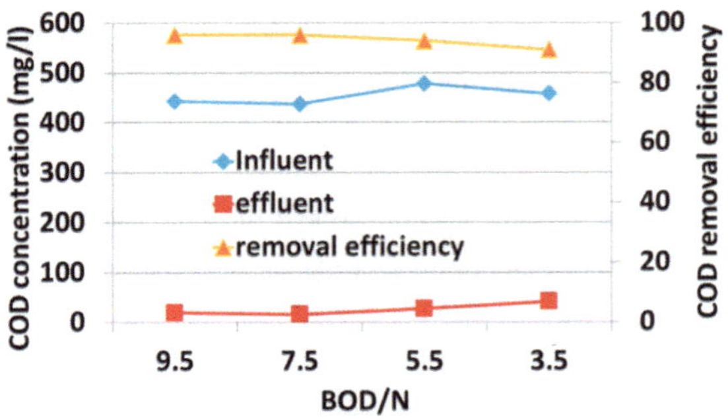

Fig. 4. COD removal of system in step 1

The source of Nitrogen used when preparing the synthetic wastewater was NH_4Cl. And the nitrogen concentration was varied in the 4 samples used. Figure 5 shows the variation of the NH_3–N in the system. According to the results, it shows a removal of NH_3–N in all four samples. However the reduction of NH_3–N increased the NO_3-N concentration of the effluent.

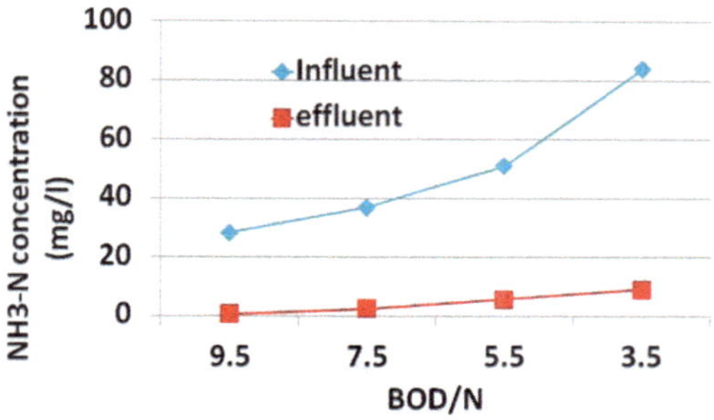

Fig. 5. NH_3–N variation of system in step 1

The results showed that the NO_3^--N in the effluent was increased after treatment. That was due to the fact that conditions of the reactor were not suitable for de-nitrification. The reactor was in high aerobic condition in the first step of the operation. The DO concentration was 5 mg/l in this step. Figure 6 shows the variation of TN in the system.

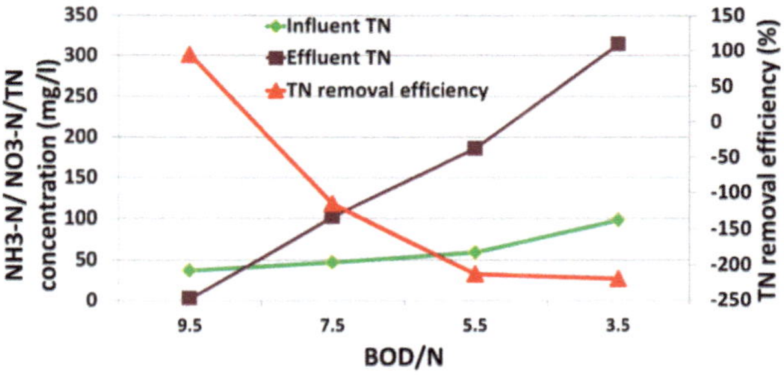

Fig. 6. TN removal of system in step 1

5.2 Observation for Step 2

During the second step of this study, the BOD/N ratio was maintained at the BOD/N ratio value of the public toilet wastewater (BOD/N = 4.2). The VSS of the reactor was also maintained at 2200 mg/l. Then the DO concentration was varied from 5 mg/l to 0.3 mg/l and the treatment efficiency of the system was tested.

The BOD removal of the system is shown in the Fig. 7. BOD removal varied from 60%–80% when the DO concentration was higher than 1 mg/l. When the DO concentration was further reduced, the BOD removal efficiency decreased. That was due to the anaerobic conditions in the reactor. Therefore it shows that the BOD removal of the system was not affected much by varying the DO concentration.

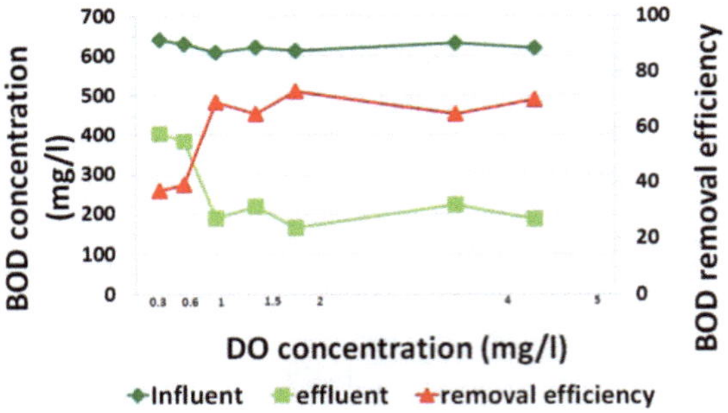

Fig. 7. BOD removal of system in step 2

The COD removal of the system is shown in the Fig. 8. COD removal had not been affected by the decreased DO concentration until it reduced to 1 mg/l. The COD removal efficiency was around 60% to 70% when the DO concentration was higher than 1 mg/l.

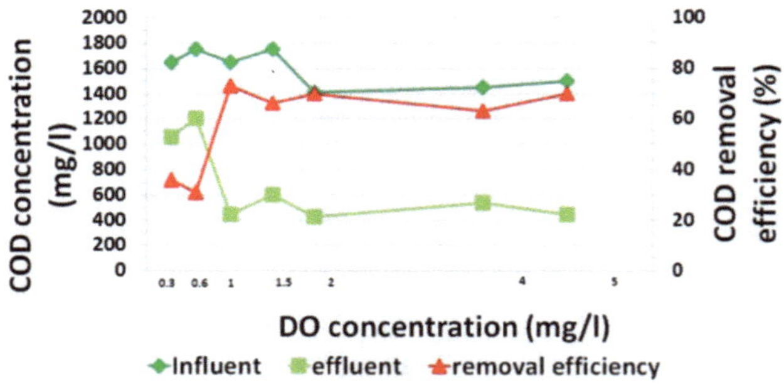

Fig. 8. COD removal of system in step 2

NH_3-N variation of the system during the step two is presented in Fig. 9. It shows that the nitrification was efficient when the DO concentration was higher than 1.5 mg/l. When the DO concentration was lower than 1.5 mg/l the anaerobic conditions occurred in the reactor. Therefore the nitrification had been reduced. The NO_3-N concentration in effluent was the highest when the DO concentration was 5 mg/l. When the DO concentration was decreased, the removal of nitrogen was increased. The optimum de-nitrification could be achieved at 1.5 mg/l DO concentration. When the DO concentration was further reduced, the de-nitrification decreased again due to low nitrification rate. The optimum DO concentration for Simultaneous Nitrification and De-nitrification for this system was obtained to be 1.5 mg/l. The Fig. 10 shows the NO_3-N variation of the system.

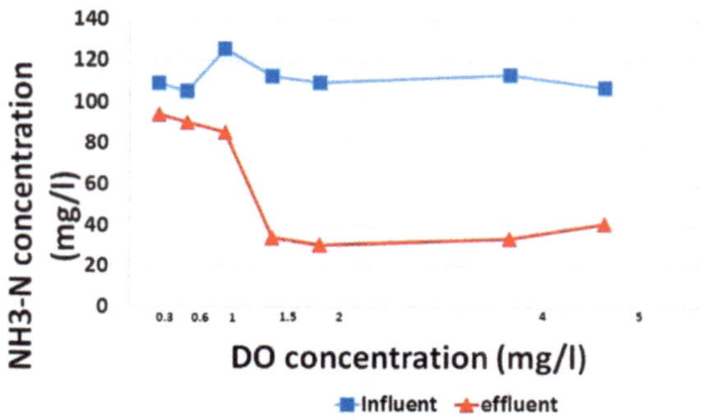

Fig. 9. NH_3-N variation of the system in step 2

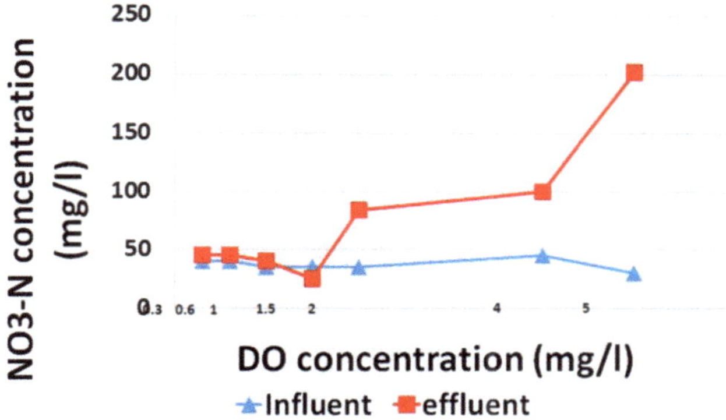

Fig. 10. NO_3-N variation of system in step 2

5.3 Observation for Step 3

In the third step, The DO concentration was kept constant at the optimum value that was obtained in the previous step. The VSS of the reactor was maintained at 2200 mg/l. Then the BOD/N ratio was changed from 3 to 7 step by step and the removal efficiencies were obtained from tests. This study was done to obtain the effect of BOD/N ratio for the SND.

The DO concentration was maintained at 1.5 mg/l. The BOD removal and COD removal increased when the BOD/N ratio was increased. Figure 11 shows the BOD removal of the system and Fig. 12 shows the COD removal of the system.

Figure 13 shows that the NH_3-N removal had not been affected by the increased BOD/N ratio. The nitrification efficiency was nearly a constant value for all 4 BOD/N ratios. The de-nitrification efficiency was increased when the BOD/N ratio increased. When the BOD/N ratio was increased, the available additional carbon source for de-nitrification increased. Adding an external carbon source is therefore favorable to maximize the nitrogen removal. Figure 14 shows the NO_3-N variation of system.

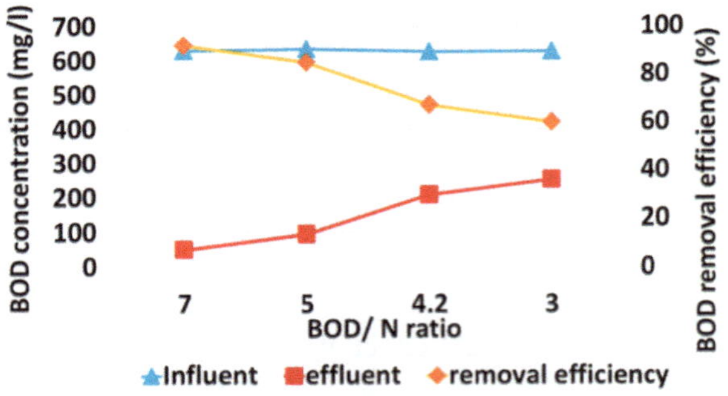

Fig. 11. BOD removal of system in step 3

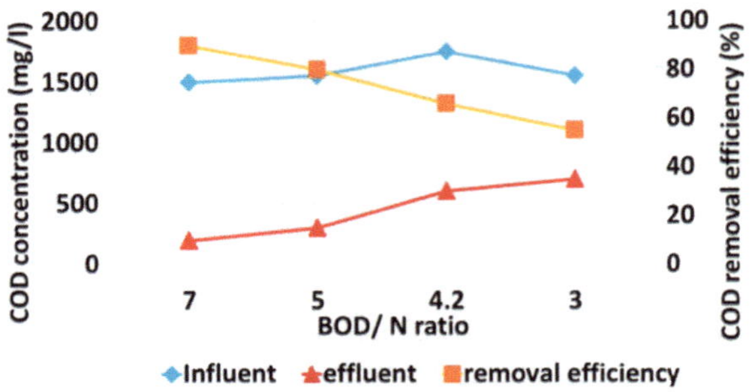

Fig. 12. COD removal of system in step 3

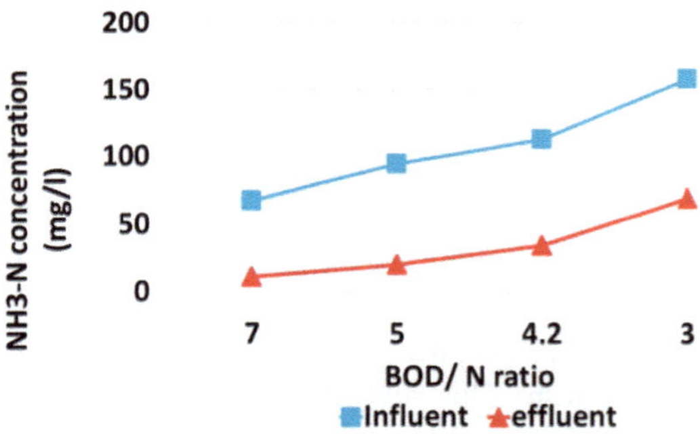

Fig. 13. NH_3-N variation of the system in step 3

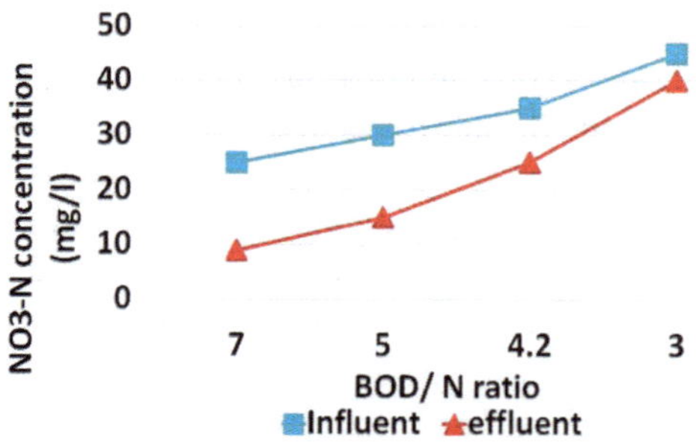

Fig. 14. NO_3-N variation of system in step 3

6 Conclusion

Treatment of public toilet wastewater with BOD/N ratio of 4.2 is viable when maintaining the DO concentration at 1.5 mg/l and VSS at 2200 mg/l. It is clear that under these operating conditions, 67% of BOD and COD removal and 60% of TN removal can be achieved. Furthermore the removal of nitrogen can be increased when the BOD/N ratio is increased.

Acknowledgements. The authors wish to extend their gratitude to the staff members of wastewater treatment plant at The Temple of the tooth relic, Kandy and the Engineers and staff of Kandy municipal council for their support, and the evaluation panel for their guidance and comments to complete this research successfully.

References

Rusten B, Eikebrokk B, Ulgenes Y, Lygren E (2004). Design and operation of Kaldnes moving bed biofilm reactor

Pochana K, Keller J (1999) Study of factors affecting simultaneous nitrification and denitrification (SND). Water Sci Technol 39:61–68

Ødegaard H (2006) Innovations in wastewater treatment: the moving bed biofilm process. Water Sci Technol 53:17–33

Zinatizadeh AAL, Ghaytooli E (2015) Simultaneous nitrogen and carbon removal from wastewater at different operating conditions in a moving bed biofilm reactor (MBBR) process modeling and optimization. J Taiwan Inst Chem Eng 53:98–111

Rosenwinkel KH, Cornelius A (2005) Deammonification in the moving bed process for the treatment of wastewater with high ammonium content. Chem Eng Technol: Ind Chem-Plant Equipment-Process Eng Biotechnol 28:49–52

Wang XJ, Xia SQ, Chen L, Zhao JF, Renault NJ, Chovelon JM (2006) Nutrients removal from municipal wastewater by chemical precipitation in a moving bed biofilm reactor. Process Biochem 41:824–828

Sumino T, Isaka K, Ikuta H, Saiki Y, Yokota T (2006) Nitrogen removal from wastewater using simultaneous nitrate reduction and anaerobic ammonium oxidation in single reactor. J Biosci Bioeng 102:346–351

Liebert MA (2015) Effective biological nitrogen removal treatment process for domestic wastewater

Metcalf E (2003) Wastewater engineering Treatment and reuse, 4th edn. McGraw-Hill, New York

Sidek LM, Mohiyaden HA, Basri H, Salih GHA, Birima AH, Ali Z, Sabri AFM, Noh MNN (2015) Experimental comparison between moving bed biofilm reactor (MBBR) and conventional activated sludge (CAS) for river purification treatment plant. Adv Mater Res 11:806–811

Experimental and Analytical Study of Residual Yield and Tensile Strengths in Partially Corroded Structural Steel

P. A. K. Karunananda(✉), K. A. D. D. T. Kariyapperuma, R. P. Weeraratne, and H. D. M. Sanjeewani

Department of Civil Engineering,
The Open University of Sri Lanka, Nawala, Nugegoda, Sri Lanka
pakar@ou.ac.lk

Abstract. Corrosion has been one of the main structural degradation processes and has been the root cause of many engineering structural failures around the world. Therefore, proper understanding of corrosion process and especially residual strength measurement is an important consideration in the engineering discipline. This study was performed to find the residual strengths of partially corroded steel using a cost effective reliable method which can be adopted easily. For that, Average Weight Ratio (AWR) was introduced instead of the Minimum Thickness Ratio (MTR), proposed by previous researchers. In this context, partially corroded steel specimens were subjected to accelerated corrosion tests in a corrosion incubator. There, correlations were built between hardness verses AWR, ultimate tensile strength verses AWR and ultimate tensile strength ratio verses AWR of partially corroded steel.

Keywords: Corrosion · Average Weight Ratio · Hardness · Residual strength

1 Introduction

The use of steel structures all over the world increased gradually due to its many advantages such as its versatility, high strength, uniformity, ease of use, toughness, ductility and many other properties. Structural steel has very high yield strength for compression and tension. Therefore, it is convenient to use structural steel as a building construction material than other construction materials such as concrete, timber, etc.

But one of the major disadvantages of using steel as a construction material is the fast deterioration in steel due to corrosion when exposed to aggressive environmental conditions and poor maintenance. Because of this, reduction in their strength capacities, corrosion wastage, fatigue cracking, or mechanical damage may occur during their service life. Therefore significant issues occur on safety, health, environment and financial costs. The catastrophic collapse of the silver bridge on Ohio River in 1967, USA is a classic example for structural failure due to corrosion.

Therefore, it is essential to carefully evaluate the remaining strength of steel structures to understand the feasibility of the steel structure for future usage. Several experimental studies and detailed investigations have been done by some researchers

R. Dissanayake and P. Mendis (Eds.): ICSBE 2018, LNCE 44, pp. 550–558, 2020.
https://doi.org/10.1007/978-981-13-9749-3_48

during the past few decades, to introduce methods of estimating the remaining strength of corroded structural steel members (Appuhamy et al. 2011).

This research study is mainly based on developing a simple and convenient method to estimate residual ultimate tensile strength and yield strength of the corroded structural steel specimens. Experimental results were used to develop and determine more reliable results. Those developed results can be used for the existing corroded structures by measuring the remaining thickness and calculating the average thickness ratio with respect to initial thickness and thereby, the remaining strength can be easily calculated.

2 Theoretical Background

It has been mentioned that regular coarse surface measurement could be sufficient for minor corrosion members as surface roughness does not affect minor corrosion. Therefore, for our experiments, average thickness ratio (ATR) was considered instead of minimum thickness ratio (MTR).

Considering that the density of steel remained unchanged and only thickness is reduced rather than length and breadth, it can be proven that average thickness ratio is equal to average weight ratio as below.

$$Density(\rho) = \frac{Mass(m)}{Volume(v)}$$

$$Initial\,Density = Final\,Density$$

$$Therefore;\frac{initial\,mass\,(wo)}{initial\,volume\,(Vo)} = \frac{final\,mass\,(wf)}{final\,volume\,(Vf)}$$

$$\frac{Wo}{L} \times B \times to = \frac{Wf}{L} \times B \times tf$$

As considering unchanged of length (L) and breadth (B)

$$\frac{tf}{to} = \frac{Wf}{Wo}$$

Hence,

$$\boldsymbol{Average\,thickness\,ratio} = \boldsymbol{Average\,Weight\,ratio}$$

2.1 Corrosion Acceleration

Corrosion accelerates due to factors such as the over potential, the aggressive medium, the temperature and the concentration. However, the materials react differently to different types of electrolytes. The accelerated corrosion test is done to obtain results in a shorter period of time than from natural exposure. In general, accelerated corrosion tests can be performed increasing temperature and potency with time to get different corrosion levels.

2.2 Techniques for Inducing Accelerated Corrosion of Steel

Corrosion of steel is a slow process in the natural environment under normal conditions. It takes reasonably long time to achieve a significant degree of corrosion in steel. Different types of techniques were introduced to accelerate corrosion in past research studies. One of it is "Impressed current technique" which is a method to induce significant corrosion in a short period of time to steel embedded in the concrete.

Further, another alternative technique for inducing accelerated reinforcement corrosion is to build up artificial climate environment in the room which controlled by a computer system. In that artificial climate room, environmental condition, temperature, relative humidity and repeated wetting and drying cycle consisted of salt water (5% NaCl solution) spraying for 1 h and infrared light shining for 7 h were controlled. This test set up is shown in Fig. 1 (Ahmad 2009)

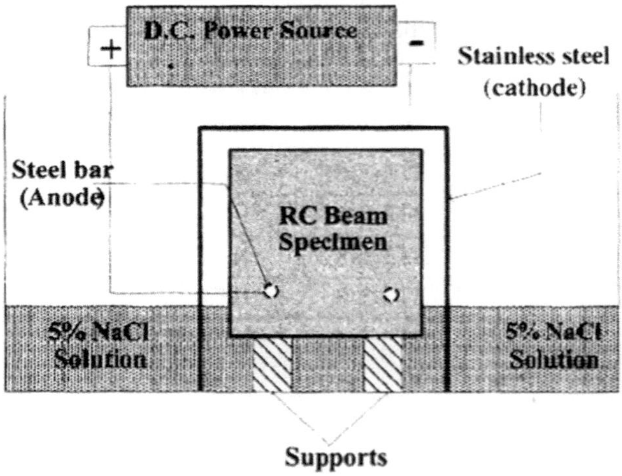

Fig. 1. Set-up for accelerating reinforcement corrosion in large-size reinforced concrete beam specimen. (Source: Ahmad 2009)

2.3 Correlations Among Mechanical and Strength Properties

In addition to yield strength and ultimate tensile strength of steel, hardness is an important mechanical property which can be measured using hardness testing machines. The above strengths can be found according to formulae among hardness, yield strength and tensile strengths, which have been developed by past researches. Of them, the following formulae are very convenient to study;

$$TS = \frac{H}{2.9}\left(\frac{n}{0.217}\right)^n \quad (1)$$

$$YS = \frac{H}{3} \times (0.1)^n \quad (2)$$

(Pavlina and Van Tyne 2008)
where, TS = Tensile strength
YS = Yield strength
H = Diamond pyramid hardness (Vickers hardness)
n = Strain-hardening exponent (n = 0.21)

Further Residual Ultimate Tensile Strength of corroded steel could be determined using the following formula by substituting Brinell hardness number.

$$TS = 3.45 \times HB \quad (3)$$

(Pavlina and Van Tyne 2008)
where,
HB = Brinell hardness

In addition to the above developed formulae, many conversion tables among hardness and tensile strengths of steel have been developed by recognized organizations.

2.4 Yield to Tensile Strength Ratio (YTR) of Steel

It is a very important parameter to determine Yield Strength of steel when the Ultimate Tensile Strength of particular steel is known. Normally YTR of range from 0.5 to 0.9 is recommended for steel (Gasko and Rosenberg 2011).

3 Experimental Set up

Two experiments were performed and the following gives the details of these experiments.

3.1 Experiment 1

Accelerated corrosion test was performed in an incubator where temperature and potency with time was controlled to check the feasibility of selected corrosion accelerated factors to be used for experiment 2 which was done to observe different corrosivity levels of specimens by changing one control factor with respect to others.

1. Two numbers of 50 × 20 × 3 mm steel specimens were prepared out of Grade S 275 structural steel (one of them was selected as a standard specimen while the other was subjected to corrosion).

2. A glass tank (300 × 200 × 150 mm), (12 V–28 V) capacity DC transformer, thermostat (30 °C–50 °C), 0.125 M Sodium Chloride solution (as a catalyst) and flexible wire pieces were used to prepare an incubator.
3. Two numbers of 50 × 20 × 3 mm structural steel specimens were submerged in the sodium chloride solution and connected to a 12 V power supply to create corrosion on surfaces of steel specimens (in 30 °C temperature).
4. The specimen connected to the cathode end acts as a standard specimen while the other specimen attached to the anode end corroded with time.
5. The above procedure was repeated twice for increasing temperature (35 °C) and increasing voltage (28 V) respectively (Fig. 2).

Fig. 2. (a) 50 × 20 × 3 mm steel specimens. (b) Arrangement of apparatus

3.2 Experiment 2 - Phase I

Experiment 2 (Phase I) was performed to get different corrosivity levels of steel. Corrosivity levels were evaluated as a measure of Average Weight Ratio instead of average thickness ratio according to the above proof.

Initial weight of each specimen (W_0) was measured using a digital scale.

The glass tank was filled with 4 L of 0.125 M NaCl solution. The electric heater coupled to the thermostat was inserted to the glass tank and temperature of the thermostat was adjusted to 30 °C.

The test specimen to corrode was connected to the anode end of the 12 V DC power supply while another specimen was connected to the cathode end of the power supply and both specimens were inserted to the solution and the power supply switched on.

After an hour, the corroded steel specimen was removed from the solution and the rust of the specimen was removed thoroughly.

- Then, the final weight of the specimen was measured.
- The above steps were repeated by increasing the time period and the solution was removed and refilled for each repetition.
- The same procedures were adopted for the other experiments by increasing the voltage and temperature as (12 V, 35 °C), (28 V, 30 °C), (28 V, 35 °C) and (28 V, 40 °C) respectively (Fig. 3).

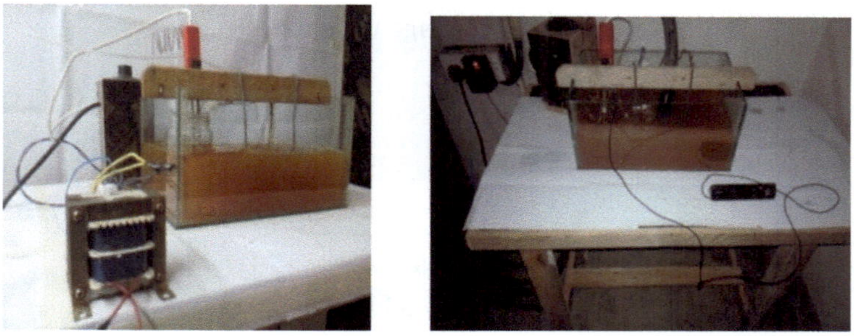

Fig. 3. Arrangement of apparatus by supplying of apparatus

4 Results

A considerable number of experiments were performed and the relationships with AWR obtained are shown in Figs. 4, 5, 6, 7 and 8.

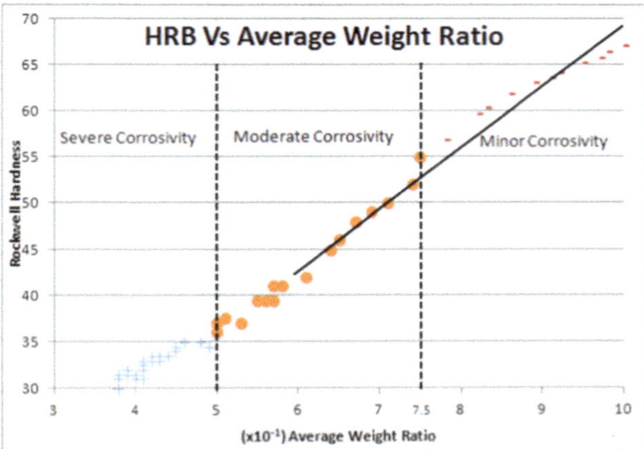

Fig. 4. The graph of Rockwell hardness (HRB) vs average weight ratio (W_f/W_0)

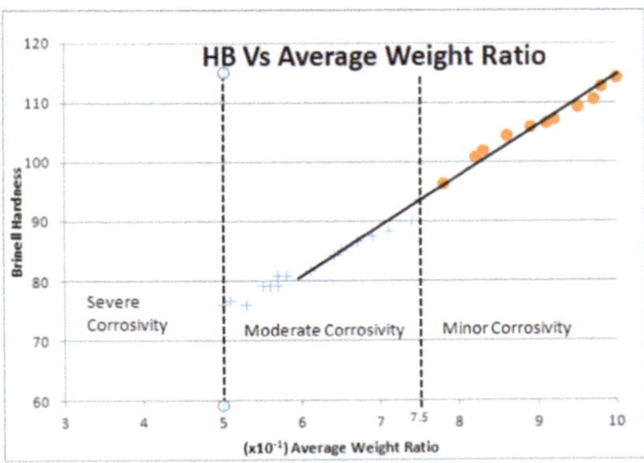

Fig. 5. The graph of Brinell hardness (HB) vs average weight ratio (W_f/W_0)

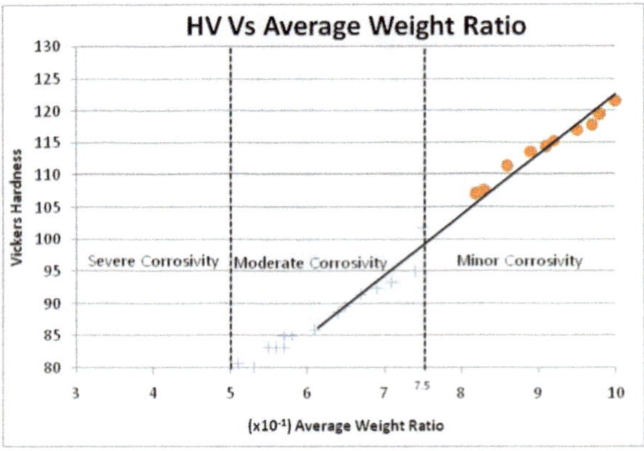

Fig. 6. The graph of Vickers hardness (HV) vs average weight ratio (W_f/W_0)

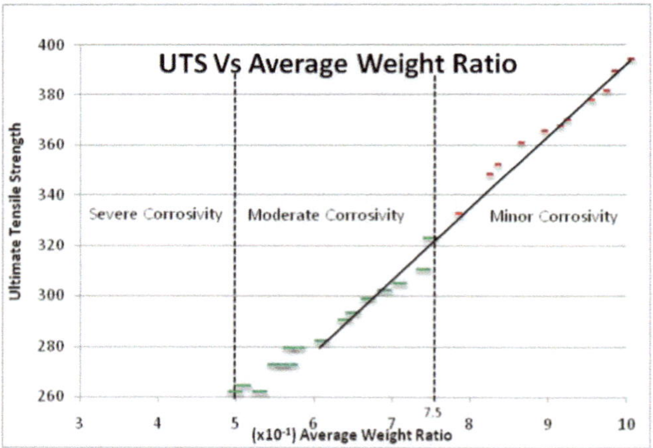

Fig. 7. The graph of Ultimate Tensile Strength (UTS) vs average weight ratio (W_f/W_0)

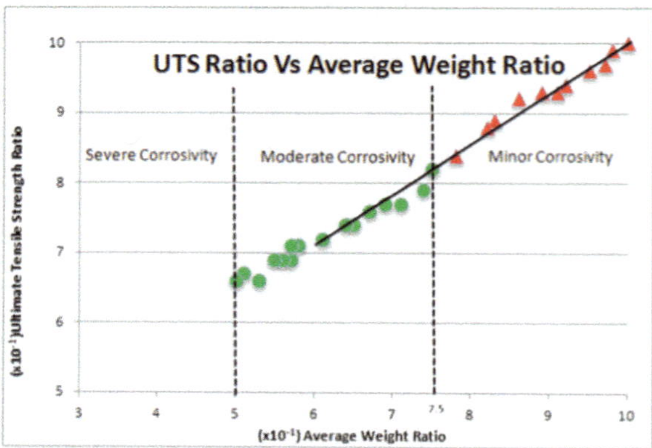

Fig. 8. The graph of UTS ratio vs average weight ratio (W_f/W_0)

4.1 Identify Correlations to Find Out Residual Strengths

From the graph of Rockwell hardness verses average weight ratio, when average weight ratio decreases, Rockwell hardness value decreases accordingly producing a linear correlation between those two parameters up to a certain extent of moderate corrosivity as shown in Fig. 4.

By scrutinizing the rest of the graphs, similar linear correlation could be identified between Brinell hardness verses average weight ratio, Vickers hardness verses average weight ratio, ultimate tensile strength verses average weight ratio and ultimate tensile strength ratio verses average weight ratio as shown in Figs. 5, 6, 7 and 8.

Residual ultimate tensile strength of partially corroded structural steel could be determined using the graph of ultimate tensile strength ratio verses average weight ratio if the initial ultimate tensile strength was known. That obtained value can justify by using Eq. (3),

$$TS = 3.45 \times HB$$

Brinell hardness value is obtained from the graph of Brinell hardness verses average weight ratio.

Residual yield strength of partially corroded structural steel could be calculated by substituting Vickers hardness value obtained from the graph of Vickers hardness verses average weight ratio to Eq. 2. That obtained value is to be multiplied by 9.81 as it is given kg_f/mm^2 and the unit of residual yield strength is N/mm^2.

$$YS = \frac{H}{3} \times (0.1)^n$$

If yield to tensile strength ratio (YTR) is known, residual yield strength value could be obtained by substituting residual ultimate tensile strength value and the above residual yield strength value could be verified.

5 Conclusion

The residual strength measurements were conducted for partially corroded structural steel and the following conclusions can be made from this study. The corrosion causes strength reduction of structural steel members and average weight ratio (AWR) can be used as the measure of the level of corrosion. The experimental and analytical results can be used to predict the residual strength of actual corroded members accurately.

Acknowledgment. Authors would like to thank to Dr. C.S. Bandara of Civil Engineering Department in University of Peradeniya for the support given in this research.

References

Ahmad S (2009) Techniques for inducing accelerated corrosion of steel in concrete. Arab J Sci Eng 34(2C):95 Department of Civil Engineering King Fahd University of Petroleum & Minerals Dhahran, Saudi Arabia

Appuhamy JMRS et al (2011) Analytical study on significance of corroded surface measurement on residual strength prediction. Proc Eng 14:2260–2268

Gasko M, Rosenberg G (2011) Correlation between hardness and tensile properties in ultra-high strength dual phase steels – short communication. Mater Eng 18:155–159

Pavlina EJ, Van Tyne CJ (2008) Correlation of yield strength and tensile strength with hardness for steels. J Mater Eng Perform 17(6):888–893

Mathematical Modeling of the Hydrolysis Phase in Anaerobic Digestion of Solid Waste

D. I. U. Edirisinghe(✉), S. M. W. T. P. K. Ariyarathna, and C. S. Kalpage

Department of Chemical and Process Engineering, Faculty of Engineering, University of Peradeniya, Peradeniya, Sri Lanka
snk_ediri@yahoo.co.uk

Abstract. Anaerobic digestion has been in use for many decades, and its range of uses keep on expanding although it has a reputation of being unreliable and unpredictable. Anaerobic digestion could provide a sustainable solution to waste disposal as well as power generation, if the process could be controlled as desired. This study was done in order to observe the behavior and to optimize the hydrolysis phase in anaerobic digestion of biodegradable solids, by developing a mathematical model for the process. Considering all aspects, a surface related model was used to describe the process since hydrolysis of complex solid substrates is dominated by surface phenomena. In order to narrow down the scope of the research, initial experimentation was focused on pure carbohydrates only. Boiled rice was selected as the substrate since it is readily available and because it is a pure form of carbohydrate. Through experimental data and the aid of the software Matlab2013, the model parameters were determined to be $K_S = -1$ and $r_{max} = 1.004$ mmol/dm^3.day under operating conditions of 1 atm pressure and temperature of 28 °C. Further developments and improvements to this model are essential to be able to predict hydrolysis of complex substrates at any given operating condition. The improved model may someday result in a capable and efficient waste management system which is sustainable, ergonomic and efficient.

Keywords: Anaerobic digestion (AD) · Mathematical modelling · Hydrolysis · Surface related two phase kinetic model · Solid waste

1 Introduction

Anaerobic digestion has been in use for over ninety years (Gene and William 1986) to treat agricultural, municipal and other waste sludge having a high organic material content. Anaerobic digestion is becoming more and more popular as a waste treatment method because of its low power requirement and ability to produce methane gas. Despite the attractiveness of the process, it has a reputation of being unstable and therefore unreliable. One of the reasons for this is the lack of fundamental knowledge on anaerobic digestion process principles. The other reasonable explanation is that originally, anaerobic digestion was used to stabilize domestic sludge, but now it is being used to treat chemically treated sludge, industrial waste and complex solid waste and therefore the treated sludge quality has changed over the period (Gene and William 1986).

R. Dissanayake and P. Mendis (Eds.): ICSBE 2018, LNCE 44, pp. 559–566, 2020.
https://doi.org/10.1007/978-981-13-9749-3_49

Anaerobic digestion takes place in four steps as hydrolysis, acidogenesis, acetogenesis and methanogenesis. Each of the stages involve different bacterial groups which result in a very complex reaction mechanism overall. Also, substrate digestibility and bio gas production are affected by process parameters such as pH, temperature and concentration of Volatile Fatty Acids (VFAs) as well. Due to this complexity, it is difficult to predict the output even though the input conditions are known. Therefore it is difficult to scale up the operation and optimize it.

Therefore, a mathematical model of the process can be very helpful in predicting the stability, understanding the behaviour under different process conditions and optimizing it. A mathematical model provides considerable insight to the process mechanism and is less time consuming and less expensive compared to a physical model. It helps in quantifying output and improving design (Andrews and Graef 1971).

Considering the facts, a mathematical model based on surface related phenomena was developed for the hydrolysis phase of carbohydrates within this study. Boiled rice was selected as the carbohydrate rich substrate and the model parameters were determined with the selected model in literature.

2 Reaction Mechanism of Carbohydrates

The first stage of anaerobic digestion is hydrolysis. Particulate carbohydrates, proteins, lipids and inert material present in a complex organic substrate are disintegrated into monosaccharides, amino acids and long chain fatty acids, solubilized and finally acted upon by hydrolytic enzymes in the hydrolysis stage. Next, these hydrolysed products are degraded into VFAs: acetate, butyrate, propionate etc. by acidogenic bacteria in the acidogenesis stage. The produced intermediate VFAs are further degraded into acetic acid by the action of acetogenic bacteria. Finally methane forming bacteria acts on the acetic acid to form methane and carbon dioxide in the methanogenesis stage.

3 Reaction Kinetics of the Hydrolysis Phase

Since the study focuses on solid waste, hydrolysis plays an important role over the other phases. Unless the complex solid substrate is hydrolysed, the acidogenic, acetogenic and methanogenic bacteria cannot digest the substrate to produce methane directly.

It is known that hydrolysis occurs in two pathways explained below;

- Micro-organisms are secreted in to the bulk of the liquid where they are adsorbed on to a particle or react with a soluble substrate (Jain et al. 1992).
- Micro-organisms attach to a particle, produce enzymes in its vicinity and benefit from soluble products released by the enzymatic reaction (Vavlin et al. 1996).

In complex substrates, hydrolysis occurs in both the mechanisms simultaneously.

A two phase surface related model coupled with Mechaelis-Menten kinetics and the Contois model are widely used in describing the hydrolysis kinetics. The surface related model is a modification of the Mechaelis-Menten kinetics coupled with 1st order rate kinetics and bacterial growth kinetics to suit realistic conditions. The Mechaelis-Menten model starts with the preliminary rate equation, and is developed gradually to suit realistic conditions.

However the parallel reaction is dominated by the surface related mechanism mentioned secondly, and the first order kinetics for the reaction can be written as;

$$\frac{dC_S}{dt} = -kC_S \tag{1}$$

$$\frac{dC_P}{dt} = \alpha kC_S \tag{2}$$

Where C_S is the substrate concentration, C_P is the product concentration, α is the conversion coefficient, k is the first order rate coefficient and t is the time.

Equations (1) and (2) considers the whole substrate to be degradable. However, in a complex substrate, only part of the substrate is degradable. Also, it has been found out that a hydrolysis reaction does not follow first order reaction kinetics all the time and the biomass concentration affects the hydrolysis rate as well (Liang et al. 2013). Therefore, the non-degradable fraction of the substrate and effect of biomass concentration have been incorporated in Eq. (3) (Liang et al. 2013);

$$\frac{dC_S}{dt} = -kX^n(C_S - \beta C_{S0}) \tag{3}$$

Where X is the biomass concentration, n is the order of the reaction and β is the non-degradable fraction of the substrate.

The Mechaelis-Menten kinetics too can be used to describe the same first order reaction as;

$$\frac{dC_S}{dt} = \frac{r_{max}C_S}{K_S(C_S - C_{S0}) + C_S} \tag{4}$$

Where C_{S0} is the initial substrate concentration, K_S is the model coefficient and r_{max} is the maximum hydrolysis rate.

All the models explained above by Eq. (1) through (4) only consider the surface related hydrolysis mechanism. It was found out that Vavlin et al. (1996) had developed the two phase surface related model which takes the colonization of waste particles by bacteria into account as well. The model equation related to this is given in Eq. (5).

$$\frac{dC_S}{dt} = r_{max}\frac{\beta X}{1 + \beta X}\frac{C_S}{K_S + C_S} \tag{5}$$

Where C_S is the substrate concentration, β is the ratio between adsorption and desorption rate constants, K_S is the half saturation coefficient, r_{max} is the maximum hydrolysis rate and t is the time.

4 Methodology

4.1 Mathematical Development of the Model

This research is based on the model expressed by Eq. (4). Since only the initial substrate concentration and variation of product concentration with time is obtained through experimentation, the rest of the parameters required for the modelling of the hydrolysis phase should be determined with the aid of mathematical equations governing hydrolysis.

Relationship shown in Eq. (6) was obtained by integrating the basic rate equations applicable to hydrolysis and using mass balance for the conversion of substrate, shown in Eqs. (1) and (2).

$$C_P = C_{P0} + \alpha C_{S0}\left(1 - e^{-kt}\right) \tag{6}$$

Once the conversion coefficient for the conversion of volatile solids to products (α) and the first order rate coefficient for the hydrolysis of carbohydrates (k), is determined, Eq. (6) and Eq. (2) can be used to obtain a relationship between the substrate and product concentration. Both these values can be determined through a simultaneous simulation.

After obtaining a relationship between substrate and product concentration, the variation of substrate concentration with time can be monitored with the aid of Eq. (2).

Once the substrate consumption rate is known, the model parameters r_{max} and K_S can be determined through a simultaneous simulation.

4.2 Experimental Procedure

The substrate was prepared by blending 680 g of boiled rice, 4.8 L of water and 30 g of sugar together. Then the blended rice slurry was transferred to a container and mixed with inoculum so that the inoculum: substrate ratio was nearly equal to 1, which is the optimum ratio according to the literature (Yoon 2014). Nine identical 500 mL digesters were filled with the prepared sample and left under anaerobic conditions. One digester was opened at a time and sampled periodically to measure the amount of acetate formed. Assuming that the rate of acidogenesis is much higher than the hydrolysis rate and acetic acid concentration is ten times higher than the rest of the VFAs, it is reasonable and accurate enough to use the acetate concentration to represent the hydrolysed product concentration (Vavlin et al. 2008). Therefore, in this study acetic acid concentration was measured with time.

5 Results and Discussion

The variation of the product concentration was determined experimentally and plotted as depicted in Fig. 1. Through a simulation, it is possible to model the variation of product concentration with time according to Eq. (6) in order to determine the conversion coefficient α and first order rate coefficient k.

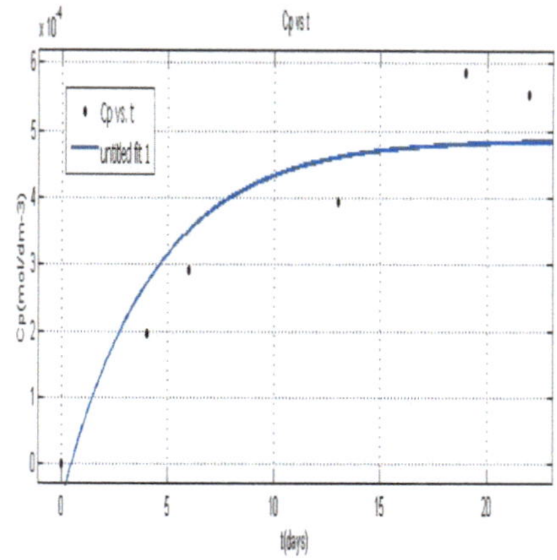

Fig. 1. Variation of product concentration with time

The model for the variation of product concentration with time is given in Eq. (7). The R^2 value of the model is 86.63%. According to Eq. (7), first order rate coefficient k = 0.2321 per day and conversion coefficient $\alpha = 0.0012$.

$$C_P = -0.000542e^{-0.2321t} + 0.0004861 \tag{7}$$

Figure 2 shows the rate of product formation, and is obtained by differentiating the above fitted model equation. Figure 3 shows the variation of substrate concentration with time which is calculated with the aid of Eq. (2).

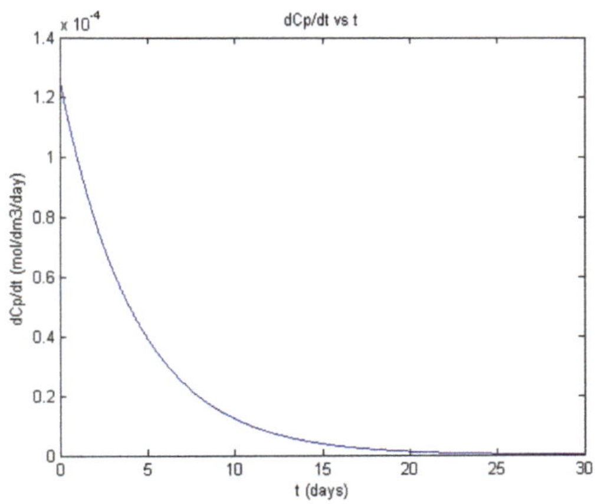

Fig. 2. Variation of the product formation rate with time

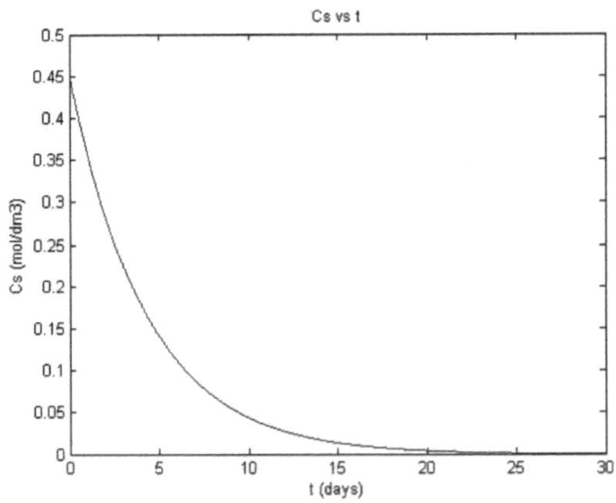

Fig. 3. Variation of the substrate concentration with time

The variation of the substrate concentration with time too can be modelled into a mathematical equation through a simulation. Using the model, the variation of the rate of substrate utilization with time and the variation of the rate of substrate utilization with substrate concentration can be obtained as shown in Figs. 4 and 5 respectively.

The model for the variation of substrate concentration is given in Eq. (8). The R^2 value of this fit is 100%, which means that it is a perfect fit.

$$C_S = 0.4517e^{-0.2321t} \tag{8}$$

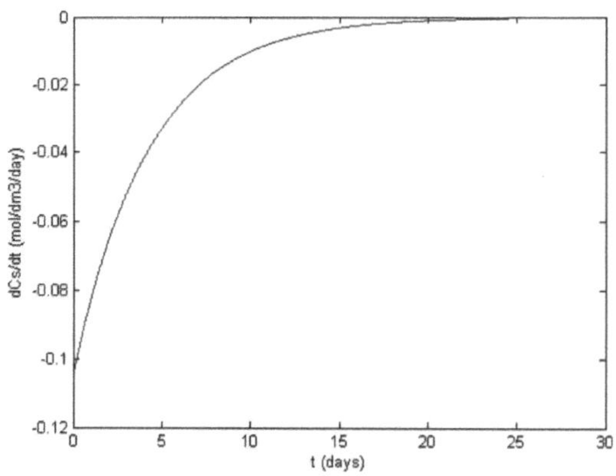

Fig. 4. Variation of substrate utilization rate with time

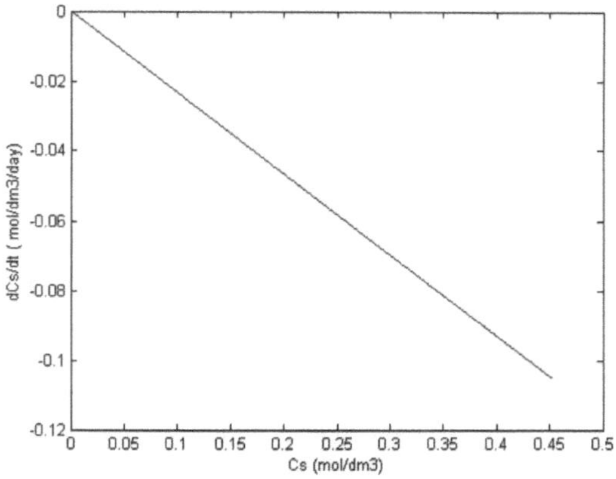

Fig. 5. Variation of substrate utilization rate with substrate concentration

After obtaining the plot in Fig. 5, the model parameters of Eq. (4) can be determined easily by linearizing the equation according to the Langmuir isotherm.

The Langmuir plot obtained by linearizing the model Eq. (4) is shown in Fig. 6. Using the plot, the model parameters were determined to be $K_S = -1$ and $r_{max} = 1.004$ mmol/dm^3.day.

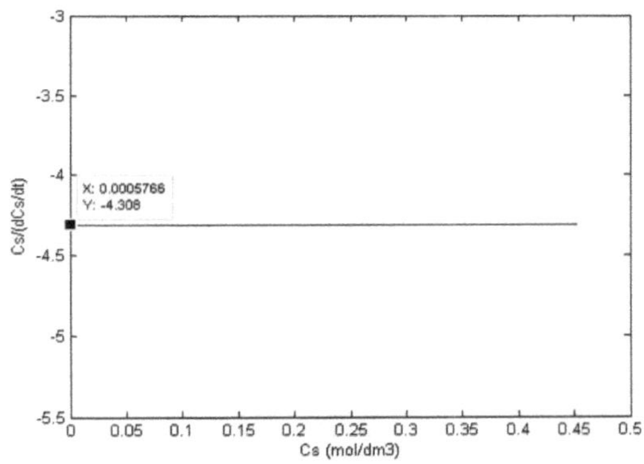

Fig. 6. Langmuir plot for the model

6 Conclusion

Even though the anaerobic digestion process is an unreliable and unpredictable method of waste treatment, given the present environmental status, sustainable and economic implementations are essential for the common benefit of the community. Therefore, during this study, effort was put in to developing a mathematical model that could be used to predict the behaviour of the hydrolysis phase in anaerobic digestion. The final model developed during this study for the hydrolysis of solid carbohydrates in anaerobic digestion is given below as Eq. (9).

$$\frac{dC_S}{dt} = \frac{1.004C_S}{(C_{S0} - C_S) + C_S} \tag{9}$$

Although the results seem to have no contradiction to theory, more improvements should be made to this model before we could entirely rely on the model as a scientific community. The behaviour of the model should be studied under various operating conditions as well as initial substrate concentrations. The approach taken during this study should be adopted to other biodegradable waste components such as proteins, lipids and ultimately to complex solid waste which would be a combination of these components.

These improvements and developments to the present model may someday result in a capable and efficient waste management system which is sustainable, ergonomic and efficient.

References

Andrews JF, Graef SP (1971) Dynamic modeling and simulation of the anaerobic digestion process. Environmental Systems Engineering Department, Clemson University, Clemson, S. C., pp 126–162

Angelidaki II, Ellegaard L, Ahring BK (1999) A comprehensive model of anaerobic bioconversion of complex substrates to biogas. Biotechnol Bioeng 63:363–372

Batstone DJ, Keller J, Angelidaki I, Kalyuzhnyi SV, Pavlostathis SG (2002) The IWA anaerobic digestion model no 1 (ADM1). Water Sci Technol 45:65–73

Gene FP, William FW (1986) Fundamentals of anaerobic digestion of waste water. J Environ Eng 112(5):867–920

Jain S, Lala AK, Bhatia SK, Kudchadker AP (1992) Modeling of hydrolysis controlled anaerobic digestion. J Chem Technol Biotechnol 53:337–344

Liang Y, Wensel PC, Jingwei M, Shulin C (2013) Mathematical modeling in anaerobic digestion. J Bioremediat Biodegradation S4

Pavlostathis SG, Giraldo-Gomez E (1991) Kinetics of anaerobic treatment. CRC Crit Rev Environ Control 21:411–490

Speece RE (1983) Anaerobic biotechnology for industrial wastewater treatment. Environ Sci Technol 17:416A–427A

Vavilin VA, Rytov SV, Lokshina LY (1996) A description of hydrolysis kinetics in anaerobic degradation of particulate organic matter. J Bioresour Technol 56:229–237

Vavlin VA, Fernandez B, Palatsi J, Flotats X (2008) Hydrolysis kinetics in anaerobic degradation of particulate organic material: an overview. J Waste Manag 28(6):939–951

Yoon YM (2014) Effects of substrate to inoculum ratio on the biochemical methane potential of piggery slaughterhouse wastes. Asian-Aust J Anim Sci 27(4):600–607

Design of a Robotic Firefighting Machine

R. Munzara[1], S. M. Ganyani[1], and T. Mushiri[1,2(✉)]

[1] Department of Mechanical Engineering, University of Zimbabwe, Mt Pleasant, P.O. Box MP167, Harare, Zimbabwe
tawanda.mushiri@gmail.com

[2] Department of Quality and Operations Management, University of Johannesburg, Bunting Road Campus, P.O. Box APB524, Johannesburg, South Africa

Abstract. The destruction of property and the environment by fires has necessitated the need for the design of an autonomous firefighting robot. Extreme cases of uncontrolled fires result in injury or even death. This paper presents a design for a firefighting robot with two degrees of freedom. The robotic arm will be able to direct a jet of water within a radius of 15 m. The ability of the machine to manoeuvre on its own, through the use of proximity sensors and flame finders reduce the risk of exposure to fire and smoke of the operator. The device is powered by two DC 40 Ah Lithium Ion batteries. At USD$3,455 initial value, the machine is estimated to depreciate to a value of USD$2,277.13 after five years, at a depreciation rate of 8%. The cost benefit analysis shows that the implementation of such a project is justifiable since having such a machine reduces the risk of losing properties and other valuables to fire. Despite the undisputed importance of fire, it can also be dangerous and costly if not handled well. Unwanted fires can result in heavy losses including properties and important documents. They can also cause injuries or even death. The use of an automatic machine (robot) to fight fire ensures less risk to firefighters at the same time increasing efficiency in putting out the fire. The use of robotics and control enhances accuracy and ease of operation of the machine. Careful design of each component guarantees fulfilment of the aim, which is putting out fire, with high efficiency and effectiveness. The cost of the machine is easily overshadowed by the privileges assumed due to its manufacture. The production of this machine thus ensures reduced financial losses and increased human safety.

Keywords: Design · Robotic · Fire · Styling · Firefighting

1 Introduction

Fire-fighting is a very imperative but hazardous occupation. A fire-fighter must be in position to access the fire safely and quickly extinguish the fire, averting further damage and lessen the possibility of fatalities. Technological advancements have finally bridged the gap that exist between firefighting and machines allowing for a more efficient and effective method to employ when extinguishing a fire. Robots are designed to locate fire, before it rages out of control and prevent it from spreading reducing the risk of injury to victims.

R. Dissanayake and P. Mendis (Eds.): ICSBE 2018, LNCE 44, pp. 567–580, 2020.
https://doi.org/10.1007/978-981-13-9749-3_50

1.1 Background

According to (Fire Service Info, 2016) the first pump designed for firefighting was created by Ctesibius of Alexandria, Egypt, around the second century BC. Thomas Lote, an engineer, built the first fire engine made in America in 1743. John Ericsson is saluted with building the first steam powered fire engine. The same self-propelled steam engine or pumper was built in New York in 1841. The self-propelled (motorized) fire apparatus did not gain acceptance until after 1900. The list of great firefighting equipment inventions through the ages can go on even up to the modern technologies like the fire sprinkler system.

Despite all the great technologies already in place to combat fire, there is need for more alternatives to put out fire as different fire scenarios need very different methods to be used. An apartment fire is so different to a veld fire. Other factors to consider are cost and ease of use of the available technologies. Not everyone can afford a fire truck. Portable fire extinguishers may never be effective in putting out fires at a burning farm. There is need for more efficient, affordable and easy to use means of putting out different fires such as an automatic fire fighting machine.

2 Literature Review

2.1 Mechanics Behind a Firefighting Device

Storage system: The reservoir tank is the agent storage facility. The water tank is the simplest as no special requirements other than less corrosive tank material are needed. For gaseous agents like CO2, high pressure tanks are need [1].

Power Supply: According to [2] a number of power sources can be used depending on design of the machine. These include 36 V or 24 V 65 Ah gel lead acid batteries, DC power via umbilical cable from 24 V electrical supply (from vehicle or socket) or diesel/petrol engines. If an engine is used then a fuel tank should be installed.

Pumping System: Motor; The motor's basic function is to convert the electrical energy from the power supply to mechanical energy (rotating shaft) so as to drive the pump.

Pump: Coupled to the motor shaft, the pump's function is to pump out the firefighting agent during the firefighting.

Agent Discharge System: Piping.
Various types and sizes of pipes are used to transfer the agent from its reservoir to the nozzle for discharge. Standardized hose pipes are used when the reservoir is separate from the machine.

Nozzle: This is the end effector of the robot [1]. The sizes vary from 5/8 to 2 in. according to the same source. Foam nozzles come in three types; high, medium and low expansion nozzles.

2.2 Robotics

Robotics is an interdisciplinary subject involving information, electronics, mechanics, automation, and control theory [3]. A robot is an electromechanical device with multiple degrees of freedom that can be programmed to execute a variety of tasks [4]. Sensors or the sensory system acquires data on the internal and external environment which involves design of sensors, data processing, signal conditioning and information retrieval [3].

2.3 General Parts of a Robot

Controller: According to [5] the controller coordinates all the movements of the mechanical system. It receives inputs from the immediate environment through sensors and also from the remote control/manipulation by an operator. According to [5] there are three levels of hierarchical control by the controller. Control signals are sent from lower levels, while feedback results are sent to the higher levels. Each level depends on the next level (above it) for instructions.

Level 1 - **Actuator Control.** This is the elementary level. Separate motions of the robot along various planes (*x, y and z*) are controlled here.
Level 2 - **Path Control.** This level coordinates the various movements along the planes determined in the first level into the desired path.
Level 3 - **Main Control.** This is the highest control level. Its function is to interpret the coded instructions from a human programmer regarding the tasks and works required to be done. Together with various environmental signals, the instructions are translated by the controller into more elementary instructions that Level 2 can understand.

Manipulator: The manipulator is an electromechanical device capable of interacting with its environment [4]. The manipulator has segments that are joined together to allow the robot to do work. It is the arm of the robot (like the human arm), providing useful work by moving materials, parts, tools, or special devices through various motions [5]. Tachometers are used in a robotic system to monitor the acceleration and deceleration movement of the manipulator [5].

End Effector: An end effector is a tool, gripper, or any other device that is mounted at the end of a manipulator for accomplishing the task. The area within reach of the robot end effector is called its work envelope [5]. In the case of the firefighting robot, the end effector will be the nozzle.

Actuators: Actuators are a motor or mechanism that converts supplied power into motion. An actuator can also be described as a motor that creates motion in robot articulations [6]. The Actuator can also be called the stimulus action of a machine or muscle of a machine. There are basically three types of actuators, each is associated with a different form of energy, be it electrical, hydraulic or pneumatic [6].

Pneumatic robots are frequently limited to tasks such as pick and place where on or off control is possible. Pneumatic energy is also implemented in gripper controls, eradicating the added motor weight [7]. Electric actuators are not as efficient as hydraulic actuators, especially in large sizes. Hydraulic actuators have the largest capability to lift and hold their final position without expending much additional power [7].

Sensors: A robot requires sensing to be an active member in the environment. Each sensor is founded on a transduction principle which is the conversion of energy from one form to another [5]. Sensors can be classified as proprioceptive and exteroceptive. Proprioceptive means sense of self or internal state in which the sensor measures values internally to the system robot for example battery level, wheel position, joint angle. Exteroceptive means the external state [8].

3 Materials and Methods

The following softwares and methods were used to design the robot

(1) Achieve number of segments of the robot.
(2) Show the mechanical anatomy of the machine.
(3) Determine the end effector and its properties (in this case the nozzle).
(4) Highlight the various softwares used in the design. The writer is to use:-
    (a) RoboAnalyzer
    (b) SolidWorks
    (c) Arduino
(5) Determine the type of controller to be used.
(6) Determine the kind of actuator to be used (AC or DC).
(7) Highlight the programming language to be run on the robot.
(8) Perform the dynamic stability of the machine.

3.1 Efficiency of Firefighting

The main aim of the design is to put out a fire effectively. More focus will be on the ability of the automated machine to perform its function efficiently before anything else. The mobility of the firefighting machine in different terrain scenarios will be of very high importance. Another area of focus is the delivery of the working fluid from the reservoir, through the nozzle to the unwanted fire. Making these two aspects efficient will increase the machine's effectiveness and enhance its functionality.

For this design research various firefighting mechanisms which already exist or have been proposed were considered and critically analyzed in a bid to develop a new and improved machine concept. In order to reach the final solution three possible solutions were generated and compared so as to obtain the best design. To determine the best possible design, the Binary Dominance Matrix (takes into account various factors including cost, functionality and weight) is going to be employed. The best was shown here and it the following technical specifications.

3.2 Design Data

The following data shows the technical specifications of the automated firefighting machine.

Table 1. Technical specifications of the robot for firefighting.

Specification	Description
Weight	800 kg
Height	1400 mm
Length	1600 mm
Width	1200 mm
Power sources	24 V rechargeable lead acid batteries or DC power via umbilical cables from 24 V vehicle supply
Fluid flow rate	700 litres per minute
Range of fluid from nozzle	15 m
Track system	Steel reinforced rubber tracks (heavy duty)
Climbing slope	25°
Speed	10 km/h

4 Results

4.1 Nozzle (End Effector) and Pipe Selection

Nozzle: The nozzle will be the end effector on the robotic arm. The purpose of the nozzle is to build fluid pressure by restricting flow. The design of the robotic arm is such that different nozzle types can be mounted and dismounted easily, depending on the operators' requirement. However, due to the smooth bore widespread and common use, this is the nozzle for use in the machine design. The specifications for this nozzle are:-

- Nozzle type - Smooth Bore
- Bore tip - 1-1/2 in. = $(\frac{3}{2} \times 25.4)$ mm = **38.1 mm**
- Discharge - 15/16 in. = $(\frac{15}{16} \times 25.4)$ mm = **23.813 mm**
- Length - **15 mm**
- Nozzle Pressure - 50 psi = (50 × 0.0689476) bar = **3.477 bar**
- Flow rate - **700.301 l/min**
- Material - Aluminum 6061 Alloy (Refer to Table 1 for properties).

Hose: The robot will make use of a hose coupled from the pump discharge to the nozzle. The pipe will be pinned to the robotic arm rather than let lose so as to prevent kinks and pressure loss of water.

- Length of hose - 0.35 m + 0.26 m + 0.39 m = **1 m**
- Hose size - $1\frac{3}{4}$ in. = **45 mm**
- Material - 100% Polyester (+80 °C heat resistant, lightweight, kink resistant, synthetic rubber coating, oil resistant)
- Mass (with water) - *Volume* × *Density* = **1.590 kg**.

For the selected nozzle, the reaction is **31.298 kg** with which the robotic arm can handle.

PUMP SIZING: A centrifugal pump will be used to pump water or foam during the firefighting operation.

The pump's specifications from calculations are as follows

Flow rate q = **0.0117 m^3/s**
H = 24.905 m + 26.117 m = **51.022**

$$Power\ required\ by\ Pump = P = \rho gHQ$$

$$P = \mathbf{5.856\,kW}$$

4.2 Determination of Length of the Robotic Arm

The arm will have to hold the nozzle in perfect position to fight fire. This is done basically by determining the length of the upper arm and forearm. The upper arm will be rigid at 30° from the horizontal and the forearm at a maximum of 50° from the horizontal at full stretch (Fig. 1).

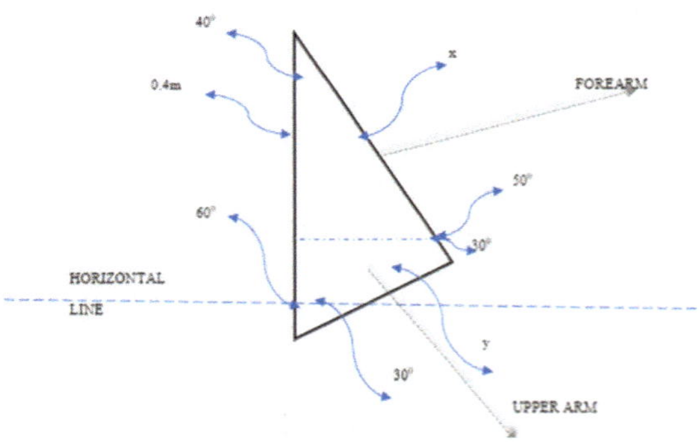

Fig. 1. Orientation of the robotic arm

The Sine Rule will be used to find x and y:-

$$Therefore\ \boldsymbol{x} = 0.35175\,\mathrm{m}$$

Also

$$\boldsymbol{y} = 0.26108\,\mathrm{m}$$

The length of the upper arm will thus be 0.26 m (**260 mm**) and the forearm 0.35 m (**350 mm**) (Fig. 2).

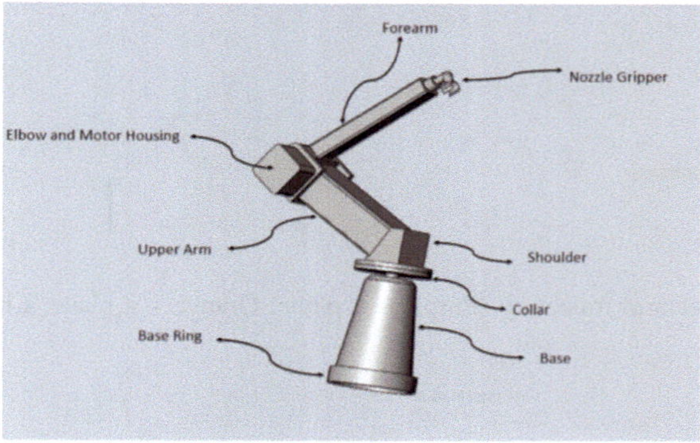

Fig. 2. Robotic arm assembly

Kinematics

The kinematics of the robotic arm was done using the RoboAnalyzer software. The position and orientation of each robotic link, with respect to the previous link, is represented by a set of four DH parameters.

The results of forward kinematics are as shown (Fig. 3).

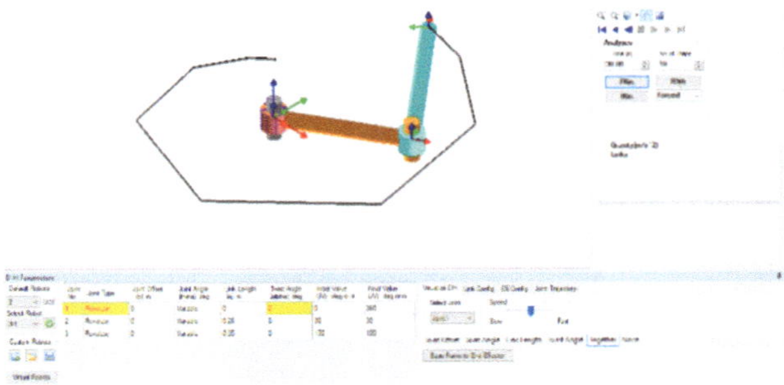

Fig. 3. Isometric view of the forward kinematics ("quantic" trajectory of the arm)

The graphical results of the upper arm and forearm trajectory and its configuration (D-H matrix) are shown below (Figs. 4, 5 and 6):-

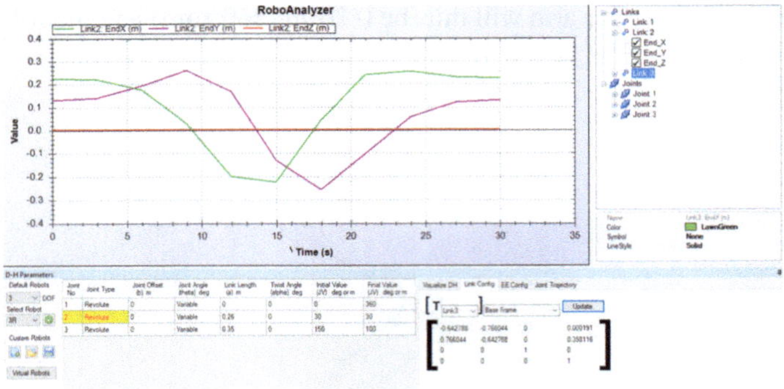

Fig. 4. Upper arm trajectory (Purple – y plane; Orange – z plane; Lime – x plane)

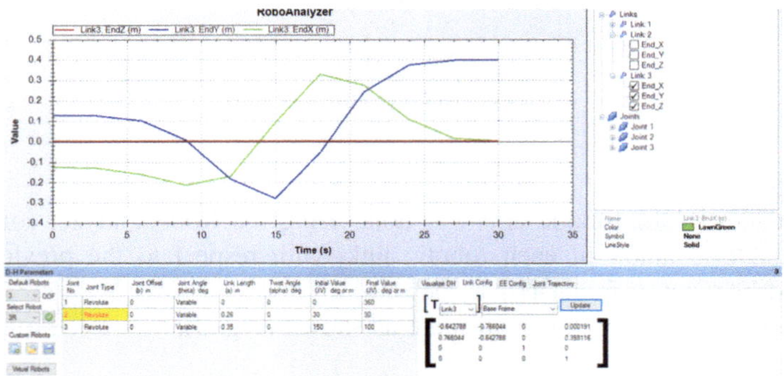

Fig. 5. Forearm trajectory (Blue – y plane; Brown – z plane; Brown – x plane)

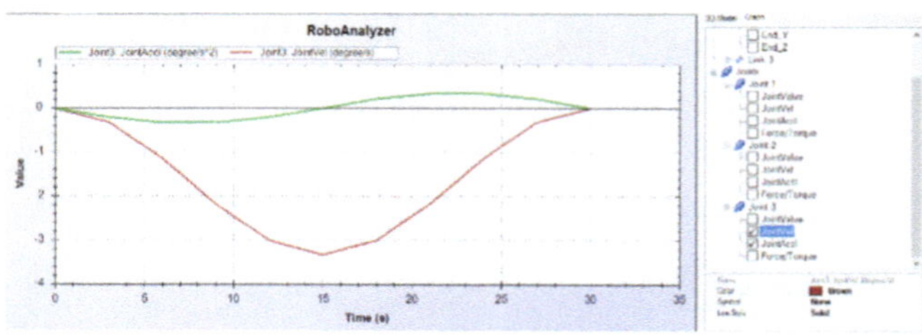

Fig. 6. Acceleration and velocity of the forearm

Sensors and Control

Encoder: Since the actuators to be used to drive the two moving joints are servo motors, built-in rotary **encoders** are to be used as the proprioceptor. The **Siemens Multi-turn Absolute Encoder** is the proprioceptor to be used.

Proximity Sensor: The robot will make use of the sharp ranger GP2D12 infrared sensor that detects from within 10 cm to 80 cm of object proximity. The machine will make use of 5 such proximity so as to provide better movement of the device. The sensors have relatively a high bandwidth that there is no need to model their dynamics.

Flame Detector: The sensor of choice is the phototransistor. The light (from flame) is detected through a transistor's base. If light from the flame is detected, the output voltage will be high and also vice versa. The SFH 309 PFA manufactured by OSRAM Opto Semiconductors will be the transistor of choice as it has a large wavelength range for the different types of flames. At 880 nm, the transistor has an acceptance angle of 150°. It also has a rise or fall time of 6 µs, which is best for the design. The device will have five flame detectors arranged the same as the proximity sensors. Three of the sensors will be at the front, and two at either side, none at the rear.

Heat Sensor: A thermistor will be used for thermal sensing. The heat sensor will be located at the front side of the machine. The thermistor will allow the machine to operate at temperatures below 90 °C, above which it will send a "high" signal to the controller and robot will move backwards.

4.3 Robot Control (Controller and Programming)

The controller acts as the computer or brain of the computer. This is where the robot program is fed and run for all the robot operations (including the arm movements and pumping system). The controller implements specific functions like motion, timing, sequencing, counting and arithmetic to regulate and control, through analogue or digital input and output modules. The control of the robot will be done using Arduino and the Arduino microcontroller. The Arduino software is used to load the control code. Signals from flame detectors, proximity sensors and the temperature sensor will be used to control the device using Arduino code (Figs. 7 and 8).

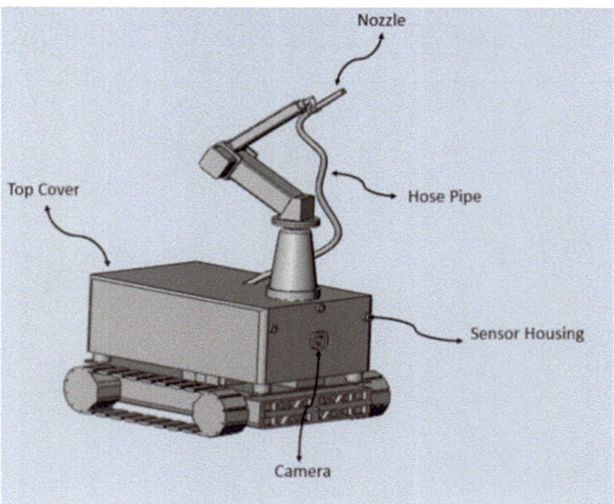

Fig. 7. Isometric view of the firefighting machine front

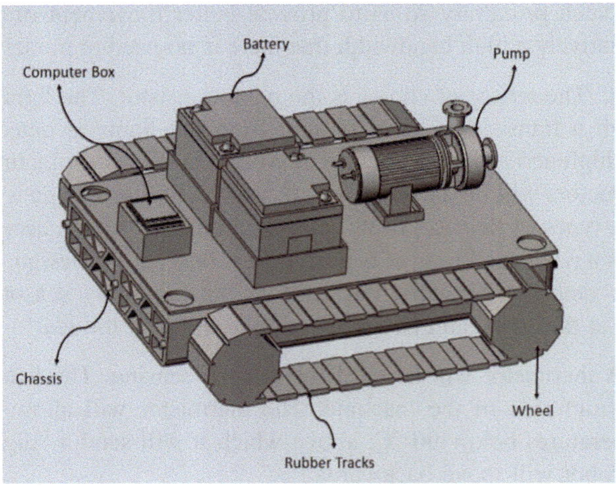

Fig. 8. Firefighting machine assembly

4.4 Stress Analysis

The stress analysis of two components was done using SolidWorks (Von Misses theory). The two components for which the analysis was performed are:-

i. Forearm - If it can withstand the nozzle reaction.
ii. Hose pipe - Pressure analysis to ensure it can withstand **60 PSI**.

Forearm: With the nozzle reaction at 369.66 N the stress, strain and displacement analysis of the robotic forearm as performed (Figs. 9, 10 and 11).

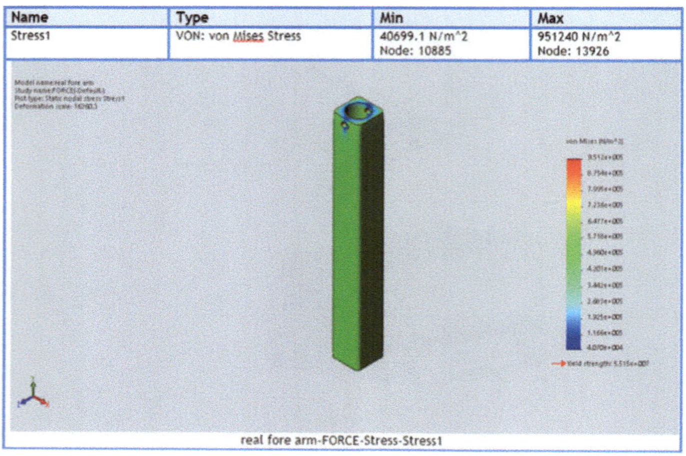

Fig. 9. Stress analysis results (forearm)

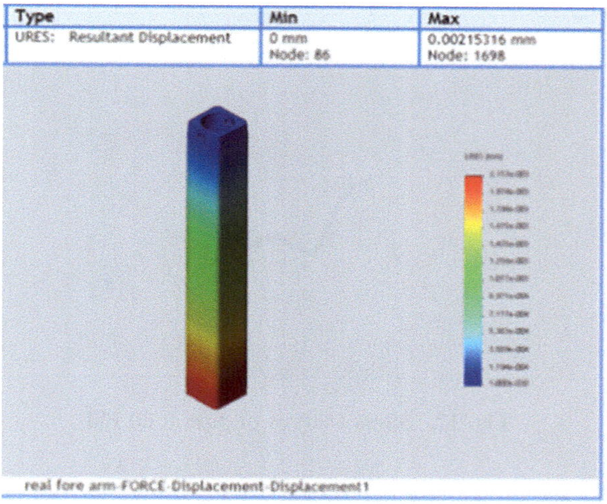

Fig. 10. Stress analysis – force displacement forearm

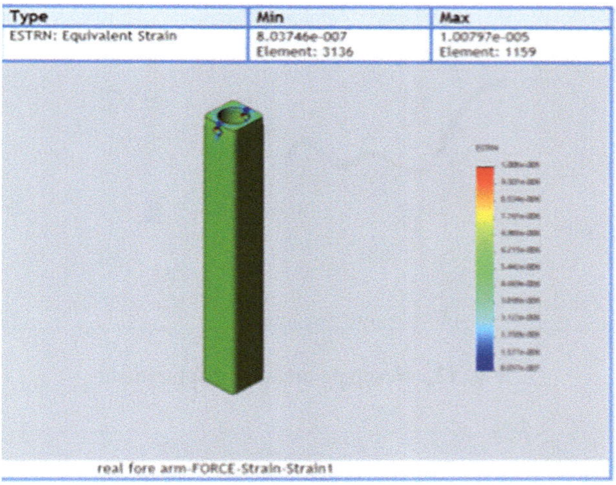

Fig. 11. Displacement and strain analysis results (forearm)

Hose pipe: The hose pipe was analyzed to see if it could withstand the set pressure of 60 PSI (Figs. 12, 13 and 14).

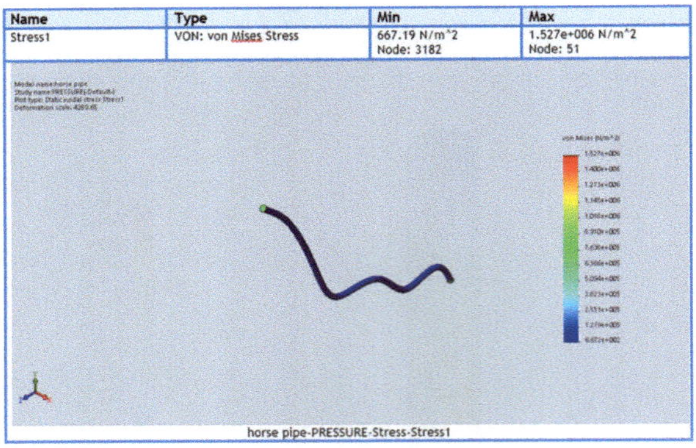

Fig. 12. Stress analysis of pipe at 60 PSI

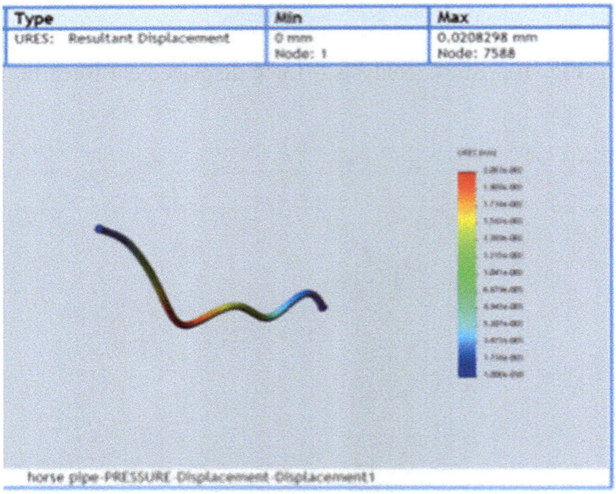

Fig. 13. Hosepipe pressure displacement

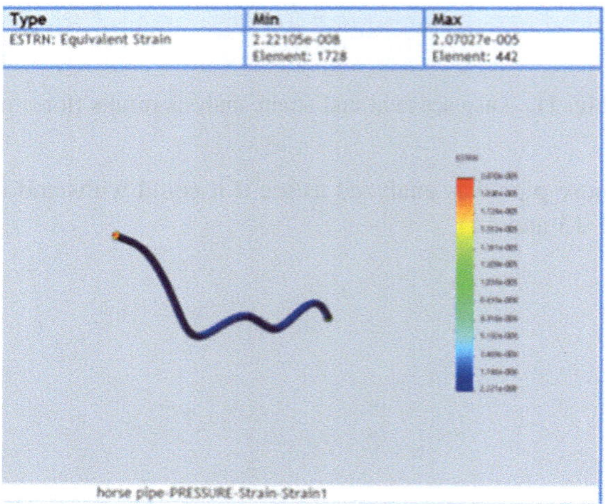

Fig. 14. Displacement and strain results of pipe assuming it has no free play

5 Recommendations and Conclusion

Despite the undisputed importance of fire, it can also be dangerous and costly if not handled well. Unwanted fires can result in heavy losses including properties and important documents. They can also cause injuries or even death. The use of an automatic machine (robot) to fight fire ensures less risk to firefighters at the same time increasing efficiency in putting out the fire. The use of robotics and control enhances accuracy and ease of operation of the machine. Careful design of each component guarantees fulfilment of the aim, which is to putting out fire, with high efficiency and effectiveness. The cost of the machine is easily overshadowed by the privileges assumed due to its manufacture. The production of this machine thus ensures reduced financial losses and increased human safety.

Acknowledgment. Would like to thank UJ for registering for this paper and funding for everything.

References

1. American Bureau of Shipping (2005) Guidance notes on fire-fighting systems. ABS, New York
2. Tan CF, Liew SM, Alkahari MR, Ranjit SSS, Said MR, Chen W, Rauterberg GWM, Sivakumar D, Sivarao (2013) Fire fighting mobile robot: state of the art and recent development. Aust J Basic Appl Sci 7:220–230
3. Koren Y (2000) Robotics for engineers. McGraw Hill Book Company, New York
4. Williams B (2014) Mechanical engineering. Ohio University, Ohio
5. Ross LT, Fardo S, Masterson JW, Towers RL (2011) Robotics: theory and industrial application, 2nd edn. The Goodheart-Willcox Co., New York
6. Chirouze MCP (1983) An introduction to robot technology. McGraw-Hill, New York
7. Poole H (1989) Fundamentals of robotics engineering, 1st edn. Van Nostrand Reinhold, New York
8. Spong MW, Hutchinson S, Vidyasagar M (2004) Robot dynamics and control, 2nd edn. Adventure Work Press, New York
9. Massachusetts Academy of Mathematics and Science (2010) Inventions that changed history. Massachusetts Academy of Mathematics and Science, Massachusetts
10. Meyers J (1993) A Short History of the Computer. www.jeremymeyers.com/img/comp/comp.pdf. Accessed 16 Sept 2016
11. Shende V (2014) Analysis of research in consumer behaviour of automobile passenger car customer. Int J Sci Res Publ IV(2):3
12. Merriam-Webster (2016) Merriam-Webster. http://www.merriam-webster.com/dictionary/sport-utility%20vehicle. Accessed 16 Sept 2016
13. DeMuro D (2013) Autotrader. http://www.autotrader.com/car-shopping/suv-vs-crossover-whats-the-difference-215843. Accessed 16 Sept 2016
14. The-Blueprints.com (2016) Artist, [Art]
15. The Great Soviet Encyclopedia (1979) The Free Dictionary. http://encyclopedia2.thefreedictionary.com/steering+system. Accessed 16 Sept 2016

16. Griffin K (2016) About Autos. http://usedcars.about.com/od/glossaryatoe/g/Crossovers.htm. Accessed 16 Sept 2016
17. Kobelev V (2014) Effect of static axial compression on the natural frequencies of helical springs. Multidiscip Model Mater Struct 10(3):379–398
18. Stander N, Du Preez R (1992) Vibration analysis of coil springs by means of isoparametric curved beam finite elements. Commun Appl Numer Methods 8(6):373–383

Design of a Portable and Sustainable Bicycle Frame

T. Mushiri[1(✉)], R. Gumbo[1], and C. Mbohwa[2]

[1] Department of Mechanical Engineering, University of Zimbabwe, P.O Box MP167, Mt Pleasant, Harare, Zimbabwe
tawanda.mushiri@gmail.com
[2] Department of Quality and Operations Management, University of Johannesburg, Bunting Road Campus, P.O Box APB524, Johannesburg, South Africa

Abstract. This paper is about designing a bicycle frame which is portable as well as sustainable for the nation of Zimbabwe. It is aimed at taking advantage of the properties offered by bamboo in the context of a bicycle frame e.g. ability to absorb vibrations when riding the bicycle on a rough terrain. Aspects of stiffness and ride quality are discussed in a detailed manner. There is also information on the three circumstances that affect a structure to work well. In this paper, frame materials like steel and carbon fiber are studied in view of how they have survived in the market and how those aspects can be incorporated in to the material bamboo to survive the market. Advantages and disadvantages of steel, titanium, aluminium and carbon fiber are viewed in a way to understand the impact of introducing bamboo as a material of choice for the nation of Zimbabwe. What makes up bamboo and its mechanical properties are also studied. The bicycle frame parts are also given attention to in order to view how they absorb force during the bicycle's transit i.e. chain-stays, down tube, bottom bracket, seat-stays, top tube and seat tubes. How the tubes are to be joined is narrated in the methodology section. This is to answer the issue that in steel and other materials there is welding to join the tubes to form the bicycle frame; what about in bamboo bicycles?

Keywords: Bamboo · Bicycle frame · Zimbabwe · Design · Portable · Sustainable

1 Introduction

Bicycles are a means of transport and use no fossil fuels to be propelled. The human energy used to propel bicycles is renewable and largely non-polluting. In a way, bicycles provide a means to lose weight in the saddle [4]. The bicycle is the primary mode of mobility for millions of people throughout many poorer parts of the world. In addition to personal transport, they see a vast number of applications including moving goods to market, the sick to hospital, run errands and delivering of mails, taking vacations and even the distributing medicines. Bicycles encourage human interaction, communication, and connectivity [1]. The cost to construct and maintain quality bike trails is much lower than the cost to construct new roads and/or widen existing roads

R. Dissanayake and P. Mendis (Eds.): ICSBE 2018, LNCE 44, pp. 581–594, 2020.
https://doi.org/10.1007/978-981-13-9749-3_51

[2]. In Zimbabwe, quite a number of people can own cars or even motorcycles and people without bicycles have to rely on inadequate and relatively undependable buses for short journey, say less than 30 km. Provision of bicycles to students can prove significant in reducing travel time and improve educational outcomes [3].

Background and Literature Review

A bicycle frame is the supreme component of a bicycle [11], onto which wheels and other bicycle accessories are fitted. The current and most common frame design for an upright bicycle is based on the safety bicycle consisting of two triangles: a frontal main triangle and a paired rear triangle.

In 1817, one Karl Barn von Drais [12, 13] devised a frame that enabled steering the front wheel. Heaviness and cumbersome was associated with the then bicycle. For many years until the present day, inventors and engineers have been busy coming up with better handling frames, which are lighter-in-weight, stiff capable of travelling long routes without disintegrating. To this day cyclists, inventors and engineers seek to make today's frames better than yesteryears in various ways [10] (Fig. 1).

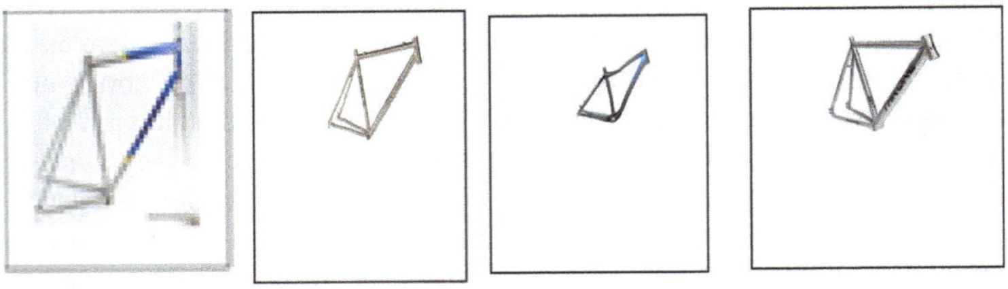

Fig. 1. a. Aluminum bicycle frame b. Carbon fiber bicycle frame c. Titanium bicycle frame d. Steel bicycle frame

Aluminum has been used in bicycle frames since the 1890's. The frame has an advantage of needing no paint [19]. Being oversized aluminum does have a very unforgiving characteristic when overstressed or cracked, it can fail very suddenly. It is of importance to inspect aluminum frames after an accident for any sign of cracking or stress, even if the bike rides perfectly. Carbon fiber have a very strong strength to weight ratio. Custom carbon fiber frames are quite expensive. Machining of titanium is difficult evidenced by requiring a lot of extra steps in the welding procedure. The material is not readily available, thus much more expensive to purchase. Titanium is more durable as a modern bicycle frame material. Steel is popular as a bicycle frame and very easy to machine. Welding steel doesn't require all of the special tooling and steps that welding titanium does. It also takes less time and less electricity than aluminum welding [19]. Steel is the standard that majority of frame materials are compared to.

1.1 Problems Associated with the Current Materials

Steel, aluminum, carbon fiber and titanium frames are very difficult to repair according to the British cycling organization. The aims and objectives of this research are to reduce the cost of bicycle frame by at least 10% and to design a sustainable bicycle frame.

1.2 Bicycle Frame Layout

The frontal, facing down, triangle and rear triangle connection are responsible for the structure of the bicycle frame. The diamond design has proven to be a great use of materials and the strength of the design comes from the triangle shapes [20]. Head tube, top tube, down tube and seat tube comprise the front triangle. Seat-tube, a pair of chain-stays and a pair of seat-stays make up the rear triangle. The two triangles are connected by use of a bottom bracket found at the lower end of the seat-tube (Fig. 2).

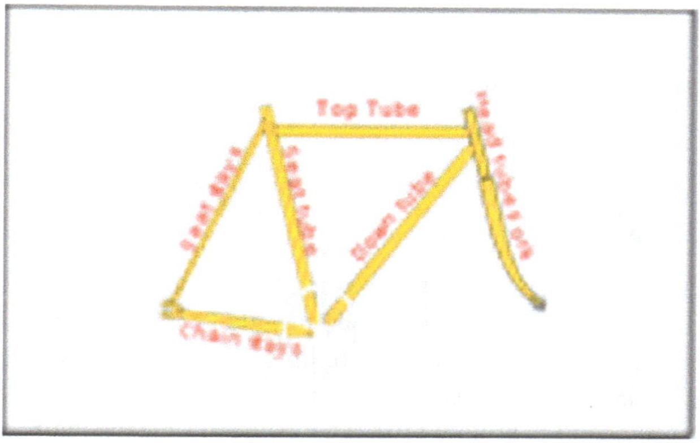

Fig. 2. The bicycle frame [15]

Every bicycle has a resonant frequency at which it oscillates in a different way. A metal tube ring can be heard when the tube is tapped with a fingernail. However, resonances below the audible range only affects the feel of a bicycle. For example, the ringing of the frame in torsion as the frontal triangle tilts relative to the rear triangle, excited by the front wheel's turning slightly to the right and left [5]. According to Allen and Brown, the front triangle is less stiff in torsion and the resonant frequency is lower. Resonances affect the feel when going over bumps. This implies that two frames which are equally stiff feel different when they resonate at different frequencies. According to the Oxford dictionary, damping is a reduction in the amplitude of a vibration or oscillation [14]. Every metal frame has a very low damping such that when they ring long enough a clear tone is produced. Carbon fiber frames will give dull sounds when tapped because carbon fiber has more damping in contrast to metal. The frame feel is much less at the low frequencies. Rubber is a material that is highly damped and so is leather. Brown and Allen alludes that the greatest damping in a bicycle is on the tires and the saddle unless the bicycle has suspension [16]. A suspension fork or frame is a very much damped resonant system [21]. If damping was absent, the frame would bounce up and down repeatedly after every bump. Suspension adds weight, hence contribute to the feel obtained from the frame. The materials of the bicycle are as shown in Table 1.

Table 1. Table of properties of bicycle frame materials [7, 8, 17, 19]

Material	Yield strength (MPa)	Ultimate tensile strength (MPa)	Density (g/cm^3)
Steel, structural ASTM a36 steel	250	400–550	7.8
Steel, 1090 mild	247	841	7.58
Chromium-vanadium steel AISI 6150	620	940	7.8
Aluminum alloy 6061-T6	241	300	2.7
Carbon fiber	N/A	4137	1.75
Bamboo		350–500	0.4
6A1/4V worked Titanium [12]	1085	1220	4.43

1.3 Bamboo as a Material

Bamboo is a natural composite material comprising cellulose fibers embedded in a lignin matrix [10]. The cellulose fibers are aligned along the length of the bamboo, thus providing maximum tensile strength and rigidity in that direction. Furthermore, the fibers are recognizably more abundant on the outermost part of the bamboo stem while the inner part of the stem is hollow [10]. This evidently means the bamboo is strong, stiff, and light (Fig. 3).

Fig. 3. a. Plantation of bamboo b. Bamboo tubes c. Bamboo sectioned tubes and way of connection [10, 18]

2 Materials and Methods Used for Design

For mechanical drawings, Solid Works 2015 was used. The specifications are shown in Table 2. Stresses were also determined by the finite element method.

Table 2. Bamboo technical specifications

Description	(N/mm^2)
Modulus of elasticity	16.17
Bending strength	20.27
Compressive strength	7.86
Tensile strength	14.96
Longitudinal shear strength	1.41

The designed bamboo frame was drawn using SolidWorks as shown in Fig. 4.

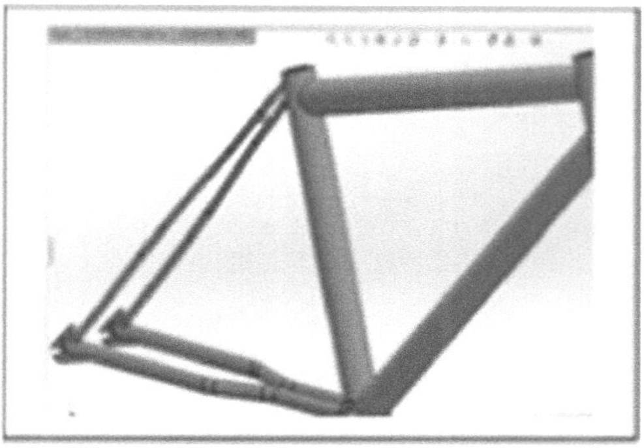

Fig. 4. Bamboo chosen shape

The length and shape of the bicycle frame tubes, together with the angles at which they are set, describes the importance of the bicycle design. The consequences would be how well the bike climbs elevated routes, how urgently it accelerates, and how stable it is, perceived at speed and through bends and corners. Importantly, geometry will have an immense influence on how comfortable the bike feels beneath different physiological types. For design the following are considered: There is need to visualize the stresses the frame will incur in each of the listed scenarios below:

(1) Sudden braking at top speed to prevent a frontal collision - fork stress or horizontal compaction.
(2) Riding uphill - out of a low part of a ridge.
(3) Jumping off a raised path.
(4) Turning aggressively i.e. leaning and shifting your body weight at an intersection.

Using anthropology data for Zimbabwe, the average human as 75 kg, the weight of frame as 2.5 kg and mass density of bamboo as 350 kg/m^3 [22].

3 Simulation and Results

For the frame: Moments about A and using support reactions in Newton's Third Law we get the forces and they are derived as follows: $+\circlearrowleft M_A = 0$ (Figs. 5, 6, 7, 8, 9, 10, 11, 12, 13 and 14) and (Tables 3 and 4);

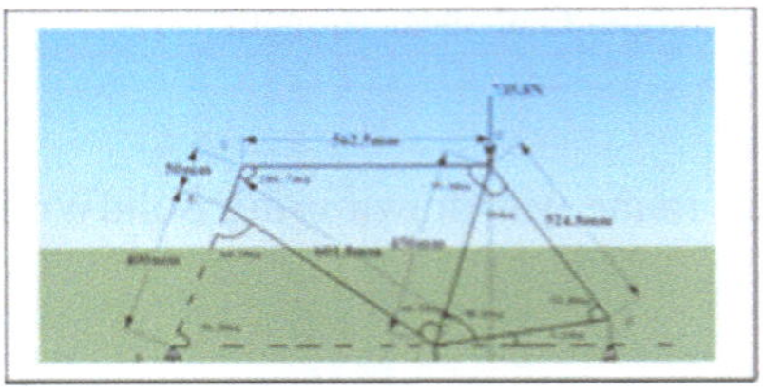

Fig. 5. Frame dimensional design.

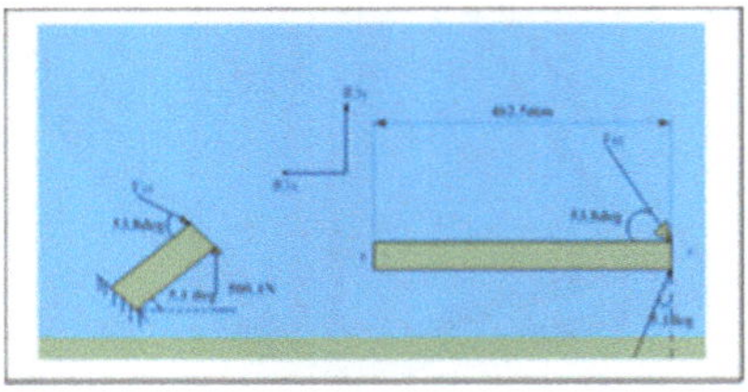

Fig. 6. Chain stay

$$24.5(0.5625) + 735.8(0.5625 + 0.45\cos 75.3^\circ) - R_2(0.5625 + 0.4975) = 0$$

$$\gg R_2 = 483.5\,\text{N} \quad + \uparrow \Sigma F_y = 0 R_1 + 483.5 = 735.8 + 24.5; \;\; R_1 = 276.8\,\text{N}$$

Let F_{SS}-be force on seat stay
F_{DT}-be force on down tube
F_{CS}-be force on chain stay
F_{TT}-be force on top tube
F_{HT}-be force on head tube
F_{ST}-be force on seat tube

Moments about E;

$$+\circlearrowleft M_E = 0;$$

$$F_{SS}\sin 53.8^\circ(0.4625) - 483.5\cos 5.1^\circ(0.4625) = 0$$

$\gg F_{SS} = 596.8\,\text{N} \quad + \rightarrow \Sigma F_x = 0; \quad 596.8\cos 53.8^\circ + 483.5\sin 5.1^\circ + R_{3x} = 0;$

$$R_{3x} = -395.5\,\text{N}$$
$$+ \uparrow \Sigma F_y = 0;$$
$$483.5cos5.1^\circ - 596.853.8^\circ + R_{3y} = 0;$$
$$R_{3y} = 0\,\text{N}$$

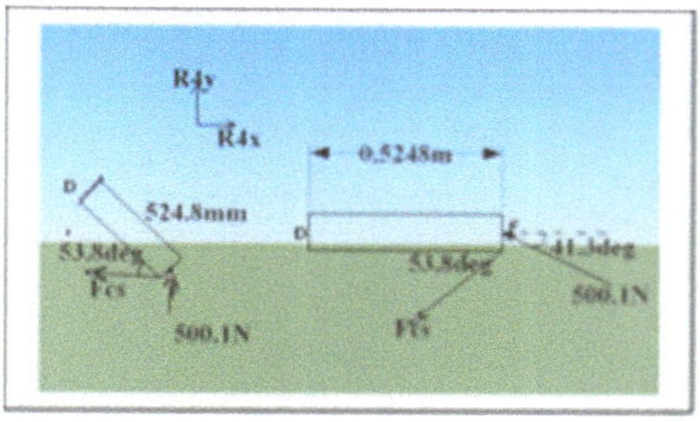

Fig. 7. Seat stay

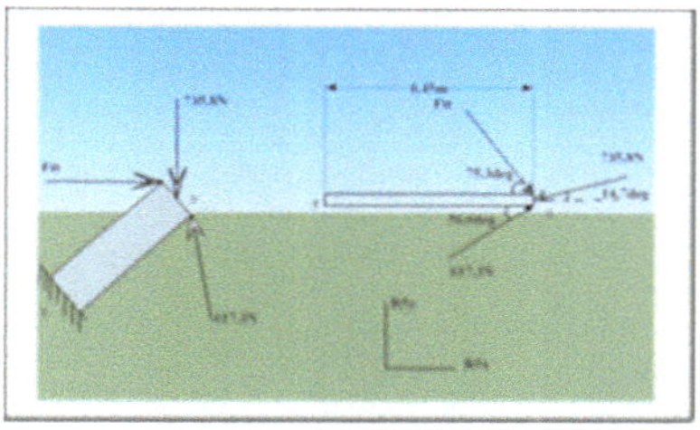

Fig. 8. Seat tube

Moments about D; $+ \circlearrowleft M_D = 0; \quad F_{CS}\sin 53.8^\circ(0.525) + 483.5\sin 41.3^\circ(0.525) = 0;$

$$F_{CS} = 395.4\,\text{N}$$
$$+ \rightarrow \Sigma F_x = 0;$$
$$-394.4\,cos53.8^\circ - 483.5\,cos41.3^\circ + R_{4x} = 0;$$
$$\gg R_{4x} = 596.8\,\text{N}$$
$$+ \uparrow \Sigma F_y = 0;$$
$$-395.4\,sin53.8^\circ + 483.5\,sin41.3^\circ + R_{4y} = 0;$$
$$\gg R_{4y} = 0\,\text{N}$$

Moments about E: $+ \circlearrowleft M_E = 0; \quad -F_{TT}\,sin75.3^\circ(0.45) - 735.8\,sin14.7^\circ(0.45) + 596.8\,sin56^\circ(0.45) = 0;$

$$\Rightarrow F_{TT} = 318\,\text{N} \quad + \rightarrow \Sigma F_x = 0;$$
$$318.5\,cos75.3° + 596.8\,cos56° - 735.8\,cos14.7° + R_{5x} = 0; \Rightarrow R_{5x} = 297.2\,\text{N}$$

$$+ \uparrow \Sigma F_y = 0;$$
$$R_{5y} + 596.8\,sin56° - 318.5\,sin75.3° - 735.8\,sin14.7 = 0; \Rightarrow R_{5y} = 0\,\text{N}$$

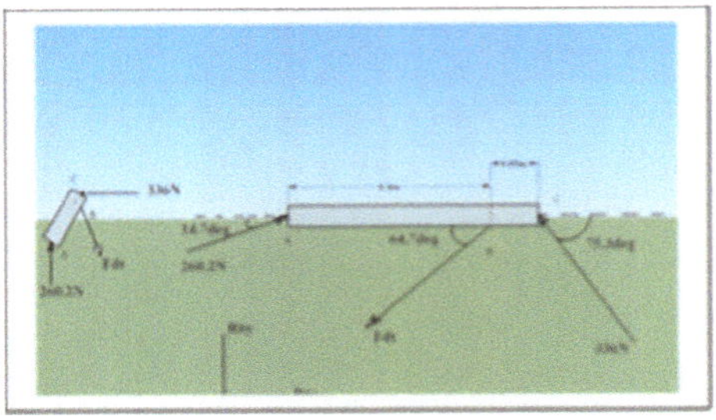

Fig. 9. Head tube

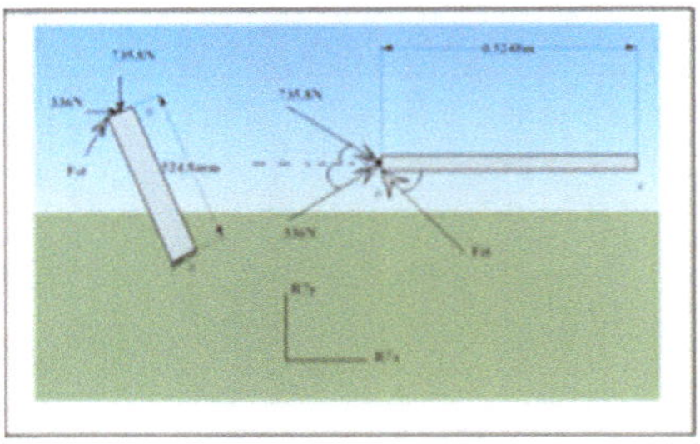

Fig. 10. Seat stay

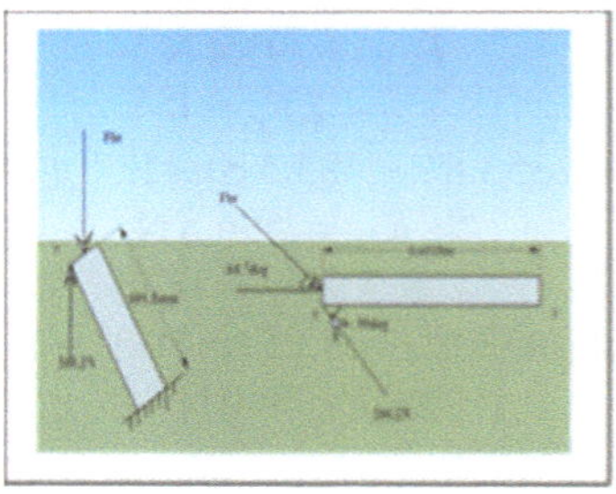

Fig. 11. Down tube

Moments about A: $+ \circlearrowleft M_A = 0; \quad 318 sin75.3°(0.45) - F_{DT} sin66.4°(0.4) = 0;$

$$\Rightarrow F_{DT} = 377.6\,\text{N} \quad + \rightarrow \Sigma F_x = 0;$$
$$276.8\,cos14.7° - 377.6\,cos66.4° 318\,cos75.3° + R_{6x} = 0;$$

$$\Rightarrow R_{6x} = -35.1\,\text{N}$$
$$+ \uparrow \Sigma F_y = 0;$$
$$276.8\,sin14.7° - 377.6\,sin66.4° + 318\,sin75.3° + R_{6y} = 0;$$
$$\Rightarrow R_{6y} = -31.8\,\text{N}$$

Moments about F: $+ \circlearrowleft M_F = 0; \quad -318.5\,sin48.7°(0.525) + 735.8\,sin41.3°(0.525) + F_{ST}\,sin56°(0.525) = 0;$

$$F_{ST} = 297.2\,\text{N} \quad + \rightarrow \Sigma F_x = 0;$$
$$R_{7x} + 735.8\,cos41.3° + 318.5\,cos48.7° + 297.2\,cos56° = 0;$$

$$\Rightarrow R_{7x} = -979.2\,\text{N} \quad + \uparrow \Sigma F_y = 0;$$
$$R_{7y} + 318.5\,sin48.7° - 735.8\,sin41.3° - 297.2\,sin56° = 0;$$

$$\Rightarrow R_{7y} = -493.1\,\text{N}$$

Moments about E: $+ \circlearrowleft M_D = 0; \quad -F_{HT}\,sin66.4°(0.625) + 276.8\,sin51.7°(0.625) = 0;$

$$\Rightarrow F_{HT} = 237.1\,\text{N} \quad + \rightarrow \Sigma F_x = 0;$$
$$R_{8x} + 237.1 cos66.4° - 276.8 cos51.7° = 0; \Rightarrow R_{8x} = 76.6\,\text{N}$$

$$+ \uparrow \Sigma F_y = 0; \quad R_{8y} - 237.1\,sin66.4° + 276.8\,sin51.7° = 0; \quad \Rightarrow R_{8y} = 0\,\text{N}$$

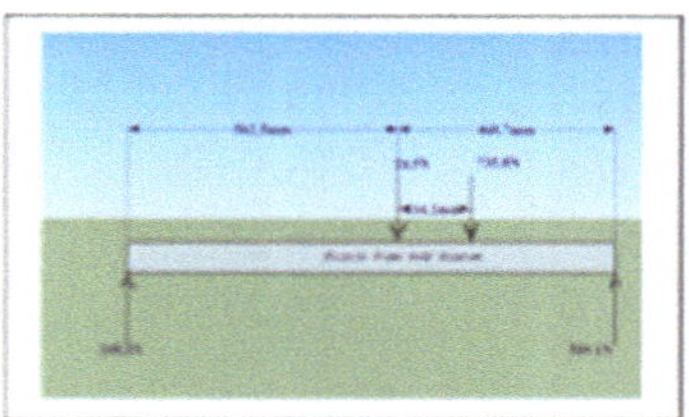

Fig. 12. Bicycle frame body diagram

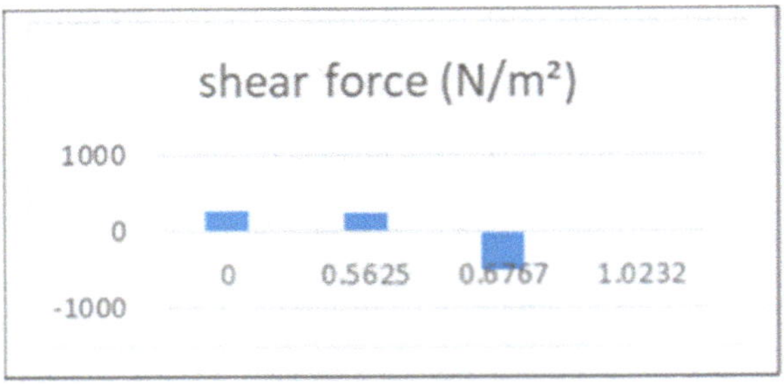

Fig. 13. Shear Force Diagram (SFD)

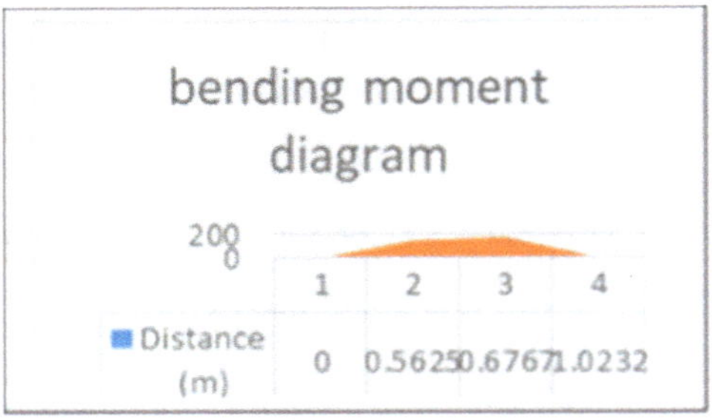

Fig. 14. Bending Moment Diagram (BMD).

Table 3. Name of frame and parameters

Name of frame	Length (m)	Diameter (mm)	Thickness (mm)	Deflection (mm)	Force (N)
F_{SS}	0.5248	25	3.1	8.2	617.3
F_{DT}	0.6018	37.5	3.1	4.1	404.4
F_{CS}	0.4625	25	3.1	4.9	409.0
F_{TT}	0.5625	37.5	3.1	3.2	336.0
F_{HT}	0.05	50	6.25	1.0	225.9
F_{ST}	0.45	37.5	6.25	1.1	281.3

Let σ_{max} = maximum stress, τ_{max} = maximum shear stress, ε = strain of the bamboo tube

D = diameter of the bamboo tube, t = thickness of the bamboo tube, $\sigma_{allowable}$ = allowable shear stress,

$\sigma_{failure}$ = ultimate strength of bamboo, E_{bamboo} = young's modulus of bamboo, δ = deflection, Choosing a safety factor (FS) of 3, $\sigma_{failure}$ = 425 MPa, E = 16.2 MPa

$FS = \frac{\sigma_{failure}}{\sigma_{allowable}}; \Rightarrow \sigma_{allowable} = \frac{425\,\text{MPa}}{3} = \mathbf{141.7\ MPa}$

Table 4. Stresses in different areas on the bicycle.

Place on the bicycle	Stress	Safe or not safe
Chain stay	$\sigma_{max} = \mathbf{1.62\ MPa} < \mathbf{141.7\ MPa} = \sigma_{allowable}$	Safe
Seat stay	$\sigma_{max} = \mathbf{2.45\ MPa} < \mathbf{141.7\ MPa} = \sigma_{allowable}$	Safe
Seat tube	$\sigma_{max} = \mathbf{0.404\ MPa} < \mathbf{141.7\ MPa} = \sigma_{allowable}$	Safe
Down tube	$\sigma_{max} = \mathbf{1.03\ MPa} < \mathbf{141.7\ MPa} = \sigma_{allowable}$	Safe
Head tube	$\sigma_{max} = \mathbf{0.241\ MPa} < \mathbf{141.7\ MPa} = \sigma_{allowable}$	Safe
Top tube	$\sigma_{max} = \mathbf{0.871\ MPa} < \mathbf{141.7\ MPa} = \sigma_{allowable}$	Safe

von Misses Stress, σ_v: $\boldsymbol{\sigma_y}\sigma_v = \frac{\sigma_y}{N(derived\ from\ factor\ of\ safety\ diagram)}$ $\sigma_v = \frac{20.26 * 10^6}{1.402 * 10^2} = \mathbf{1.445} * \mathbf{10^5} \mathrm{N/m^2}$

Since $\boldsymbol{\sigma_v} < \boldsymbol{\sigma_y}$; **design is safe** (Figs. 15, 16, 17 and 18) (Table 5).

Table 5. Costing related to the bamboo

Material	Units	Thickness	Dimensions	Cost ($)
Bamboo tube	1	6.25 mm	Dia. 50 mm * length 300 mm	3.00
Bamboo tube		3.1 mm	Dia. 37.5 mm * length 475 mm	4.00
Bamboo tube		3.1 mm	Dia. 37.5 mm * length 575 mm	5.00
Bamboo tube		6.25 mm	Dia. 37.5 mm * length 500 mm	5.00
Steel drop outs	2	5 mm	Dia. 25 mm	1.00
Bottom-bracket (steel)	1	6.25 mm	Dia. 50 mm * 40 mm	3.00
Bamboo tube		3.1 mm	Dia. 37.5 mm * 600 mm	5.00
Bamboo tube		3.1 mm	Dia. 25 mm * 525 mm	4.00
Wrapping tape	12		15 mm * 5000 mm	0.30
Hemp fiber	20		5000 mm * 10 mm * 2 mm	0.25
Total				**39.60**

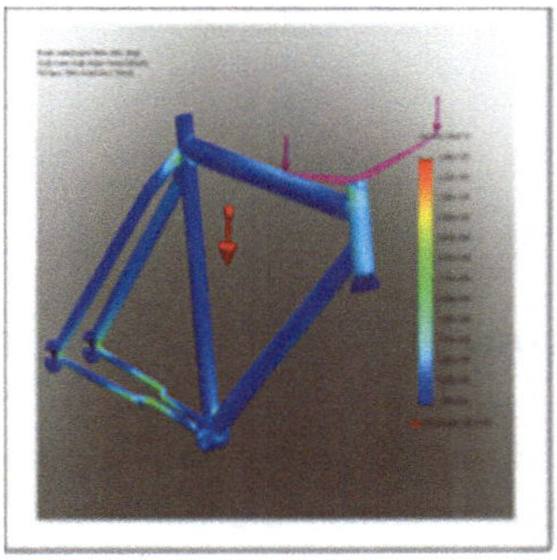

Fig. 15. von Mises Stress

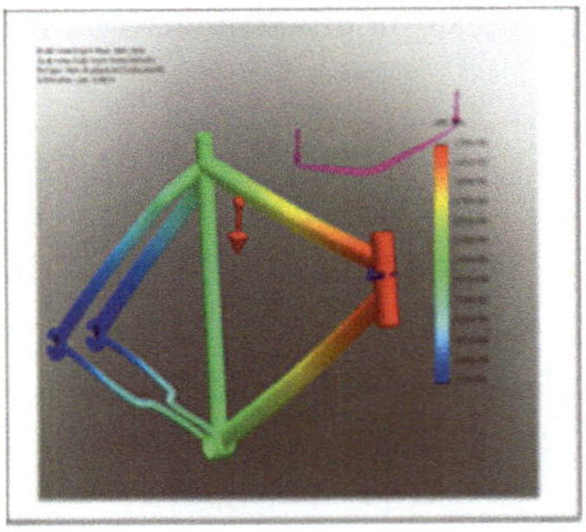

Fig. 16. Displacement analysis

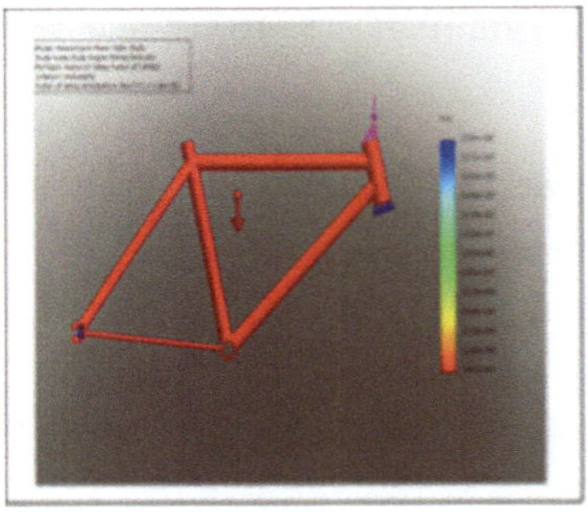

Fig. 17. Factor of Safety (FOS).

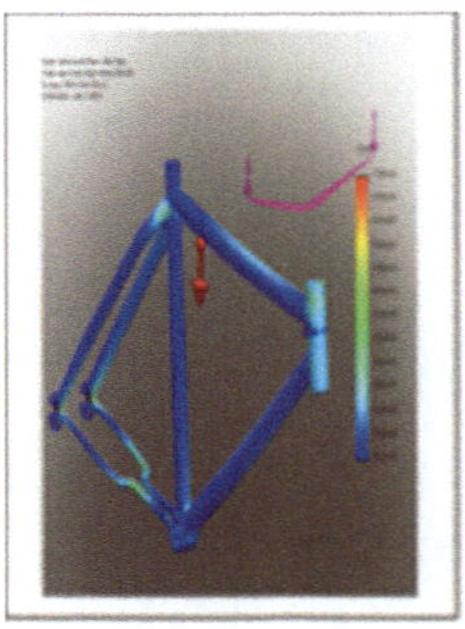

Fig. 18. Strain

4 Discussion and Recommendations for Further Work

People who seek to be their own employers should see this project as an idea to acquire finances. It is recommended to establish the bamboo frame building factory on the bamboo plantation to cut down costs mentioned in the bill of materials i.e. tubes will be acquired for free. For further savings, the government of Zimbabwe should reduce importation of bicycle from the Far East and start harvesting more bamboo plants in Zimbabwe, especially in provinces where rainfall is limited. It is recommended that maximum loading on the frame is not to exceed 100 kgs. For better enjoyment, store frame in a dry storeroom. Since some bamboo will be cut, afforestation should be done more often and policies of plantation should be encouraged. Zimbabwe has a National Tree Planting Day which will support this project of bamboos as raw materials for bicycles in abundance and well replaced [23]. To cater for comfort, it will be necessary to check the reach and make it long, handle height and add spacers under the stem, make the saddle comfortable by covering with canvas and adjust height and angle and lastly tyre pressure should not be too much but average.

5 Conclusion

The bamboo bicycle frame is a successful project. Objectives of the frame were realized; being portable as well as the use of sustainable material in bicycle frame make. The cost was reduced in the forming of the frame including cutting cost on importation. This can be used but one thing to be noted is that forces are less than those of steel hence average people should cycle general to make it durable. This research is very useful to Zimbabwe.

Acknowledgment. Would like to thank UJ for paying the registration fees and avail resources for paper to be presented in Sri Lanka.

References

1. Agur L (2013) Carbon fiber vs Aluminium vs Steel vs Titanium i love bicycling. http://www.ilovebicycling.com. Accessed 9 Oct 2016
2. Bamboo bicycle club (n.d.) Frequently Asked Questions About Bamboo Bikes. http://bamboobicycleclub.org/. Accessed 16 Mar 2017
3. Bamboo Bike Supplies (2015) Composite Lug Wrapping with Hemp Fibers. https://bamboobikesupplies.com. Accessed 19 Apr 2017
4. Barbour M (2016) 30 reasons to take up cycling. www.bikeradar.com. Accessed 15 Sept 2016
5. Brown S, Allen J (2010) Frame Materials for the Touring Cyclist. http://www.sheldonbrown.com/framematerials. Accessed 21 Feb 2017
6. DeMerchant C (2017) Youngs Modulus is a Measure of Stiffness. http://www.christinedemerchant.com/youngmodulus.html. Accessed 21 Mar 2017
7. Gerguri J, Broughton J (2016)
8. Hearn EJ (1997) Mechanics of materials, 3rd edn. Butterworth Heinemann, United Kingdom
9. Herlihy D (2004) Bicycle: the history. Yale University Press, New Haven
10. Instron (2012) Bamboo Bikes - A "Growth" Industry. http://www.instron.us/. Accessed 20 Mar 2017
11. Kolin MJ, de la Rosa DM (1979) The bicycle frame. In: Lepley TA (ed) the custom bicycle, p 1
12. Maleque MA, Hossain MS, Dyuti S (2014) Material properties and design aspects of folding bicycle frame. https://www.researchgate.net/publication. Accessed 7 Mar 2017

13. Madyda K, Nixon N (n.d.) Bike history. http://nickandkevin.weebly.com/. Accessed 10 Mar 2017
14. Mancone M (2008) Brief bicycle history. iml.jou.ufl.edu. Accessed 15 Sept 2016
15. Moyo J (2016) Zimbabweans spot green gold in bamboo that spares forests. http://www.reuters.com/article/us-zimbabwe-environment-forests-bamboo-idUSKCN10L1DP. Accessed 7 Mar 2017
16. Mozer D (n.d.) Bicycle history timeline. www.ibike.org. Accessed 15 Sept 2016
17. Palermo E (2014) Who Invented the Bicycle? http://www.livescience.com/. Accessed 10 Mar 2017
18. Reiss-Davis Z (2012) What are the benefits of a bamboo bicycle frame? www.quora.com. Accessed 16 Mar 2017
19. Rodriguez D (2009) Custom bicycles and frame materials. http://www.rodbikes.com. Accessed 22 Sept 2016
20. Salvagione P (n.d.) Science of Cycling: Bicycle Frame Design—Exploratorium. https://www.exploratorium.edu/cycling/frames4.html. Accessed 21 Feb 2017
21. The Engineering Toolbox (n.d.) Modulus of Elasticity or Young's Modulus - and Tensile Modulus for common Materials. http://www.engineeringtoolbox.com/young-modulus-d_417.html. Accessed 21 Mar 2017
22. Walker R (2011) Mass, weight, density or specific gravity of wood. https://www.simetric.co.uk/si_wood.htm. Accessed 10 May 2017
23. Bulawayo24, 3 December 2013, Zimbabwe National Tree Planting Day. https://bulawayo24.com/index-id-news-sc-national-byo-39716-article-Zimbabwe+National+Tree+Planting+Day.html. Accessed 17 July 2018

Design for Manufacture and Assembly of an Automated Dish Washing Machine

Tawanda Mushiri[1(✉)], Rutendo F. Solomon[2], and Charles Mbohwa[3]

[1] Department of Quality and Operations Management, Faculty of Engineering and the Built Environment, University of Johannesburg, Auckland Park, 2006 Johannesburg, South Africa
tawanda.mushiri@gmail.com

[2] Department of Mechanical Engineering, University of Zimbabwe, P.O Box MP167, Mt Pleasant, Harare, Zimbabwe

[3] Faculty of Engineering and the Built Environment, University of Johannesburg, Auckland Park, 2006 Johannesburg, South Africa

Abstract. This paper was a research done at a local university in Zimbabwe. Dishwashing by hand involves scrubbing, rinsing and drying of dishes which is time consuming and laborious when done on a large scale. Each dining hall at the university uses about one thousand plates per meal and the university is understaffed thus the dishes are not thoroughly cleaned. A dishwashing machine is thus a necessity for the university. Researching of literature was done and information on the current designs of dishwashers, their operations, merits and demerits was gathered. A case study on the university was carried out using interviews and questionnaire. The dishwasher for the university was designed and it uses pressurised water jets to clean the plates. The pressurised water jets are produced by a pump installed in the dishwasher. The plates are put into the dishwashing using a rake conveyer system which is automatically controlled. The von Mises stresses of the system were analysed using Solidworks and they were in the desired range. Deflections were also done using Solidworks. Specifications and drawings of the dishwashing machine were developed. Materials of the design were selected concentrating on reduction of cost through minimising the number of parts retaining its function.

Keywords: Design for manufacture and assembly (DFMA), automated · Dish washing machine

1 Introduction

Dish washing machines are devices that are used to clean and dry dishes and kitchen utensils. When using a dish washing machine the dishes are put into the racks and the machine is closed, the dishes are showered by pressurised, hot and soapy water jets until they are clean. This machine can clean any type of kitchen utensil. The cleaning of dishes by hand involves scrubbing, rinsing and drying of the dishes. There is use of detergents in the washing and some of the detergents have long time effects on the hands [1]. The process of hand dish washing takes a lot of time and consumes a lot of energy and water [6]. This has resulted in the design and manufacture of dish washing

R. Dissanayake and P. Mendis (Eds.): ICSBE 2018, LNCE 44, pp. 595–603, 2020.
https://doi.org/10.1007/978-981-13-9749-3_52

machines that can clean dishes using minimum energy, minimum water and taking less time [9]. The existing dish washing machines are expensive and the maintenance costs are high thus there are not common in Zimbabwe.

1.1 Background of the Study

The first dishwashing machine was manufactured in the middle of the 19th century. A patent of the first device was signed in 1850 by Joel Houghton [2]. The machine was crank operated and built from wood. When cranked, water would make its way through the wood plumbing and spray over the dishes. A woman named Josephine Cochrane who was very wealthy improved on that invention and manufactured the first successful dish washing machine [7]. These dishwashers could not dry the dishes thus the dishes were dried by hand. The water for washing was also heated manually using wood thus they were not effective. They also wasted a lot of water.

Dish washing technology continued improving and the recent dish washing machines are far much better than those in the 19th and 20th centuries. The recent dishwashing machines are now more helpful than ever. They have a water efficiency of about 4 to 6 gallons of water per cycle [4]. The electricity usage has reduced by about 15%. These dishwashers produce less sound than the previous ones. Some of them even have soil sensors and innovative dish rack designs. Due to the increase in water and energy efficiency, the recent machine take a lot of time as compared to those of the 20th century. The 20th century machines took 60 to 80 min per cycle but the recent ones take about 120 to 140 min. All the improvements have led to the increase of the cost of dishwashing machines thus they are not common in developing countries.

Dishwashers are a reproducing ground for possibly unsafe organisms, new research has uncovered. Scientists found that the warm, soggy condition in dishwashers is perfect for the development of growths including Exophiala, Rhodotorula, and Candida parapsilosis – yeast-like species that can be hurtful to individuals. On the off chance that breathed in, these parasites can colonize the lungs and cause contaminations that can be hard to treat. The elastic seals in dishwasher entryways are an ideal rearing ground for organisms and that a few sorts, for example, Exophiala.

About 500 plates, 1000 pieces of cutlery and 500 cups are used per meal in a dining at the university, there is no enough manpower for dishwashing thus dishes are not thoroughly cleaned.

The major objectives of the research are:

- To design a dish washing machine for the university that washes 500 plates, 1000 pieces of cutlery and 300 cups per hour.
- To design a machine which requires minimum or zero human assistance.
- To design a machine that can be manufactured by locally available resources.

2 Literature Review

Dish washing involves preparation, filling, washing, rinsing and drying (Fig. 1).

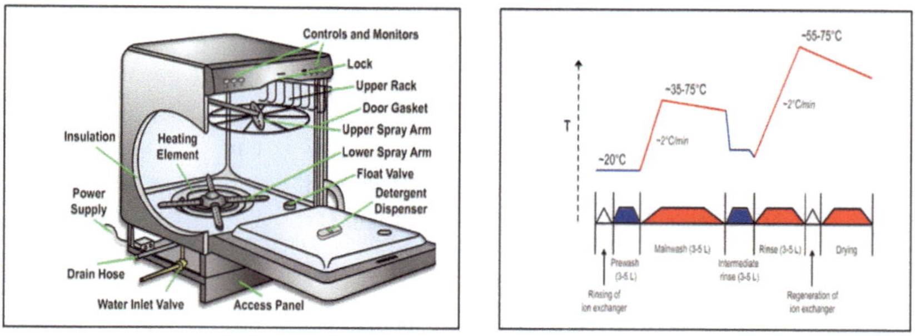

Fig. 1. a. Parts of a dishwasher b. Dishwasher cleaning cycle

Despite the fact that part of human exercises are robotized in the present aggressive world. There is a slack in robotized dish washer. A few machines are as of now planned with the assistance of high speed water jet; henceforth there is a possibility of uncleanliness and not evacuating extreme strains in dishes. Keeping in mind the end goal to defeat the above issues an exceptional machine called—Automatic Dishwasher Machine" with rollers, brush and water fly presented in this work. By consolidating dishwasher with roller component utilizing brush and water stream, it is conceivable to clean the articles viably than alternate machines and with no labor [11].

3 Materials and Methods Used for Design

For mechanical drawings, Solid Works 2015 software was used. SolidWorks (stylized as SOLIDWORKS) is a solid modeling Computer-Aided Design (CAD) and Computer-Aided Engineering (CAE) computer program that was run on Microsoft Windows in this research. It is a solid modeler, and utilizes a parametric feature-based approach to create models and assemblies. The dish washer designed has the following parameters as in Table 1. The selection was done as steps in SolidWorks. The response of a section relies upon the material selected to the part. SimulationXpress must know the flexible properties of the material of the part. You allocate material to the part by picking a material from a material library. Materials in SOLIDWORKS have two arrangements of properties, visual and physical (mechanical). SimulationXpress utilizes the physical properties of the materials characterized in the SOLIDWORKS Material library. Materials can be isotropic, orthotropic, or anisotropic. SimulationXpress bolsters isotropic materials as it were (Table 2).

Table 1. Parameters used in the design

Parameter	Quantity
Water efficiency	10 gallons per cycle
Time per cycle	15 min
Number of plates per cycle	300 plates
Type of plates	Melamine
Energy efficiency	10 joules per circuit
Number of rakes	2
Dimension	$0.7 \times 0.7 \times 1$ m
Plate area	0.0625 m^2
Plate thickness	3 mm

Table 2. Components and quantities to be designed

Component	Quantity
Centrifugal pumps	2
Spray arms	2
Water collector pipes	1
Racks	2
Housing	1

To Assign/Modify a Material to a Part:

1. Click Choose Material (Material tab of the SimulationXpress wizard).
2. In the Material dialog box, expand the class of materials and select a material.
3. Click Apply.
4. Click Close
5. Click Next.

You can also assign a material by right-clicking the part in the SimulationXpress study tree and clicking Apply/Edit Material (Fig. 2).

Approximately 12,000 students need to eat three times a day.

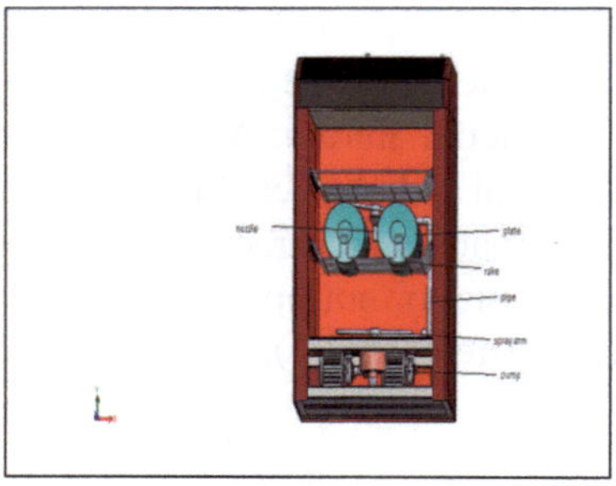

Fig. 2. Designed dish washing machine

4 Simulation and Results

4.1 Water Jet Arrangement

The arrangement of the jet of water for design was shown below in Fig. 3.

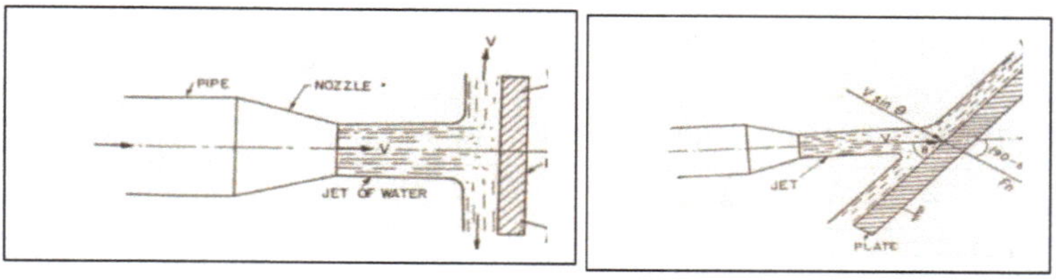

Fig. 3. a. Water jets on horizontal plate b. Inclined plate

$F_x = \rho a V^2 = 1000 \times 3.14 \times 10^{-6} \times 3.98^2 = 0.0497$ N (Horizontal plane). This is the force acting in horizontal plane.

$F_x = \rho A v^2 \sin^2 \theta = 1000 \times 3.14 \times 10^{-6} \times 3.98^2 \times \sin^2 60°$; $F_x = 0.0373$ N for inclined plane.

From the Darcy equations the researchers calculated the losses due to friction, $h = \frac{4flv2}{2dg}$; but Re $= \frac{1030.4*1.141}{1.14*10} = 14035$ and the flow is turbulent, so the Darcy equation must be used. To determine the value of friction factor:

Relative roughness = k/d = 0.000 08*0.04 = 0.002. From the Moody chart in appendix 1, for Re $= 1.4 \times 10^{-4}$ and relative roughness of 0.002.

f = 0.008; $h = \frac{4x0.008x1.65x3.98x3.98}{2x9.81x0.030}$; $h_f = 1.1$ m, $h_i = 1.1 + 0.55 + 1.25 + 1.5$; $h_i = 4.5$ m; Energy = ρghQ = 1000*9.81*4.5*5 $\times 10^{-5}$ = 0.720725 KW. 1 HP pump with a flow rate of 30 litres per minute with a head of 4.5 m and pressure of 0.2 bar was selected (Fig. 4).

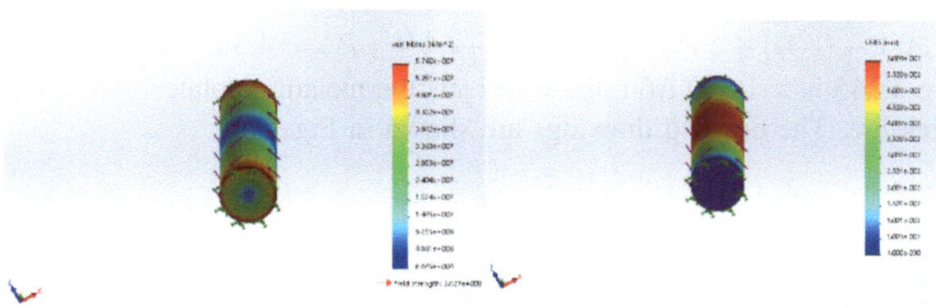

Fig. 4. a. The stress analysis of the shaft b. The displacement analysis of the shaft

4.2 Stress Analysis

$$\sigma_x = \frac{p}{\pi r2} = \frac{100 * 10^{-3}}{4 * 30 * 10^{-3}} = 18.7\text{MPa}; \sigma_y = \frac{p}{2\pi rl}; = \frac{120 * 10^{-3}}{2 * 30 * 10^{-3}} = 33.3\text{MPa}$$

$$\sigma_{1,2} = \frac{\sigma_x + \sigma_y}{2} + \sqrt{\frac{(\sigma_x - \sigma_x)^2}{2} + \tau_{xy}^2};$$

$$= \frac{(16.7 + 33.3) * 10^6}{2} + \sqrt{\frac{(16.7 * 10^6 - 33.3 * 10^6)^2}{2} + 200 * 10^{5^2}} = 48.2\text{MPa}$$

$$\sigma_v = \sqrt{\sigma_x^2 + \sigma_y^2 - \sigma_{L\sigma_x^2} + 3\tau_{xy}^2};$$

$$\sqrt{(16.6 * 10^6)^2 + (33.6 * 10^6)^2 - (16.8 * 10^6 * 33.5 * 10^6) + (3 * (10 * 10^6)^2)}$$

$$= 49.1\text{MPa}; \sigma_d = \frac{S_y}{n}; = \frac{350}{3}$$

$$= 116\text{MPa}.$$

4.3 Stress Analysis on the Melamine Plate

Calculation of slenderness ratio is as follows: $S = \frac{Le}{r}$; But $r = \sqrt{\frac{I}{A}}$; $r = \sqrt{\frac{0.03}{0.0067}}$; $r = 12.33$

$$S = \frac{0.03}{12.33}; S = 0.0045; S_{cr} = \sqrt{\frac{2\pi^2 E}{\sigma_y}}; S_{cr} = \sqrt{\frac{2\pi^2 24 * 10^9}{200 * 10^6}}; S_{cr} = 204.56$$

Calculation of the critical load: Since S = 0.0045 < S_{cr} = 204.6 the melamine plate is a short column thus the following equation is used for the critical load $P_{cr} = \sigma_y A(1 - (\frac{\sigma_y}{4\pi^2 E})\frac{Le}{r}^2$; $P_{cr} = \sigma_y A(1 - (\frac{\sigma_y}{4\pi^2 E})\frac{Le}{r}^2$; $P_{cr} = 4.345\ kN$.

Since 4.345 is > 1.020 from the water jets the melamine plates will fail due to the water pressure. The detailed drawings are shown in Fig. 5.

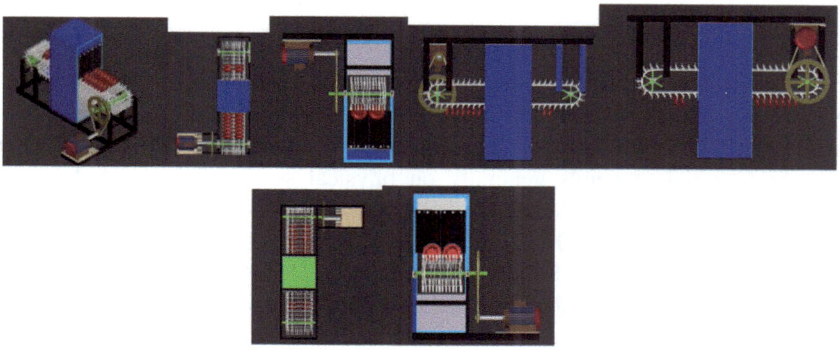

Fig. 5. a. Detailed full drawing b. Top view c. Front view d. Right hand side e. Left hand side f. Bottom side g. Back view

5 Discussion and Recommendations

The number of pumps was reduced from four to two since the dirty water from the dishes will be drained by gravity the water collector then to the sewer system thus no need for the two pumps to pump dirty water from the washer. The following were allocated materials as follows and suggested prices to buy the materials were put across. The dishwasher uses a lot of detergents to disinfect the plates. This can be improved by using pressurised heated water. Heated water will also resulting in the dishes drying faster (Table 3).

Table 3. Part list and materials

Part name	Description	Quantity	Unit price (US$)	Cost
Drive shaft	Carbon steel	1	$2.00	$2.00
Motor		1	$90.00	$90.00
Pump		1	$100.00	$100.00
Bearing	Gray cast Iron SN	1	$11.00	$11.00
Belt	Steel	1	$30.00	$30.00
Pulley	Brass	2	$11.00	$22.00
Spray arms	Copper	2	$5.00	$10.00
Pipes	Copper	30	$1.00	$30.00
Housing	Cast stainless steel	1	$10.00	$10.00
Water collector	Gray cast iron SN	1	$2.00	$2.00
Idler	Gray cast iron SN	2	$2.50	$5.00
Stand	Cast stainless steel	1	$4.50	$4.50
Pipe connectors	Copper	20	$0.65	$13.00
Total				**$329.50**

6 Conclusion

The research was a success, next step is to install this dishwasher at the university. The university community will benefit a lot in coming up with this dishwasher. A dishwasher is a mechanical gadget for cleaning dishware and cutlery. Not at all like manual dishwashing, which depends generally on physical scouring to expel dirtying, the mechanical dishwasher cleans by showering high temp water, ordinarily in the vicinity of 45 and 75 °C (110 and 170 °F), at the dishes, with bring down temperatures utilized for sensitive items. A blend of water and dishwasher cleanser is pumped to at least one turning shower arms, which shoot the dishes with the cleaning blend. Once the wash is done, the water is depleted, more heated water enters the tub by methods for an electromechanical solenoid valve, and the flush cycle starts. After the flush cycle completes and the water is depleted, the dishes are dried utilizing one of a few drying techniques. Ordinarily a flush guide, a synthetic to decrease surface pressure of the water, is utilized to lessen water spots from hard water or different reasons. Plates will be very much clean than manual washing.

Notwithstanding local units, mechanical dishwashers are accessible for use in business foundations, for example, lodgings and eateries, where a substantial number of dishes must be cleaned. Washing is led with temperatures of 65–71 °C (149–160 °F) and sanitation is accomplished by either the utilization of a promoter warmer that will give a 82 °C (180 °F) "last flush" temperature or using a substance sanitizer.

The new age best-in-class machine beat the earlier age unit by a vast edge. It did likewise measure of work however utilized significantly less vitality and water. The advancements on the new designed dishwasher at university are at present available and accessible for manufacture and use. These advancements increment the productivity of the dishroom and take care of numerous upkeep issues, and likely increment the lifetime of the machine while guaranteeing appropriate machine task. The generous cost investment funds and low straightforward payback time influence a sensible business to case to resign even high-proficiency machines early.

Appendix 1: Moody Diagram

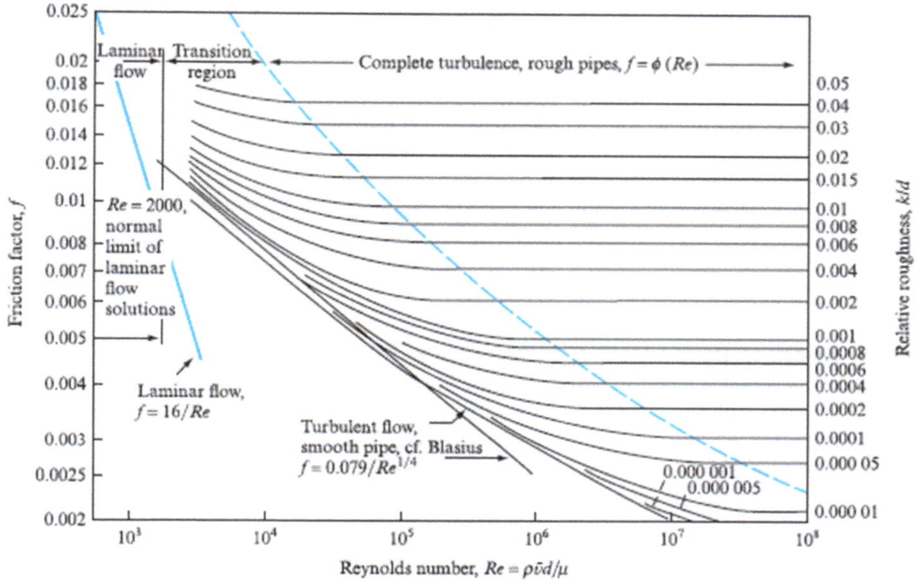

References

1. Barro N, Bello AR, Savadogo A, Ouattara CAT, Iiboudo A, Traoré AS (2006) Hygienic status assessment of dish washing waters, utensils, hands and pieces of money from street food processing sites in Ouagadougou (Burkina Faso). Afr J Biotechnol 5
2. Emmel J, Parott K, Beamish J (2003) Dishwashing and water conservation: an opportunity for environmental education. J Ext 41
3. Flannery JP (1956) Dishwashers. Google Patents
4. Gram M, Schweier P (2016) Dishwasher with a dynamic filling sequence. Google Patents
5. Pruehs H, Schenker G, Altenschoepfer T, Piorr R (1990) Use of hydroxyalkyl polyethylene glycol ethers as surfactants in rinse aids for dishwashing machines. Google Patents
6. Stamminger R (2011) Modelling resource consumption for laundry and dish treatment in individual households for various consumer segments. Energ Effi 4:559–569
7. von Hippel E, Ogawa S, de Jong JP (2011) The age of the consumer-innovator. MIT Sloan Manage Rev 53:27
8. William D, Andrews DC (1974) Technology and the housewife in nineteenth-century America. Women's Stud Interdisc J 2:309–328
9. Wood G, Newborough M (2003) Dynamic energy-consumption indicators for domestic appliances: environment, behaviour and design. Energ Build 35:821–841
10. Venkatesh RB, (2017) Design and fabrication of automatic dishwasher machine. In: International Conference on Emerging trends in Engineering, Science and Sustainable Technology (ICETSST), E-ISSN: 2348 – 8360, pp 26–31

Integration of BIM and Construction Supply Chain Through Supply Chain Management; An Information Flow Model

A. P. Rathnasinghe(✉), M. K. C. S. Wijewickrama, U. Kulatunga, and H. S. Jayasena

Department of Building Economics, University of Moratuwa, Moratuwa 10400, Sri Lanka
akilaprl993@gmail.com

Abstract. Building Information Modelling (BIM) is a revolutionary stride of technology in the orthodox construction and procedures in conventional Architecture Engineering and Construction (AEC) industry which is stressed upon for its delay on embracing new technologies. The strength of BIM has been disvalued in the eyes of many professionals, to think of BIM as just another 'software' which is in reality, a 'process' of attaining an outstanding collaboration among each and every stakeholder throughout a project's life cycle. Hence, the philosophy of Supply Chain Management (SCM) in a construction project can be achieved effectively with the BIM's promise of flexible and transparent interaction among Construction Supply Chain (CSC) contributors. However, the complexities in CSC of a conservative setup has been more composite with the intervention of BIM. Consequently, the BIM project stakeholders have encountered complications on the effective application of BIM on CSC while reaching the envisioned goals of BIM. Thus, the intention of this study is to develop a foreseeable information flow model related to construction supply chain in a BIM aided project. In order to attain the aim, an extensive literature synthesis was piloted to develop a conceptual informational flow model among CSC stakeholders of a BIM project. This contemporary research outlines that the CSC of a BIM project is much interactive and flexible with its collaborative effort of stakeholders when compared to the clashes among professionals in traditional setups. In a practical context, the roles and duties of BIM project stakeholders identified in BIM standards, have slightly differed due to real-world complications in construction industry. Besides that, BIM standards have laid down charismatic arrangements on information flows among BIM project stakeholders, which is in reality, a complex and random setup. Hence, the research outcome has successfully answered the complications by laying down guidelines for any outsider to the BIM field to recognise what would be the contribution of each stakeholder throughout the BIM project's life cycle.

Keywords: Building Information Modelling (BIM) · PAS1192:2 · Construction Supply Chain (CSC) · BIM project stakeholders · Supply chain information flows

R. Dissanayake and P. Mendis (Eds.): ICSBE 2018, LNCE 44, pp. 604–614, 2020.
https://doi.org/10.1007/978-981-13-9749-3_53

1 Introduction

The success of a construction project is usually due to its effective and efficient management of various knowledge areas. The areas identified are the major management knowledge areas in Organizational Management. Within the modern trends in global construction, Building Information Modelling (BIM) helps to substitute the customary procedures and implement innovation to orthodox construction. BIM also acts as an 'Integrator' for all those aspects of management. Moreover, such quality of BIM would be more effective only with a clearly defined information flow among the project stakeholders. The development of a BIM model is based on coherent information feedings by said stakeholders. However, due to the vagueness in Construction Supply Chain (CSC), the defined information sharing, feeding or collection is difficult. This may result in reducing the productivity of BIM application too.

The paper comprehends the preliminary findings of the literature synthesis on how the information management among the BIM CSC stakeholders has been affected with the intervention of BIM standards as with special particulars to BIM Level 2 introduced by the United Kingdom (UK). Accordingly, the outcomes of the literature synthesis is illustrated in a conceptual model so as to express the information flows of BIM environment in a much structured manner.

2 Influence of BIM Towards the CSC

Construction projects establish the critical portion of all disciplines in projects focusing on their budget, multiplicity and cost (Rokooei 2015). Moreover, Rokooei (2015) has highlighted the necessity of managing it, regardless of the financial scale of a construction project. Rokooei (2015) further extended the success criterion of construction project management into information management, where the successfulness could have been assured through aiding BIM.

Accordingly, modern Architecture Engineering and Construction (AEC) trade is multifaceted due to critical issues such as split nature in resources production, complexity in supply processes, constraints to construction and the outsized number of project stakeholders as of latest procurement pathways (Zavadskas et al. 2010). In view of that, the information supply chain concept is being used in investigating the contribution and dependence of stakeholders within project environment.

The concept of Supply Chain (SC) has been defined as "the network of organizations that are involved, through upstream and downstream linkages, in the different processes and activities that produce value in the form of products and services in the hands of the ultimate customer" (Chirstopher 2011, p. 3). Consequently, Supply Chain Management (SCM) envisions the entire supply chain as a process of production without being restricted to a single level, and expects to increase the transparency and logical alignment among the SC actors upon their interconnections which will not be limited on practical or commercial aspects (Cooper and Ellram 1993). Besides, as Mentzer et al. (2001) further clarifies, SCM is a successful supervisory mechanism to manage the various flows of information and products within the project boundaries.

The implementation of BIM in the construction industry is appreciated by Robson et al. (2014) as an expedition through the intelligence of building information up to the 'maturity' from the conventional 2D drawings at level 0, to the exclusively amalgamated 3D modelling in criterions of design, time and cost. However, Goh et al. (2014) underlined that BIM was conceived as a technological solution for problems related to the informational SC processes.

However, the traditional SCM intents at achieving naive and obvious interactions among the various stakeholders in SC, common profits and controlling the aforementioned project uncertainties (London 2009). With the mandating of BIM in construction projects by the most government authorities, the traditional SC of the construction projects was drastically changed due to the intervention of the digital world to orthodox construction and planning techniques. Accordingly, the British government has made a significant contribution towards introducing uniformed and consistent specification series, which is commonly practised in the global context. Therefore, it is worth investigating how the information management among the stakeholders is performed with this new technological interference of BIM as introduced by the British Standards on BIM implementation. This paper presents a literature synthesis aimed to identify a suitable information flow model for information management of a CSC of a construction project.

3 Research Method

This research seeks to address the research problem of "how the information management among BIM project stakeholders is to be performed with the technological intervention of BIM in CSC" through a qualitative approach, as qualitative methods contribute in conducting in-depth investigations on emerging concepts and is more appropriate, when the research has a trivial base of literature. Accordingly, this research is mainly focussed on a thorough investigation of literature sources such as; BIM standards published by The British Standards Institution and other BIM related books, periodicals, and conference proceedings. Subsequently, the outcomes have been illustrated in a conceptual model which is totally based on BIM related literature.

4 Standard Practices of BIM and Their Effect on Traditional Construction Phenomena

As identified by Wickersham (2009), existing BIM-related standards provide a successful initiative to establish a collaborative BIM method. Further, McGraw Hill (2014) surveys suggested that the high implementation of BIM should be effectively managed and controlled by introducing relevant standards depending on the project context.

However, these BIM standards are mainly paper-based standards (e.g. BS1192: 2007 and PAS 1192-1, 2 and 3), and IT related data exchange standards (e.g. COBie and IFC) (Smith 2014). BIM implementation in the UK has improved gradually until the introduction of BIM policy by the UK government's BIM Task Group in May 2011 (HM Government 2012). This transitional position has been called by the scholars as the BIM 'light bulb moment' for UK constructions (NBS 2012). Henceforth, the UK is presently considered to be the most leading and influencing nation on global BIM resourcefulness (Saxon 2013).

4.1 Publicly Available Specification (PAS) 1192-2:2013

This UK based specification is formed to manage the information with BIM which may arise during the supply stages of resources. Accordingly, it provides a guideline to implement a BIM project and standard requisites on the Employer's Information Requirements (EIR) and the BIM Execution Plan (BEP) (Rock 2018). Further, PAS 1192:2 describes the CDE based on its collaborative features to perform as a uniform stage for information distribution in between the task teams of different professional codes (Applecore Designs 2016).

4.2 Significant Changes Due to BIM Standards

Arayici et al. (2012) argued on BIM standards as it has made significant changes in predefined design instances regarding the documentation and management procedures of those design scenarios. Meanwhile, it is considered that those BIM values denote a mutual concern on achieving utmost goals of the time, cost and quality triangle and employer's gratification (Takim et al. 2013). Lee (2008) highlights a different outlook on BIM, on reinventing the standard construction procedures, through a simulated environment which enables greater productivity, corporation and communication. Therefore, Takim et al. (2013) further appraised the implementation of BIM in the AEC industry as a turnover point, which has improved the quality of life among stakeholders.

Even though many authors suggest achieving a greater efficiency through BIM implementation within the AEC industry, some have identified the limitations and issues in this scenario. Arayici et al. (2012) found out that one such limitation of BIM is in the information congregation process where the final design proposal is solely based on the deliberation and user requirements which are poorly collected. As a remedy, Howard and Bjork (2008) highlighted on the importance of developing BIM standards which are compatible with the industry's stakeholder requirements.

5 BIM Common Data Environment (CDE) and Its Information Flows

The main pathway to achieving a well-organized data collection in a BIM-aided project is to have a CDE; a virtual platform managed by project teams for assembling, handling and distributing digital information among themselves upon requirement. The layout of a CDE is defined in the PAS 1192-2 specification which is the regulatory manual for public projects done in BIM Level 2. This guidance is based on the core principles of BS 1192: 2007.

As identified in PAS 1192:2, there are four main phases of a CDE. The first phase of CDE is 'Work in progress (WIP)' where data is currently in fabrication and has not yet been checked and verified for use outside of the authoring team (Architectural, Engineering and Construction Council (AEC) 2018).

Once the WIP data has been checked, and approved by the particular stakeholder who does the uploading those would be moved into the 'Shared' area. In this phase, other stakeholders may have in access and can use that information in the creation and development of their contributions. Under BIM Level 2, while the other stakeholders re-use the preliminary information, the ability to alter the information remains with its creator.

Once an individual WIP has gone over the 'Shared' process, all other stakeholders of the BIM team may have already made their comments where the work has been matched and detected for any clashes with other personnel. Such pieces of work will be suitable for publishing. 'Published' means the specific piece of information is the outcome to a particular client's information requirement. The outcomes of the Published or Issued area would deliver as 2D design format or Portable Document Format (PDF), drawings or data sheets, which would also be kept in the Published Area of the folder structure.

6 BIM Project Stakeholders

It is the common viewpoint that, BIM has established a rigid commitment based environment among the project members. Sebastian (2011), explained that the ideal status in BIM utilization can only be achieved through an effective collaborative environment consisting of multi-disciplinary professionals. Furthermore, Sebastian (2011) argued that this collaborative environment requires clear administrative definitions on key roles rather than modifying them. Moreover, such environments will be successful through the rescheduled contractual relationships and re-engineered integration process.

6.1 Characteristics of BIM Project Stakeholders

Latest surveys revealed the necessity of engaging BIM skilled professionals for a successful BIM implementation (Allen Consulting Group 2010). As mentioned by Olatunji (2011), experts with required professional skills can be gained with certain modifications and training to the conventional project professionals. Accordingly, the training of stakeholders should be compatible to comply with the specialized needs of BIM software.

Arayici et al. (2012), highlighted that BIM stakeholders should possess modern management skills with special concentration to external party support. Henceforward, research studies aim at an effective redevelopment of consultants within the BIM team who support the completion milestones in the model.

As Ashcraft (2008) suggested, when producing a BIM model, undivided commitment and care from the BIM project stakeholders is most important. The main aim of such a collaborative environment should be to develop a comprehensive BIM model with less disputes among the project members. This point of view has been discussed again under the BIM overlay to the Royal Institute of British Architects (RIBA 2012) as expected job profiles and services to be performed by respective stakeholders within a BIM project.

6.2 Roles and Duties of BIM Project Stakeholders

The arrangement of the BIM team and its members' copyrights will be settled by the end of project. However, on other hand, the expected usage of technologies and approaches may be considered at the time of creating the model. Consequently, the administration of BIM projects of such nature will be done by a panel of BIM experts consisting of BIM architects, BIM managers, model developers, and draftsmen (Olatunji 2011). Even though specific professional titles in BIM environment are still untried in the conventional setup, traditional professionals with BIM knowledge are still valuable as substitutes to the BIM job profiles.

As identified by Aranda-Mena et al. (2009), 'project manager' is one of the critical BIM project team members who is distinct with the BIM and technological knowledge. Moreover, studies of Kassem et al. (2014), highlighted the role of 'BIM coordinator' as the main professional who acts on behalf of the client with professional liability and the job profile has been recognized by the BIM protocol as a central role in CDE.

CIC BIM protocol (2013) suggested that the lead designer and BIM coordinator should work together so as to embrace the contractual obligations related to the BIM model work distribution. Moreover, the protocol recommended a 'BIM manager' on regulating the administrative and procedural rules related to BIM model management so as to preserve the collaborative nature and the duties attached to the BIM concept. BIM overlay to RIBA (2012), introduced a number of professional profiles associated with the BIM project team as in Table 1 which are significantly different when compared to the traditional project team stakeholders.

Table 1. Identified duties of BIM project stakeholders

Job title	Key duties within BIM environment
Main designer/Architect	Lead architect is the person who ensures inter-relationships among various designs of task teams used in BIM environment. Moreover, lead designer is the main regulator on the accuracy of the information fed by the task team members to the shared model
Client representative	Briefly, a personality who represents the client at each phase of BIM project lifespan. Accordingly, the role is granted with limited decision making authority on project milestones, and power to assign minor project staff
Technical advisors	Technical advisors facilitate the construction professional staff and the client in understanding the technological aspects behind the BIM concept. Furthermore, the title holder is responsible with the upgrading of intelligent brief on the project plan
Delivery manager	Delivery managers are accommodated with the monitoring and supervising on the supply chain of the project whether the project is going along the intended progress of work. However, in practical context, the client's representative will replace with the role of delivery manager within the project team

Source: Adapted from (BIM overlay to RIBA 2012)

7 Information Flows in Construction Supply Chain Partnerships

A successful case study on communication amongst project and SC stakeholders with the inspiration of BIM was conducted by Papadonikolaki et al. (2017). According to their outcomes, the communication among the SC stakeholders is directed through the pre-defined channels as identified in their particular contract. The architect, the engineers and other identified main stakeholders would be in continuous communication with the contractor, where they are obliged to confer on decision making through their respective employment contracts.

Meanwhile, the supplier, who is obliged with a 'chain contract' (SC framework agreement) with the contractor, may control a part of direct channels with subsidiary suppliers or engineers in behalf of the contractor for the ease of solving problems within an integrated environment. However, such communication is exclusively practiced among the engineers or suppliers correspondingly, where the consultants (architect, structural and mechanical) may not be involved in. In contrast, the supplier and consultant often communicate mostly on fiscal indecisions and build trust.

Further Papadonikolaki et al. (2017) recognised that the contractor's role towards successful CSCM information establishment would be impossible if it has been addressed from the monetary view instead of building trust. Moreover, lead designers hold a vital role in such information flows as s/he holds the main outcome on clarifying agreements on design and the material perspectives with the contractor.

8 Integration of BIM and Construction Supply Chain

As identified by Zavadskas et al. (2010), the AEC industry is having a challenging SC because of the split supply of resources, building challenges and the increasing of project participants. Further, they suggested the deployment of BIM may neutralize such executive and bureaucratic impediments in the AEC SC.

Rezgui and Miles (2010) further commented on the use of inter-organisational IT, such as BIM to perform an effective CSCM. Therefore, they suggested that the integration among BIM may cause an effective management of information flows of the SC, because BIM is a designed data model of building information (Eastman et al. 2011).

However, while the integration among those two concepts are vital to the effectiveness of the AEC industry, the organizational challenges due to this integration are not yet thoroughly reviewed by the scholars. In contemporary studies, BIM implementation and SCM integration are discussed only by means of the theoretical aspects while not significantly involved in the research process (Papadonikolaki et al. 2015). Accordingly, Flows of material, information, and the network of stakeholders are the main components to be considered in SCM.

8.1 BIM Integration with Information Flows and CSC Stakeholders

The definition of BIM by the National BIM Standard Project Committee itself states the BIM's relationship in assisting and controlling the information flow of a building (BuildingSMART 2016). Accordingly, BIM provides with solutions for 3D modelling, interoperability and transparency in project decision making throughout the SC stakeholders. Furthermore, it provides an accessible podium for each project member to improve the harmonization among the stakeholders' decision making and feedback mechanisms while improving the traditional data management (Demian and Walters 2013). However, while the impacts of the BIM models are still doubtful, Van Berlo (2012) recommends workarounds that provide reasonable BIM interoperability. Therefore, the above certifies BIM's effort on managing the information flow, a core feature of SCM.

9 Development of Flow Model for Construction Supply Chain's Information Flows

In developing the conceptual model as in Fig. 1, the concern was on the information management principles offered in PAS 1192:2 specifications, where it clearly presented the relationships between the documents and stakeholders in information management of a BIM project.

According to the readings, during the tender stage (Pre-contract stage), a forthcoming contractor to the project may develop a BIM Execution Plan (BEP) while validating their competence, and aptitude with evidence to fulfil Employer's Information Requirements (EIR). Accordingly, Pre-contract BEP makes the pathway to Project Implementation Plan (PIP), which consists of project goals for collaboration, information modelling and key project milestones and where they fit with the broader project programme.

However, when the most suitable contractor has been selected, s/he is required to develop a more comprehensive BEP while confirming SC's proficiencies. This is indicated in the conceptual model as the 'Post-contract BIM Execution Plan'. Furthermore, this Post-contact BEP arouses the need of 'Master Information Delivery Plan (MIDP)' which signifies when and how the information is being produced. Meanwhile, it represents the responsible party's involvement in producing a piece of information and their probable use of standard protocols and procedures.

In general terms, BEP is the set of plans prepared by the respective suppliers of SC to signify on how the information modelling features of the BIM project are being sustained. Furthermore, the BEP presents the solutions for the problems raised in EIR; the document consisting of standards, procedures and information transmissions to be embraced by the respective contractors (suppliers) as a part of continuing the project.

Furthermore, the final outcome as a MIDP is established on a compacted chain of individual Task Information Delivery Plans (TIDP), consisting of each team's tasks with the layout, date and duties.

Accordingly, the combination of MIDP, BEP and the PIP, will form the BIM model based on the design and specification data provided by the lead architect and other task teams to the CDE. Henceforward, the traditional SC relationships among the contractor, suppliers and BIM task teams commence as illustrated in the bottom part of Fig. 1.

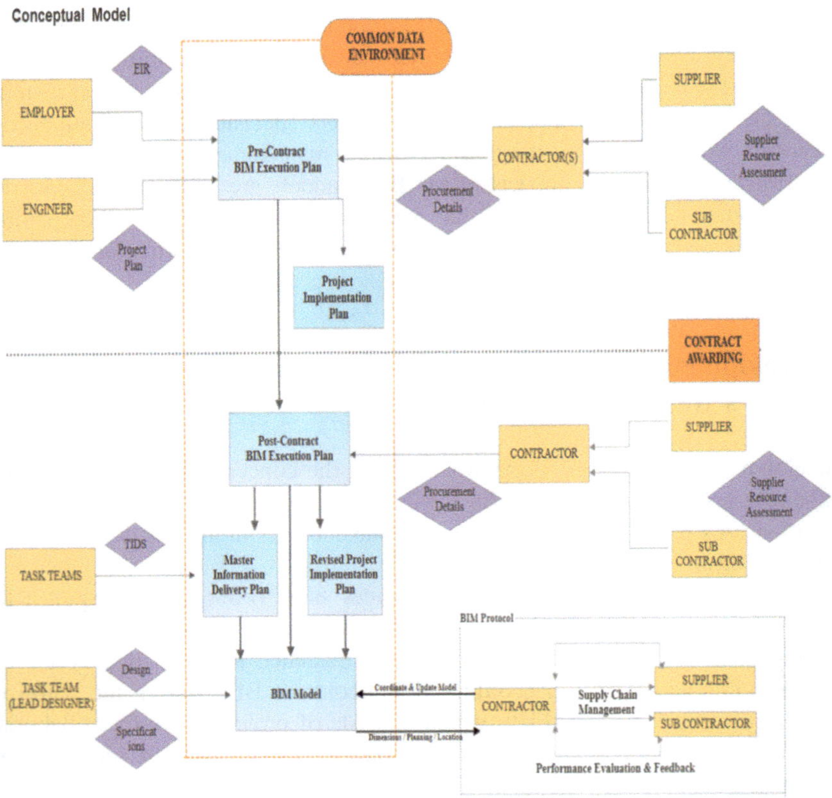

Fig. 1. Conceptual information flow model

10 Conclusion

The comprehensive literature synthesis revealed the complexity of the CSC due to numerous dynamics and the intervention of BIM as an IT solution to the construction industry which is a late embracer on such contexts.

Furthermore, the paper discusses the standard roles and duties of BIM project stakeholders and the composition of BIM project teams as identified by the regulatory institutions. Accordingly, BIM related literature exposed BIM specific job profiles such as; lead architect, client's representative, delivery manager and technical advisors.

Subsequently, as a BIM collaborative environment is technically called as the CDE, the literature proved that there are different flows among each and every stakeholder which makes it more complex to examine. Moreover, the standards have suggested the technical stages of CDE and upon each stage standards have been presented with the flow of information from the working stage to the published stage. Moreover, the works of scholars proved that the effective management of material and information flow can be ascertained through the integration of BIM with SC information flows.

Therefore, upon the outcomes of BIM project stakeholders of CSC and the information flows among them, a structure of the anticipated information flows for CSC of a BIM project was proposed in Fig. 1. Consequently, it was further developed with the assistance of researches in field of SC in BIM projects. As the model was totally based upon the literature outcomes, it was acknowledged as a conceptual model.

In view of that, the model illustrated with main BIM project stakeholders and their anticipated information flows accordance with the informational outputs.

References

Allen Consulting Group (2010) Productivity in the buildings network: assessing the impacts of building information models. Built Environment Innovation and Industry Council. Sydney: Allen Consulting Group

Applecore Designs (2016) The Eight Pillars of BIM Level 2 - Part One - Applecore Designs. https://www.applecoredesigns.co.uk/coreblog/pillars-bim-level2-part-one. Accessed 9 Nov 2018

Architectural, Engineering and Construction Council (UK) (2018) AEC (UK) BIM Standard for Autodesk Revit. https://aecuk.files.word.press.com/2010/06/aecukbimstandardforrevit-v1-0.pdf. Accessed 9 Nov 2018

Aranda-Mena G, Crawford J, Chevez A, Froese T (2009) Building information modelling demystified: does it make business sense to adopt BIM? Int J Manag Proj Bus 2(3):419–434

Arayici Y, Egbu C, Coates P (2012) Building Information Modelling (BIM) implementation and remote construction projects: issues, challenges and critiques. J Inf Technol Constr (ITcon) **17**, pp 72–59. http://www.itcon.org/data/works/att/2012_5.content.03794.pdf. Accessed 9 Nov 2018

Ashcraft H (2008) Building information modelling: a framework for collaboration. Construction Lawyer **28**(3). http://www.hansonbridgett.com/About-Hanson-Bridgett/press-and-news/~/media/Files/News/CL_HowardAshcraft_BIM.pdf. Accessed 9 Nov 2018

BuildingSMART (2016) Frequently asked questions about the national BIM standard-United States™. https://www.nationalbimstandard.org/faqs. Accessed 9 Nov 2018

Christopher M (2011) Logistics and Supply Chain Management. Prentice Hall, London

CIC BIM Protocol (2013) CIC BIM Protocol. Construction Industry Council, London

Cooper M, Ellram L (1993) Characteristics of supply chain management and the implications for purchasing and logistics strategy. Int J Logist Manage 4(2):13–24

Demian P, Walters D (2013) The advantages of information management through building information modelling. Constr Manage Econ 32(12):1153–1165

Eastman C, Teicholz P, Sacks R, Kathleen L (2008) BIM Handbook: A guide to Building Information Modeling for owners, managers, designers, engineers and contractors. Constr Econ Build 12(3):101–102

Goh K, Goh H, Toh S, Peniel Ang S (2014) Enhancing communication in construction industry through BIM. In: Proceedings of the 11th International Conference of Innovation and Management, pp 23–35. http://eprints.uthm.edu.my/6696/1/Enhancing_Communication_in_Construction_Industry_through_BIM.pdf. Accessed 9 Nov 2018

HM Government (2012) Building Information Modelling, Industrial Strategy – Government and Industry in partnership. London: Government Report

Howard R, Bjork B (2008) Building information modelling – Experts' views on standardisation and industry deployment. Adv Eng Inf 22(2):271–280

Kassem M, Iqbal N, Kelly G, Lockley S, Dawood N (2014) Building information modelling: protocols for collaborative design processes. J Inf Technol Constr (ITcon) **19**, pp 126–149. http://www.itcon.org/data/works/att/2014_7.content.00672.pdf. Accessed 9 Nov 2018

Lee C (2008) BIM: changing the AEC industry. In: Proceedings of PMI Global Congress 2008. Project Management Institute, Denver, Colorado, USA

London K (2008) Industrial organization object-oriented project model of the facade supply chain cluster. Construction Supply Chain Management Handbook, pp 13-1–13-46

McGraw-Hill Construction (2010) The business value of BIM in Europe. Smart Market Report. Bedford: McGraw-Hill Construction. Available at: http://images.autodesk.com/adsk/files/business_value_of_bim_in_europe_smr_final.pdf. Accessed 9 Nov 2018

Mentzer J, De Witt W, Keebler J, Min S, NIx N, Smith C, Zacharia Z (2001) Defining supply chain management. J Bus Logist **22**(2). https://www.biblioteca.fundacionicbc.edu.ar/images/e/e4/Conexion_y_logistica_2.pdf. Accessed 9 Nov 2018

NBS (2012) BIM Legal Roundtable Discussion. NBS Website. http://www.thenbs.com/roundtable/legal/bimLegalRoundtable_2012.asp. Accessed 9 Nov 2018

Olatunji O (2011) Modelling the costs of corporate implementation of building information modelling. J Financ Manage Prop Constr 16(3):211–231

Papadonikolaki E, Verbraeck A, Wamelink H (2017) Formal and informal relations within BIM-enabled supply chain partnerships. Constr Manage Econ 35(8-9):531–552. https://doi.org/10.1080/01446193.2017.1311020

Papadonikolaki E, Vrijhoef R, Wamelink H (2015) Supply chain integration with BIM: a graph-based model. Struct Surv 33(3):257–277

Rezgui Y, Miles J (2010) Exploring the potential of SME alliances in the construction sector. J Constr Eng Manage 136(5):558–567

Robson A, Boyd D, Thurairajah N (2014) UK Construction Supply Chain Attitudes to BIM. In: 50th ASC Annual International Conference Proceedings. Birmingham: Birmingham City University. http://ascpro.ascweb.org/chair/paper/CPRT224002014.pdf. Accessed 9 Nov 2018

Rock S (2018) Construction: The eight pillars of Level 2 BIM. Herbert Smith Freehills - Real estate development notes. https://hsfnotes.com/realestatedevelopment/2016/05/04/the-eight-pillars-of-level-2-bim/. Accessed 9 Nov 2018

Rokooei S (2015) Building information modeling in project management: necessities, challenges and outcomes. Proc Soc Behav Sci 210:87–95

Royal Institute of British Architects [RIBA] (2012) BIM Overlay to the RIBA Outline. London: Author

Saxon R (2013) Growth through BIM. Construction Industry Council, London

Sebastian R (2011) Changing roles of the clients, architects and contractors through BIM. Eng Constr Archit Manage 18(2):176–187

Smith P (2014) BIM implementation – global strategies. Proc Eng 85:482–492

Takim R, Harris M, Nawawi A (2013) Building Information Modelling (BIM): a new paradigm for quality of life within architectural, engineering and construction (AEC) industry. In: Proceedings of AMER International Conference on Quality of Life. Langkawi, Malaysia: Universiti Teknologi MARA (UiTM), pp 23–32

Van Berlo L (2012) Collaborative engineering with IFC: new insights and technology. In: Proceedings of 9th European Conference on Product and Process Modelling. Iceland: Blackwell Publishers

Wickersham J (2009) Legal and Business Implications of Building Information Modelling (BIM) and Integrated Project Delivery (IPD). Rocket Publishing Ltd, USA

Zavadskas E., Turskis Z, Tamošaitienė J (2010) J Civil Eng Manage **16**(1):33–46. https://journals.vgtu.lt/index.php/JCEM/article/view/5898

Removal of Heavy Metals from Industrial Wastewater Through Minerals

U. A. S. L. Muthukalum(✉), C. A. Gunathilake, and C. S. Kalpage

University of Peradeniya, Peradeniya, Sri Lanka
muthukaluml@gmail.com

Abstract. Heavy metals are toxic to the health and environment and causes harmful consequences to human beings. It is essential to take effective measures to remove metals from contaminated water. In this research, naturally available brick materials with different compositions collected from Kandy (type A), Gampaha (type B), and Embilipitiya (type C) areas were used to remove Cu, Cr and Pb ions. These brick materials are cost effective and are a readily available alternative to conventional heavy metal removal. The characterization of brick material is performed using X-Ray Fluorescence (XRF) and Nitrogen Adsorption – Desorption analyser. Calculated surface area and total pore volume and pore width of type A, B, and C are lie respectively, in the range of 128–154 m^2/g, 0.24–0.45 cm^3/g, and 5.6–16.7 nm. The main objective of this research is to investigate how the characteristics of adsorbents influences the adsorption process and to identify the best model to describe the kinetic and equilibrium adsorption to purify metal contaminated water. Results indicate that Pseudo – first – order kinetics model properly described the adsorption of Cu^{2+} to the brick type C, which has maximum adsorption capacity of 497 mg g^{-1}. The adsorption process of Pb^{2+} to the brick type A and Cu^{2+} and Pb^{2+} to brick type B and C were well-fitted with Pseudo – second – order kinetics model. In equilibrium studies, the Langmuir isotherm showed a better fitness in adsorption of Cu^{2+} into brick type A and C, Pb^{2+} into brick type B, whereas the Freundlich isotherm well represented the adsorption characteristics of Cu^{2+} into brick type A, Pb^{2+} into brick type B and Cr^{6+} into all brick types. The comparison of results indicated that the use of brick types A, B, C can be used as a potential nan sorbent to remove heavy metals from industrial waste water.

Keywords: Heavy metals · Minerals · Adsorption · Specific surface area · Ion concentration

1 Introduction

In the current world, heavy metals are a major source of pollution in the aquatic and soil environments. Due to certain Industrial activities, a lot of companies generate waste water containing different metals; mainly Zn, Cr, Cu, Ni and Cd. According to previously conducted studies, it has been revealed that high levels of particular heavy metals in streams and soil can be identified due to the activities such as mining societies' effluents, agricultural activities, discharge from Industrial parks and sewage work. The conventional way of removing heavy metal is usually performed by the

R. Dissanayake and P. Mendis (Eds.): ICSBE 2018, LNCE 44, pp. 615–632, 2020.
https://doi.org/10.1007/978-981-13-9749-3_54

addition of a base to precipitate hydroxides. But the expected dissolved concentrations of heavy metals cannot be achieved due to the complexation with inorganic and organic ligands of the water (Danielsson 2013).

Considering the adsorption capacity of minerals containing oxide, they have the capability of removing many minerals over a wider pH range and lowering dissolved levels of heavy metals. As an effective method of removing toxic heavy metals, adsorption is an important physicochemical process that occurs at the solid liquid interface (Garcia Sanchez 1999). Natural minerals have been investigated as adsorbents in many researches for the removal of various toxic and hazardous pollutants. The most important parameters that control heavy metal adsorption and their distribution have been identified as mineral type, metal speciation, metal concentration, soil pH, solid: solution mass ratio and contact time (Pare 2012).

In this research, different types of bricks which are made up of soil compositions from 3 locations in the country are tested for the ability to remove heavy metals through adsorption. The adsorption of selected heavy metals such as Pb, Cr, Cu etc. are investigated for kinetic adsorption and equilibrium adsorption. According to the variation of the factors in the reaction medium and the characteristics of the adsorbent, the adsorption process is observed and the best model which fits the observed data is used to obtain the adsorption capacity for each heavy metal.

2 Materials and Methods

2.1 Materials

2.1.1 Adsorbent

Three different kinds of bricks from different locations in the country (Kandy, Gampaha, Embilipitiya) were selected as shown in Table 1 and they were ground into powder form.

Table 1. Brick materials obtained from different places

Brick type	Area
A	Kandy
B	Gampaha
C	Embilipitiya

2.1.2 Adsorbate

Metal ions, subject to study are Pb^{2+}, Cr^{6+} and Cu^{2+}. Salt solutions of $Pb(NO_2)_2$, $CuSO_4$ and K_2CrO_4 were used as sources of heavy metal ions. Dilute HNO_3 and $NaHCO_3$ were used for pH adjustment.

2.2 Methods

To determine the ion concentration in each sample, an atomic adsorption spectrophotometer is used. In AAS, free atoms are generated by the sample (to be analysed) in the atomizer through a flame, and a light beam is directed to the flame. The amount of light absorbed by the atomized elements is measured eventually and the unknown concentration is determined through a calibration curve which is obtained using known concentrations of solutions.

Elementary material composition is investigated using X-Ray Fluorescence (XRF) method.

The specific surface area is determined using Nitrogen Adsorption Desorption Analyzer. The specific surface area and the pore size distribution of a sample is determined by the physical adsorption of a gas on the surface of the solid and by calculating the amount of adsorbate gas corresponding to a monomolecular layer on the surface. Initially, the mineral sample is placed in the degassing unit to be cleaned and removed of impurities. The sample with the actual weight is then moved to the surface area analyser. Liquid nitrogen is used to make the inert gas adsorbed onto the mineral surface. The data processing unit observes the sample and the isotherm is obtained with the parameters under BET summary (Kruk *et al.* 1997).

3 Adsorption Studies and Isotherms

3.1 Kinetic Studies

3.1.1 Kinetic Adsorption Experiments

Kinetic adsorption studies were carried out using the initial metal ion concentration for all 3 solutions (Pb^{2+}, Cu^{2+}, Cr^{6+}) of 100 ppm. 0.05 g of brick material samples (adsorbent) were prepared. Initially, for one brick material sample selected and one ion solution, 50 ml of ion solution was taken into a beaker and the adsorbent mass was placed while stirring (150 rpm, 25 °C, pH = 5). 0.2 ml of solution was withdrawn before mixing the adsorbent and ion solution (t = 0 min) at pre-determined intervals (t = 10 to 120 min) (Gunathilake *et al.* 2015). The amount of adsorbed ions per unit mass of the adsorbent at a time t (q_t) was calculated by the following equation.

$$q_t = \frac{(C_0 - C_t)V}{m} \tag{1}$$

Where C_0 (ppm) is the initial concentration of metal ions, V (mL) is the volume of metal ion solution and m (g) is the adsorbent mass. C_t (ppm) is the metal ion concentration at a time t. Metal ion concentration was analysed by atomic adsorption spectroscopy.

3.1.2 Analysis of Adsorption Kinetics Data

Two models were used to analyse the adsorption kinetics data which is mentioned below. The first order adsorption kinetics model can be expressed by the following equation.

$$\frac{dQ}{dt} = K_1(q_e - q_t) \tag{2}$$

Where q_e and q_t are the amount of adsorbed metal ions at the equilibrium and time t, respectively, K_1 is the rate constant of the first order adsorption reaction. Integration of above equation gives the following linear model.

$$\log(q_e - q_t) = \frac{K_1 t}{2.303} + \log q_e \tag{3}$$

The pseudo second order kinetic equation is as following.

$$\frac{dQ}{dt} = K_2(q_e - q_t)^2 \tag{4}$$

where K_2 is the pseudo second order rate constant of adsorption (g mg^{-1} min^{-1}). By integrating this equation, following linear model can be obtained.

$$\frac{t}{q_t} = \frac{1}{K_2 q_e^2} + \frac{t}{q_e} \tag{5}$$

3.2 Equilibrium Studies

All the adsorption studies are carried out in 200 ml flasks by subjecting a given 50 mg dose of adsorbent to a period of stirring with 50 ml of solution on a magnetic stir.

To obtain the adsorption isotherm, solutions with metal concentrations ranging between 50–300 ppm are used and stirred for 6 h as it was optimized in kinetic studies, under 25 °C, 5 pH and 150 rpm.

The adsorption isotherms under Langmuir model and Freundlich model are developed and the most suitable model for each adsorbent is evaluated by comparing the coefficient of determination and the model constants are also calculated

Langmuir equation,

$$Q_e = \frac{K Q_m C_e}{1 + K C_e} \tag{6}$$

In linear form,

$$\frac{1}{Q_e} = \frac{1}{Q_m} + \frac{1}{K Q_m} \frac{1}{C_e} \tag{7}$$

Where,

$$q_e = (C_0 - C_e) \times \frac{V}{m} \tag{8}$$

Freundlich equation,

$$Q_e = KC_e^{\frac{1}{n}} \tag{9}$$

The linear model is expressed as,

$$\log Q_e = \log K + (1/n) \times \log C_e \tag{10}$$

In these models, the used notations are, Q_e - Equilibrium concentration of adsorbate in the solid phase in mg/g, Ce - Equilibrium concentration of adsorbate in the liquid phase in mg/l, Co - Initial concentration of adsorbate in the liquid phase in mg/l, m - Initial mass of the adsorbent in g, V - Volume of the liquid phase in ml, K - Langmuir Constant, Q_m - Maximum adsorption capacity and 1/n - Adsorption Intensity.

4 Result and Discussion

4.1 Characterization of Raw Brick Samples

The composition of each brick material was investigated using X-Ray Fluorescence (XRF) and the results are presented in Figs. 1, 2 and 3. It can be seen that the brick samples A, B and C contains approximately equal amounts of Fe, but the content of Ti, Pb and Cu in brick sample A is much less than that of brick sample B. The XRF pattern for brick sample A reveals that it contains 89.6% Fe, 2.3% Ba, 1.9% Mn and 1.4% Ca whereas brick sample B contains 87.4% Fe, 4.8% Cu and 4.1% Ti.

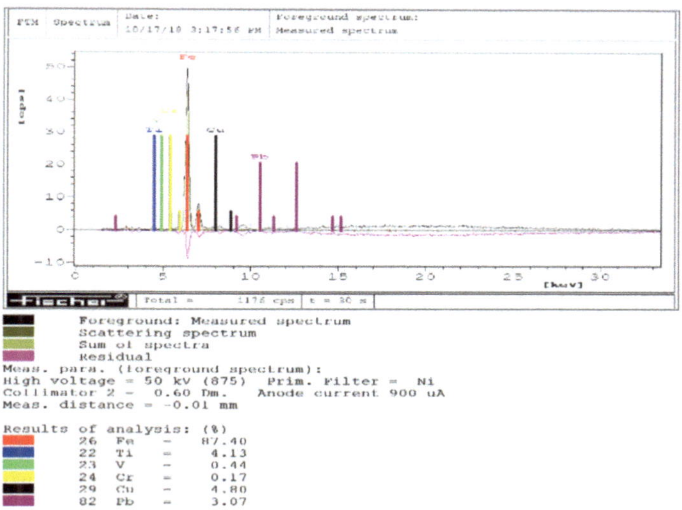

Fig. 1. XRF pattern for brick sample A

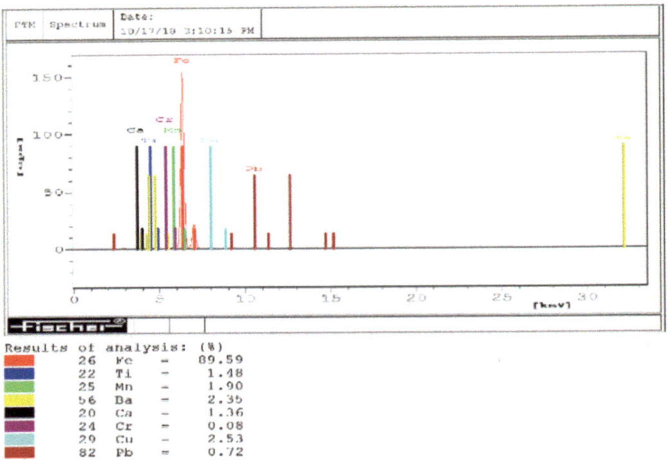

Fig. 2. XRF pattern for brick sample B

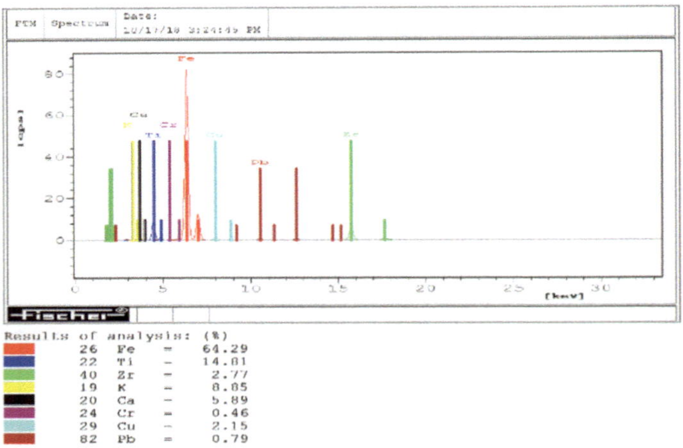

Fig. 3. XRF pattern for brick sample C

4.2 Calculation of Surface Properties

The Brunauer–Emmett–Teller (BET) specific surface area (S_{BET}) of each sample was calculated by following its N_2 adsorption isotherm in the relative pressure range of 0.05 to 0.20. The single-point pore volume (V_{sp}) was determined based on the amount N_2 gas adsorbed at a relative pressure of ~0.98. The pore size distributions (PSD) were calculated using adsorption branches of N_2 adsorption/desorption isotherms and applying the improved Kruk–Jaroniec–Sayari (KJS) method developed for cylindrical pores. The total pore volume (V_t) was calculated by integrating the entire PSD curve and the volume of micropores (V_{mi}) was calculated by integrating the PSD curve up to 2 nm. The mesopore volume (V_{meso}) was determined by integrating the PSD curve from 2 nm to the end of the PSD curve. In addition to the PSD curves that display the overall distribution of pores, the pore widths (W_{max}) corresponding to the maxima of these distributions are also tabulated to show the characteristic pore sizes of the samples studied.

Figures 4 shows comparisons of N_2 adsorption/desorption isotherms for the type A brick, type B brick and type C brick. All materials exhibit type IV nitrogen adsorption isotherms and H1 type hysteresis loop with capillary condensation–evaporation steps beginning at a relative pressure of around 0.70. Insets in Fig. 4 shows the pore size distribution (PSD) curves corresponding to the aforementioned adsorption isotherms obtained by the KJS method.

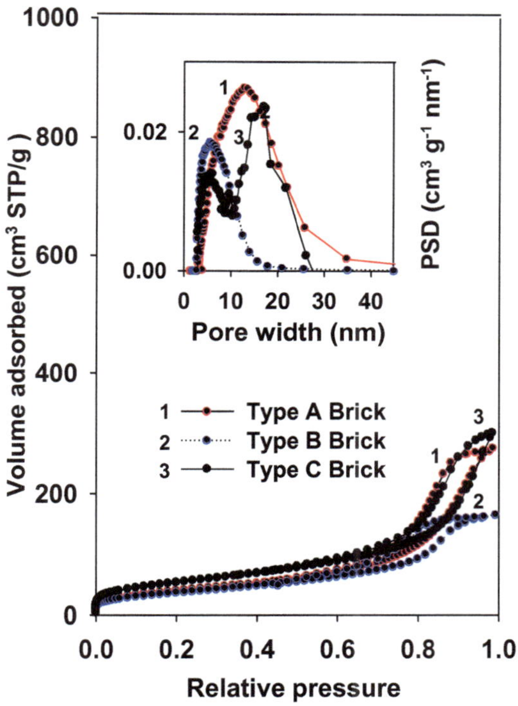

Fig. 4. N_2 adsorption isotherms and their corresponding PSD curves (insets) for the type A brick and type B brick.

Surface properties calculated using N_2 adsorption isotherms reveals that the type C brick sample has better surface properties, as evidenced by higher surface area, total pore volume (V_t), and pore width (W_{max}), when compared to the type B brick sample (Table 2). The notations used are S_{BET}—specific surface area, V_{sp}—single point pore volume, V_t—total pore volume, V_{mi}—volume of micropores, W_{max}—pore width.

Table 2. Adsorption properties obtained for the A, B and C brick samples studied.

Samples	S_{BET} (m^2/g)	V_{sp} (cc/g)	V_t (cc/g)	V_{mi} (cc/g)	W_{max} (nm)
Type A brick	146	0.39	0.42	0.03	12.9
Type B brick	128	0.22	0.24	0.02	5.6
Type C brick	154	0.42	0.45	0.03	16.7

4.2.1 Kinetic Studies of Metal Ion Adsorption

Kinetic studies of metal ion adsorption process revealed the rate of metal ion uptake by the brick samples. Brick sample – A, brick sample – B and brick sample – C were considered and each cation solution (Pb^{2+}, Cu^{2+}, Cr^{6+}) was tested for the time dependence of adsorption, as shown in Figs. 5, 6 and 7.

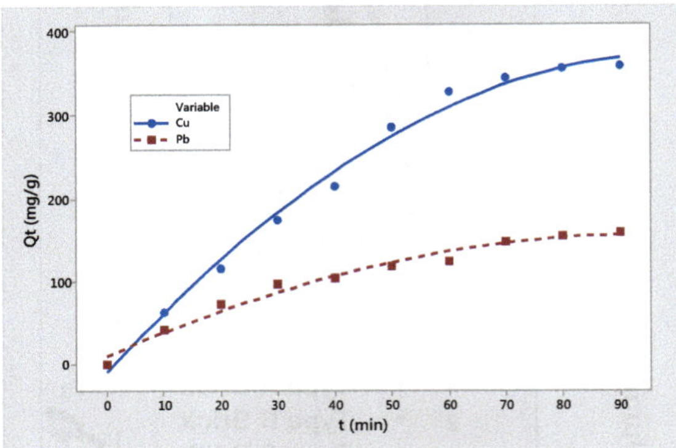

Fig. 5. Experimental adsorption kinetics of metal ions on to brick type A

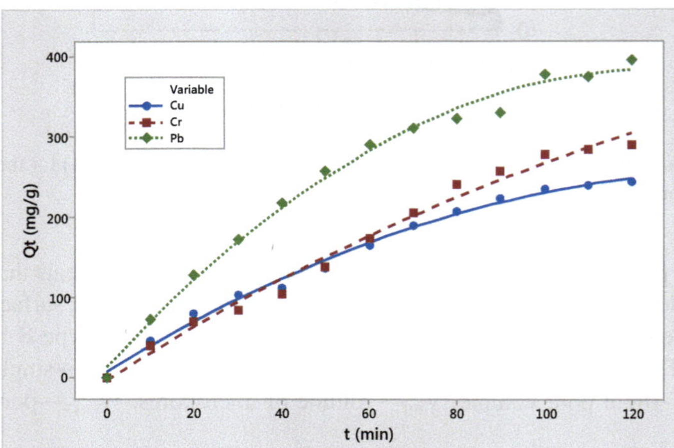

Fig. 6. Experimental adsorption kinetics of metal ions on to brick type B

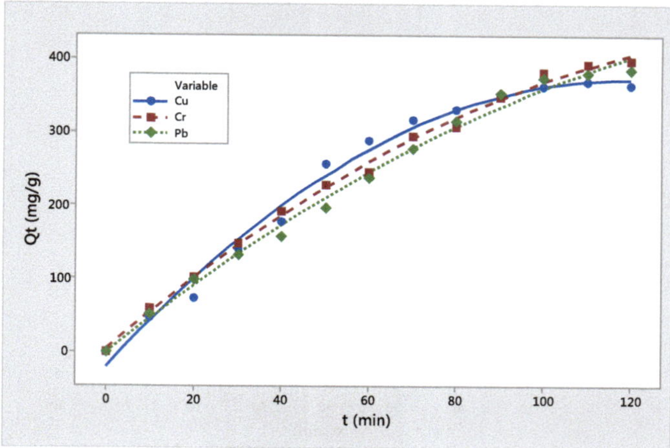

Fig. 7. Experimental adsorption kinetics of metal ions on to brick type C

The linear form of Pseudo – first – order and Pseudo – second – order kinetic equations were applied to obtain the rate of ion adsorption. These forms of linear equations were used to model the adsorption process and the validity of each equation was checked using the correlation coefficient (R^2) for each brick type in each ion solution. Cu^{2+} ion uptake for brick type A fit well with Pseudo – first – order kinetics model as shown in Fig. 8 with 95.8% data fitting the model. In the case of Pb^{2+} ion uptake, brick type A was more consistent with Pseudo – second – order kinetic model which represented 96.4% data in the plot as shown in the Fig. 9. But Cr^{6+} ion concentration was observed in the solution as gradually increasing with time passing, and therefore it could not fit with any kinetic model. The uptake of Pb^{2+} into brick type B was investigated and experimental data could be fitted with Pseudo – second – order kinetics model which had a correlation coefficient of 92.4%. Cu^{2+} ion adsorption into brick type B was properly fitted with Pseudo – second – order kinetics with 98.9% data represented by the plot as shown in Fig. 10. In the case of adsorbing into brick type B, Cr^{6+} shows a better fitness of 96.3% with Pseudo – second – order kinetics model with 352 mg/g adsorption capacity. Cu^{2+} was fit well in adsorbing into brick type C with Pseudo – first – order kinetics model having 497 mg/g, whereas Cr^{6+} and Pb^{2+} showed a best fit with Pseudo – second – order kinetic model as shown in Figs. 7, 8 and 9 with adsorption capacities 191.8 mg/g 240.6 mg/g (Figs. 11, 12, 13, 14 and 15).

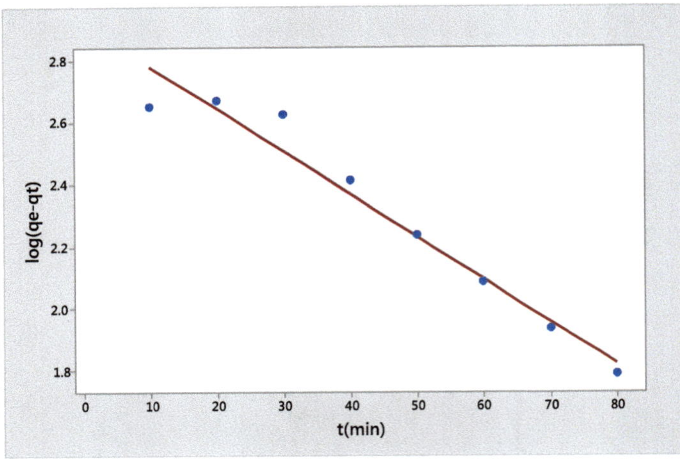

Fig. 8. Linear form of adsorption kinetics equation for Pseudo - first - order kinetic model for Cu^{2+} adsorption on brick type A

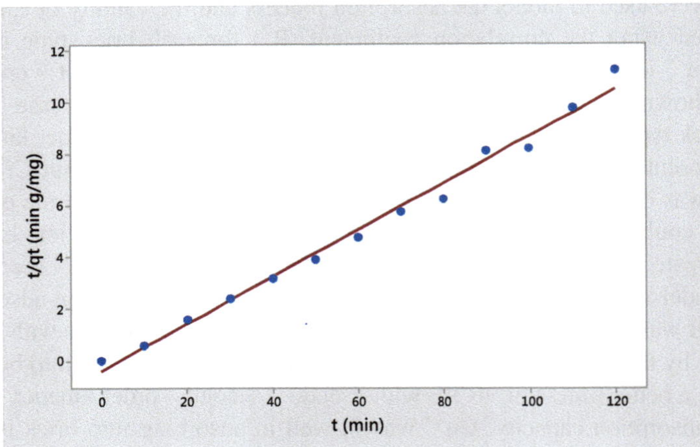

Fig. 9. Linear form of adsorption kinetics equation for Pseudo - second – order kinetic model for Pb^{2+} adsorption on brick type A

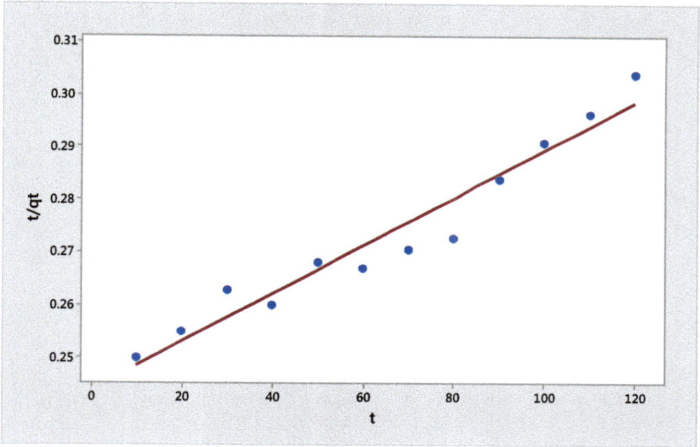

Fig. 10. Linear form of adsorption kinetics equation for Pseudo - second - order kinetic model for Cu^{2+} adsorption on brick type B

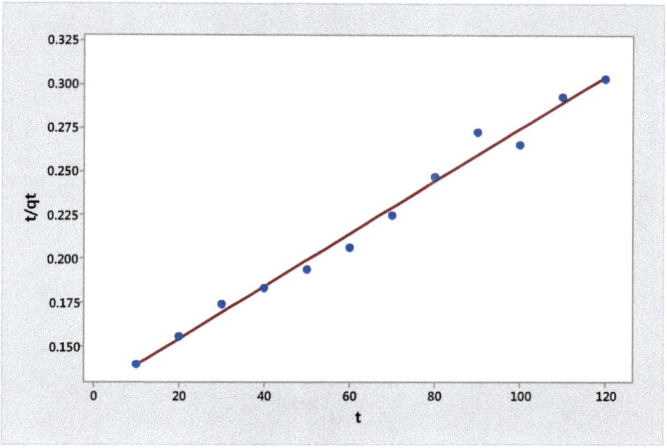

Fig. 11. Linear form of adsorption kinetics equation for Pseudo - second - order kinetic model for Cr^{6+} adsorption on brick type B

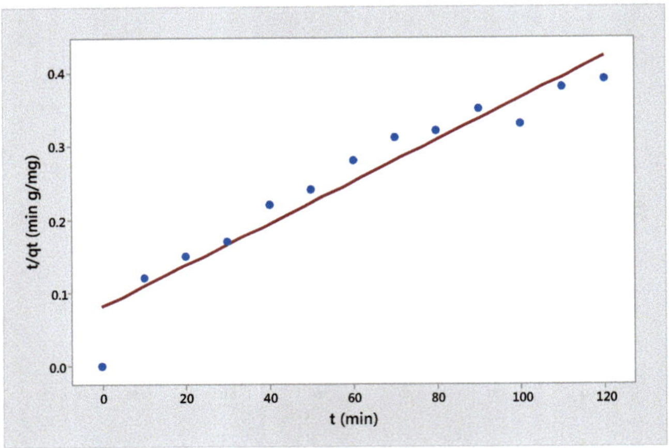

Fig. 12. Linear form of adsorption kinetics equation for Pseudo - second - order kinetic model for Cu^{2+} adsorption on brick type B

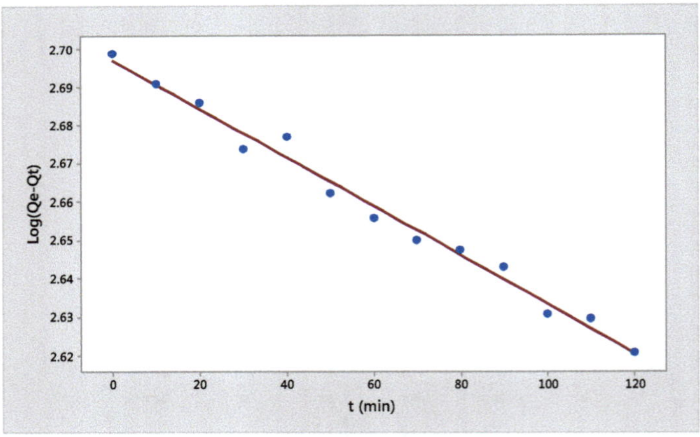

Fig. 13. Linear form of adsorption kinetics equation for Pseudo - first - order kinetic model for Cu^{2+} adsorption on brick type C

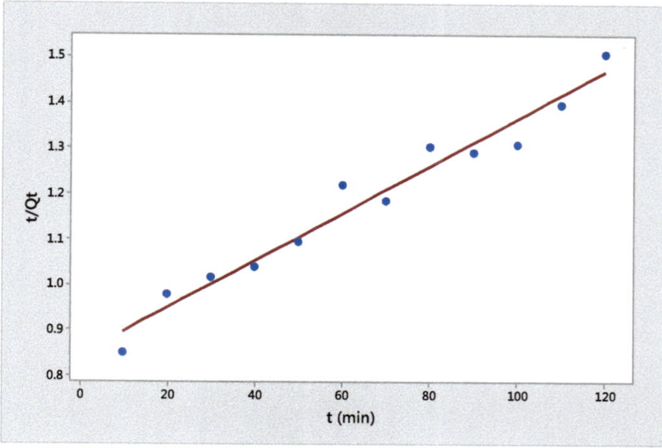

Fig. 14. Linear form of adsorption kinetics equation for Pseudo - second - order kinetic model for Cr^{6+} adsorption on brick type C

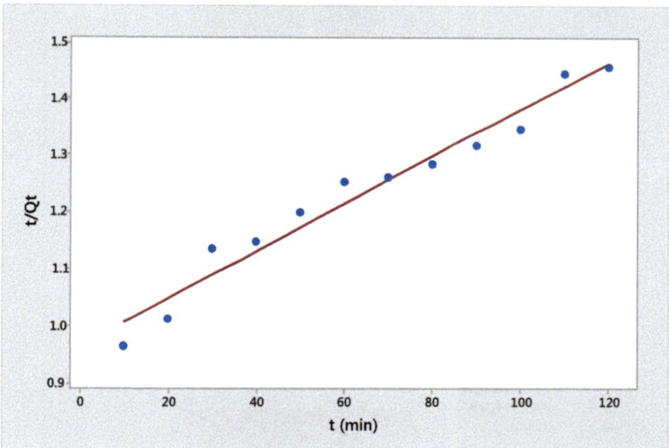

Fig. 15. Linear form of adsorption kinetics equation for Pseudo - second - order kinetic model for Pb^{2+} adsorption on brick type C

Figures 16, 17 and 18 shows the possibility of up-taken of each metal ion into each brick type.

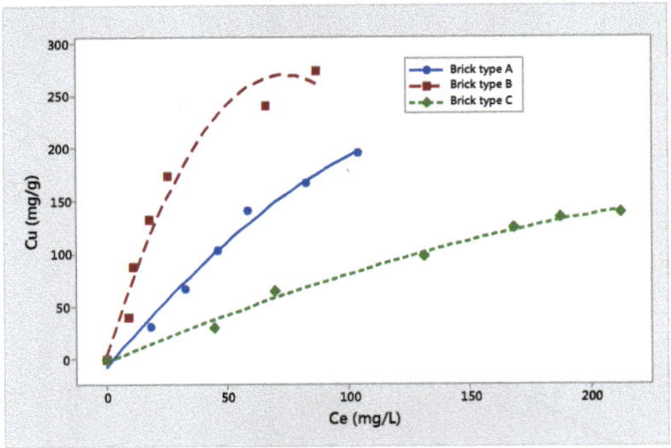

Fig. 16. Cu^{2+} uptake into brick materials

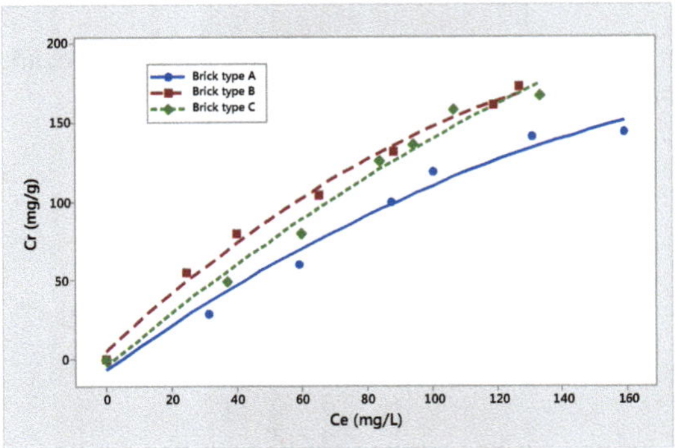

Fig. 17. Cr^{6+} uptake into brick materials

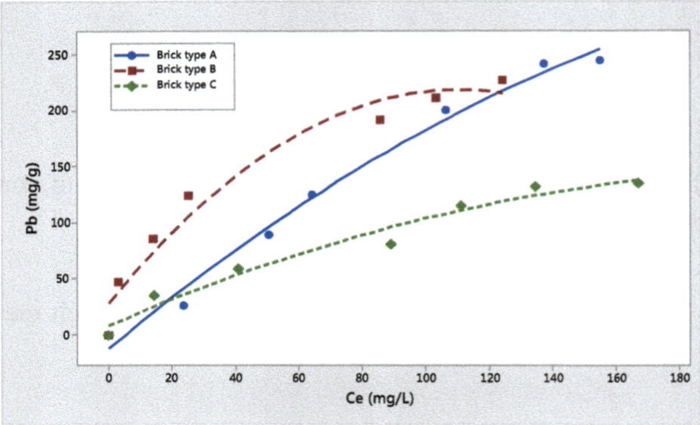

Fig. 18. Pb^{2+} uptake into brick materials

According to the Figs. 16, 17 and 18, brick type B shows a higher possibility of uptaking Cu^{2+} while brick type A shows ascending adsorption capacity as it leads to predict a higher possibility of adsorption in higher Cu^{2+} concentrations. Each brick type shows almost similar adsorption capacities of up-take of Cr^{6+} in lower concentration whereas brick type A shows increasing adsorption capacity in higher concentration. Brick type A shows a significant adsorption of Cr^{6+} compared to other brick types.

The considered ion uptakes into each brick type is listed in the table. According to the results, the amount of Cu^{2+} uptake to brick type A (106.6 mg/g) is significantly higher compared to the amount of Cu^{2+} uptake to brick type B (11.04 mg/g). Pb^{2+} uptake value obtained for adsorbing to brick type A (753 mg/g) is considerably higher compared to brick type B (25.08 mg/g) and it is relatively competitive with other values as shown in Table 3. Brick type C shows relatively higher adsorption possibility compared to brick type B but not as high as brick type A.

Table 3. Kinetic parameters obtained for A, B and C brick samples

Brick type	Metal ion	K_1 (/min)	q_e (mg/g)	R^2
A	Cu^{2+}	0.036	106.6	95.8%
C	Cu^{2+}	0.004	497.0	97.2%
Brick type	Metal ion	K_2 (g/ mg min)	q_e (mg/ g)	R^2
A	Pb^{2+}	0.001	753	94.5%
B	Cu^{2+}	0.022	11.0	92.4%
B	Pb^{2+}	0.004	25.1	98.9%
C	Cr^{2+}	0.032	191.8	97.8%
C	Pb^{2+}	0.008	240.6	99.6%

4.2.2 Equilibrium Studies of Metal Ion Adsorption

Equilibrium studies were investigated based on two models which are Langmuir isotherm and Freundlich isotherm, for the uptake of three metal ions; Cu^{2+}, Cr^{6+} and Pb^{2+}, for each brick type. The fitness of each model to the corresponding experiment was observed and plotted as shown in Figs. 19, 20 and 21.

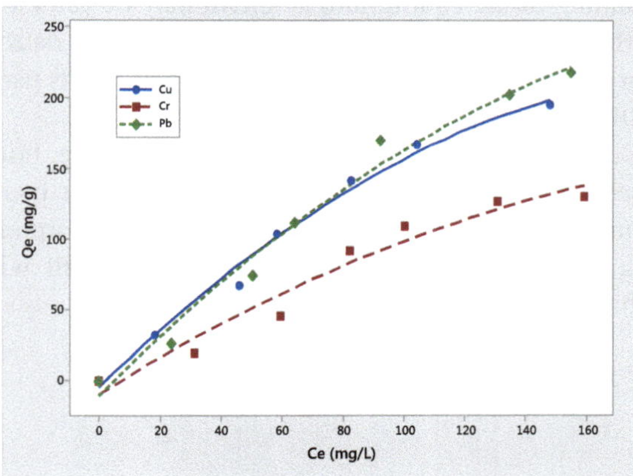

Fig. 19. Equilibrium studies of different metal ions into brick material A

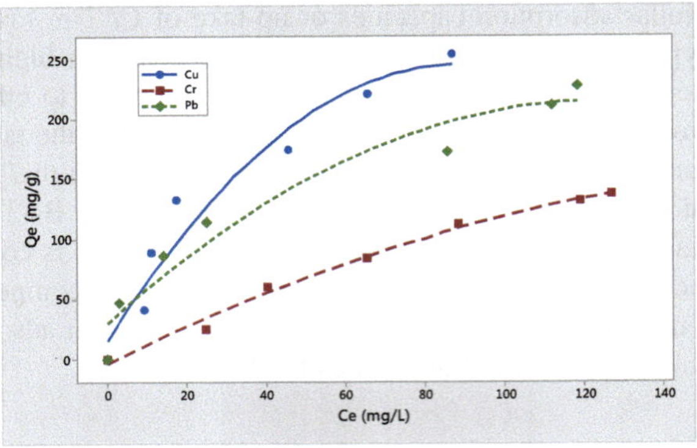

Fig. 20. Equilibrium studies of different metal ions into brick material B

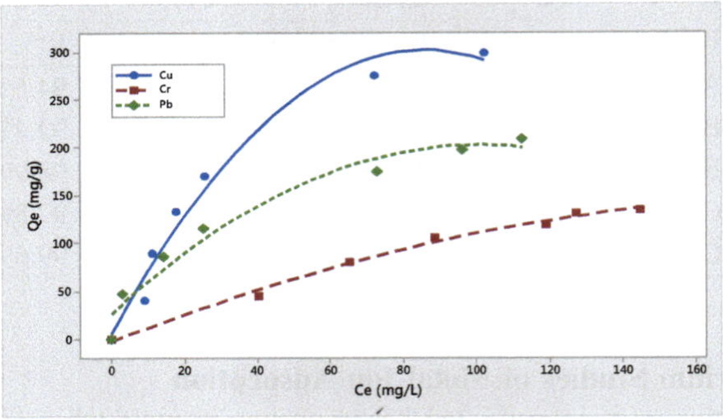

Fig. 21. Equilibrium studies of different metal ions brick material C

These figures show that adsorption of heavy metals initially increases with increasing metal concentration and then levels off. The high metal ion uptake of brick type C at the beginning of the equilibrium isotherm may be caused by the availability higher specific surface area and pore volume. The experimental data were correlated by the nonlinear form of Langmuir and Freundlich isotherms and its parameters calculated from the curve fitting the results are listed in Table 4.

Considering Cu^{2+} ion uptakes measured for each brick type, brick type A shows a significantly higher value (802.4 mg/g) compared to the other two brick types. Pb^{2+} uptake is higher into brick type C having 312.3 mg/g of adsorption capacity. In the case of adsorbing Cr^{6+} into each brick type, which fits well with the Freundlich isotherm, shows higher uptake in adsorbing brick type A.

Table 4. Equilibrium parameters obtained for A, B and C brick samples

Brick type	Metal ion	Langmuir isotherm		R^2
		K (l/mg)	Q_{max} (mg/g)	
A	Cu^{2+}	0.003	802.4	98.9%
A	Pb^{2+}	0.024	522.3	94.5%
B	Pb^{2+}	0.042	187.2	99.8%
C	Cu^{2+}	0.002	300.0	96.4%
C	Pb^{2+}	0.004	312.3	97.9%
Brick type	Metal ion	Freundlich isotherm		R^2
		K	n	
A	Cr^{6+}	4.22	0.84	97.3%
B	Cu^{2+}	3.36	0.65	99.7%
B	Cr^{6+}	1.44	0.36	99.4%
C	Cr^{6+}	2.84	0.58	98.2%

5 Conclusion

Brick materials have space inside their porous structure, which is the main reason they adsorb toxic heavy metals that are presented in the wastewater molecules. The ability of brick materials to accommodate those ionic species increases their adsorption capacity. The adsorption capacity of the investigated two brick types for selected metal ions (Cu^{2+}, Pb^{2+}, Cr^{6+}) show considerably significant features. Considering Cu^{2+} ion, the adsorption capacities, brick type A > brick type B > brick type C and for Pb^{2+} ion, the adsorption capacities were brick type A > brick type C > brick type B. Cu^{2+} adsorbed to brick type A fit well with Pseudo – first – order kinetic model and, Cu^{2+} adsorbed to brick type B and Cu^{2+} and Pb^{2+} adsorbed to brick type B show a good fit with Pseudo – second – order kinetic model. Brick type C is not much significantly different from other brick types which shows relatively average adsorption capacities. In the case of uptake possibility of the Cu^{2+}, brick type B > brick type A > brick type C. Further Cr^{6+} brick type B $\sim$ brick type A > brick type C. It is evident that the presence of meso-structure in brick materials allows it to effectively adsorb metal ions leading to a substantial enlargement of the adsorption capacity.

Acknowledgements. The BET analysis data were obtained at department of chemistry, California State University, Stanislaus, and the authors thanks to Dr. Amanpreet Manchanda for the support with BET experiments. Further, thanks to Dr. Buddhika Dassanayake from Faculty of Science, university of Peradeniya for the technical support with the XRF measurements.

References

Conseil National de Recherches du Canada (1980) Can J 64:1–8

Danielsson L (2013) Removal of chromium in wastewater with natural clays in southern Malawi

Garcia Sanchez A (1999) Sorption of heavy metals from industrial waste water by low-cost mineral silicates. Clay Miner 34(3):469–477

Gunathilake C et al (2015) Adsorption of lead ions from aqueous phase on mesoporous silica with P-containing pendant groups. ACS Appl Mater Interfaces 7(41):23144–23152

Kruk M, Jaroniec M, Sayari A (1997) Application of large pore MCM-41 molecular sieves to improve pore size analysis using nitrogen adsorption measurements. Langmuir 13:6267–6273

Pare S (2012) Heavy metal removal from aqueous solutions by sorption using natural clays from Burkina Faso. Af J Biotechnol 11(45):10395–10406

Rahman MM et al (2014) Removal of heavy metal ions with acid activated carbons derived from oil palm and coconut shells. Materials 7(5):3634–3650

Rashed MN (2001) Lead removal from contaminated water using mineral adsorbents. Environmentalist 21(3):187–195

Talaat HA, El Defrawy NM, Abulnour AG (2011) Evaluation of heavy metals removal using some Egyptian clays, vol 6, pp 37–42

Tripathi A, Rawat Ranjan M (2015) Heavy metal removal from wastewater using low cost adsorbents. J Bioremediat Biodegradation 6(6)

Uddin MK (2017) A review on the adsorption of heavy metals by clay minerals, with special focus on the past decade. Chem Eng J 308:438–462

Wingenfelder U et al (2005) Removal of heavy metals from mine waters by natural zeolites. Environ Sci Technol 39(12):4606–4613

Author Index

R. Dissanayake and P. Mendis (Eds.): ICSBE 2018, LNCE 44, pp. 633–635, 2020.
https://doi.org/10.1007/978-981-13-9749-3

MIX
Papier aus verantwortungsvollen Quellen
Paper from responsible sources
FSC® C105338

If you have any concerns about our products,
you can contact us on
ProductSafety@springernature.com

In case Publisher is established outside the EU,
the EU authorized representative is:
Springer Nature Customer Service Center GmbH
Europaplatz 3, 69115 Heidelberg, Germany

Printed by Libri Plureos GmbH
in Hamburg, Germany